METHODOLOGIES FOR CONTROL OF JUMP TIME-DELAY SYSTEMS

Methodologies for Control of Jump Time-Delay Systems

by

Magdi S. Mahmoud
Faculty of Engineering,
Arab Academy for Sciences & Technology, Egypt

and

Peng Shi
Weapons Systems Division,
Defense and Technology Organization, Australia

KLUWER ACADEMIC PUBLISHERS
BOSTON / DORDRECHT / LONDON

A C.I.P. Catalogue record for this book is available from the Library of Congress.

ISBN 1-4020-7489-1

Published by Kluwer Academic Publishers,
P.O. Box 17, 3300 AA Dordrecht, The Netherlands.

Sold and distributed in North, Central and South America
by Kluwer Academic Publishers,
101 Philip Drive, Norwell, MA 02061, U.S.A.

In all other countries, sold and distributed
by Kluwer Academic Publishers,
P.O. Box 322, 3300 AH Dordrecht, The Netherlands.

Printed on acid-free paper

Printed in the Netherlands.

To The Memory of my Parents and
To My Family (Salwa, Medhat, Monda, Mohamed)
MSM

To The Memory of my Father and
To My Family (Fengmei, Lisa, Michael)
PS

Contents

Preface

In many industrial and engineering applications we deal with dynamical systems subject to frequently occuring and unpredictable structural changes and a convenient representation of these systems would be via *piece-wise deterministic models* in order to make efficient use of well-established theories. The underlying system dynamics assume different forms, that is multiple-modes, depending on the value of an associated Markov chain process (random jump process) known as *form* process associated with the system under consideration. More often, these dynamical systems have been termed *jump systems.* The usefulness of such model representation is quite apparent since it permits the decision-maker to properly treat the discrete-events that significantly change the normal operation by exploiting the knowledge of their occurrence and the statistical patterns about their arrival information. From a different aspect, the operation of a wide class of industrial system applications experience time-delays be inherently as a result of physical properties, externally due to the use of special equipment or due to finite capabilities of information processing and data flow amongst various parts of the system. Most of the time delays have crucial impact on the plant performance. The employment of functional differential equations (FDEs) rather than ordinary differential equations (ODEs) in the modeling effort becomes the rule not the exception. Putting them together, a new system configuration readily emerges which, from now onwards, we call **jump time-delay systems (JTDS)**.

This book is about the recent advances in control analysis and design methodologies for such a new class of systems. This class possesses the main ingredients of multi-modes of operation, nominally inherent time-delay model and parametric uncertainties and external disturbances. Indeed, this class reflects several important features on the performance analysis and control design and emphasizes the existence of a hybrid system: state-space dynamics and Markov chain dynamics. There are numerous applications that can cast in the framework of such JTDS. Examples include, but not limited to, water quality control, electric power systems, productive manufacturing systems and cold steel rolling mills. For obvious reasons, JTDS can best represented in the time-domain by a hybrid state-space formalism the major part of which is a state-space hereditary model and a random process model forming the remaining part.

In dealing with JTDS, we follow a systematic modeling approach in that a convenient representation of the system state would be by observing a finite-dimensional vector at a particular instant of time and then examining the subsequent behavior. Looked at in this light, the primary objective of this book is to present an introductory, yet comprehensive, treatment of JTDS by jointly combining the two fundamental attributes: the system dynamics possesses an inherent time-delay and the system parameters may undergo jump behavior. While each attribute has been examined individually in several texts, the integration of both attributes is quite unique and deserves special consideration. Additionally, JTDS are nowadays receiving increasing attention by numerous investigators as evidenced by the number of articles appearing in journal and conference proceedings.

The material contained in this book not only organized to focus on the new developments in the analysis and control methodologies for such JTD systems, but it also integrates the impact of the delay factor on important issues like stochastic stability and control design. After an introductory chapter, it is intended to split the book into seven self-contained chapters with each chapter being equipped with illustrative examples, problems and questions. The book will be supplemented by an extended bibliography, appropriate appendices and indexes.

It is planned while organizing the material that this book would be appropriate for use either as graduate-level textbook in applied mathematics as well as different engineering disciplines (electrical, mechanical, civil, chemical, systems), a good volume for independent study or a reference for practicing engineers, interested readers, researchers and students.

Acknowledgment

Although the material contained in this volume is an outgrowth of our academic research activities over the past several years, the idea of jointly writing the book arouse and developed only while first author was visiting the second one at the **University of South Australia**, Adelaide in fall 2001. We are deeply appreciated to the Center for Applied Mathematics at the University of South Australia for supporting this visit.
In writing this volume, we took the approach of referring within the text to papers and/or books which we believed taught us some ideas and methods. We further complement this by adding some notes and questions at the end of each chapter to shed some light on other related results. We apologize in advance in case we committed injustice and assure our colleagues that any mistake was unintentional.

The first author owes a measure of gratitude to the **Arab Academy for Sciences & Technology**, Egypt for providing the permission of sabbatical leave and extending excellent opportunity of academic collaboration during the year (2001-2002). In particular, the continuous encouragements of Professor Omar Abdel-Aziz, Dean of Engineering and Professor Ahmed Amer are wholeheartedly acknowledged.
The process of fine tuning and producing the final draft was pursued at **UAE University**, United Arab Emirates and special thanks must go to my colleagues Dr Abdulla Ismail, Dr Naser Abdel-Rahim and Dr Habib-Ur Rehman of the electrical engineering department. In this regard, the overall scientific environment at UAEU is gratefully recorded.

We are immensely pleased for many stimulating discussions with colleagues, students and friends throughout our technical career which have definitely enriched our knowledge and experience.

The great support and enthusiasm of Mark de Jongh, Electrical Engineering Editor of Kluwer Academic Publishers were instrumental to the success of this project

Most of all however, we would like to thank our families. Without their constant love, incredible amount of patience and (mostly) enthusiastic support this volume would not have been finished.

Magdi S. Mahmoud

Peng Shi

Chapter 1

Introduction

1.1 Overview

In engineering practice, there is a wide class of dynamical systems whose parameters change in an unpredictable manner. This could result from abrupt phenomena such as component and interconnection failures, parameters shifting, tracking and variation in the time frame of measurements [32, 47, 154]. To make use of the wealth theories of deterministic systems, one convenient representation of this class of systems would be via **piece-wise deterministic models**. In this way, the underlying system dynamics assume different forms (multiple modes) depending on the value of an associated Markov chain process, which is known as form process associated with the system under consideration. More often, these dynamical systems have been termed **jump systems**. The usefulness of such model representation is apparent since it permits the decision maker to properly treat the discrete events that significantly change the normal operation by making use of the knowledge of their occurrence and the statistical patterns about their arrival information. Looked at from alternative perspective, the operations of a wide class of industrial system applications experience

time-delays due to various reasons including inherent physical properties (mass transport flow, recycling), data transmission delays or finite capabilities of information exchange. In most of these cases time-delays cause harmful effects to the plant performance [61, 63, 84, 85, 86, 89, 98, 126]. In this situation, the employment of functional differential equations (FDEs) rather than ordinary differential equations (ODEs) in the modeling effort becomes the rule not the exception. Thus we have two basic attributes: a) the Markovian behavior of the parameters, and b) the time-delay factor. Putting them together, a new system configuration readily emerges which, from now onwards, we call **jump time-delay systems (JTDS)**, a block-diagram of which is sketched below. This book is about the recent advances in the analysis and control methodologies for JTD systems which possess multiple modes of operation and in the normal situation at a particular mode, it behaves like a time-delay system. Meanwhile its parameters behave in accordance to Markov stochastic process (random jump process). In turn, this emphasizes the existence of a hybrid system: state space dynamics and Markov chain dynamics. We observe that there are numerous applications that can cast into the framework of such JTD systems. Examples include, but not limited to, water quality management, electric power systems, productive manufacturing systems and cold steel rolling mills. Throughout the book, we are going to represent JTD systems in the time-domain by a hybrid state-space formalism the major part of which is a state-space hereditary model and a random process model forming the remaining part .

1.2 Historical Perspectives

Interest in jump linear systems **JLS**, as an important class of dynamical systems, arouse around the mid 1960s [138] since the system model allows the decision maker to cope adequately with the discrete events that disrupt and /or change significantly the normal operation of a system, by using the knowledge of their occurrence and the statistical information on the rate at which the Markovian events take place. Research activities into the class of JLS and their applications into manufacturing management span several decades. Some representative references in this area are [42, 66, 67, 45, 4, 59, 153, 157]

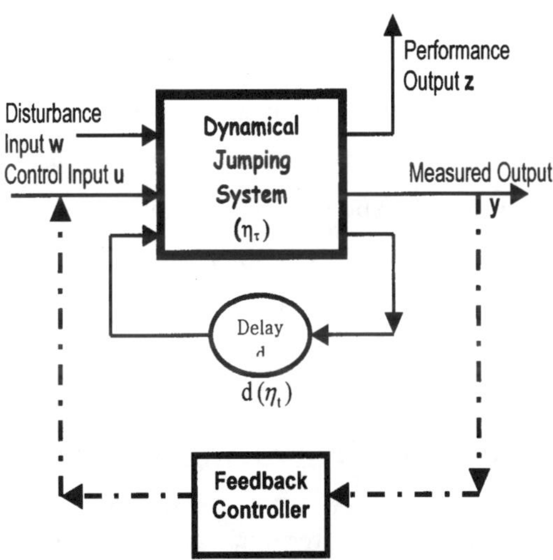

Fig 1.1: A Block-Diagram of Jump Time-Delay System There has been dramatic progress in jump linear quadratic **JLQ** control theory since the pioneering work of [76]. The JLQ control problem was solved by [155] using the stochastic maximum principle for state-feedback in the finite-horizon case. Later on, [159] obtained the same results using dynamic programming for both the finite-horizon and infinite-horizon cases. [130] provided an approach to the output-feedback JLQ control problem. The continuous-time partially observable situation was studied by [51] and an analysis of the discrete-time version of JLQ control problem was conducted by [28] for the case without driving noise. The concept of stochastic Lyapunov function was introduced and developed in [78].

In recent years, research into the control problems of JLS in the presence of unknown continuous disturbances has been initiated in [12, 33, 139]. This opens up the paradigm of $\mathcal{H}_\infty$ optimal control, where there is an additional (discrete) element, which is the stochastic (piecewise constant) Markov process disturbance, that causes structural changes. Further results can be found in [20, 22, 133, 149, 152]. On the filtering problem of JLS, results of some research investigations are contained in [34, 38, 39, 151].

In this volume, Chapter 2 will be devoted to some examples and models of JTDS and Chapter 3 will focus on stochastic stability and stabilization. The control design will be divided into two chapters: Chapter 4 deals with state and output feedback and Chapter 5 treats simultaneous $\mathcal{H}_2/\mathcal{H}_\infty$ control. In Chapter 6, we examine the filtering problem and in Chapter 7, we investigate the control and filtering neutral jumping system (NJS). Chapter 8 will contain results on interconnected JTDS.

1.3 A Glossary of Terminology and Notations

In this section, we assemble the terminologies and notations to be adopted throughout the book with the objective of paving the way to the technical development of subsequent chapters. These terminologies and notations are quite standard are in the scientific media and only vary in form or character.

1.3.1 General Terms

As a start, matrices as $n \times m$ dimensional arrays of elements with n-rows and m-columns are represented by capital letters while vectors as n-tuples or columns (unless otherwise specified)and scalars (single elements) are represented by lower case letters. We use $\mathbb{R}, \mathbb{R}_+$, $\mathbb{R}^n$ and $\mathbb{R}^{n\times m}$ to denote the set of real , positive real numbers , real n-tuples (vectors) and real $n \times m$ matrices, respectively.

Alternatively, $\Re^n$ is called the Euclidean space and is equipped with the vector-norm as $||x|| \stackrel{\Delta}{=} \sqrt{[x_1^2 + \cdots x_n^2]}$. The terms $f(t), g(s)$ denote, respectively, scalar-valued functions of the real variables t and s. The quantities $\dot{x}, \ddot{x}$ are the first and the second derivative of x with respect to time, respectively. The symbols $[.,.]$, $(.,.]$, $(.,.)$ denote, respectively, closed, semiclosed, and open intervals; that is $t \in (a,b] \Rightarrow a < t \leq b$. The open left-half $(\equiv \{s : Re(s) \leq 0\}$, the open proper left-half $(\equiv \{s : Re(s) < 0\}$ and the open proper right-half $(\equiv \{s : Re(s) > 0\}$ of the complex plane are represented by $\mathbb{C}^{\dagger}$, $\mathbb{C}^{-}$ and $\mathbb{C}^{+}$, respectively. Let $\bar{\mathcal{C}} \stackrel{\Delta}{=} \bigcup_{j \in \mathcal{S}} \mathcal{C}[-\tau_j, 0] \times \{j\}$. The Lebsegue space $\mathcal{L}_2[0,\infty)$ consists of square integrable functions on the interval $[0,\infty)$ and equipped with the norm

$$||p||_2 \stackrel{\Delta}{=} \left(\int_0^{\infty} p^t(\tau)\, p(\tau)\, d\tau \right)^{1/2} \tag{1.1}$$

For any square matrix W of arbitrary dimension $n \times n$, let W^t, W^{-1}, $\lambda(W)$, $tr(W)$, $r(W)$, $det(W)$, $\rho(W)$ and $||W||$ denote, respectively, the transpose, the inverse, the spectrum, rank, the trace , the determinant, the spectral radius and the induced norm defined by

$$||W|| \stackrel{\Delta}{=} \lambda(W\, W^t)^{1/2} \tag{1.2}$$

We use $W > 0$ $(\geq, <, \leq 0)$ to denote a symmetric positive definite (positive semidefinite, negative, negative semidefinite matrix W with $\lambda_m(W)$ and $\lambda_M(W)$ being the minimum and maximum eigenvalues of W. Frequently, I stands for the identity matrix with appropriate dimension, $W^{\dagger}$ denotes the pseudo-inverse of W and $diag(W_1, \cdots, W_p)$ stands for the block-diagonal matrix

$$\begin{bmatrix} W_1 & 0 & \cdots & 0 \\ 0 & W_2 & \cdots & 0 \\ 0 & \cdots & \cdots & 0 \\ 0 & \cdots & \cdots & W_p \end{bmatrix} \tag{1.3}$$

Throughout the book in portions dealing with multi-state-delay models, we let $J_r = \{1, ..., r\}$ be the set of the r first positive integers. Given a set of constant matrices, $A_{d_1}, ..., A_{d_q} \in \Re^{n\times n}$, we denote $\mathcal{E}_t = \left[A_{d_1} .... A_{d_q}\right] \in \Re^{n\times nq}$ and $\mathcal{E}_s = \left[A_{d_1} + ... + A_{d_q}\right] \in \Re^{n\times n}$. Also, for some positive constants $(a_1, ..., a_q)$ with the vector $x(t - a_j) \in \Re^n$, $\forall j \in J_q$, we let, $\chi(t, a, q) = [x^t(t - a_1),\ x^t(t - a_2),\ ....x^t(t - a_q)]^t \in \Re^{nq\times 1}$. Also, we use $\mathcal{T} = (\tau_1, ..., \tau_q)^t$ and in the sequel, we let $\mathcal{Z}_r \stackrel{\Delta}{=} \{1, \cdots, r\}$ be the set of the r first positive integers.

1.3.2 Functional Differential Equations

Let $\mathbb{C}_{n,\tau} = \mathbb{C}([-\tau, 0], \mathbb{R}^n)$ denotes the Banach space of continuous vector functions mapping the interval $[-\tau, 0]$ into $\mathbb{R}^n$ with the topology of uniform convergence and designate the norm of an element $\phi \in \mathbb{C}_{n,\tau}$ by

$$||\phi||_* \stackrel{\Delta}{=} \sup_{\theta\in[-\tau,0]} ||\phi(\theta)||_2 \tag{1.4}$$

If $\alpha \in \mathbb{R}, d \geq 0$ and $x \in \mathbb{C}([\alpha-\tau, \alpha+d], \mathbb{R}^n)$ then for any $t \in [\alpha, \alpha+d]$, we let $x_t \in \mathbb{C}$ be defined by $x_t(\theta) := x(t+\theta),\ -\tau \leq \theta \leq 0$. If $\mathcal{D} \subset \Re \times \mathbb{C}$, $f : \mathbb{D} \to \mathbb{R}^n$ is a given function , the relation $\dot{x}(t) \ = \ f(t, x_t)$ is a retarded functional differential equation (RFDE) on $\mathbb{D}$ where $x_t(t), t \geq t_o$ denotes the restriction of $x(\cdot)$ to the interval $[t-\tau, t]$ translated to $[-\tau, 0]$. Here, $\tau > 0$ is termed the *delay factor*. In the sequel, if $\alpha \in \mathbb{R},\ d \geq 0$ and $x \in \mathbb{C}([\alpha - \tau, \alpha + d], \Re^n)$ then for any $t \in [\alpha, \alpha + d]$, we let $x_t \in \mathbb{C}$ be defined by $x_t(\theta) \stackrel{\Delta}{=} x(t + \theta),\ -\tau \leq \theta \leq 0$. In addition, if $\mathbb{D} \subset \mathbb{R} \times \mathbb{C}, f : \mathbb{D} \to \mathbb{R}^n$ is given function, then the relation

$$\dot{x}(t) \ = \ f(t, x_t) \tag{1.5}$$

is a retarded functional differential equation (RFDE) on $\mathbb{D}$ where $x_t, t \geq t_0$ denotes the restriction of $x(.)$ on the interval $[t-\tau, t]$ translated to $[-\tau, 0]$. Here $\tau > 0$ is termed the **state-delay factor**. A function x is said to be a **solution** of (1.5) on $[\alpha - \tau, \alpha + d]$ if there $\alpha \in \mathbb{R}$ and $d > 0$ such that

$$x \in \mathbb{C}([\alpha - \tau, \alpha + d], \mathbb{R}^n), \quad (t, x_t) \in \mathbb{D}, \quad t \in [\alpha, \alpha + d] \tag{1.6}$$

and $x(t)$ satisfies (1.5) for $t \in [\alpha, \alpha + d]$. For a given $\alpha \in \mathbb{R}$, $\phi \in \mathbb{C}$, $x(\alpha, \phi, f)$ is said to be a solution of (1.5) with **initial value** ϕ at α.

1.3.3 Markov Processes

In terms of the theory of stochastic processes [35], let $\mathbb{E}[\cdot]$ stands for mathematical expectation and the sample space be denoted by Ω with generic point ω. The range of the process is in the Euclidean space $\mathbb{R}^n$ with $\mathcal{B}$ being a Borel field of sets in $\mathbb{R}^n$, $\mathbb{F}$ is the algebra of events and $\mathbf{P}$ is the probability measure defined on $\mathbb{F}$. Then the triplet $(\Omega, \mathbb{F}, \mathbf{P})$ represents a probability space.

Let $x_1, \cdots$ be a discrete-time parameter stochastic process with the associated transition function $\mathbf{P}(s, x; n+s, \Gamma)$ which can be interpreted as the probability that x_{n+s} is in $\Gamma \in \mathcal{B}$, given that $x_s = x$. Suppose that the conditional distribution function satisfies

$$\mathbf{P}\{x_{n+s} \in \Gamma | x_1, \cdots, x_s\} \;=\; \mathbf{P}\{x_{n+s} \in \Gamma\} \tag{1.7}$$

for all Γ in $\mathcal{B}$ and nonnegative n and s, with probability one. Then the process is termed a *Markov process* and, in addition, the Chapman-Kolmogorov

$$\mathbf{P}(s, x; n+m+s, \Gamma) \;=\; \int_{\mathbb{R}^n} \mathbf{P}(s, x; s+m, dy)\mathbf{P}(s+m, y; n+m+s, \Gamma) \tag{1.8}$$

holds.

In the sequel, we suppose that time is not a component of the state and that the process x_s be homogeneous. Then $\mathbf{P}_x(x_t \in \Gamma) \;=\; \mathbf{P}(s, x; s+t, \Gamma) \;\equiv\; \mathbf{P}(x_t, \Gamma)$ with probability one for any $s \geq 0$. Also $\mathbb{E}_x f(x_t) \;=\; \int f(y)\mathbf{P}(x, t, dy)$. On the other hand, if the process is not homogeneous, we let $\mathbf{P}_{x,s}(x_{t+s} \in \Gamma) \;=\; \mathbf{P}(s, x; s+t, \Gamma)$ and $\mathbb{E}_{x,s} f(x_{t+s}) = \int f(y)\mathbf{P}(s, x, s+t, dy)$ where x is the initial value of the process at initial time s, $\mathbf{P}_x(d\omega)$ and $\mathbf{P}_{x,s}(d\omega)$ are the probability measures corresponding to the homogeneous and the nonhomogeneous case, respectively.

The process x_t is said to be *stochastically continuous at the point* x if

$$\mathbf{P}_x\{||x_\delta - x|| \geq \epsilon\} \rightarrow 0$$

as $\delta \rightarrow 0$, for any $\epsilon > 0$. If

$$\mathbf{P}_x\{\sup_{\delta \geq \Delta \geq 0} ||x_\Delta - x|| \geq \epsilon\} \rightarrow 0$$

uniformly for x in a set M, as $\delta \rightarrow 0$, for any $\epsilon > 0$, then the process is *uniformly stochastically continuous* in the set M.

Let $\aleph_t$ be the minimum $\sigma-$algebra on Ω determined by conditions on $x_s,\ 0 \leq s \leq t,\ x_0 = x$. A random variable τ taking values in $[0, \infty]$ and depending also on x is called a *Markov time* if the event $\{\tau \leq t\}$ is contained in $\aleph_t$ for each $t < \infty$ and fixed τ. Note in general that τ is a functional of the sample paths. The process x_t is termed a *strong Markov process* if, for any Markov time τ and any $t \geq 0,\ \Gamma \in \mathcal{B},$ and $x \in \Re^n$, the conditional probability satisfies

$$\mathbf{P}_x\{x_{t+\tau} \in \Gamma | \aleph_t\} = \mathbf{P}(t, x_\tau, \Gamma) \tag{1.9}$$

with probability one. An interpretation of (1.9) is that the probability of $\{x_{t+\tau} \in \Gamma\}$, conditioned upon the history up to τ, equals the probability of $\{x_{t+\tau} \in \Gamma\}$, conditioned upon x_τ only. Observe that (1.9) holds for τ equal to any finite constant, any stong Markov process is also a Markov process.

A function $f(x)$ is said to be in the domain of the weak infinitesimal operator $\tilde{A}$ of the process x_s, and we write $\tilde{A}f(x) = h(x)$, if the limit

$$\lim_{\delta \rightarrow 0} \frac{\mathbb{E}_x f(x_\delta) - f(x)}{\delta} = h(x) \tag{1.10}$$

exists pointwise in $\Re^n$ and satisfies

$$\lim_{\delta \to 0} \quad \mathbb{E}_x h(x_\delta) \;=\; h(x) \tag{1.11}$$

It is readily seen that $\tilde{A}$ is linear. To shed more light, let x_t be the solution of the differential equation

$$\dot{x} \;=\; g(x)$$

. The phrase "$f(x) is in the domain of \tilde{A}$" implies that $f(x)$ has continuous first partial derivatives and that $f(\dot{x_t})$ is continuous in t. Under these conditions

$$\tilde{A}f(x) \;=\; \frac{\partial f}{\partial x(x)}\dot{x} \;=\; \dot{f}(x)$$

and in general, $\tilde{A}f(x)$ is the average time rate of change of the process $f(x_s)$ at time s, given that $x_s = x$.

Consider the case where x_t is a right continuous strong Markov process and τ is a random time with $\mathcal{E}_x\tau \;<\; \infty$. Let $f(x)$ be in the domain of $\tilde{A}$, with $\tilde{A}f(x) \;=\; h(x)$. Then

$$\mathbb{E}_x f(x_\tau) \;-\; \mathbb{E}_x \int_0^\tau h(x_s)\, ds \;=\; \mathbb{E}_x \int_0^\tau \tilde{A}f(x_s)\, ds \tag{1.12}$$

Equation (1.12) is the well-known *Dynkin's formula* and it plays a major role in the construction of stochastic stability.

The *differential generator* of the process x_t is defined by the operator

$$\Upsilon \;=\; \sum_j f_j(x,t)\, \frac{\partial}{\partial x_j} \;+\; \frac{1}{2}\sum_{j,k} S_{j,k}(x,t)\, \frac{\partial^2}{\partial x_j \partial x_k} \;+\; \frac{\partial}{\partial t} \tag{1.13}$$

Finally, let the random form process $\{\eta_t, t \in [0,\infty)\}$ be a homogeneous, finite-state Markovian process with right continuous trajectories and taking values in a finite set $\mathcal{S} \;=\; \{1, 2, ..., s\}$ with transition probability from mode i at time t to mode j at time $t + \delta$, $i, j \in \mathcal{S}$:

$$\begin{aligned} p_{ij} \;&=\; Pr(\eta_{t+\delta} = j \mid \eta_t = i) \\ &=\; \begin{cases} \alpha_{ij}\delta + o(\delta), & if \;\; i \neq j, \\ 1 + \alpha_{ij}\delta + o(\delta), & if \;\; i = j \end{cases} \end{aligned} \tag{1.14}$$

with transition probability rates $\alpha_{ij} \geq 0$ for $i, j \in \mathcal{S}, i \neq j$ and

$$\alpha_{ii} = - \sum_{m=1, m \neq i}^{s} \alpha_{im} \tag{1.15}$$

where $\delta > 0$ and $\lim_{\delta \downarrow 0} o(\delta)/\delta = 0$. Let $\hat{\alpha} \stackrel{\Delta}{=} \max_j \{|\alpha_{jj}|, j \in \mathcal{S}\}$ and note that the set $\mathcal{S}$ comprises the various operational modes of the system under study.

Interestingly enough, the processes that we are going to deal with throughout the entire book are right continuous strong Markov processes since they are eventually models of physical problems.

Sometimes in different places, the arguments of a function will be omitted in the analysis when no confusion can arise.

1.4 Main Features of the Book

In writing this book, we endeavored to make the material coherent, systematic and readable in order to achieve the two-fold objective: educate students and impress colleagues. Therefor, we believe that our book possesses the following outstanding features:

1) It provides an in-depth treatment of continuous-time jump time-delay systems through systematic presentation, ease and complete coverage of the analytical subjects matter.

2) It treats the various topics related to JTDS with a balanced and logical compromise between mathematical rigor, simplicity in exposition and engineering interpretations.

3) It complements the mathematical analysis and theoretical results by adding illustrative remarks and solved examples.

4) It recognizes the practical aspects of system analysis and design by focusing on the time-delay system as the "nominal system" and exposing the delay dependence as a key analytical tool.

5) It appends the technical developments with user-friendly computational algorithms based on the MATLAB-software.

6) It stimulates researchers, practicing engineers and students by adding questions and problems at the end of each chapter to encourage them to further prop into the different areas of interests. It inserts some hints as well to motivate the readers and to help in tackling the problems.

By and large, our approach is to focus on time-delay systems (at the core) and to accommodate the stochastic variations as well as the parametric uncertainties within the analytical framework (see Fig. 1.1). Therefore, it is fair to state the available books are either emphasize one aspect of the topics covered by the proposed book or follow closely one approach.

1.5 Notes and References

In addition to the numerous papers and articles on time-delay systems and/or jumping systems, there are a number of reference books which might have some connection to the topics to be discussed in this book. This includes [24, 32, 78, 98, 126]. It is fair to say that the first book by Malek and Jamshidi [126] provides the rudimentary basics for time-delay systems using the tools of the late seventies and early eighties. It treats the topics of analysis, elementary design and optimization of constant time-delay (time-lag) systems based on the developments up to the mid eighties. The second book by Mahmoud [98] upgrades this effort in various ways and expands it extensively to the late nineties and beyond. It focuses on the fundamental role exhibited by the delay factor on system dynamics and performance behavior while discussing different notions of stability , stabilization and robustness as applied to single and interconnected systems.

The third book by Kushner [78] presents mathematical tools of stochastic systems to be used extensively throughout the book and the fourth book by Davis [32] deals with models with Markovian jump parameters. Both of them have nothing to do with time-delay systems. The most recent book by Boukas and Liu [24] seems to adopt the author's own approach and in which some parts overlap with the second book by Mahmoud [98]. Indeed, it covers stochastic stability, stabilization of JTD systems but from totally different perspectives. It does not cover however, neutral systems and interconnected systems. Our book have a unified methodology for the subject matter not shared by Boukas and Liu [24].

Chapter 2

Jump Time-Delay Systems

The primary objective of this chapter is to prepare the reader to the types of systems that the book is concerned with. Most of the examples are physically based and for convenience, we split them into time-delay systems, jump systems and jump time-delay systems with each one being treated in a different section.

2.1 Examples of Time-Delay Systems

In this section, we present some typical examples of physical systems that exhibit time-delay phenomena. The examples selected in this section fit nicely into the model (1.5).

2.1.1 Economic Systems

The existence of delays (or gestation lags) in economic systems is quite natural since there must be finite period of time following a decision for its effects to appear. In one model [61] of aggregate economy, we let $Y(t)$ be the income which can split into consumption $C(t)$, investment $I(t)$ and autonomous expenditure

E. Thus

$$Y(t) = C(t) + I(t) + E(t) \tag{2.1}$$

Define

$$C(t) = c\,Y(t) \tag{2.2}$$

where c is a consumption coefficient. From (2.1) we get

$$Y(t) = \frac{I(t) + E}{1 - c} \tag{2.3}$$

It is assumed that there is finite interval of time h between ordering and delivery of capital equipment following a decision to invest $D(t)$. In terms of the stock of capital assets $K(t)$, we have

$$\dot{K}(t) = D(t - h) \tag{2.4}$$

$$I(t) = \frac{1}{h}\int_{t-h}^{t} D(r)dr \tag{2.5}$$

Economic rationale implies that $D(t)$ is determined by the rate of saving (proportional to $Y(t)$) and by the capital stock $K(t)$. This means that

$$D(t) = \alpha(1 - c)Y(t) - \beta K(t) + \varepsilon \tag{2.6}$$

where $\alpha > 0$, $\beta > 0$ and ε is a trend factor. Combining (2.4)-(2.5), we obtain:

$$I(t) = \frac{1}{h}\big[K(t + h) - K(t)\big] \tag{2.7}$$

By (2.3) and (2.7), we arrive at

$$Y(t) = \frac{1}{h(1 - c)}\big[K(t + h) - K(t)\big] + \frac{E}{1 - c} \tag{2.8}$$

Finally, it follows from (2.5), (2.6) and (2.8) that

$$\dot{K}(t) = \frac{\alpha}{h}K(t) \; - \; \left(\beta + \frac{\alpha}{h}\right)K(t-h) + [\alpha E + \varepsilon] \tag{2.9}$$

which expresses the formation of the rate of delivery of the new equipment. This is a typical functional differential equation (FDE) of retarded type.

2.1.2 Nuclear Reactors

In modeling the dynamics of nuclear reactors, it turns out [73] that the resulting equations are described by FDEs and the delay factors are due to finite time of heat transport through different elements, warming up time of the reactor , snapping time of the control system, to name a few. In terms of
$x(t) =$ the relative change of neutron density,
$y(t) =$ proportional to the relative change in the reactor temperature,
$r_1(t) =$ proportional to the relative change in the fuel temperature,
$r_2(t) =$ proportional to the relative change in the temperature of deccelaration devices.
one model that takes into account the delayed neutrons is given by:

$$\begin{aligned}
\dot{x}(t) &= \Big[\alpha_1 r_1(t) + \alpha_2 r_2(t)\Big]\Big[1 + x(t)\Big] - \alpha_3\Big[x(t) - y(t)\Big] \\
\dot{y}(t) &= \alpha_4\Big[x(t) - y(t)\Big] \\
\dot{r_1}(t) &= \frac{1-\beta}{\gamma}x(t) - \theta\Big[r_1(t) - r_2(t)\Big] \\
\dot{r_2}(t) &= \sigma r_2(t-h) - r_2(t) + \beta x(t) + \theta\Big[r_1(t) - r_2(t)\Big]
\end{aligned} \tag{2.10}$$

where $\alpha_1, \cdots, \theta$ are known coefficients and h is the system delay corresponding to the time of liquid fuel transportation along the circulation contour.

2.1.3 Predator-Prey Models

Predator-prey interactions abound in the biological world and offer an interesting area of ecological studies. One simple logistic growth model is due to Volterra [135] is given in terms of the population sizes of the prey $x_1(t)$ and the predator $x_2(t)$, respectively. The model is described by

$$\begin{aligned} \dot{x}_1(t) &= \alpha_1 x_1(t) - \alpha_2 x_1(t) x_2(t) - \alpha_3 x_1^2(t) \\ \dot{x}_2(t) &= -\alpha_4 x_2(t) + \alpha_5 x_1(t-h) x_2(t-h) \end{aligned} \tag{2.11}$$

where $\alpha_1, cdots, \alpha_5$ are constants and the delay $h > 0$ is the average time between death of a prey and the birth of subsequent number of predators.

There are numerous extensions of model (2.11) to take into account factors of environmental inhomogeneity, competition, age structure, etc see [26].

2.2 Examples of Jump Systems

In this book, we are interested in the kind of stochastic models arising in characterizing many important physical systems subject to random failures and structure changes. This includes, but not limited to, electric power systems [157], control systems of a solar thermal central receiver [42], communications systems [32], aircraft flight control [131] and manufacturing systems [154]. It is known that hybrid stochastic control systems have been used extensively in the modeling of many dynamic planning models and in particular of manufacturing systems [60]. Our purpose in the ensuing sections is to provide representative models of jump systems.

2.2.1 Manufacturing Flow Control

In a typical model of manufacturing flow control, the state of the system is described by a pair $\mathbf{s} = (\xi, x)$ where $\xi \in \mathbf{E}$ is discrete and $x \in \mathbb{R}^n$ is a continuous.

The material flow is described by a simple state equation

$$\dot{x}(t) = u(t) \; - \; d(t) \tag{2.12}$$

where the vector $x(t)$ measures the cumulative difference between production and demand for parts, $u(t)$ is the production rate vector and $d(t)$ is the demand rate, sometimes taken as a constant. The modal changes act on the admissible control set through the constraints

$$u(t) \; \in \; \Omega(\xi(t)) \tag{2.13}$$

Over a prescribed tine horizon $[0, T]$, the system cost is usually represented by a functional

$$J \; = \; \mathbb{E}\left[\int_0^T \; g(x(s)) \; d\, s \middle| x(0), \xi(0) \right] \tag{2.14}$$

The above framework has been used to derive a closed-loop solution to the problem of dispatching parts to machines in a failure-prone flexible manufacturing systems. Using hierarchical structures [60], mean time to failure and mean time to repair are much longer than operation times. The flow control level then determines the short-time production rates u_j of each member j of the part family and the mix of parts being produced is adjusted continuously to take into account the random failure state (γ_{ij}) of the work stations.

Extensions of the foregoing modeling effort to production systems in which the modal transition rates being state and/or control dependent have been examined in [49, 16, 1, 18, 50].

2.2.2 Optimal Inventory/Production Control

In this example [46], an inventory/production problem is considered within a stochastic manufacturing system where the machine capacity is a birth-death process and the demand is uncertain. The associate cost criterion is discounted with the demand, the production capacity and the processing time per unit are being random variables.

A brief description of the underlying system is now given. The manufacturing system is subject to random machine breakdowns such that when the system is up, the machine can produce at a rate $r > 0$ where r is the machine capacity. The time duration of making a unit of the product is exponentially distributed. When the machine is down, the system is not producing. If the machine recovers from the breakdown, it will continue the incomplete work on the unit left from the last breakdown. The durations of up and down periods are exponentially distributed with rates q_1 and q_o, respectively. The demand process forms a homogeneous Poisson flow with a constant rate d. The associated cost function $c(x)$ is:

$$c(x) = \begin{cases} c^+ \, x, & if \;\; x \geq 0 \\ -\, c^- \, x, & otherwise \end{cases} \tag{2.15}$$

where c^+ and c^- are the holding and backlog cost per unit of item over unit of time, respectively. In this model the production rate is piece-wise constant, that is no rate adjustment is allowed until either the machine is broken down or the current unit is completed. Under the assumption of sufficient capacity, the average capacity is no less than the average demand:

$$\frac{r \, q_o}{q_1 + q_o} = \geq d \tag{2.16}$$

Observe that the capacity-demand gap

$$\delta \triangleq \frac{r \, q_o}{q_1 + q_o} - d \geq 0$$

Define the system states as $(x(t), i(i))$, $x(t) \in I; i(t) \in \{0, 1\}$ where $x(t)$ denotes the inventory state and $i(t)$ is the machine state for $\forall t \geq 0$

$$i(t) = \begin{cases} 1, & if \; the \; system \; is \; up \\ 0, & if \; the \; system \; is \; down \end{cases} \tag{2.17}$$

The purpose here is to minimize the discounted operational cost

$$J^u(x,i) \;=\; \mathbb{E}\Big[\int_0^{\infty} e^{\gamma\, t}\big[c^+ \; x_u^+(t) \;+\; c^- \; x_u^-(t)\big] d\, t \;, i \;=\; 0, 1\Big] \tag{2.18}$$

where $x_u(t)$ is the level of inventory at $t \geq 0$ under control u, $x^+ = \max(0, x)$, $x^- = \max(0, -x)$ and γ is the discount factor.

In [162], a variation of the forgoing model is considered to represent a manufacturing firm facing stochastic demand where the inventory level varies according to (2.12) with a unit production capacity constraint $0 \geq u(t) \geq 1$ and the marketing (advertising) rate $w(t) \in \{0, w_d\}$, $w_d > 0$. Here, the demand rate is assumed to follow a two-state Markov chain with generator

$$Q(w(t)) = \begin{pmatrix} -k\, w(t) & k\, w(t) \\ 0 & 0 \end{pmatrix}$$

2.3 Classes of Jump Time-Delay Systems

Recall from section (1.3) that the relation

$$\dot{x}(t) \;=\; f(t, x_t) \tag{2.19}$$

is a retarded functional differential equation (RFDE) on $\mathcal{D}$ where $x_t, t \geq t_0$ denotes the restriction of $x(.)$ on the interval $[t-\tau, t]$ translated to $[-\tau, 0]$ where $\tau > 0$ is termed the **state-delay factor**. For a given $\alpha \in \Re$, $\phi \in \mathcal{C}$, $x(\alpha, \phi, f)$ is said to be a solution of (2.19) with **initial value** ϕ at α.

On the other hand, given a probability space $(\Omega, \mathcal{F}, \mathbf{P})$ let the random form process $\{\eta_t, t \in [0, \mathcal{T}]\}$ be a homogeneous, finite-state Markovian process with right continuous trajectories and taking values in a finite set $\mathcal{S} = \{1, 2, ..., s\}$ with transition probability p_{ij} from mode i at time t to mode j at time $t+\delta$, $i, j \in \mathcal{S}$ as described by equation (1.7).

In the sequel, we study a class of stochastic uncertain time-delay systems with Markovian jump parameters described over the space $(\Omega, \mathcal{F}, \mathbf{P})$ by:

$$
\begin{aligned}
(\Sigma_G): \quad \dot{x}(t) &= A_{\Delta o}(t,\eta_t)x(t) + A_{\Delta d}(t,\eta_t)x(t-\tau) + B_{\Delta o}(t,\eta_t)u(t) \\
&+ \Gamma_{\Delta o}(\eta_t)w(t), \; t \geq 0, \qquad (2.20) \\
x(t) &= \phi(t), \; t \in [-\tau, 0], \; \eta_o = i \qquad (2.21) \\
y(t) &= C_{\Delta o}(t,\eta_t)x(t) + D_{\Delta d}(t,\eta_t)x(t-\tau) \\
&+ \Psi_{\Delta o}(\eta_t)w(t) \qquad (2.22) \\
z(t) &= G_{\Delta o}(t,\eta_t)x(t) + F_{\Delta o}(t,\eta_t)x(t-\tau) + \Phi_{\Delta o}(\eta_t)w(t) \qquad (2.23)
\end{aligned}
$$

where the system variables

$x(t) \in \Re^n$ is the state vector;

$u(t) \in \Re^m$ is the control input;

$w(t) \in \Re^q$ is the disturbance input which belongs to $\mathcal{L}_2[0,\infty)$;

$y(t) \in \Re^p$ is the measured output;

$z(t) \in \Re^r$ is the controlled output which belongs to $\mathcal{L}_2\big[(\Omega, \mathcal{F}, \mathbf{P}), [0,\infty)\big]$. and the systems matrices have the linear perturbed form

$$
\begin{bmatrix} A_{\Delta o}(t,\eta_t) & A_{\Delta d}(t,\eta_t) & \Gamma_{\Delta o}(t,\eta_t) \\ C_{\Delta o}(t,\eta_t) & D_{\Delta d}(t,\eta_t) & \Psi_{\Delta o}(t,\eta_t) \end{bmatrix} = \begin{bmatrix} A_o(\eta_t) & A_d(\eta_t) & \Gamma_o(\eta_t) \\ C_o(\eta_t) & D_d(\eta_t) & \Psi_o(\eta_t) \end{bmatrix}
$$

$$
\begin{bmatrix} \Delta A_o(t,\eta_t) & \Delta A_d(t,\eta_t) & \Delta \Gamma_o(t,\eta_t) \\ \Delta C_o(t,\eta_t) & \Delta D_d(t,\eta_t) & \Delta \Psi_o(t,\eta_t) \end{bmatrix} \qquad (2.24)
$$

The delay factor τ plays a major role throughout this book. To examine its impact on other variables and system behavior, it is considered to have one of the following possible cases:

1) **Case of Weak-dependence**

in which τ is treated as unknown and time-varying with given bounds such that

$$\tau \in (0, \tau^*] \quad , \quad \dot{\tau} \leq \tau^+ < 1$$

where τ^*, τ^+ being finite known constants.

Obviously the case of $\dot{\tau} \equiv 0$ represents constant delay for which $\tau \equiv \tau^*$ at all times. This is a bit trivial case.

2)**Case of Strong-dependence**

in which τ is treated as known time-varying quantity and most of the time independent of the mode of operation.

3)**Case of Mode-dependence**

in this case the time delay depends on the operational mode and we write $\tau \equiv \tau(\eta_t)$. This could be obtained via look-up tables and represented in functional form. For this latter case, we call it functional time-delays.

In the sequel, for each possible value (mode) $\eta_t = i, \ i \in \mathcal{S},$ we will denote the system matrices of (2.20)-(2.23) associated with mode i by

$$\begin{aligned} A_o(\eta_t) &\triangleq A_o(i) \ , \ A_d(\eta_t) \triangleq A_d(i) \ , \ \Gamma(\eta_t) \triangleq \Gamma(i) \\ C_o(\eta_t) &\triangleq C_o(i) \ , \ D_d(\eta_t) \triangleq D_d(i) \ , \ \Psi(\eta_t) \triangleq \Psi(i) \\ B_o(\eta_t) &\triangleq B_o(i) \ , \ G_o(\eta_t) \triangleq G_o(i) \ , \ F(\eta_t) \triangleq F_j(i) \\ \Phi(\eta_t) &\triangleq \Phi(i) \end{aligned} \tag{2.25}$$

where $A_o(i), B_o(i), A_d(i), C_o(i), D_d(i), G(i), F(i), \Gamma(i), \Psi(i)$ and $\Phi(i)$ are known real constant matrices of appropriate dimensions which describe the nominal system of (2.20)-(2.23).

2.3.1 Model of Uncertainties

The matrices $\Delta A_o(t, \eta_t)$, $\Delta A_d(t, \eta_t)$ and $\Delta B_o(t, \eta_t)$ are real, time-varying matrix functions representing parameter uncertainties. The admissible uncertainties are assumed to be modeled in the norm-bounded form:

$$\begin{bmatrix} \Delta A_o(t,\eta_t) & \Delta A_d(t,\eta_t) & \Delta \Gamma_o(t,\eta_t) \\ \Delta C_o(t,\eta_t) & \Delta D_d(t,\eta_t) & \Delta \Psi_o(t,\eta_t) \end{bmatrix} = \begin{bmatrix} H_a(\eta_t) \\ H_c(\eta_t) \end{bmatrix} \Delta_a(t,\eta_t) \begin{bmatrix} E_a(\eta_t) & E_d(\eta_t) & E_s(\eta_t) \end{bmatrix} \tag{2.26}$$

$$\Delta B_o(t,\eta_t) = H_a(\eta_t)\Delta_b(t,\eta_t)E_b(\eta_t) \tag{2.27}$$

$$\begin{bmatrix} \Delta G_o(t,\eta_t) \\ \Delta F_o(t,\eta_t) \\ \Delta\Phi_o(t,\eta_t) \end{bmatrix} = H_z(\eta_t)\Delta_c(t,\eta_t) \begin{bmatrix} E_g(\eta_t) \\ E_f(\eta_t) \\ E_h(\eta_t) \end{bmatrix} \tag{2.28}$$

where for $\eta_t = i$, $H_a(\eta_t) \equiv H_a(i) \in \Re^{n\times\alpha_a}$, $H_c(\eta_t) \equiv H_c(i) \in \Re^{p\times\alpha_c}$, $H_z(\eta_t) \equiv H_z(i) \in \Re^{r\times\alpha_z}$, $E_a(\eta_t) \equiv E_a(i) \in \Re^{\beta\times n}$, $E_d(\eta_t) \equiv E_d(i) \in \Re^{\beta\times n}$, $E_s(\eta_t) \equiv E_s(i) \in \Re^{\beta\times q}$, $E_b(\eta_t) \equiv E_b(i) \in \Re^{\beta_b\times m}$, $E_g(\eta_t) \equiv E_g(i) \in \Re^{\beta_c\times n}$, $E_f(\eta_t) \equiv E_f(i) \in \Re^{\beta_c\times n}$, and $E_h(\eta_t) \equiv E_h(i) \in \Re^{\beta_c\times q}$ and are known real constant matrices, which designates the way how uncertain parameters in $\Delta_a(t,\eta_t) \in \Re^{\alpha_a\times\beta}$, $\Delta_b(t,\eta_t) \in \Re^{\alpha_a\times\beta_b}$, $\Delta_c(t,\eta_t) \in \Re^{\alpha_z\times\beta_c}$ affect the nominal system with $\Delta(t,\eta_t)$, $\eta_t = i \in \mathcal{S}$ and being unknown, time-varying matrix functions satisfying

$$\begin{aligned} &||\Delta_a(t,\eta_t)||_2 \le 1\,,\ ||\Delta_b(t,\eta_t)||_2 \le 1\,, \\ &||\Delta_c(t,\eta_t)||_2 \le 1 \qquad \eta_t = i \in \mathcal{S} \end{aligned} \tag{2.29}$$

where the elements of $\Delta_a(t,\eta_t) \equiv \Delta(t,i)$, $\Delta_b(t,\eta_t) \equiv \Delta(t,i)$, $\Delta_c(t,\eta_t) \equiv \Delta(t,i)$ are Lebesgue measurable for any $\eta_t = i \in \mathcal{S}$.

It is worthwhile to observe that system (2.20)-(2.23) is a hybrid system in which one state $x(t)$ takes values continuously, and another "state" $\eta(t)$ takes values discretely. We should also note that the uncertainty structure of (2.26)-(2.28) satisfying (2.29) has been widely used in robust control and filtering for uncertain systems for both deterministic and stochastic cases, see, for example, [16,17,20] and the references therein. By and large, the matrices $\Delta_a(t,i)$, $\Delta_b(t,i)$, and $\Delta_c(t,i)$ are allowed to be state-dependent, i.e., $\Delta_a(t,i) = \Delta_a(t,x,i)$ and $\Delta_d(t,i) = \Delta_d(t,x,i)$, as long as (2.29) is satisfied along all possible state trajectories. Also, observe that the unit overbound for $\Delta_a(t,i)$, $\Delta_a(t,i)$ and $\Delta_c(t,i)$ does not cause any loss of generality. Indeed, $\Delta_a(t,i)$, $\Delta_a(t,i)$ and $\Delta_c(t,i)$ can be always normalized, in the sense of (2.26)-

(2.28), by appropriately choosing the matrices $H_a(i), \cdots, E_h(i)$.

2.4 Relevant Special Cases

Admittedly, system (2.20)-(2.23) along with associated matrices (2.24)-(2.28) is quite general to encompass almost all classes of jump time-delay models under consideration. To simplify the analysis in the subsequent chapters, we extract hereafter several models as special cases from system (2.20)-(2.23). Each model of the extracted cases will be appropriately used in a particular task.

2.4.1 Nominal Models

These models are free of uncertainties and are essentially derived from system (2.20)-(2.23) by setting

$$(\Delta_a(t,i) \equiv 0\ ,\ \Delta_b(t,i) \equiv 0\ ,\ (\Delta_c(t,i) \equiv 0 \quad , \quad i \in \mathcal{S}$$

These models are further subdivided into two groups: the first group comprised of models 1)-3) and will be used for weak delay-dependent studies (only extreme values of the delay factor are needed) and the other one comprised of models 4)-6) and will be used for strong delay-dependent studies where the delay factor is known at every instant of time.

1) The free nominal jump system:

$$\begin{aligned} (\Sigma_f): \quad \dot{x}(t) &= A_o(i)x(t) + A_d(i)x(t-\tau) \ , \quad t \geq 0 \\ x(t) &= \phi(t),\ t \in [-\tau, 0],\ \ \eta_o = i \end{aligned} \tag{2.30}$$

2) The controlled nominal jump system:

$$\begin{aligned} (\Sigma_c): \quad \dot{x}(t) &= A_o(i)x(t) + A_d(i)x(t-\tau) + B_o(i)u(t) \ , \quad t \geq 0 \\ x(t) &= \phi(t),\ \ t \in [-\tau, 0],\ \ \eta_o = i \end{aligned} \tag{2.31}$$

3)The nominal jump system

$$(\Sigma_n): \quad \dot{x}(t) = A_o(i)x(t) + A_d(i)x(t-\tau) + \Gamma(i)w(t) \ , \quad t \geq 0$$

$$x(t) = \phi(t), \ \ t \in [-\tau, 0], \ \ \eta_o = i, \qquad (2.32)$$

$$z(t) = G(i)x(t) + \Phi(i)w(t) \qquad (2.33)$$

It should be remarked that systems (2.30), (2.31) and (2.32)-(2.33) are readily usable for stochastic stability, stabilization and disturbance attenuation, respectively, with weak delay-dependence. To derive the models for strong delay-dependent studies, we recall the well-known Leibniz-Newton formula to express the delayed state as:

$$\begin{aligned} x(t-\tau) &= x(t) - \int_{\tau}^{0} \dot{x}(t+\theta)d\theta \\ &= x(t) - \int_{\tau}^{0} A_o(i)x(t+\theta)d\theta - \int_{\tau}^{0} A_d(i)x(t-\tau+\theta)d\theta \\ &- \int_{\tau}^{0} B_o(i)u(t+\theta)d\theta - \int_{\tau}^{0} \Gamma(i)w(t+\theta)d\theta \end{aligned} \qquad (2.34)$$

which when substituted back into (2.20)-(2.23) and suppressing the uncertainties for $\eta_t = i \in \mathcal{S}$, it yields the respective remaining models:

4) The free nominal jump system:

$$\begin{aligned} (\Sigma_{fs}): \ \dot{x}(t) &= [A_o(i) + A_d(i)]x(t) - A_d(i)\Big[\int_{-\tau}^{0} A_o(i)x(t+\theta)d\theta \\ &+ \int_{-\tau}^{0} A_d(i)x(t-\tau+\theta)d\theta\Big] \ , \ t \geq 0 \\ x(t) &= \phi(t), \ \ t \in [-2\tau, 0], \ \ \eta_o = i \end{aligned} \qquad (2.35)$$

5) The controlled nominal jump system:

$$(\Sigma_{cs}): \ \dot{x}(t) = [A_o(i) + A_d(i)]x(t) - A_d(i)\Big[\int_{-\tau}^{0} A_o(i)x(t+\theta)d\theta$$

$$
\begin{aligned}
&+ \int_{-\tau}^{0} A_d(i)x(t-\tau+\theta)d\theta \\
&+ \int_{-\tau}^{0} B_o(i)u(t+\theta)d\theta\Big] \ , \quad t \geq 0 \\
x(t) &= \phi(t), \quad t \in [-2\tau, 0], \quad \eta_o = i
\end{aligned}
\tag{2.36}
$$

6) The nominal jump system:

$$
\begin{aligned}
(\Sigma_{ns}) : \dot{x}(t) &= [A_o(i) + A_d(i)]x(t) - A_d(i)\Big[\int_{-\tau}^{0} A_o(i)x(t+\theta)d\theta \\
&+ \int_{-\tau}^{0} A_d(i)x(t-\tau+\theta)d\theta + \int_{-\tau}^{0} B_o(i)u(t+\theta)d\theta \\
&+ \int_{\tau}^{0} \Gamma(i)w(t+\theta)d\theta\Big] \ , \ t \geq 0 \\
x(t) &= \phi(t), \quad t \in [-2\tau, 0], \quad \eta_o = i
\end{aligned}
\tag{2.37}
$$

It should be observed that models 4)-6) require initial data over the period $[-2\tau, 0]$.

2.4.2 Uncertain Models

In a similar way and by preserving the uncertainties for $\eta_t = i \in \mathcal{S}$, we deal with the following related systems:

1) The free uncertain jump system:

$$
\begin{aligned}
(\Sigma_{\Delta f}) : \quad \dot{x}(t) &= [A_o(\eta_t) + \Delta A_o(t, \eta_t)]x(t) + [A_d(\eta_t) + \Delta A_d(t, \eta_t)]x(t-\tau) \\
&= A_{\Delta o}(t, \eta_t)x(t) + A_{\Delta d}(t, \eta_t)x(t-\tau) \ , \ t \geq 0 \\
x(t) &= \phi(t) \ \ , t \in [-\tau, 0] \ , \ \eta_o \ = \ i
\end{aligned}
\tag{2.38}
$$

2) The controlled uncertain jump system:

$$
\begin{aligned}
(\Sigma_{\Delta c}) : \dot{x}(t) &= [A_o(\eta_t) + \Delta A_o(t, \eta_t)]x(t) + [A_d(\eta_t) + \Delta A_d(t, \eta_t)]x(t-\tau) \\
&+ [B_o(\eta_t) + \Delta B_o(t, \eta_t)]u(t)
\end{aligned}
$$

$$
\begin{aligned}
&= A_{\Delta o}(t,\eta_t)x(t) + A_{\Delta d}(t,\eta_t)x(t-\tau) + B_{\Delta o}(t,\eta_t)u(t)\ ,\ t \geq 0 \\
x(t) &= \phi(t)\ , t \in [-\tau, 0]\ ,\ \eta_o = i \qquad (2.39)
\end{aligned}
$$

3) The uncertain jump system

$$
\begin{aligned}
(\Sigma_{\Delta n}):\ \dot{x}(t) &= [A_o(\eta_t) + \Delta A_o(t,\eta_t)]x(t) + [A_d(\eta_t) + \Delta A_d(t,\eta_t)]x(t-\tau) \\
&+ \Gamma(\eta_t)w(t) \\
&= A_{\Delta o}(t,\eta_t)x(t) + A_{\Delta d}(t,\eta_t)x(t-\tau) + \Gamma(t,\eta_t)w(t)\ ,\ t \geq 0 \\
x(t) &= \phi(t)\ , t \in [-\tau, 0]\ ,\ \eta_o = i\ , \qquad (2.40) \\
z(t) &= G(\eta_t)x(t) + \Phi(\eta_t)w(t) \qquad (2.41)
\end{aligned}
$$

Likewise, systems (2.38), (2.39) and (2.40)-(2.41) are used in examining the problems of robust stability, robust stabilization, and robust disturbance attenuation of the uncertain system (2.20)-(2.23) under weak-delay dependence. Finally, to examine the system behavior under strong delay-dependence we employ the Leibniz-Newton formula again to express the delayed state as:

$$
\begin{aligned}
x(t-\tau) &= x(t) - \int_{\tau}^{0} \dot{x}(t+\theta)d\theta \\
&= x(t) - \int_{\tau}^{0} A_{\Delta o}(t,\eta_t)x(t+\theta)d\theta - \int_{\tau}^{0} A_{\Delta d}(t,\eta_t)x(t-\tau+\theta)d\theta \\
&- \int_{\tau}^{0} B_{\Delta o}(t,\eta_t)u(t+\theta)d\theta - \int_{\tau}^{0} \Gamma_{\Delta d}(t,\eta_t)w(t+\theta)d\theta \qquad (2.42)
\end{aligned}
$$

Substituting (2.42) back into system (2.20)-(2.23) for $\eta_t = i \in \mathcal{S}$, it yields the following related models:

4) The free uncertain jump system:

$$
\begin{aligned}
(\Sigma_{\Delta fs}):\ \dot{x}(t) &= [A_{\Delta o}(t,i) + A_{\Delta d}(t,i)]x(t) \\
&- A_{\Delta d}(t,i)\Big[\int_{-\tau}^{0} A_{\Delta o}(t,i)x(t+\theta)d\theta
\end{aligned}
$$

$$
\begin{aligned}
& + \int_{-\tau}^{0} A_{\Delta d}(t,i)x(t-\tau_{\eta_t}+\theta)d\theta\Big] \ , \ t \geq 0 \\
x(t) &= \phi(t) \ , t \in [-2\tau, 0] \ , \ \eta_o = i \ , \qquad (2.43)
\end{aligned}
$$

5) The controlled uncertain jump system:

$$
\begin{aligned}
(\Sigma_{\Delta cs}): \ \dot{x}(t) &= [A_{\Delta o}(t,i) + A_{\Delta d}(t,i)]x(t) \\
&- A_{\Delta d}(t,i)\Big[\int_{-\tau}^{0} A_{\Delta o}(t,i)x(t+\theta)d\theta \\
&+ \int_{-\tau}^{0} A_{\Delta d}(t,i)x(t-\tau_{\eta_t}+\theta)d\theta \\
&+ \int_{-\tau}^{0} B_{\Delta o}(t,i)u(t+\theta)d\theta + \int_{-\tau}^{0} \Gamma_{\Delta o}(t,i)w(t+\theta)d\theta\Big] \ , \ t \geq 0 \\
x(t) &= \phi(t) \ , t \in [-2\tau, 0] \ , \ \eta_o = i \ , \qquad (2.44)
\end{aligned}
$$

6) The uncertain jump system:

$$
\begin{aligned}
(\Sigma_{\Delta ns}): \ \dot{x}(t) &= [A_{\Delta o}(t,i) + A_{\Delta d}(t,i)]x(t) \\
&- A_{\Delta d}(t,i)\Big[\int_{-\tau}^{0} A_{\Delta o}(t,i)x(t+\theta)d\theta \\
&+ \int_{-\tau}^{0} A_{\Delta d}(t,i)x(t-\tau_{\eta_t}+\theta)d\theta \\
&+ \int_{-\tau}^{0} B_{\Delta o}(t,i)u(t+\theta)d\theta + \int_{-\tau}^{0} \Gamma_{\Delta o}(t,i)w(t+\theta)d\theta\Big] \ , \ t \geq 0 \\
x(t) &= \phi(t) \ , t \in [-2\tau, 0] \ , \ \eta_o = i \ , \qquad (2.45) \\
z(t) &= G(\eta_t)x(t) + \Phi(\eta_t)w(t) \qquad (2.46)
\end{aligned}
$$

Throughout the book we will adopt a dual notational frame of reference in the sense that we either refer to the general jump time-delay system as system (2.20)-(2.23) or simply system (Σ_G). Similarly, we refer to the nominal jump systems as (2.32)-(2.33) or simply (Σ_n).

Chapter 3

Stochastic Stability and Stabilization

In this chapter we provide a closer look at the fundamental problems of stochastic stability and stabilizability of continuous-time jump time-delay systems (JTDS) while taking into consideration all the system ingredients: model uncertaintes, jumping parameters and time-delay factors. Our rational throughout this chapter is to give a short account explaining the concepts and subject matter which then followed by a mathematical definition and finally establish a testable criteria in the form of family of linear matrix inequalities (LMIs).

3.1 Introduction

In control engineering, stability problem is the key problem in almost all research studies, development and design. Depending on the model representation, system stability can be formulated in many different ways. Common to all of these ways is the major objective of guaranteeing asymptotic behavior of JTDS irrespective of external disturbances or internal changes. In this section, we focus on the stability issues of and by examining various factors we seek to reach computable criteria to be used in simulation.

Recall the probability space $(\Omega, \mathcal{F}, \mathbf{P})$ described in Chapter 1 over which all class of dynamical systems with Markovian jump parameters described in the sequel are defined. In addition, we consider the classes of models presented in Chapter 2, section 2.4.2 and focus initially on the model given by (2.38) which describes a mode-independent model delay. It is repeated hereafter for easy reference.

$$\begin{aligned} \dot{x}(t) &= [A_o(\eta_t) + \Delta A_o(t, \eta_t)]x(t) + [A_d(\eta_t) + \Delta A_d(t, \eta_t)]x(t - \tau) \\ &= A_{\Delta o}(t, \eta_t)x(t) + A_{\Delta d}(t, \eta_t)x(t - \tau) \end{aligned} \tag{3.1}$$

All the associated matrices are defined there and the subsequent analysis will be carried out for every $\eta_t = i \in \mathcal{S}$. The initial vector function is specified as $\xi_0 \equiv \langle x(0), x(s) \rangle = \langle x_o, \phi(s) \rangle$, where $\phi(\cdot) \in \mathcal{L}_2[-\tau, 0]$ and will be assumed, throughout this paper, that it is independent of the process $\{\eta_t,\ t \in [0, \mathcal{T}]\}$. In the sequel, we let $\mathsf{X}(\xi_0, \eta_0)$ denote the state trajectory in system (3.1) from the initial state (ξ_0, η_0).

In our endeavor to examine the notions of robust stochastic stability and analyze the subsequent behavior of the JTDS under consideration, we subdivide our effort into four distinct categories, each of which has its own definition and testing condition. The distinction between these cases arises from the information set available to the designer in relation to the time-delay factor. The categories are:

(1) Robust stochastic stability with weak delay-dependence (RSSWDD)

In this case, the time-delay $\tau(t)$ is unknown and time-varying quantity satisfying

$$0 \le \tau(t) \le \tau^*, \qquad 0 \le \dot{\tau}(t) \le \tau^+$$

where the bounds τ^*, and $\tau^+ < 1$ are known constants. It is readily seen that the information size is small and hence the link between stability and time-delay is rather weak.

(2) Robust stochastic stability with strong delay-dependence (RSSSDD)

Here the link between stability and time-delay is rather strong since the information size is big due to the requirement that the time-delay $\tau(t)$ is time-varying quantity known at every instant of time (continuously measurable)

(3) Robust stochastic stability with functional time-delay (RSSFTD)

In this case, the time-delay varies with the mode of operation expressed as

$$\tau_{\eta_t} \equiv \tau_j, \; j \in \mathcal{S}$$

and the information set contains only the maximum and minimum bounds on the functional dependence

$$\hat{\tau} \stackrel{\Delta}{=} \max_j \{\tau_j, j \in \mathcal{S}\} \;\; , \;\; \check{\tau} \stackrel{\Delta}{=} \min_j \{\tau_j, j \in \mathcal{S}\}$$

(4) Robust strong stochastic stability with functional time-delay (RSSS-FTD)

Here time-delay τ_{η_t} is strongly associated with the mode of operation which, in turn, requires a look-up table or stored relation to retrieve the amount of delay at each operational mode.

3.2 Mode-Independent Stochastic Stability

In the sequel, we examine mode-independent robust stochastic stability with weak and strong delay-dependence corresponding to categories **(1)** and **(2)**. The theorems established in the sequel show that the stochastic stability behavior of system (3.1) is related to the existence of a positive definite solution of a family of algebraic Riccati linear matrix inequalities (LMIs) or inequalities (ARIs) thereby providing a clear key to designing feedback controllers later on.

3.2.1 Weak Delay-Dependence

Initially, we have the following definition :

Definition 3.1 *System (3.1) is said to be* **robustly stochastically stable and weakly delay-dependent (RSSWDD)** *given the bounds* τ^* *and* $\tau^+ < 1$ *if for all finite initial vector function* $\xi_0 \in \Re^n$ *defined on the interval* $[-\tau^*, 0]$ *and any initial mode* $\eta_o \in \mathcal{S}$, *the following inequality*

$$\mathbb{E}\left\{ \int_0^\infty \|X(\xi_0, \eta_0)\|^2 \ dt|\xi_0, \eta_0 \right\} < \infty$$

holds for any admissible parameter uncertainties satisfying (2.26)-(2.29).

Before establishing the stability result, we consider that given a sequence of matrices $Q(i) = Q^t(i) > 0, \ i \in \mathcal{S}$ we define

$$\begin{aligned} \bar{Q}(i) &= (1 - \tau^+)Q(i) - \xi^{-1}(i) \sum_{m=1}^{s} \alpha_{im} Q(m) \\ \hat{Q}(i) &= Q(i) + \xi(i) \sum_{m=1}^{s} \alpha_{im} Q(m), \qquad i \in \mathcal{S} \end{aligned} \tag{3.2}$$

for some scalars $\xi(i) > 0, \ i \in \mathcal{S}$ where $\bar{Q}(i) > 0, \ i \in \mathcal{S}$ by selection of $\xi(i) \, , i \in \mathcal{S}$.

Now we establish the following result:

Theorem 3.1 *System (3.1) is* **RSSWDD** *if given the bounds* τ^* *and* $\tau^+ < 1$ *and matrices* $\bar{Q}(i) = \bar{Q}^t(i) > 0, \ \hat{Q}(i) = \hat{Q}^t(i) > 0, \ i \in \mathcal{S}$ *there exist matrices* $P(i) = P^t(i) > 0$ *and scalars* $\epsilon(i) > 0$, $\rho(i) > 0$, $i \in \mathcal{S}$, *satisfying the system of LMIs for all* $i \in \mathcal{S}$

$$\begin{bmatrix} \Sigma_a(i) & \Sigma_b(i) & P(i)A_d(i) \\ \Sigma_b^t(i) & -\Sigma_c(i) & 0 \\ A_d^t(i)P(i) & 0 & \begin{matrix} -\bar{Q}(i) + \\ \rho(i)N_d^t(i)N_d(i) \end{matrix} \end{bmatrix} < 0 \tag{3.3}$$

where

$$\begin{aligned}
\Sigma_a(i) &= P(i)A_o(i) + A_o^t(i)P(i) + \sum_{m=1}^{s} \alpha_{im} P(m) \\
&+ \hat{Q}(i) + \epsilon(i) N_a^t(i) N_a(i) \\
\Sigma_b(i) &= [P(i)M_a(i) \quad P(i)M_a(i)] \\
\Sigma_c(i) &= diag[\epsilon(i)I \quad \rho(i)I]
\end{aligned} \tag{3.4}$$

Proof: Let $\mathbf{x}_s(t) \stackrel{\Delta}{=} x(s+t),\ t-\tau \ \leq \ s \ \leq \ t$ and define the process $\{(\mathbf{x}(t), \eta_t),\ t \ \geq \ 0\}$ over the state space $\bar{\mathcal{C}}$. It should be observed that $\{(\mathbf{x}(t), \eta_t),\ t \ \geq \ 0\}$ is strong Markovian [78]. For $\eta_t = i, i \in \mathcal{S}$, and given $Q(i) = Q^t(i) > 0$, let the Lyapunov functional $V(\cdot) : \Re^n \times \Re_+ \times \mathcal{S} \to \Re_+$ be selected as

$$\begin{aligned}
V(x, \eta_t = i) &= V(x, i) \\
&= x^t(t)P(i)x(t) + \int_{-\tau}^{0} x^t(t+\theta)Q(i)x(t+\theta)d\theta
\end{aligned} \tag{3.5}$$

The weak infinitesimal operator $\Im_1^x[\cdot]$ of the process $\{x(t), \eta_t, t \geq 0\}$ for system (3.1) at the point $\{t, x, \eta_t\}$ is given by:

$$\Im_1^x[V] = \partial V/\partial t + \dot{x}^t(t)\ \partial V/\partial x \mid_{\eta_t = i} + \sum_{m=1}^{s} \alpha_{im} V(t, x, i, m) \tag{3.6}$$

Using (3.1) into (3.5)-(3.6) and manipulating the terms we get:

$$\begin{aligned}
\Im_1^x[V] &= x^t(t)\Big\{ A_o^t(i)P(i) + P(i)A_o(i) + \sum_{m=1}^{s} \alpha_{im} P(m) \\
&+ Q(i) \Big\} x(t) - (1-\dot{\tau}) x^t(t-\tau) Q(i) x(t-\tau) \\
&+ \sum_{m=1}^{s} \alpha_{im} \int_{-\tau}^{0} x^t(t+\theta) Q(m) x(t+\theta) d\theta \\
&+ x^t(t-\tau) A_d^t(i) P(i) x(t) + x^t(t) P(i) A_d(i) x(t-\tau)
\end{aligned}$$

$$
\begin{aligned}
&+ \quad x^t(t)\Big\{\Delta A_o^t(i)P(i) + P(i)\Delta A_o(i)\Big\}x(t) \\
&+ \quad x^t(t-\tau)\Delta A_d^t(i)P(i)x(t) \\
&+ \quad x^t(t)P(i)\Delta A_d(i)x(t-\tau)
\end{aligned} \tag{3.7}
$$

First, it is easy to see that the following inequality holds

$$
\begin{aligned}
&\sum_{m=1}^{s} \alpha_{im} \int_{-\tau}^{0} x^t(t+\theta)Q(m)x(t+\theta)d\theta \\
&\le \xi(i)\; x^t(t) \sum_{m=1}^{s} \alpha_{im} Q(m)x(t) + \\
&\xi^{-1}(i)\; x^t(t-\tau) \sum_{m=1}^{s} \alpha_{im} Q(m)x(t-\tau)
\end{aligned} \tag{3.8}
$$

for some scalars $\xi(i) > 0,\ i \in \mathcal{S}$. Now we focus on the uncertainties and use **Fact 1** from the appendix to yield:

$$
\begin{aligned}
&\Delta A_o^t(i)P(i) + P(i)\Delta A_o(i) \\
&\le \epsilon^{-1}(i)P(i)M_a(i)M_a^t(i)P(i) + \epsilon(i)N_a^t(i)N_a
\end{aligned} \tag{3.9}
$$

for some scalars $\epsilon(i) > 0,\ i \in \mathcal{S}$. Next, combining (3.2) and (3.8)-(3.9), applying the argument of 'completing the squares' and over-bounding the result using **Fact 2**, we get:

$$
\begin{aligned}
\Im_1^x[V] \quad &\le \quad x^t(t)\Big\{ A_o^t(i)P(i) + P(i)A_o(i) + \sum_{m=1}^{s} \alpha_{im} P(m) \\
&+ \quad \epsilon(i)N_a^t(i)N_a + \epsilon^{-1}(i)P(i)M_a(i)M_a^t(i)P(i) + \hat{Q}(i) \\
&+ \quad \rho^{-1}(i)P(i)M_a(i)M_a^t(i)P(i) \\
&+ \quad P(i)A_d(i)[\bar{Q}(i) - \rho(i)N_d^t(i)N_d(i)]^{-1}A_d^t(i)P(i)\Big\}x(t) \\
&\triangleq \quad x^t(t)\Pi(i)x(t)
\end{aligned} \tag{3.10}
$$

In view of (3.3) and the Schur complements (See Appendix), it follows that $\Pi(i) < 0 \; \forall \; i \in \mathcal{S}$. Therefore we conclude that $\Im_1^x[V] \; < \; 0$ for all $x \neq 0$ and $\Im_1^x[V] \; \leq \; 0$ for all x.

Recalling from [85] that

$$||x(t+\beta)|| \; \leq \varphi||x(t)|, \; \forall \beta \in [-\tau, 0]$$

and some $\varphi > 0$. It follows from (3.5) that

$$V(x,i) \leq x^t(t)P(i)x(t) + \mu \; ||x||^2$$

where

$$\mu = \varphi\tau\big(\max_i \lambda_M[P(i)] + \lambda_M[Q(i)]\big)$$

. Therefore, for all $x \neq 0$, we have

$$\begin{aligned} \frac{\Im_1^x[V]}{V(x,i)} \;\; &\leq \;\; \frac{x^t\Pi(i)x}{x^tP(i)x + \mu||x||^2} \\ &\leq \;\; -\zeta \; \triangleq \; -\min_{i \in \mathcal{S}} \left\{ \frac{\lambda_m[-\Pi(i)]}{\lambda_M[P(i)] + \mu} \right\} \end{aligned} \tag{3.11}$$

It is readily seen from (3.11) that $\zeta > 0$ and hence we get

$$\Im_1^x[V] \;\; \leq -\zeta \, V(t,x,i)$$

It follows from [78] by using the Gronwall-Bellman lemma [98] and letting $x(t = 0, \phi, \eta_o) = x_o$, one has

$$\mathbb{E}[V(x,i)|\phi, \eta_o] \;\; \leq \;\; e^{-\zeta \, t} \, V(x_o, i) \tag{3.12}$$

Since

$$\mathbb{E}\bigg\{ \int_{-\tau}^{0} x^t(t+\theta)Q(i)x(t+\theta)d\theta|\phi, \eta_o \bigg\} \geq 0$$

it is easy to see from (3.5) that

$$\mathbb{E}\bigg\{ x^t(t)P(i)x(t)|\phi, \eta_o \bigg\} \leq e^{-\zeta \, t}V(x_o, i) \Longrightarrow$$

$$
\begin{aligned}
&\mathbb{E}\Big\{ \int_0^{\mathcal{T}} x^t(t)P(i)x(t)dt|\phi, \eta_o = i \Big\} \\
&\le \Big[\int_0^{\mathcal{T}} e^{-\zeta\, t} dt \Big] V(x_o, i) = \frac{1}{\zeta}[e^{-\zeta\, \mathcal{T}} \; - \; 1] \; V(x_o, i) \\
&\Longrightarrow \lim_{\mathcal{T}\to\infty} \; \mathbb{E}\Big\{ \int_0^{\mathcal{T}} x^t(t)P(i)x(t)dt|\phi, \eta_o = i \Big\} \\
&\le \; \frac{1}{\zeta} x_o^t P(\eta_o)x_o + \frac{\tau^*}{\zeta}[Q(\eta_o)]||x(t+\theta)||_*^2, \;\; \forall \theta \in [-\tau, 0] \qquad (3.13)
\end{aligned}
$$

where $||x(t+\theta)||_*^2 \stackrel{\Delta}{=} \sup_{\theta\in[-\tau,0]} ||x(t+\theta)||_2^2$. Letting

$$
\bar{P}(i) \; = \; \max_{i \in \mathcal{S}} \left\{ \frac{P(\eta_o)||x_o||^2 + \tau^*[Q(\eta_o)]||x(t+\theta)||_*^2}{\zeta[P(\eta_o)]||x_o||^2} \right\}
$$

it follows from (3.13) for $i \in \mathcal{S}$ that

$$
\begin{aligned}
&\lim_{\mathcal{T}\to\infty} \; \mathbb{E}\Big\{ \int_0^{\mathcal{T}} x^t(t)x(t)dt|\phi, \eta_o = i \Big\} \\
&\le x_o^t \lambda_M(\bar{P}(i))x_o < +\infty
\end{aligned}
$$

which, in the light of **Definition** 3.1, shows that system (3.1) is **RSSWDD**. $\nabla\nabla\nabla$

Remark 3.1 *One the significant features of* **Theorem** *3.1 is that in the delay-less case ($A_d(.) \equiv 0$), it recovers the basic result of [45] and thus it generalizes the results of [45] to JTD systems.*

Next consider the jump dynamical system for $\eta_t = i \in \mathcal{S}$

$$
\begin{aligned}
\dot{x}(t) \;\; &= \;\; A_{\Delta o}(t,i)x(t) + A_{\Delta d}(t,i)x(t-\tau) + \Gamma(i)w(t) \\
z(t) \;\; &= \;\; G_o(i)x(t) + \Phi(i)w(t) \qquad (3.14)
\end{aligned}
$$

For system (3.14), we have the following definition

Definition 3.2 *System (3.14) is said to be* **RSSWDD with a disturbance attenuation** γ *if given the bounds* τ^* *and* $\tau^+ < 1$, *for zero initial vector function* $\phi \equiv 0$ *defined on the interval* $[-\tau, 0]$ *and initial mode* $\eta_o \in \mathcal{S}$ *the following inequality*

$$||z(t)||_{E_2} \triangleq I\!\!E\left[\int_0^\infty z^t(t)z(t)dt\right]^{1/2} < \gamma\,||w(t)||_2$$

holds for all $0 \neq w(t) \in \mathcal{L}_2[0, \infty)$ *and for all admissible parameter uncertainties satisfying (2.26)-(2.29).*

Now we are in a position to present the following result

Theorem 3.2 *System (3.14) is* **RSSWDD with a disturbance attenuation** γ, *if given the bounds* τ^* *and* $\tau^+ < 1$ *and matrices* $\bar{Q}(i) = \bar{Q}^t(i) > 0$, $\hat{Q}(i) = \hat{Q}^t(i) > 0$, $i \in \mathcal{S}$ *there exist matrices* $P(i) = P^t(i) > 0$ *and scalars* $\gamma > 0$, $\epsilon(i) > 0$, $\rho(i) > 0$, $i \in \mathcal{S}$, *satisfying the system of LMIs for all* $i \in \mathcal{S}$

$$\begin{bmatrix} \Sigma_g(i) & \Sigma_b(i) & P(i)A_d(i) & \begin{matrix} P(i)\Gamma(i)+ \\ G_o^t(i)\Phi(i) \end{matrix} \\ \Sigma_b^t(i) & -\Sigma_c(i) & 0 & 0 \\ A_d^t(i)P(i) & 0 & \begin{matrix} -\bar{Q}(i)+ \\ \rho(i)N_d^t(i)N_d(i) \end{matrix} & 0 \\ \begin{matrix} \Gamma^t(i)P(i)+ \\ \Phi^t(i)G_o(i) \end{matrix} & 0 & 0 & -\gamma^2 I \end{bmatrix} < 0 \quad (3.15)$$

where $\Sigma_g(i) = \Sigma_a(i) + G_o^t(i)G_o(i)$

Proof. The stochastic stability of system (3.14) follows as a result of **Theorem** 3.1. What we need at this stage is to show that system (3.14) has a disturbance attenuation γ. Let the Lyapunov functional $V(t, x, \eta_t)$, for $\eta_t = i$ be given by (3.5). By evaluating the weak infinitesimal operator $\Im_2^x[\cdot]$ of the process $\{x(t), \eta_t, t \geq 0\}$ for system (3.14) at the point $\{t, x, \eta_t\}$ using (3.10) and manipulating we get

$$\Im_2^x[V] \leq \Im_1^x[V] + x^t(t)P(i)\Gamma(i)w(t) + w^t(t)\Gamma^t(i)P(i)x(t)$$

$$\triangleq \Im_{+}^{x}[V]$$

Now, we introduce

$$\mathcal{J}(x) := \mathbb{E}\Big\{ \int_0^{\infty} [z^t(t)z(t) \; - \; \gamma^2 w^t(t)w(t) \;]dt \Big\}$$

By Dynkin's formula [78], one has

$$\mathbb{E}\Big\{ \int_0^{\infty} \Im_{+}^{x}[V]dt \Big\} = \mathbb{E}\{V(t,x,\eta_t)|_{t=\infty}\} - V(t,x,\eta_t)|_{t=0} \geq 0$$

With some standard manipulations using (3.14), we obtain:

$$\begin{aligned}
&\mathcal{J}(x) = \mathbb{E}\Big\{ \int_0^{\infty} [z^t(t)z(t) \; - \; \gamma^2 w^t(t)w(t) \; + \Im_{+}^{x}[V] - \Im_{+}^{x}[V]]dt \Big\} \\
&\leq \mathbb{E}\Big\{ \int_0^{\infty} [z^t(t)z(t) \; - \; \gamma^2 w^t(t)w(t) \; + \Im_{+}^{x}[V]]dt \Big\} \\
&< \mathbb{E}\Big\{ \int_0^{T} x^t(t)\Big\{ A^t(i)P(i) + P(i)A(i) + \sum_{m=1}^{s} \alpha_{im}P(m) + \\
&P(i)A_d(i)Q^{-1}(i)A_d^t(i)P(i) + G^t(i)G(i) + \\
&\gamma^{-2}[P(i)\Gamma(i) + G^t(i)\Phi(i)][\Gamma^t(i)P(i) + \Phi^t(i)G(i)] + Q(i)\Big\}x(t)\Big\} \quad (3.16)
\end{aligned}$$

By using (3.15) via the Schur complements and the results of **Theorem** 3.1, it follows from inequality (3.16) that $\mathcal{J}(x) < 0$ and by **Definition** 3.2, the proof is completed. $\nabla\nabla\nabla$

Remark 3.2 *It should be remarked that the foregoing two theorems provide sufficient stochastic stability criteria expressed as LMI-feasibility conditions These conditions are standard linear matrix inequalities in the variables* $P(i)$, $\mu(i)$, $\epsilon(i)$, $i \in \mathcal{S}$ *[25] and therefore can be conveniently solved by the software environment [57]. In those cases when the result gives infeasible solution, we adjust* $Q(i)$, $\xi(i)$, $i \in \mathcal{S}$ *and* $\xi(i), i \in \mathcal{S}$ *before repeating the process.*

Now by suppressing the uncertainties $\Delta(.,.) \equiv 0$ in (3.14), we readily obtain the nominal dynamical system

$$\begin{aligned} \dot{x}(t) &= A_o(i)x(t) + A_d(i)x(t-\tau) + \Gamma(i)w(t) \\ z(t) &= G_o(i)x(t) + \Phi(i)w(t) \end{aligned} \tag{3.17}$$

The following two corollaries as special cases of **Theorems** 3.1-3.2.

Corollary 3.1 *System (3.17) with $w(t) \equiv 0$ is* **stochastically stable and weakly delay-dependent (SSWDD)** *if given the bounds τ^*, and $\tau^+ < 1$ and matrices $\bar{Q}(i) = \bar{Q}^t(i) > 0$, $\hat{Q}(i) = \hat{Q}^t(i) > 0$, $i \in \mathcal{S}$ there exist matrices $P(i) = P^t(i) > 0$ satisfying the system of LMIs for all $i \in \mathcal{S}$*

$$\begin{bmatrix} \Sigma_{ao}(i) & P(i)A_d(i) \\ A_d^t(i)P(i) & -\bar{Q}(i) \end{bmatrix} < 0 \tag{3.18}$$

where

$$\Sigma_{ao}(i) = P(i)A_o(i) + A_o^t(i)P(i) + \sum_{m=1}^{s} \alpha_{im}P(m) + \hat{Q}(i) \tag{3.19}$$

Corollary 3.2 *System (3.17) is* **SSWDD with a disturbance attenuation** *γ, if given the bounds τ^*, and $\tau^+ < 1$ and matrices $\bar{Q}(i) = \bar{Q}^t(i) > 0$, $\hat{Q}(i) = \hat{Q}^t(i) > 0$, $i \in \mathcal{S}$ there exist matrices $P(i) = P^t(i) > 0$ and scalar $\gamma > 0$, satisfying the system of LMIs for all $i \in \mathcal{S}$*

$$\begin{bmatrix} \Sigma_{ao}(i) & P(i)A_d(i) & \begin{matrix} P(i)\Gamma(i)+ \\ G_o^t(i)\Phi(i) \end{matrix} \\ A_d^t(i)P(i) & -\bar{Q}(i) & 0 \\ \begin{matrix} \Gamma^t(i)P(i)+ \\ \Phi^t(i)G_o(i) \end{matrix} & 0 & -\gamma^2 I \end{bmatrix} < 0 \tag{3.20}$$

3.2.2 Strong Delay-Dependence

Sometimes, it can be argued that weak delay-dependent criteria of stability, stabilization and disturbance attenuation are generally conservative [98] since they have limited information about the delay factor τ. This was the case with the first category **1**. Turning now to the second category **2** and focus attention on the strong delay-dependent stochastic stability.

To proceed further and since we rely on the availability of the delay factor τ all the time, we need to bring the factor $A_d(.)$ into the main dynamics of the system. This in turn calls for an appropriate system transformation. There are several methods to accomplish this task and they will be discussed in subsequent chapters. Here we utilize a simple transformation based on the standard the Leibniz-Newton formula

$$\int_a^b \dot{f}(t)\, dt = f(b) - f(a)$$

It follows for $t \geq \tau$ and $\eta_t = i \in \mathcal{S}$ into system (3.14) that

$$\begin{aligned} x(t-\tau) &= x(t) - \int_{-\tau}^{0} \dot{x}(t+\theta)d\theta \\ &= x(t) - \int_{-\tau}^{0} A_{\Delta o}(t,i)x(t+\theta)d\theta \\ &- \int_{-\tau}^{0} A_{\Delta d}(t,i)x(t-\tau+\theta)d\theta x(t-\tau) \\ &- \int_{-\tau}^{0} \Gamma(i)w(t+\theta)d\theta \end{aligned} \tag{3.21}$$

Upon substituting (3.21) back into (3.14) it yields:

$$\begin{aligned} \dot{x}(t) &= [A_{\Delta o}(t,i) + A_{\Delta d}(t,i)]x(t) \\ &- A_{\Delta d}(t,i)\left\{\int_{-\tau}^{0} A_{\Delta o}(t,i)x(t+\theta)d\theta\right\} \\ &+ A_{\Delta d}(t,i)\left\{\int_{-\tau}^{0} A_{\Delta d}(t,i)x(t-\tau+\theta)d\theta\right\} \\ &- A_{\Delta d}(t,i)\int_{-\tau}^{0} \Gamma(i)w(t+\theta)d\theta \\ x(t) &= \phi(t),\ t \in [-2\tau, 0],\ \eta_o = i \end{aligned} \tag{3.22}$$

Definition 3.3 *System (3.22) with $w(.) \equiv 0$ is said to be* **robustly stochastically stable and strongly delay-dependence (RSSSDD)** *for any time-delay τ satisfying $0 \leq \tau \leq \bar{\tau}$ if for all finite initial vector function $\xi_0 \in \Re^n$ defined on the interval $[-\bar{\tau}, 0]$ and any initial mode $\eta_o \in \mathcal{S}$, the following inequality*

$$\mathbb{E}\left\{ \int_0^\infty \|\mathsf{X}(\xi_0, \eta_0)\|^2 \; dt | \xi_0, \eta_0 \right\} \; < \; \infty$$

holds for any admissible parameter uncertainties satisfying (2.26)-(2.29).

Definition 3.4 *System (3.22) is said to be* **robustly stochastically stable and strongly delay-dependence (RSSSDD) with a disturbance attenuation γ for any time-delay** τ *satisfying $0 \leq \tau \leq \bar{\tau}$ if for all finite initial vector function $\xi_0 \in \Re^n$ defined on the interval $[-\bar{\tau}, 0]$ and any initial mode $\eta_o \in \mathcal{S}$, the following inequality*

$$||z(t)||_{E_2} \; \stackrel{\Delta}{=} \; \mathbb{E}\left[\int_0^\infty z^t(t) z(t) dt \right]^{1/2} \; < \; \gamma \, ||w(t)||_2$$

holds for all $0 \neq w(t) \in \mathcal{L}_2[0, \infty)$ and for all admissible parameter uncertainties satisfying (2.26)-(2.29).

Theorem 3.3 *Given a scalar $\bar{\tau} > 0$, system (3.22) with $w(.) \equiv 0$ is* **RSSSDD for any time-delay τ satisfying** $0 \; \leq \; \tau \; \leq \; \bar{\tau}$ *if there exist matrices $0 < X(i) = X^t(i) \in \Re^{n \times n}$, $i \in \mathcal{S}$ and scalars $\sigma_1(i) > 0$, $\sigma_2(i) > 0$, $\sigma_3(i) > 0$, $\sigma_4(i) > 0$, $\sigma_5(i) > 0$, $\sigma_6(i) > 0$, $\sigma_7(i) > 0$, $i \in \mathcal{S}$ satisfying the system of LMIs for all $i \in \mathcal{S}$:*

$$\begin{bmatrix} \Pi_a(i) & \Lambda_a(i,\tau) & \Lambda_b(i,\tau) \\ \Lambda_a^t(i,\tau) & -\Theta_a(i,\tau) I & 0 \\ \Lambda_b^t(i,\tau) & 0 & -\Theta_b(i,\tau) I \end{bmatrix} < 0,$$

$$\begin{bmatrix} -\omega_1(i) I & \sigma_2(i) N_d^t(i) \\ \sigma_2(i) N_d(i) & -\sigma_2(i) I \end{bmatrix} < 0 \, , \begin{bmatrix} -\omega_2(i) I & \sigma_3(i) M_a(i) \\ \sigma_3(i) M_a(i) & -\sigma_3(i) I \end{bmatrix} < 0$$

$$\begin{bmatrix} -\sigma_6(i)I & \sigma_4(i)M_a(i) \\ \sigma_4(i)M_a^t(i) & -\sigma_4(i)I \end{bmatrix} < 0 \, , \begin{bmatrix} -\sigma_7(i)I & \sigma_5(i)M_d(i) \\ \sigma_5(i)M_d^t(i) & -\sigma_5(i)I \end{bmatrix} < 0$$

$$\begin{bmatrix} X(i) & I \\ I & Y(i) \end{bmatrix} \geq 0 \ , \ \begin{bmatrix} \sigma_6(i) & I \\ I & \omega_1(i) \end{bmatrix} \geq 0 \ , \ \begin{bmatrix} \sigma_7(i) & I \\ I & \omega_2(i) \end{bmatrix} \geq 0 \quad (3.23)$$

for all admissible uncertainties satisfying (2.26)-(2.29) where

$$\begin{aligned}
\Pi_a(i) &= X(i)[A_o(i) + A_d(i)] + [A_o(i) + A_d(i)]^t X(i) \\
&+ X(i)\{\sum_{m=1}^{s} \alpha_{im} Y(m)\} X(i) + \sigma_1(i)[N_a^t(i)N_a(i) + N_d^t(i)N_d(i)] \\
\Lambda_a(i,\tau) &= [\Lambda_{a1}(i,\tau) \;\; \Lambda_{a2}(i,\tau)] \\
\Lambda_{a1}(i,\tau) &= [X(i)M_a(i) \;\; X(i)M_d(i) \;\; \tau X(i)M_d(i)] \\
\Lambda_{a2}(i,\tau) &= [\tau X(i)M_d(i) \;\; \tau N_a^t(i) \;\; \tau N_d^t] \\
\Lambda_b(i,\tau) &= [\tau X(i)A_d^t(i) \;\; \tau X(i)A_d^t(i) \;\; \tau A_o^t(i) \;\; \tau A_d^t(i)] \\
\Theta_a(i,\tau) &= diag[\Theta_{a1}(i,\tau) \;\; \Theta_{a2}(i,\tau) \;\; \Theta_{a3}(i,\tau) \;\; \Theta_{a4}(i,\tau)] \\
\Theta_{a1}(i,\tau) &= diag[\sigma_1(i)I \;\; \sigma_1(i)I \;\; \tau\sigma_2(i)I] \\
\Theta_{a2}(i,\tau) &= diag[\tau_{(i)}\sigma_3(i)I \;\; \tau\sigma_4(i)I \;\; \tau\sigma_5(i)I] \\
\Theta_b(i,\tau) &= diag[\Theta_{b1}(i,\tau) \;\; \Theta_{b2}(i,\tau)] \\
\Theta_{b1}(i,\tau) &= \tau[\omega_1(i)I - \sigma_2(i)N_d^t(i)N_d(i)] \\
\Theta_{b2}(i,\tau) &= \tau[\omega_1(i)I - \sigma_3(i)N_d^t(i)N_d(i)] \\
\Theta_{b3}(i,\tau) &= \tau[\sigma_6(i)I - \sigma_4(i)M_a(i)M_a^t(i)] \\
\Theta_{b3}(i,\tau) &= \tau[\sigma_7(i)I - \sigma_5(i)M_d(i)M_d^t(i)] \qquad (3.24)
\end{aligned}$$

Proof: Introduce the following Lyapunov-Krasovskii functional for $\eta_t = i \in \mathcal{S}$, $\mathbf{x} \in \mathcal{C}[-\tau, 0]$, thus:

$$V(\mathbf{x}, t, i) \;=\; x^t(t)X(i)x(t) + W(\mathbf{x}, t, i) \qquad (3.25)$$

where $0 < X(i) = X^t(i) \in \Re^{n \times n}$, $i \in \mathcal{S}$ and $W(\mathbf{x}, t, i) > 0$ is a quadratic form to defined later on. The weak infinitesimal operator $\Im_s^x[\cdot]$ of the process

$\{(\mathbf{x}(t), \eta_t),\ t\ \geq\ 0\}$ for system (3.22) at the point $\{t, \mathbf{x}, \eta_t\}$ is given by :

$$
\begin{aligned}
\Im_s^x[V(.,.)] &= x^t(t)\Big\{X(i)[A_{\Delta o}(t,i) + A_{\Delta d}(t,i)] + [A_{\Delta o}(t,i) + A_{\Delta d}(t,i)]^t X(i) \\
&+ \sum_{k=1}^{s} \alpha_{ik} X(k)\Big\} x(t) + \varrho(t,i) + \varphi(t,i) + \Im_s^x[W(.,.)] \qquad (3.26)
\end{aligned}
$$

where

$$
\begin{aligned}
\varrho(t,i) &= -2\int_{-\tau}^{0} x^t(t)X(i)A_{\Delta d}(t,i)A_{\Delta o}(t,i)x(t+\theta)d\theta \\
\varphi(t,i) &= -2\int_{-\tau}^{0} x^t(t)X(i)A_{\Delta d}(t,i)A_{\Delta d}(t,i)x(t-\tau_{(i)}+\theta)d\theta \qquad (3.27)
\end{aligned}
$$

It follows on using **Fact 1** that

$$
\begin{aligned}
\varrho(t,i) &\leq \omega_1^{-1}(i)\int_{-\tau}^{0} \Big[x^t(t)X(i)A_{\Delta d}A_{\Delta d}^t X(i)x(t)\Big] d\theta \\
&+ \omega_1(i)\int_{-\tau}^{0} \Big[x^t(t+\theta)A_{\Delta o}^t A_{\Delta o}(t,i)x(t+\theta)\Big] d\theta \\
&= \omega_1^{-1}(i)\tau x^t(t)X(i)A_{\Delta d}(t,i)A_{\Delta d}^t(t,i)X(i)x(t) \\
&+ \omega_1(i)\int_{-\tau}^{0} \Big[x^t(t+\theta)A_{\Delta o}^t(t,i)A_{\Delta o}(t,i)x(t+\theta)\Big] d\theta \qquad (3.28) \\
\varphi(t,i) &\leq \omega_2^{-1}(i)\int_{-\tau}^{0} \Big[x^t(t)X(i)A_{\Delta d}(t,i)A_{\Delta d}^t(t,i)X(i)x(t)\Big] d\theta \\
&+ \omega_2(i)\int_{-\tau}^{0} \Big[x^t(t-\tau+\theta)A_{\Delta d}^t(t,i)A_{\Delta d}(t,i)x(t-\tau+\theta)\Big] d\theta \\
&= \omega_2(i)\int_{-\tau}^{0} \Big[x^t(t-\tau+\theta)A_{\Delta d}^t(t,i)A_{\Delta d}(t,i)x(t-\tau+\theta)\Big] d\theta \\
&+ \omega_2^{-1}(i)\tau x^t(t)X(i)A_{\Delta d}(t,i)A_{\Delta d}^t(t,i)X(i)x(t) \qquad (3.29)
\end{aligned}
$$

for some scalers $\omega_1(i) > 0$, $\omega_2(i)$, $i \in \mathcal{S}$. Now define

$$
\begin{aligned}
W(\mathbf{x},t,i) &= \int_{-\tau}^{0}\int_{t+\theta}^{t} \omega_1(i)[x^t(s)A_{\Delta o}^t(t,i)A_{\Delta o}(t,i)x(s)]dsd\theta \\
&+ \int_{-\tau}^{0}\int_{t-\tau+\theta}^{t} \omega_2(i)[x^t(s)A_{\Delta d}^t(t,i)A_{\Delta d}(t,i)x(s)]dsd\theta \qquad (3.30)
\end{aligned}
$$

The weak infinitesimal generator $\Im_s^x[W(.,.)]$ is given by:

$$\begin{aligned}
\Im_s^x[W(.,.)] &= \frac{1}{\delta}\lim_{\delta\to 0}\mathcal{E}[W(\mathbf{x},t,\eta_t)|\eta_o = i, \mathbf{x}_o = \phi] - W(\mathbf{x},t,\eta_t) \\
&= \omega_1(i)\tau\Big[x^t(t)A_{\Delta o}^t(t,i)A_{\Delta o}(t,i)x(t)\Big] \\
&- \omega_1(i)\int_{-\tau}^{0}\Big[x^t(t+\theta)A_{\Delta o}^t(t,i)A_{\Delta o}(t,i)x(t+\theta)\Big]d\theta \\
&- \omega_2(i)\int_{-\tau}^{0}\Big[x^t(t-\tau+\theta)A_{\Delta d}^t(t,i)A_{\Delta d}(t,i)x(t-\tau+\theta)\Big]d\theta \\
&+ \omega_2(i)\tau\Big[x^t(t)A_{\Delta d}^t(t,i)A_{\Delta d}(t,i)x(t)\Big] \qquad (3.31)
\end{aligned}$$

By taking into account (3.28)-(3.31), it follows from (3.24) that

$$\begin{aligned}
\Im_s^x[V(.,.)] &\le x^t(t)\Big\{X(i)[A_{\Delta o}(t,i) + A_{\Delta d}(t,i)] + [A_{\Delta o}(t,i) + A_{\Delta d}(t,i)]^t X(i) \\
&+ \sum_{k=1}^{s}\alpha_{ik}X(k) + \omega_1^{-1}(i)\tau X(i)A_{\Delta d}(t,i)A_{\Delta d}^t(t,i)X(i) \\
&+ \omega_2^{-1}(i)\tau X(i)A_{\Delta d}(t,i)A_{\Delta d}^t(t,i)X(i) \\
&+ \omega_2(i)\tau A_{\Delta d}^t(t,i)A_{\Delta d}(t,i)\Big\}x(t) + \omega_1(i)\tau A_{\Delta o}^t(t,i)A_{\Delta o}(t,i) \qquad (3.32)
\end{aligned}$$

Direct application of **Facts 1** and **2** using the uncertainty representation of chapter 2 yields:

$$\begin{aligned}
&X(i)[A_{\Delta o}(t,i) + A_{\Delta d}(t,i)] + [A_{\Delta o}(t,i) + A_{\Delta d}(t,i)]^t X(i) \le \\
&X(i)[A_o(t,i) + A_d(t,i)] + [A_o(t,i) + A_d(t,i)]^t X(i) + \\
&\theta_1^{-1}(i)X(i)[M_a(i)M_a^t(i) + M_d(i)M_d^t(i)]X(i) + \\
&\theta_1(i)[N_a^t(i)N_a(i) + N_d^t(i)N_d(i)] \qquad (3.33) \\
&X(i)A_{\Delta d}(t,i)A_{\Delta d}^t(t,i)X(i) \le X(i)\Big[\theta_2^{-1}(i)M_d(i)M_d^t(i) + \\
&A_d(i)[I - \theta_2(i)N_d^t(i)N_d(i)]^{-1}A_d^t(i)\Big]X(i) \qquad (3.34)
\end{aligned}$$

$$
\begin{aligned}
& A^t_{\Delta o}(t,i)A_{\Delta o}(t,i) \ \leq \theta_3^{-1}(i)N^t_a(i)N_a(i) + \\
& A^t_o(i)[I - \theta_3(i)M_a(i)M^t_a(i)]^{-1}A_o(i) \qquad (3.35) \\
& A^t_{\Delta d}(t,i)A_{\Delta d}(t,i) \ \leq \theta_4^{-1}(i)N^t_d(i)N_d(i) + \\
& A^t_d(i)[I - \theta_4(i)M_d(i)M^t_d(i)]^{-1}A_d(i) \qquad (3.36)
\end{aligned}
$$

for some scalars $\theta_1(i) > 0, \cdots, \theta_4(i) > 0, \ i \in \mathcal{S}$ satisfying

$$
\begin{aligned}
& [I - \theta_2(i)N^t_d(i)N_d(i)] > 0 \ , \ [I - \theta_3(i)M_a(i)M^t_a(i)] > 0 \\
& [I - \theta_4(i)M_d(i)M^t_d(i)] > 0 \qquad (3.37)
\end{aligned}
$$

Since $\omega_1(i), \omega_2(i), \ i \in \mathcal{S}$ are given weights, we introduce the following change of variables

$$
\begin{aligned}
\sigma_1(i) &= \theta_1(i) \ , \ \sigma_2(i) = \omega_1(i)\theta_2(i) \ , \ \sigma_3(i) = \omega_2(i)\theta_2(i) \\
\sigma_4(i) &= \omega_1^{-1}(i)\theta_3(i) \ , \ \sigma_5(i) = \omega_2^{-1}(i)\theta_2(i) \\
\sigma_6(i) &= \omega_1^{-1}(i) \ , \ \sigma_7(i) = \omega_2^{-1}(i) \qquad (3.38)
\end{aligned}
$$

It then follows from (3.32)-(3.38) that

$$
\Im^x_s[V(.,.)] \leq \ x^t(t) \ \Omega(X,\sigma,\omega,\tau,i) \ x(t) \qquad (3.39)
$$

with

$$
\begin{aligned}
\Omega(X,\sigma,\omega,\tau,i) &= X(i)[A_o(i) + A_d(i)] + [A_o(i) + A_d(i)]^t X(i) \\
&+ X(i)\{\sum_{m=1}^{s} \alpha_{im} Y(m)\}X(i) + \sigma_1^{-1} X(i)M_a(i)M^t_a(i)X(i) \\
&+ \sigma_1^{-1} X(i)M_d(i)M^t_d(i)X(i) \\
&+ \sigma_1(i)[N^t_a(i)N_a(i) + N^t_d(i)N_d(i)] \\
&+ \tau X(i)\left[\sigma_1^{-1}(i)M_d(i)M^t_d(i) + \sigma_3^{-1}(i)M_d(i)M^t_d(i)\right] \\
&+ \tau X(i)A_d(i)\left[\omega_1(i)I - \sigma_2(i)N^t_d(i)N_d(i)]^{-1}A^t_d(i)\right]X(i)
\end{aligned}
$$

$$
\begin{aligned}
&+ \quad \tau X(i)\Big[A_d(i)[\omega_2(i)I - \sigma_3(i)N_d^t(i)N_d(i)]^{-1}A_d^t(i)\Big]X(i) \\
&+ \quad \sigma_4^{-1}(i)\tau_{(i)}N_a^t(i)N_a(i) + \sigma_5^{-1}(i)\tau N_d^t(i)N_d(i) \\
&+ \quad \tau A_o^t(i)[\omega_1^{-1}(i)I - \sigma_4(i)M_a(i)M_a^t(i)]^{-1}A_o(i) \\
&+ \quad \tau A_d^t(i)[\omega_2^{-1}(i)I - \sigma_5(i)M_d(i)M_d^t(i)]^{-1}A_d(i) \qquad (3.40)
\end{aligned}
$$

where $\sigma(i)$ denotes $\{\sigma_1(i), \cdots, \sigma_7(i)\}$ and $\omega(i)$ denotes the combined weights $\{\omega_1(i), \omega_2(i)\}$. In view of the monotonic nondecreasing behavior of $\Omega(X, \theta, \omega, \tau, i)$ and using the Schur complements, it can be readily verified that LMIs (3.23) ensure that $\Omega(X, \sigma, \omega, \tau, i) < 0$. The remaining part of the proof follows parallel development to **Theorem** 3.1. $\nabla\nabla\nabla$

To end this section, we list without proof the counterpart of **Theorem** 3.2 and **corollaries** 3.1-3.2.

Theorem 3.4 *Given a scalar* $\bar{\tau} > 0$, *system (3.22) is* **RSSSDD with a disturbance attenuation** γ **for any time-delay** τ **satisfying** $0 \le \tau \le \bar{\tau}$ *if there exist matrices* $0 < X(i) = X^t(i) \in \Re^{n\times n}$, $i \in \mathcal{S}$ *and scalars* $\sigma_1(i) > 0$, $\sigma_2(i) > 0$, $\sigma_3(i) > 0$, $\sigma_4(i) > 0$, $\sigma_5(i) > 0$, $\sigma_6(i) > 0$, $\sigma_7(i) > 0$, $i \in \mathcal{S}$ *satisfying the system of LMIs for all* $i \in \mathcal{S}$:

$$
\begin{bmatrix}
\Pi_g(i) & \Lambda_a(i,\tau) & \Lambda_b(i,\tau) & \begin{matrix} P(i)\Gamma(i)+ \\ G_o^t(i)\Phi(i) \end{matrix} \\
\Lambda_a^t(i,\tau) & -\Theta_a(i,\tau)I & 0 & 0 \\
\Lambda_b^t(i,\tau) & 0 & -\Theta_b(i,\tau)I & 0 \\
\begin{matrix} \Gamma^t(i)P(i)+ \\ \Phi^t(i)G_o(i) \end{matrix} & 0 & 0 & -\gamma^2 I
\end{bmatrix} < 0,
$$

$$
\begin{bmatrix} -\omega_1(i)I & \sigma_2(i)N_d^t(i) \\ \sigma_2(i)N_d(i) & -\sigma_2(i)I \end{bmatrix} < 0\,, \quad
\begin{bmatrix} -\omega_2(i)I & \sigma_3(i)M_a(i) \\ \sigma_3(i)M_a(i) & -\sigma_3(i)I \end{bmatrix} < 0
$$

$$
\begin{bmatrix} -\sigma_6(i)I & \sigma_4(i)M_a(i) \\ \sigma_4(i)M_a^t(i) & -\sigma_4(i)I \end{bmatrix} < 0\,, \quad
\begin{bmatrix} -\sigma_7(i)I & \sigma_5(i)M_d(i) \\ \sigma_5(i)M_d^t(i) & -\sigma_5(i)I \end{bmatrix} < 0
$$

$$
\begin{bmatrix} X(i) & I \\ I & Y(i) \end{bmatrix} \ge 0\,, \quad
\begin{bmatrix} \sigma_6(i) & I \\ I & \omega_1(i) \end{bmatrix} \ge 0\,, \quad
\begin{bmatrix} \sigma_7(i) & I \\ I & \omega_2(i) \end{bmatrix} \ge 0 \qquad (3.41)
$$

for all admissible uncertainties satisfying (2.26)-(2.29) where

$$\Pi_g(i) = \Pi_a(i) + G_o^t(i)G_o(i) \tag{3.42}$$

Corollary 3.3 *Given a scalar* $\bar{\tau} > 0$, *system (3.17) with* $w(.) \equiv 0$ *is* **SSSDD for any time-delay** τ **satisfying** $0 \leq \tau \leq \bar{\tau}$ *if there exist matrices* $0 < X(i) = X^t(i) \in \Re^{n\times n}$, $i \in \mathcal{S}$ *satisfying the system of LMIs for all* $i \in \mathcal{S}$:

$$\begin{bmatrix} \Pi_{ao}(i) & \Lambda_b(i,\tau) \\ \Lambda_b^t(i,\tau) & -\Theta_{bo}(i,\tau)I \end{bmatrix} < 0 \begin{bmatrix} X(i) & I \\ I & Y(i) \end{bmatrix} \geq 0 \tag{3.43}$$

for all admissible uncertainties satisfying (2.26)-(2.29) where

$$\begin{aligned} \Pi_{ao}(i) &= X(i)[A_o(i) + A_d(i)] + [A_o(i) + A_d(i)]^t X(i) \\ &+ X(i)\{\sum_{m=1}^{s} \alpha_{im} Y(m)\} X(i) \\ \Theta_{bo}(i,\tau) &= diag[\Theta_{bo1}(i,\tau) \;\; \Theta_{bo2}(i,\tau)] \\ \Theta_{bo1}(i,\tau) &= diag\Big[\tau\omega_1(i)I \;\; \tau\omega_1(i)I\Big] \\ \Theta_{bo2}(i,\tau) &= diag\Big[\tau\omega_1^{-1}(i)I \;\; \tau\omega_2^{-1}(i)I\Big] \end{aligned} \tag{3.44}$$

Corollary 3.4 *Given a scalar* $\bar{\tau} > 0$, *system (3.17) is* **RSSSDD with a disturbance attenuation** γ **for any time-delay** τ **satisfying** $0 \leq \tau \leq \bar{\tau}$ *if there exist matrices* $0 < X(i) = X^t(i) \in \Re^{n\times n}$, $i \in \mathcal{S}$ *satisfying the system of LMIs for all* $i \in \mathcal{S}$:

$$\begin{bmatrix} \Pi_{go}(i) & \Lambda_b(i,\tau) & \begin{matrix} P(i)\Gamma(i)+ \\ G_o^t(i)\Phi(i) \end{matrix} \\ \Lambda_b^t(i,\tau) & -\Theta_{bo}(i,\tau)I & 0 \\ \begin{matrix} \Gamma^t(i)P(i)+ \\ \Phi^t(i)G_o(i) \end{matrix} & 0 & -\gamma^2 I \end{bmatrix} < 0$$

$$\begin{bmatrix} X(i) & I \\ I & Y(i) \end{bmatrix} \geq 0 \tag{3.45}$$

for all admissible uncertainties satisfying (2.26)-(2.29) where

$$\Pi_{go}(i) = \Pi_{ao}(i) + G_o^t(i)G_o(i) \tag{3.46}$$

3.2.3 Examples

For the purpose of demonstrating the developed analytical results, we provide hereafter two examples.

Example 3.1

We consider a pilot-scale single-reach water quality system which can fall into the models discussed in Chapter 2, section 2.4.2 (the reader is referd to (2.38) or (2.40)-(2.41) for description). In simulation, we take $\tau^* = 0.95$, $\tau^+ = 0.5$. Let the Markov process governing the mode switching has generator

$$\Im = \begin{bmatrix} -4 & 4 \\ 3 & -3 \end{bmatrix}$$

For the two operating conditions (modes), the associated date are:

Mode 1:

$$\begin{aligned}
A_o(1) &= \begin{bmatrix} -3 & -2 \\ 1 & 0 \end{bmatrix}, \ A_d(1) = \begin{bmatrix} 0 & 0.3 \\ -0.3 & -0.2 \end{bmatrix} \\
B_o(1) &= \begin{bmatrix} 1 & 0 \\ 0 & 2 \end{bmatrix}, \ M_a(1) = \begin{bmatrix} 0.2 \\ 0.1 \end{bmatrix} \\
\Gamma(1) &= \begin{bmatrix} 0.1 \\ 0.2 \end{bmatrix}, \ \Phi(1) = \begin{bmatrix} 0.2 \\ 0.1 \end{bmatrix} \\
G(1) &= [0.1 \ \ 0.1] \, , \ N_a(1) = [0.2 \ \ 0.4] \\
N_d(1) &= [0.1 \ \ 0.3] \, , \ N_b(1) = [0.3 \ \ 0.3]
\end{aligned}$$

Mode 2:

$$\begin{aligned}
A_o(2) &= \begin{bmatrix} -1 & 0 \\ 2 & -2 \end{bmatrix}, \ A_d(2) = \begin{bmatrix} -0.5 & -0.6 \\ -0.2 & -0.1 \end{bmatrix} \\
B_o(2) &= \begin{bmatrix} 1 & 0 \\ 0 & 1 \end{bmatrix}, \ M_a(2) = \begin{bmatrix} 0.1 \\ 0.2 \end{bmatrix} \\
\Gamma(2) &= \begin{bmatrix} 0.2 \\ 0.1 \end{bmatrix}, \ \Phi(2) = \begin{bmatrix} 0.1 \\ 0.2 \end{bmatrix} \\
G(2) &= [0.1 \ \ 0.1] \, , \ N_a(2) = [0.4 \ \ 0.2]
\end{aligned}$$

$$N_d(2) = [0.3\ \ 0.1]\ ,\ N_b(2) = [0.1\ \ 0.1]$$

Using the initial data for $i = 1, 2$:

$$Q(1) = \begin{bmatrix} 2 & 0 \\ 0 & 2 \end{bmatrix}\ ,\ Q(2) = \begin{bmatrix} 2 & 0 \\ 0 & 2 \end{bmatrix}$$

and selecting $\xi(1) = 10$, $\xi(2) = 12.5$ ensures that $\bar{Q}(1) > 0$, $\bar{Q}(2) > 0$. Invoking the software environment [57], the feasible solutions of LMIs (3.3) are given by:

$$\begin{aligned} P(1) &= \begin{bmatrix} 2.9930 & 2.3143 \\ 2.3143 & 5.2018 \end{bmatrix}\ ,\ \epsilon(1) = 0.3112\ ,\ \rho(1) = 0.5009 \\ P(2) &= \begin{bmatrix} 3.1172 & 1.2844 \\ 1.2844 & 6.0789 \end{bmatrix}\ ,\ \epsilon(2) = 1.2632\ ,\ \rho(2) = 1.9743 \end{aligned}$$

Since $P(1) > 0$, $P(2) > 0$, **Theorem** 3.1 is validated and in turn confirms the robust stochastic stability with weak delay-dependece.

On solving the LMIs (3.15), we get

$$\begin{aligned} P(1) &= \begin{bmatrix} 3.0124 & 2.1850 \\ 2.1850 & 7.3122 \end{bmatrix}\ ,\ \epsilon(1) = 1.2548\ ,\ \rho(1) = 0.7598 \\ P(2) &= \begin{bmatrix} 3.4520 & 2.0843 \\ 2.0843 & 8.1065 \end{bmatrix}\ ,\ \epsilon(2) = 3.4398\ ,\ \rho(2) = 2.4784 \\ \gamma &= 2.9762 \end{aligned}$$

which verifies **Theorem** 3.2. Using

$$Q(1) = \begin{bmatrix} 1 & 0 \\ 0 & 1 \end{bmatrix}\ ,\ Q(2) = \begin{bmatrix} 1 & 0 \\ 0 & 1 \end{bmatrix}$$

and selecting $\xi(1) = 5$, $\xi(2) = 5$ ensures that $\bar{Q}(1) > 0$, $\bar{Q}(2) > 0$. We then obtain

$$P(1) = \begin{bmatrix} 0.5032 & 0.5920 \\ 0.5920 & 2.2628 \end{bmatrix}\ ,\ P(2) = \begin{bmatrix} 0.5149 & 0.5128 \\ 0.5128 & 1.9912 \end{bmatrix}$$

This verifies **corollary** 3.1 and in turn confirms the stochastic stability with weak delay-dependece. Next, we solve the LMIs (3.20) to get

$$\begin{aligned} P(1) &= \begin{bmatrix} 0.4574 & 0.4180 \\ 0.4180 & 1.9415 \end{bmatrix}, \quad P(2) = \begin{bmatrix} 0.5636 & 0.5628 \\ 0.5628 & 2.2182 \end{bmatrix} \\ \gamma &= 3.1428 \end{aligned}$$

Once again $P(1) > 0,\ P(2) > 0$,**corollary** 3.2 is validated as expected.

Focusing on the strong delay-dependent stability, we solve inequality (3.23) with $\omega_1(1) = 1,\ \omega_2(1) = 2,\ \omega_1(2) = 1,\ \omega_2(2) = 2$. The feasible results are:

$$\begin{aligned} P(1) &= \begin{bmatrix} 4.7985 & 2.1054 \\ 2.1054 & 2.1168 \end{bmatrix} \\ \beta_1(1) &= 0.2987\ ,\ \beta_2(1) = 1.0374,\ \beta_3(1) = 0.8117 \\ \beta_4(1) &= 1.1577,\ \beta_5(1) = 2.9487,\ \beta_6(1) = 5.2066 \\ \beta_7(1) &= 0.2987,\ \delta_5(1) = 11.0321,\ \delta_7(1) = 9.6108 \\ P(2) &= \begin{bmatrix} 5.4325 & 0.5546 \\ 0.5546 & 1.9065 \end{bmatrix} \\ \beta_1(2) &= 0.3164,\ \beta_2(2) = 0.7374,\ \beta_3(2) = 1.2870 \\ \beta_4(2) &= 1.5063,\ \beta_5(2) = 6.0094,\ \beta_6(2) = 4.1286 \\ \beta_7(2) &= 0.3358,\ \delta_5(2) = 1.9987,\ \delta_7(1) = 7.5671,\ \bar{\tau} = 0.3098 \end{aligned}$$

Next, we solve inequality (3.43) with $\omega_1(1) = 1,\ \omega_2(1) = 2,\ \omega_1(2) = 1,\ \omega_2(2) = 2$. The feasible results are:

$$\begin{aligned} P(1) &= \begin{bmatrix} 4.9377 & 2.2076 \\ 2.2076 & 2.1177 \end{bmatrix}, \quad P(2) = \begin{bmatrix} 5.5537 & 0.6743 \\ 0.6743 & 1.8289 \end{bmatrix} \\ \tau^* &= 0.2105 \end{aligned}$$

This reads that the water quality system is stochastically stable for any constant time-delay τ satisfying $0 \le \tau \le 0.2105$.

On comparing these results with the foregoing ones finds that the upper bound on τ for weak delay-dependent stability was set at 0.95 where it is determined for strong delay-dependent stability as 0.2105 for the nominal case and 0.3098 for the uncertain case, respectively. This clearly emphasizes the fact that stochastic stability independent of delay is more conservative that delay-dependent stochastic stability.

Example 3.2

This is again a pilot-scale single-reach water quality system having three reaches with $\tau^* = 0.95$, $\tau^+ = 0.5$. The first two-reaches have charractristic values similar to **Example 1**. Let the Markov process governing the mode switching has generator

$$\Im = \begin{bmatrix} -6 & 4 & 2 \\ 2 & -5 & 3 \\ 3 & 1 & -4 \end{bmatrix}$$

For the three operating conditions (modes), the associated date are:

Mode 1:

$$\begin{aligned} A_o(1) &= \begin{bmatrix} -3 & -2 \\ 1 & 0 \end{bmatrix}, \ A_d(1) = \begin{bmatrix} 0 & 0.3 \\ -0.3 & -0.2 \end{bmatrix}, \ B_o(1) = \begin{bmatrix} 1 & 0 \\ 0 & 2 \end{bmatrix} \\ \Gamma(1) &= \begin{bmatrix} 0.1 \\ 0.2 \end{bmatrix}, \ \Phi(1) = \begin{bmatrix} 0.2 \\ 0.1 \end{bmatrix}, \ M_a(1) = \begin{bmatrix} 0.2 \\ 0.1 \end{bmatrix}, \ N_b(1) = [0.3 \ \ 0.3] \\ G_o(1) &= [0.1 \ \ 0.1] \, , \ N_a(1) = [0.2 \ \ 0.4] \, , \ N_d(1) = [0.1 \ \ 0.3] \end{aligned}$$

Mode 2:

$$\begin{aligned} A_o(2) &= \begin{bmatrix} -1 & 0 \\ 2 & -2 \end{bmatrix}, \ A_d(2) = \begin{bmatrix} -0.5 & -0.6 \\ -0.2 & -0.1 \end{bmatrix}, \ B_o(2) = \begin{bmatrix} 1 & 0 \\ 0 & 1 \end{bmatrix} \\ \Gamma(2) &= \begin{bmatrix} 0.2 \\ 0.1 \end{bmatrix}, \ \Phi(2) = \begin{bmatrix} 0.1 \\ 0.2 \end{bmatrix}, \ M_a(2) = \begin{bmatrix} 0.1 \\ 0.2 \end{bmatrix}, \ N_b(2) = [0.1 \ \ 0.1] \\ G_o(2) &= [0.1 \ \ 0.1] \, , \ N_a(2) = [0.4 \ \ 0.2] \, , \ N_d(2) = [0.3 \ \ 0.1] \end{aligned}$$

Mode 3:

$$A_o(3) = \begin{bmatrix} -4 & -3 \\ 1 & 0 \end{bmatrix}, \ A_d(3) = \begin{bmatrix} 0.1 & 0.2 \\ -0.4 & 0 \end{bmatrix}, \ B_o(3) = \begin{bmatrix} 1 & 0 \\ 0 & 1 \end{bmatrix}$$

$$\begin{aligned}
\Gamma(3) &= \begin{bmatrix} 0.1 \\ 0.1 \end{bmatrix}, \ \Phi(3) = \begin{bmatrix} 0.1 \\ 0.2 \end{bmatrix}, \ M_a(3) = \begin{bmatrix} 0.1 \\ 0.1 \end{bmatrix}, \ N_b(3) = [0.2 \ \ 0.3] \\
G_o(1) &= [0.1 \ \ 0.1], \ N_a(3) = [0.2 \ \ 0.3], \ N_d(3) = [0.1 \ \ 0.3]
\end{aligned}$$

Using the initial data for $i = 1, 2, 3$:

$$Q(1) = \begin{bmatrix} 1.3 & 0 \\ 0 & 1.3 \end{bmatrix}, \ Q(2) = \begin{bmatrix} 1.5 & 0 \\ 0 & 1.5 \end{bmatrix}, \ Q(3) = \begin{bmatrix} 1.1 & 0 \\ 0 & 1.1 \end{bmatrix}$$

and selecting $\xi(1) = 20$, $\xi(2) = 25$, $\xi(3) = 30$ ensures that $\bar{Q}(1) > 0$, $\bar{Q}(2) > 0$, $\bar{Q}(3) > 0$. Invoking the software environment [57], we solve LMIs (3.3) and the feasible solutions are given by:

$$\begin{aligned}
P(1) &= \begin{bmatrix} 0.4732 & 0.5920 \\ 0.5920 & 1.8968 \end{bmatrix}, \ P(2) = \begin{bmatrix} 0.5149 & 0.5128 \\ 0.5128 & 1.9912 \end{bmatrix} \\
P(3) &= \begin{bmatrix} 0.5066 & 0.5437 \\ 0.5437 & 2.2129 \end{bmatrix}
\end{aligned}$$

This verifies **Theorem** 3.1 and in turn confirms the robust stochastic stability with weak-delay dependence. Next, we solve the LMIs (3.15) to get

$$\begin{aligned}
P(1) &= \begin{bmatrix} 0.4474 & 0.4180 \\ 0.4180 & 1.8415 \end{bmatrix}, \ P(2) = \begin{bmatrix} 0.5436 & 0.5628 \\ 0.5628 & 2.1182 \end{bmatrix} \\
P(3) &= \begin{bmatrix} 0.4636 & 0.4628 \\ 0.4628 & 1.9176 \end{bmatrix}, \ \gamma = 6.1428
\end{aligned}$$

Since $P(1) > 0$, $P(2) > 0$, $P(3) > 0$, **Theorem** 3.2 is validated. Turning to LMIs (3.23), we use

$$Q(1) = \begin{bmatrix} 2 & 0 \\ 0 & 2 \end{bmatrix}, \ Q(2) = \begin{bmatrix} 2.2 & 0 \\ 0 & 2.2 \end{bmatrix}, \ Q(3) = \begin{bmatrix} 2.2 & 0 \\ 0 & 2.2 \end{bmatrix}$$

and selecting $\xi(1) = 20$, $\xi(2) = 25$, $\xi(3) = 30$ ensures that $\bar{Q}(1) > 0$, $\bar{Q}(2) > 0$, $\bar{Q}(3) > 0$. The feasible solution is:

$$\begin{aligned}
P(1) &= \begin{bmatrix} 2.8930 & 2.3143 \\ 2.3143 & 5.0018 \end{bmatrix}, \ \epsilon(1) = 1.2112, \ \mu(1) = 1.5009 \\
P(2) &= \begin{bmatrix} 3.1172 & 1.2844 \\ 1.2844 & 6.3789 \end{bmatrix}, \ \epsilon(2) = 2.2632, \ \mu(2) = 2.9743 \\
P(3) &= \begin{bmatrix} 2.9945 & 1.1564 \\ 1.1564 & 5.7839 \end{bmatrix}, \ \epsilon(2) = 2.2345, \ \mu(2) = 2.3497
\end{aligned}$$

Once again, since $P(1) > 0,\ P(2) > 0,\ P(3) > 0$, **Theorem** 3.3 is validated. On solving the LMIs (3.41), we get

$$
\begin{aligned}
P(1) &= \begin{bmatrix} 3.0124 & 2.1850 \\ 2.1850 & 7.3122 \end{bmatrix}, \quad \epsilon(1) = 1.2548,\ \mu(1) = 0.7598 \\
P(2) &= \begin{bmatrix} 3.4520 & 2.0843 \\ 2.0843 & 8.1065 \end{bmatrix}, \quad \epsilon(2) = 3.4398,\ \mu(2) = 2.4784 \\
P(3) &= \begin{bmatrix} 3.1520 & 2.0423 \\ 2.0423 & 6.9865 \end{bmatrix}, \quad \epsilon(2) = 2.1134,\ \mu(2) = 1.7864
\end{aligned}
$$

which verifies **Theorem** 3.4.

3.3 Mode-Dependent Stochastic Stability

In the previous section we dealt with two categories of stochastic stability of JTDS in which the delays are independnet of the operational modes. The developed results are rigorous extensions of the time-delay reeults [87, 85, 89, 92, 93, 94, 95, 97, 165, 96, 105, 106, 110, 99, 166]. Bearing in mind that time-delays and operational modes are basic ingredients of JTDS, it would them seem desirable to explore the impact of the joint ingredients on the stochastic stability and the subsequent behavior. This comes in line of the recent developments on time-delay systems [53]. It turns out that among the central issues of interest are the degree of conservativeness of the stability results, the available information related to the delay factors and the choice of the Lypunov functional. Some recent development to resolve these issues have been addressed in [54] for time-delay systems and in [22, 114] for classes of jumping systems.

The purpose of this section is to extend the results of [22, 95, 97, 96, 165, 105, 110, 106, 99, 114, 166] further by developing criteria of stochastic stability and stabilization of a class of uncertain Markovian jump systems with functional time-delays (mode-dependent delays). Needless to say that the ensuing results complement those of the previous section.

3.3.1 Mode-Dependent Model

Given a probability space $(\Omega, \mathcal{F}, \mathbf{P})$ where Ω is the sample space, $\mathcal{F}$ is the algebra of events and $\mathbf{P}$ is the probability measure defined on $\mathcal{F}$. Let the random form process $\{\eta_t, t \in [0, \mathcal{T}]\}$ be a homogeneous, finite-state Markovian process with right continuous trajectories and taking values in a finite set $\mathcal{S} = \{1, 2, ..., s\}$ with generator $\Im = (\alpha_{ij})$ and transition probability from mode i at time t to mode j at time $t + \delta$, $i, j \in \mathcal{S}$:

$$
\begin{aligned}
\mathbf{p}_{ij} &= Pr(\eta_{t+\delta} = j \mid \eta_t = i) \\
&= \begin{cases} \alpha_{ij}\delta + o(\delta), & if \;\; i \neq j \\ 1 + \alpha_{ij}\delta + o(\delta), & if \;\; i = j \end{cases}
\end{aligned} \tag{3.47}
$$

with transition probability rates $\alpha_{ij} \geq 0$ for $i, j \in \mathcal{S}, i \neq j$ and

$$
\alpha_{ii} = -\sum_{m=1, m \neq i}^{s} \alpha_{im} \;, \quad \hat{\alpha} \stackrel{\Delta}{=} \max_i \{|\alpha_{ii}|, i \in \mathcal{S}\} \tag{3.48}
$$

where $\delta > 0$ and $\lim_{\delta \downarrow 0} o(\delta)/\delta = 0$. Note that the set $\mathcal{S}$ comprises the various operational modes of the system under study.

We consider a class of stochastic uncertain systems with Markovian jump parameters and functional state-delay described over the space $(\Omega, \mathcal{F}, \mathbf{P})$ by:

$$
\begin{aligned}
(\Sigma_J): \quad \dot{x}(t) &= [A_o(\eta_t) + \Delta A_o(t, \eta_t)]x(t) + [A_d(\eta_t) + \Delta A_d(t, \eta_t)]x(t - \tau_{\eta_t}) \\
&+ [B_o(\eta_t) + \Delta B_o(t, \eta_t)]u(t) + \Gamma(\eta_t)w(t) \\
&= A_{\Delta o}(t, \eta_t)x(t) + A_{\Delta d}(t, \eta_t)x(t - \tau_{\eta_t}) + B_{\Delta o}(t, \eta_t)u(t) \\
&+ \Gamma(\eta_t)w(t), \; x(t) = \phi(t), \; t \in [-\tau, 0], \; \eta_o = i
\end{aligned} \tag{3.49}
$$

$$
z(t) = G_o(\eta_t) + \Phi(\eta_t)w(t) \tag{3.50}
$$

where $x(t) \in \Re^n$ is the state vector; $u(t) \in \Re^m$ is the control input; $w(t) \in \Re^q$ is the disturbance input which belongs to $\mathcal{L}_2[0, \mathcal{T}]$; $z(t) \in \Re^r$ is the controlled output which belongs to $\mathcal{L}_2\left[(\Omega, \mathcal{F}, \mathbf{P}), [0, \mathcal{T}]\right]$ and τ_{η_t} denotes the time-delay in the jumping system when the mode is in η_t with

$$\hat{\tau} \triangleq \max_j\{\tau_j, j \in \mathcal{S}\} \quad , \quad \check{\tau} \triangleq \min_j\{\tau_j, j \in \mathcal{S}\}$$

Note that, in general, the functional relationship could be expressed analytically or presented in tabular form. For simplicity in exposition, the matrices associated with the i-th mode will be denoted in the sequel by

$$\begin{aligned} A_o(i) &\triangleq A_o(\eta_t),\ \Delta A_o(t,i) \triangleq \Delta A_o(t,\eta_t),\ A_d(i) \triangleq A_d(\eta_t) \\ \Delta A_d(t,i) &\triangleq \Delta A_d(t,\eta_t),\ B_o(i) \triangleq B_o(\eta_t),\ \Delta B_o(t,i) \triangleq \Delta B_o(t,\eta_t) \\ \Gamma(i) &\triangleq \Gamma(\eta_t),\ \Phi(i) \triangleq \Phi(\eta_t),\ G_o(i) \triangleq G_o(\eta_t) \end{aligned} \tag{3.51}$$

where $A_o(i), A_d(i), B_o(i), \Gamma(i), G_o(i)$ and $\Phi(i)$ are constant matrices with compatible dimensions that describe the nominal system (3.47)-(3.51) for every $\eta_t = i,\ i \in \mathcal{S}$ while the matrices $\Delta A_o(t,\eta_t)$, $\Delta A_d(t,\eta_t)$ and $\Delta B_o(t,\eta_t)$ are real time-varying matrix functions representing the norm-bounded parameter uncertainties. For $\eta_t = i$, the admissible uncertainties are assumed to be modeled in the form:

$$[\Delta A_o(t,i) \quad \Delta A_d(t,i) \quad \Delta B_o(t,i)] = M_a(i)\ \Delta(t,i)\ [N_a(i) \quad N_d(i) \quad N_b(i)] \tag{3.52}$$

where $M_a(i)$, $M_g(i)$, $N_a(i)$, $N_b(i)$ and $N_d(i)$ are known real constant matrices, with appropriate dimensions, and $\Delta(t,i)$ being unknown time-varying matrix function satisfying

$$||\Delta(t,i)||_2 \ \leq \ 1 \tag{3.53}$$

where the elements of $\Delta(t,i)$ are Lebesgue measurable for any $i \in \mathcal{S}$.

Remark 3.3 *It should be noted that system (3.47)-(3.51) encompasses many state space models of delay systems and generalizes the models put forward in*

Chapter 2. Therefore it can be used to represent many important physical systems; for example, power systems, cold rolling mills, wind tunnel and water resources systems, see for example [98] and the references therein. Also it should be remarked that the parameter uncertainty structure (3.52) has been widely used in robust control and filtering, for example, [92, 96, 98, 105, 149, 152, 151], which can be used to describe many real systems with modeling uncertainties. Also, it is worthwhile to mention that this parameter uncertainty structure covers the usual so-called matching condition [36] as a special case.

3.3.2 Weak-Delay Dependence

Now we direct attention to the stochastic stability of system (3.47)-(3.51) in which the time-delay varies with the mode of operation. Form now onwards, we use the terms functional time-delay or mode-dependent delay on an equivalent basis. Initially, we consider the case where the bounds of the functional dependence (maximum and minimum) are the only available information. This corresponds to category **3**. The related free, uncertain jump system is given by

$$\begin{aligned}(\Sigma_{\Delta f}): \quad \dot{x}(t) &= [A_o(\eta_t) + \Delta A_o(t,\eta_t)]x(t) + [A_d(\eta_t) + \Delta A_d(t,\eta_t)]x(t-\tau_{\eta_t}) \\ &= A_{\Delta o}(t,\eta_t)x(t) + A_{\Delta d}(t,\eta_t)x(t-\tau_{\eta_t}) \end{aligned} \tag{3.54}$$

Definition 3.5 *System $\Sigma_{\Delta f}$ is said to be* **robustly stochastically stable with functional time-delays (RSSFTD)** *if there exist matrices $W^t(i) = W(i) > 0, P(i) = P^t(i) > 0,\ i \in \mathcal{S}$ such that the LMIs for all $\eta_t = i \in \mathcal{S}$*

$$\Upsilon(\eta_t,t) \triangleq \begin{bmatrix} \Pi_t(\eta_t,t) & P(\eta_t)A_{\Delta d}(t,\eta_t) \\ A^t_{\Delta d}(t,\eta_t)P(\eta_t) & -\hat{W}(i) \end{bmatrix} < 0 \tag{3.55}$$

$$\begin{aligned}\Pi_t(\eta_t,t) &= P(\eta_t)A_{\Delta o}(t,\eta_t) + A^t_{\Delta o}(t,\eta_t)P(\eta_t) + \sum_{k=1}^{s}\alpha_{ik}P(k) + \beta\, W(i) \\ \beta &= \hat{\tau} + \hat{\alpha}(\hat{\tau} - \check{\tau})(\hat{\tau} + \check{\tau}) \quad , \quad \hat{W}(i) = \hat{\tau}\, W(i) \end{aligned} \tag{3.56}$$

hold for all admissible uncertainties satisfying (3.52)-(3.53).

The following theorem establishes an LMI-based sufficient for robust stochastic stability with functional time-delays

Theorem 3.5 *System* $\Sigma_{\Delta f}$ *is* **RSSFTD**, *if there exist matrices* $W^t(i) = W(i) > 0, P(i) = P^t(i) > 0,\ i \in \mathcal{S}$ *and scalars* $\epsilon(i) > 0,\ \mu(i) > 0,\ i \in \mathcal{S}$ *satisfying the LMIs for all* $i \in \mathcal{S}$

$$\begin{bmatrix} \Pi_s(i) & P(i)M_a(i) & P(i)M_d(i) & P(i)A_d(i) \\ M_a^t(i)P(i) & -\epsilon(i)I & 0 & 0 \\ M_d^t(i)P(i) & 0 & -\mu(i)I & 0 \\ A_d^t(i)P(i) & 0 & 0 & \begin{matrix} -\hat{W}(i)+ \\ \mu(i)N_d^t(i)N_d(i) \end{matrix} \end{bmatrix} < 0 \tag{3.57}$$

$$\Pi_s(i) = A_o^t(i)P(i) + P(i)A_o(i) + \sum_{k=1}^{s} \alpha_{ik}P(k) + \beta\, W(i) + \epsilon(i)N_a^t(i)N_a \tag{3.58}$$

Proof: Let $\mathbf{x}_s(t) \stackrel{\Delta}{=} x(s+t),\ t - \tau_{\eta_t} \leq s \leq t$ and define the process $\{(\mathbf{x}(t), \eta_t),\ t \geq 0\}$ over the state space $\bar{\mathcal{C}}$. It should be observed that $\{(\mathbf{x}(t), \eta_t),\ t \geq 0\}$ is strong Markovian [78]. Now introduce the following Lyapunov-Krasovskii functional:

$$\begin{aligned} V_t(\mathbf{x}(t), \eta_t) &= V_n(\mathbf{x}(t), \eta_t) + V_d(\mathbf{x}(t), \eta_t) + V_f(\mathbf{x}(t), \eta_t) \\ &+ V_h(\mathbf{x}(t), \eta_t) \\ V_n(\mathbf{x}(t), \eta_t) &= x^t(t)P(\eta_t)x(t) \\ V_d(\mathbf{x}(t), \eta_t) &= \int_{-\tau_{\eta_t}}^{0} \int_{t+\beta}^{t} x^t(r)W(\eta_t)x(r)\, dr\, d\beta \\ V_h(\mathbf{x}(t), \eta_t) &= \hat{\alpha} \int_{-\hat{\tau}}^{-\check{\tau}} \int_{t+\beta}^{t} x(r-t+\hat{\tau})W(\eta_t)x(r-t+\hat{\tau})\, dr\, d\beta \\ V_f(\mathbf{x}(t), \eta_t) &= \hat{\alpha}\left(\hat{\tau} - \check{\tau}\right) \int_{-\check{\tau}}^{0} \int_{t+\beta}^{t} x^t(r)W(\eta_t)x(r)\, dr\, d\beta \end{aligned} \tag{3.59}$$

The weak infinitesimal operator $\Im_t^x[\cdot]$ of the process $\{(\mathbf{x}(t), \eta_t),\ t \ \geq\ 0\}$ for system $(\Sigma_{\Delta f})$ at the point $\{t, x, \eta_t\}$ is given by [78, 154]:

$$\begin{aligned}
\Im_k^x[V_k] &= \frac{\partial V_k}{\partial t} + \dot{x}^t(t)\frac{\partial V_k}{\partial x} \mid_{\eta_t = i} + \sum_{m=1}^{s} \alpha_{im} V_k(t, x, i, m) \\
\Im_t^x[V_t] &= \sum_{k=n,d,f,h} \Im_m^x[V_m] \qquad (3.60)
\end{aligned}$$

Select $\eta_t = i \in \mathcal{S},\ \mathbf{x} \in \mathcal{C}[-\tau_{(i)}, 0]$. Upon applying (3.59)-(3.60) to system (3.54) it yields:

$$\begin{aligned}
\Im_n^x[V_n] &= x^t(t)\Big\{P(\eta_t)A_{\Delta o}(t,\eta_t) + A^t_{\Delta o}(t,\eta_t)P(\eta_t) + \sum_{k=1}^{s} \alpha_{\eta_t k} P(k)\Big\}x(t) \\
&+ 2x^t(t)P(\eta_t)A_{\Delta d}(t,\eta_t)x(t-\tau_{\eta_t}) \qquad (3.61)
\end{aligned}$$

$$\begin{aligned}
\Im_d^x[V_d] &= \tau_{\eta_t}\Big[x^t(t)W(i)x(t) \ - \ x^t(t-\tau_{\eta_t})W(i)x(t-\tau_{\eta_t})\Big] \\
&+ \sum_{m=1}^{s} \alpha_{im} \int_{-\tau_{\eta_t}}^{0} \int_{t+\beta}^{t} x^t(r)W(i)x(r)\,dr\,d\beta \qquad (3.62)
\end{aligned}$$

Standard algebraic manipulations lead to

$$\begin{aligned}
&\sum_{m=1}^{s} \alpha_{im} \int_{-\tau_{\eta_t}}^{0} \int_{t+\beta}^{t} x^t(r)W(i)x(r)\,dr\,d\beta \\
&= \alpha_{ii} \int_{-\tau_{\eta_t}}^{0} \int_{t+\beta}^{t} x^t(r)W(i)x(r)\,dr\,d\beta \\
&+ \sum_{m=1, m\neq i}^{s} \alpha_{im} \int_{-\tau_{\eta_t}}^{0} \int_{t+\beta}^{t} x^t(r)Wx(r)\,dr\,d\beta \\
&\leq -|\alpha_{ii}| \int_{-\hat{\tau}}^{0} \int_{t+\beta}^{t} x^t(r)W(i)x(r)\,dr\,d\beta \\
&+ \sum_{m=1, m\neq i}^{s} \alpha_{im} \int_{-\check{\tau}}^{0} \int_{t+\beta}^{t} x^t(r)W(i)x(r)\,dr\,d\beta \\
&= |\alpha_{ii}| \int_{-\check{\tau}}^{-\hat{\tau}} \int_{t+\beta}^{t} x^t(r)W(i)x(r)\,dr\,d\beta
\end{aligned}$$

$$\begin{aligned}
&\leq \hat{\alpha}\int_{-\check{\tau}}^{-\hat{\tau}}\int_{t+\beta}^{t} x^t(r)W(i)x(r)\,dr\,d\beta \\
&= \hat{\alpha}\Big(\hat{\tau}-\check{\tau}\Big)\int_{t-\check{\tau}}^{t} x^t(r)W(i)x(r)\,dr \\
&+ \hat{\alpha}\int_{t-\check{\tau}}^{t-\hat{\tau}} x^t(t+\beta)W(i)x(t+\beta)\,dr
\end{aligned} \tag{3.63}$$

Therefore

$$\begin{aligned}
\Im_d^x[V_d] &\leq \tau_{\eta_t}\Big[x^t(t)W(i)x(t) \; - \; x^t(t-\tau_{\eta_t})W(i)x(t-\tau_{\eta_t})\Big] \\
&+ \hat{\alpha}\Big(\hat{\tau}-\check{\tau}\Big)\int_{t-\check{\tau}}^{t} x^t(r)W(i)x(r)\,dr \\
&+ \hat{\alpha}\int_{t-\check{\tau}}^{t-\hat{\tau}} x^t(r+\hat{\tau})W(i)x(r+\hat{\tau})\,dr
\end{aligned} \tag{3.64}$$

In a similar way, it is easy to show that:

$$\begin{aligned}
\Im_f^x[V_f] &= \hat{\alpha}\check{\tau}\Big(\hat{\tau}-\check{\tau}\Big)x^t(t)W(i)x(t) \\
&- \hat{\alpha}\Big(\hat{\tau}-\check{\tau}\Big)\int_{t-\check{\tau}}^{t} x^t(\beta)W(i)x(\beta)\,d\beta \\
\Im_h^x[V_h] &= \hat{\alpha}\hat{\tau}\Big(\hat{\tau}-\check{\tau}\Big)x^t(t)W(i)x(t) \\
&- \hat{\alpha}\int_{t-\check{\tau}}^{t-\hat{\tau}} x^t(r+\hat{\tau})W(i)x(r+\hat{\tau})\,dr
\end{aligned} \tag{3.65}$$

Now by combining (3.61) through (3.65) and arranging terms, we obtain

$$\begin{aligned}
\Im_t^x[V_t] &\leq x^t(t)\Big\{P(\eta_t)A_{\Delta o}(t,\eta_t) + A_{\Delta o}^t(t,\eta_t)P(\eta_t) \\
&+ \sum_{k=1}^{s}\alpha_{\eta_t k}P(k) + \beta\, W(i)\Big\}x(t) \\
&+ 2x^t(t)P(\eta_t)A_{\Delta d}(t,\eta_t)x(t-\tau_{\eta_t}) \\
&- x^t(t-\tau_{\eta_t})\hat{W}(i)x(t-\tau_{\eta_t})
\end{aligned} \tag{3.66}$$

which is negative from (3.55)-(3.56). Application of the Schur complements to yields

$$[x^t(t) \quad x^t(t-\tau_{\eta_t})]\Upsilon(i,t)\begin{bmatrix} x(t) \\ x(t-\tau_{\eta_t}) \end{bmatrix} =$$

$$x^t(t)\Big\{P(\eta_t)A_{\Delta o}(t,\eta_t) + A^t_{\Delta o}(t,\eta_t)P(\eta_t) + \sum_{m=1}^{s} \alpha_{im}P(m) + \beta\, W +$$

$$P(\eta_t)A_{\Delta d}(t,\eta_t)\hat{W}^{-1}(i)A^t_{\Delta d}(t,\eta_t)P(\eta_t)\Big\}x(t) \stackrel{\Delta}{=} x^t(t)\Lambda(i,t)x(t)$$

$$< 0\,, \quad \forall\, \eta_t = i \in \mathcal{S}. \tag{3.67}$$

Considering (3.52)-(3.53) and applying **Facts 1-2** to (3.67) with some algebraic manipulations, we get for some scalars $\epsilon(i) > 0,\ \mu(i) > 0,\ i \in \mathcal{S}$:

$$\begin{aligned} x^t(t)\Lambda(i)x(t) &= x^t(t)\Bigg\{A^t_o(i)P(i) + P(i)A_o(i) + \sum_{m=1}^{s} \alpha_{im}P(m) \\ &+ \epsilon(i)N^t_a(i)N_a + \epsilon^{-1}(i)P(i)M_a(i)M^t_a(i)P(i) \\ &+ P(i)A_d(i)[\hat{W}(i) - \mu(i)N^t_d(i)N_d(i)]^{-1}A^t_d(i)P(i) \\ &+ \mu^{-1}(i)P(i)M_d(i)M^t_d(i)P(i) + \beta\, W(i)\Bigg\}x(t) < 0 \end{aligned} \tag{3.68}$$

This is equivalent to the LMIs (3.57) via the Schur complements. Hence, we conclude that $\Im^x_t[V_t] < 0$ for all $x \neq 0$ and $\Im^x_t[V_t] \leq 0$ for all x.

Since $||x(t+v)|| \leq \varphi||x(t)||,\ \forall v \in [-\tau_\eta, 0]$ and some $\varphi > 0$ [85], it follows from (3.59) that $V_t(x,i) \leq x^t(t)P(i)x(t) + \mu||x||^2$ where $\mu = \varphi\kappa(\max_i \lambda_M[W(i)])$, $\kappa = \hat{\tau} + \hat{\alpha}[\hat{\tau}^2 + \hat{\tau} - \check{\tau}(1+\hat{\tau})]$. Therefore, for all $x \neq 0$, we have

$$\begin{aligned} \frac{\Im^x_t[V_t]}{V_t(x,i)} &\leq \frac{x^t\Lambda(i)x}{x^tP(i)x + \mu||x||^2} \leq -\,\zeta \\ &\stackrel{\Delta}{=} -\min_{i\in\mathcal{S}}\left\{\frac{\lambda_m[-\Lambda(i)]}{\lambda_M[P(i)] + \mu}\right\} \end{aligned} \tag{3.69}$$

It is readily seen from (3.69) that $\zeta > 0$ and hence we get $\Im^x_t[V_t] \leq -\zeta\, V_t(x,i)$. It follows from [78] by using the Gronwall-Bellman lemma [98] and letting $x(t =$

$0, \phi, \eta_o) = x_o$, one has

$$\mathbb{E}[V_t(x,i)|\phi,\eta_o] \;\leq\; e^{-\zeta\, t}\, V_t(x_o,i) \tag{3.70}$$

Recall that

$$\mathbb{E}\Big\{ \int_{-\tau_\eta}^{0} x^t(t+\theta)W(i)x(t+\theta)d\theta|\phi,\eta_o \Big\} \geq 0$$

It is then easy to see from (3.59) that

$$\begin{aligned}
&\mathbb{E}\Big\{x^t(t)P(i)x(t)|\phi,\eta_o\Big\} \leq e^{-\zeta\, t}V_t(x_o,i) \Longrightarrow \\
&\mathbb{E}\Big\{\int_0^{\mathcal{T}} x^t(t)P(i)x(t)dt|\phi,\eta_o = i\Big\} \\
&\leq \Big[\int_0^{\mathcal{T}} e^{-\zeta\, t}dt\Big]\, V_t(x_o,i) = \frac{1}{\zeta}[e^{-\zeta\, \mathcal{T}} \; - \; 1]\; V_t(x_o,i) \Longrightarrow \\
&\lim_{\mathcal{T}\to\infty} \; \mathbb{E}\Big\{\int_0^{\mathcal{T}} x^t(t)P(i)x(t)dt|\phi,\eta_o = i\Big\} \\
&\leq \; \frac{1}{\zeta}x_o^t P(\eta_o)x_o + \frac{\kappa}{\zeta}[W(\eta_o)]||x(t+\theta)||_*^2, \quad \forall \theta \in [-\tau_\eta, 0]
\end{aligned} \tag{3.71}$$

where

$$||x(t+\theta)||_*^2 \triangleq \sup_{\theta\in[-\tau,0]} ||x(t+\theta)||_2^2$$

Let

$$\bar{P}(i) \;=\; \max_{i\in\mathcal{S}} \left\{ \frac{P(\eta_o)||x_o||^2 + \kappa[W(\eta_o)]||x(t+\theta)||_*^2}{\zeta[P(\eta_o)]||x_o||^2} \right\}$$

Finally it follows from (3.71) for $i \in \mathcal{S}$ that

$$\begin{aligned}
&\lim_{\mathcal{T}\to\infty} \; \mathbb{E}\Big\{\int_0^{\mathcal{T}} x^t(t)x(t)dt|\phi,\eta_o = i\Big\} \leq \\
&x_o^t\lambda_M(\bar{P}(i))x_o < +\infty
\end{aligned} \tag{3.72}$$

which shows that system $(\Sigma_{\Delta f})$ is **RSSFTD**. $\nabla\nabla\nabla$

Consider the JTD system

$$
\begin{aligned}
(\Sigma_{\Delta n}): \quad \dot{x}(t) &= [A_o(\eta_t) + \Delta A_o(t,\eta_t)]x(t) + [A_d(\eta_t) + \Delta A_d(t,\eta_t)]x(t-\tau_{\eta_t}) \\
&+ \Gamma(\eta_t)w(t) \\
&= A_{\Delta o}(t,\eta_t)x(t) + A_{\Delta d}(t,\eta_t)x(t-\tau_{\eta_t}) + \Gamma(\eta_t)w(t) \\
x(t) &= \phi(t), \ t \in [-\tau, 0], \ \eta_o = i \\
z(t) &= G_o(\eta_t) + \Phi(\eta_t)w(t)
\end{aligned}
$$

Extending on **Theorem** 3.5 in the manner of the foregoing section, the following result could be readily established for system $(\Sigma_{\Sigma n})$:

Theorem 3.6 *System* $\Sigma_{\Delta n}$ *is* **RSSFTD with a disturbance attenuation** γ, *if there exist matrices* $W^t(i) = W(i) > 0, P(i) = P^t(i) > 0, \ i \in \mathcal{S}$ *and scalars* $\epsilon(i) > 0, \ \mu(i) > 0, \ i \in \mathcal{S}$ *satisfying the LMIs for all* $i \in \mathcal{S}$

$$
\begin{bmatrix}
\Pi_{sg}(i) & \Lambda_w(i) & P(i)A_d(i) & P(i)\Gamma(i) + G_o^t(i)\Phi(i) \\
\Lambda_w^t(i) & -\Lambda_d(i) & 0 & 0 \\
A_d^t(i)P(i) & 0 & -\hat{W}(i) + \mu(i)N_d^t(i)N_d(i) & 0 \\
\Gamma^t(i)P(i) + \Phi^t(i)G_o(i) & 0 & 0 & -\gamma^2 I + \Phi^t(i)\Phi(i)
\end{bmatrix} < 0 \quad (3.73)
$$

$$(3.74)$$

$$
\begin{aligned}
\Pi_{sg}(i) &= A_o^t(i)P(i) + P(i)A_o(i) + \sum_{m=1}^{s} \alpha_{im}P(m) + \beta\, W(i) \\
&+ \epsilon(i)N_a^t(i)N_a + G_o^t(i)G_o(i) \\
\Lambda_w(i) &= [P(i)M_a(i) \quad P(i)M_d(i)] \\
\Lambda_d(i) &= diag[\epsilon(i)I \quad \mu(i)I] \qquad (3.75)
\end{aligned}
$$

Had we suppressed the uncertainties in systems $\Sigma_{\Delta f}$ and $\Sigma_{\Delta n}$ we would have obtained systems Σ_f and Σ_n:

$$(\Sigma_f): \quad \dot{x}(t) = A_o(\eta_t)x(t) + A_d(\eta_t)x(t-\tau_{\eta_t}) \qquad (3.76)$$

$$
\begin{aligned}
(\Sigma_n): \quad \dot{x}(t) &= A_o(t,\eta_t)x(t) + A_d(t,\eta_t)x(t-\tau_{\eta_t}) + \Gamma(\eta_t)w(t) \quad (3.77)\\
z(t) &= G_o(\eta_t) + \Phi(\eta_t)w(t) \quad (3.78)
\end{aligned}
$$

for which the following results stand as corollaries from **Theorem** 3.5 and **Theorem** 3.6, respectively.

Corollary 3.5 *System* Σ_f *is* **stochastically stable with functional time-delays (SSFTD)**, *if there exist matrices* $W^t(i) = W(i) > 0, P(i) = P^t(i) > 0,\ i \in \mathcal{S}$ *and scalars* $\epsilon(i) > 0,\ \mu(i) > 0,\ i \in \mathcal{S}$ *satisfying the LMIs for all* $i \in \mathcal{S}$

$$
\begin{bmatrix} \Pi_s(i) & P(i)A_d(i) \\ A_d^t(i)P(i) & -[\hat{W}(i) - \mu(i)N_d^t(i)N_d(i)] \end{bmatrix} < 0
$$

$$
\Pi_{so}(i) = A_o^t(i)P(i) + P(i)A_o(i) + \sum_{m=1}^{s} \alpha_{im}P(m) + \beta\, W(i)
$$

Corollary 3.6 *System* Σ_n *is* **SSFTD with a disturbance attenuation** γ, *if there exist matrices* $W(i) > 0, P(i) = P^t(i) > 0,\ i \in \mathcal{S}$ *satisfying the LMIs for all* $i \in \mathcal{S}$

$$
\begin{bmatrix} \Pi_s(i) & P(i)A_d(i) & \begin{matrix} P(i)\Gamma(i)+ \\ G_o^t(i)\Phi(i) \end{matrix} \\ A_d^t(i)P(i) & \begin{matrix} -\hat{W}(i)+ \\ \mu(i)N_d^t(i)N_d(i) \end{matrix} & 0 \\ \begin{matrix} \Gamma^t(i)P(i)+ \\ \Phi^t(i)G_o(i) \end{matrix} & 0 & \begin{matrix} -\gamma^2 I+ \\ \Phi^t(i)\Phi(i) \end{matrix} \end{bmatrix} < 0
$$

$$
\begin{aligned}
\Pi_{swo}(i) &= A_o^t(i)P(i) + P(i)A_o(i) + \sum_{m=1}^{s} \alpha_{im}P(m) \\
&+ \beta\, W(i) + G_o^t(i)G_o(i)
\end{aligned}
$$

3.3.3 Strong Delay-Dependence

In this section, we focus attention on the case of strong stochastic stability with mode-dependent delays and develop corresponding criteria for system (Σ_J). In this case, the information about the functional dependence τ_{η_t} is known in terms of a given relation or look-up table. To this end, we start by system (3.49) with $(u(.) \equiv 0)$ and use the Leibniz-Newton formula

$$\int_a^b \dot{f}(t)\,dt \;=\; f(b) \;-\; f(a)$$

It follows for $t \geq \tau_{\eta_t}$ and $\eta_t = i \in \mathcal{S}$ that

$$\begin{aligned}
x(t - \tau_{\eta_t}) &= x(t) - \int_{\tau(i)}^{0} \dot{x}(t+\theta)d\theta \\
&= x(t) - \int_{\tau_{\eta t}}^{0} A_{\Delta o}(t,i)x(t+\theta)d\theta \\
&- \int_{\tau(i)}^{0} A_{\Delta d}(t,i)x(t - \tau_{(i)} + \theta)d\theta x(t - \tau_{(i)}) \\
&- \int_{\tau(i)}^{0} \Gamma(i)w(t+\theta)d\theta
\end{aligned} \tag{3.79}$$

Upon substituting (3.79) back into (3.49) with $(u(.) \equiv 0)$ it yields:

$$\begin{aligned}
(\Sigma_{JS}): \quad \dot{x}(t) &= [A_{\Delta o}(t,i) + A_{\Delta d}(t,i)]x(t) \\
&- A_{\Delta d}(t,i)\int_{-\tau_{\eta t}}^{0} A_{\Delta o}(t,i)x(t+\theta)d\theta \\
&+ A_{\Delta d}(t,i)\int_{-\tau_{\eta t}}^{0} A_{\Delta d}(t,i)x(t - \tau_{\eta_t} + \theta)d\theta \\
&- A_{\Delta d}(t,i)\int_{-\tau_{\eta t}}^{0} \Gamma(i)w(t+\theta)d\theta \\
x(t) &= \phi(t), \; t \in [-2\tau, 0], \; \eta_o = i
\end{aligned} \tag{3.80}$$

The following theorem summarizes the main result:

Theorem 3.7 *Consider the jumping system* (Σ_{JS}) *with* $(w(.) \equiv 0)$*. Given scalars* $\omega_1(i) > 0,\ \omega_2(i) > 0,\ i \in \mathcal{S}$ *this system is* **robustly strong stochastically stable (RSSS) for any mode-dependent time-delay** $\tau_{\eta_t} \equiv \tau_{(i)},\ i \in \mathcal{S}$ *if there exist matrices* $0 < X(i) = X^t(i) \in \Re^{n\times n},\ i \in \mathcal{S}$ *and scalars* $\sigma_1(i) >$

0, $\sigma_2(i) > 0$, $\sigma_3(i) > 0$, $\sigma_4(i) > 0$, $\sigma_5(i) > 0$, $\sigma_6(i) > 0$, $\sigma_7(i) > 0$, $i \in \mathcal{S}$ *satisfying the system of LMIs for all* $i \in \mathcal{S}$*:*

$$\left[\begin{array}{ccc} \Pi(i) & \Lambda_a(i,\tau_{(i)}) & \Lambda_b(i,\tau_{(i)}) \\ \Lambda_a^t(i,\tau_{(i)}) & -\Theta_a(i,\tau_{(i)})I & 0 \\ \Lambda_b^t(i,\tau_{(i)}) & 0 & -\Theta_b(i,\tau_{(i)})I \end{array}\right] < 0$$

$$\left[\begin{array}{cc} -\omega_1(i)I & \sigma_2(i)N_d^t(i) \\ \sigma_2(i)N_d(i) & -\sigma_2(i)I \end{array}\right] < 0\,, \left[\begin{array}{cc} -\omega_2(i)I & \sigma_3(i)M_a(i) \\ \sigma_3(i)M_a(i) & -\sigma_3(i)I \end{array}\right] < 0$$

$$\left[\begin{array}{cc} -\sigma_6(i)I & \sigma_4(i)M_a(i) \\ \sigma_4(i)M_a^t(i) & -\sigma_4(i)I \end{array}\right] < 0\,, \left[\begin{array}{cc} -\sigma_7(i)I & \sigma_5(i)M_d(i) \\ \sigma_5(i)M_d^t(i) & -\sigma_5(i)I \end{array}\right] < 0$$

$$\left[\begin{array}{cc} X(i) & I \\ I & Y(i) \end{array}\right] \geq 0\,,\ \left[\begin{array}{cc} \omega_1(i) & I \\ I & \sigma_6(i) \end{array}\right] \geq 0\,,\ \left[\begin{array}{cc} \omega_2(i) & I \\ I & \sigma_7(i) \end{array}\right] \geq 0 \quad (3.81)$$

for all admissible uncertainties satisfying (3.52-3.53) where

$$\begin{aligned}
\Pi(i) &= X(i)[A_o(i) + A_d(i)] + [A_o(i) + A_d(i)]^t X(i) \\
&+ X(i)\{\sum_{m=1}^{s} \alpha_{im} Y(m)\} X(i) + \sigma_1(i)[N_a^t(i)N_a(i) + N_d^t(i)N_d(i)] \\
\Lambda_a(i,\tau_{(i)}) &= [\Lambda_{a1}(i,\tau_{(i)}) \ \ \Lambda_{a2}(i,\tau_{(i)})] \\
\Lambda_{a1}(i,\tau_{(i)}) &= [X(i)M_a(i) \ \ X(i)M_d(i) \ \ \tau X(i)M_d(i)] \\
\Lambda_{a2}(i,\tau_{(i)}) &= [\tau X(i)M_d(i) \ \ \tau N_a^t(i) \ \ \tau N_d^t] \\
\Lambda_b(i,\tau) &= [\tau X(i)A_d^t(i) \ \ \tau X(i)A_d^t(i) \ \ \tau A_o^t(i) \ \ \tau_{(i)}A_d^t(i)] \\
\Theta_a(i,\tau_{(i)}) &= diag[\Theta_{a1}(i,\tau_{(i)}) \ \ \Theta_{a2}(i,\tau_{(i)})] \\
\Theta_{a1}(i,\tau_{(i)}) &= diag[\sigma_1(i)I \ \ \sigma_1(i)I \ \ \tau\sigma_2(i)I] \\
\Theta_{a2}(i,\tau_{(i)}) &= diag[\tau_{(i)}\sigma_3(i)I \ \ \tau\sigma_4(i)I \ \ \tau\sigma_5(i)I] \\
\Theta_b(i,\tau_{(i)}) &= diag[\Theta_{b1}(i,\tau_{(i)}) \ \ \Theta_{b2}(i,\tau_{(i)}) \ \ \Theta_{b3}(i,\tau_{(i)}) \ \ \Theta_{b4}(i,\tau_{(i)})] \\
\Theta_{b1}(i,\tau_{(i)}) &= \tau[\omega_1(i)I - \sigma_2(i)N_d^t(i)N_d(i)] \\
\Theta_{b2}(i,\tau_{(i)}) &= \tau[\omega_1(i)I - \sigma_3(i)N_d^t(i)N_d(i)] \\
\Theta_{b3}(i,\tau_{(i)}) &= \tau[\sigma_6(i)I - \sigma_4(i)M_a(i)M_a^t(i)] \\
\Theta_{b4}(i,\tau_{(i)}) &= \tau[\sigma_7(i)I - \sigma_5(i)M_d(i)M_d^t(i)] \qquad (3.82)
\end{aligned}$$

Proof: Introduce the following Lyapunov-Krasovskii functional for $\eta_t = i \in \mathcal{S}$, $\mathbf{x} \in \mathcal{C}[-\tau_{(i)}, 0]$, thus:

$$V(\mathbf{x}, t, i) \;=\; x^t(t)X(i)x(t) + W(\mathbf{x}, t, i) \tag{3.83}$$

where $0 < X(i) = X^t(i) \in \Re^{n\times n}$, $i \in \mathcal{S}$ and $W(\mathbf{x}, t, i) > 0$ is a quadratic form to defined later on. The weak infinitesimal operator $\Im_s^x[\cdot]$ of the process $\{(\mathbf{x}(t), \eta_t),\ t \geq 0\}$ for system (3.120) at the point $\{t, \mathbf{x}, \eta_t\}$ is given by :

$$\begin{aligned} \Im_s^x[V(.,.)] &= x^t(t)\Big\{X(i)[A_{\Delta o}(t,i) + A_{\Delta d}(t,i)] + [A_{\Delta o}(t,i) + A_{\Delta d}(t,i)]^t X(i) \\ &+ \sum_{k=1}^{s} \alpha_{ik} X(k)\Big\}x(t) + \varrho(t,i) + \varphi(t,i) + \Im_s^x[W(.,.)] \end{aligned} \tag{3.84}$$

where

$$\begin{aligned} \varrho(t,i) &= -2\int_{-\tau(i)}^{0} x^t(t)X(i)A_{\Delta d}(t,i)A_{\Delta o}(t,i)x(t+\theta)d\theta \\ \varphi(t,i) &= -2\int_{-\tau(i)}^{0} x^t(t)X(i)A_{\Delta d}(t,i)A_{\Delta d}(t,i)x(t-\tau_{(i)}+\theta)d\theta \end{aligned} \tag{3.85}$$

It follows on using **Fact 1** that

$$\begin{aligned} \varrho(t,i) &\leq \omega_1^{-1}(i)\int_{-\tau(i)}^{0} \Big[x^t(t)X(i)A_{\Delta d}A_{\Delta d}^t X(i)x(t)\Big]d\theta \\ &+ \omega_1(i)\int_{-\tau(i)}^{0} \Big[x^t(t+\theta)A_{\Delta o}^t A_{\Delta o}(t,i)x(t+\theta)\Big]d\theta \\ &= \omega_1^{-1}(i)\tau_{(i)}x^t(t)X(i)A_{\Delta d}(t,i)A_{\Delta d}^t(t,i)X(i)x(t) \\ &+ \omega_1(i)\int_{-\tau_{\eta_t}}^{0} \Big[x^t(t+\theta)A_{\Delta o}^t(t,i)A_{\Delta o}(t,i)x(t+\theta)\Big]d\theta \end{aligned} \tag{3.86}$$

$$\begin{aligned} \varphi(t,i) &\leq \omega_2^{-1}(i)\int_{-\tau(i)}^{0} \Big[x^t(t)X(i)A_{\Delta d}(t,i)A_{\Delta d}^t(t,i)X(i)x(t)\Big]d\theta \\ &+ \omega_2(i)\int_{-\tau(i)}^{0} \Big[x^t(t-\tau_{(i)}+\theta)A_{\Delta d}^t(t,i)A_{\Delta d}(t,i)x(t-\tau_{(i)}+\theta)\Big]d\theta \end{aligned}$$

$$
\begin{aligned}
&= \omega_2(i) \int_{-\tau_{(i)}}^{0} \Big[x^t(t - \tau_{(i)} + \theta) A^t_{\Delta d}(t,i) A_{\Delta d}(t,i) x(t - \tau_{(i)} + \theta) \Big] d\theta \\
&+ \omega_2^{-1}(i) \tau_{(i)} x^t(t) X(i) A_{\Delta d}(t,i) A^t_{\Delta d}(t,i) X(i) x(t) \qquad (3.87)
\end{aligned}
$$

for some scalers $\omega_1(i) > 0,\ \omega_2(i),\ i \in \mathcal{S}$. Now define

$$
\begin{aligned}
W(\mathbf{x}, t, i) &= \int_{-\tau_{(i)}}^{0} \int_{t-\tau_{(i)}+\theta}^{t} \omega_2(i)[x^t(s) A^t_{\Delta d}(t,i) A_{\Delta d}(t,i) x(s)] ds d\theta \\
&+ \int_{-\tau_{(i)}}^{0} \int_{t+\theta}^{t} \omega_1(i)[x^t(s) A^t_{\Delta o}(t,i) A_{\Delta o}(t,i) x(s)] ds d\theta \qquad (3.88)
\end{aligned}
$$

The weak infinitesimal generator $\Im^x_s[W(.,.)]$ is given by:

$$
\begin{aligned}
\Im^x_s[W(.,.)] &= \frac{1}{\delta} \lim_{\delta \to 0} \mathcal{E}[W(\mathbf{x}, t, \eta_t) | \eta_o = i, \mathbf{x}_o = \phi] - W(\mathbf{x}, t, \eta_t) \\
&= \omega_1(i) \tau_{(i)} \Big[x^t(t) A^t_{\Delta o}(t,i) A_{\Delta o}(t,i) x(t) \Big] \\
&- \omega_1(i) \int_{-\tau_{(i)}}^{0} \Big[x^t(t+\theta) A^t_{\Delta o}(t,i) A_{\Delta o}(t,i) x(t+\theta) \Big] d\theta \\
&+ \theta) A^t_{\Delta d}(t,i) A_{\Delta d}(t,i) x(t - \tau_{(i)} + \theta) \Big] d\theta \\
&+ \omega_2(i) \tau_{(i)} \Big[x^t(t) A^t_{\Delta d}(t,i) A_{\Delta d}(t,i) x(t) \\
&- \omega_2(i) \int_{-\tau_{(i)}}^{0} \Big[x^t(t - \tau_{(i)} \Big] \qquad (3.89)
\end{aligned}
$$

On considering (3.86)-(3.89), it follows from (3.84) that

$$
\begin{aligned}
\Im^x_s[V(.,.)] &\leq x^t(t) \Big\{ X(i)[A_{\Delta o}(t,i) + A_{\Delta d}(t,i)] + [A_{\Delta o}(t,i) + A_{\Delta d}(t,i)]^t X(i) \\
&+ \omega_1^{-1}(i) \tau_{(i)} X(i) A_{\Delta d}(t,i) A^t_{\Delta d}(t,i) X(i) + \sum_{k=1}^{s} \alpha_{ik} X(k) \\
&+ \omega_2^{-1}(i) \tau_{(i)} X(i) A_{\Delta d}(t,i) A^t_{\Delta d}(t,i) X(i) \\
&+ \omega_2(i) \tau_{(i)} A^t_{\Delta d}(t,i) A_{\Delta d}(t,i) \Big\} x(t) \\
&+ \omega_1(i) \tau_{(i)} A^t_{\Delta o}(t,i) A_{\Delta o}(t,i) \qquad (3.90)
\end{aligned}
$$

Direct application of **Facts 1** and **2** using (3.52) yields:

$$
\begin{aligned}
& X(i)[A_{\Delta o}(t,i) + A_{\Delta d}(t,i)] + [A_{\Delta o}(t,i) + A_{\Delta d}(t,i)]^t X(i) \ \leq \\
& X(i)[A_o(t,i) + A_d(t,i)] + [A_o(t,i) + A_d(t,i)]^t X(i) + \\
& \theta_1^{-1}(i) X(i)[M_a(i)M_a^t(i) + M_d(i)M_d^t(i)]X(i) + \\
& \theta_1(i)[N_a^t(i)N_a(i) + N_d^t(i)N_d(i)] \qquad (3.91) \\
& X(i)A_{\Delta d}(t,i)A_{\Delta d}^t(t,i)X(i) \ \leq X(i)\Big[\theta_2^{-1}(i)M_d(i)M_d^t(i) + \\
& A_d(i)[I - \theta_2(i)N_d^t(i)N_d(i)]^{-1}A_d^t(i)\Big]X(i) \qquad (3.92) \\
& A_{\Delta o}^t(t,i)A_{\Delta o}(t,i) \ \leq \theta_3^{-1}(i)N_a^t(i)N_a(i) + \\
& A_o^t(i)[I - \theta_3(i)M_a(i)M_a^t(i)]^{-1}A_o(i) \qquad (3.93) \\
& A_{\Delta d}^t(t,i)A_{\Delta d}(t,i) \ \leq \theta_4^{-1}(i)N_d^t(i)N_d(i) + \\
& A_d^t(i)[I - \theta_4(i)M_d(i)M_d^t(i)]^{-1}A_d(i) \qquad (3.94)
\end{aligned}
$$

for some scalars $\theta_1(i) > 0, \cdots, \theta_4(i) > 0$, $i \in \mathcal{S}$ satisfying

$$
\begin{aligned}
& [I - \theta_2(i)N_d^t(i)N_d(i)] > 0 \ , \ [I - \theta_3(i)M_a(i)M_a^t(i)] > 0 \\
& [I - \theta_4(i)M_d(i)M_d^t(i)] > 0 \qquad (3.95)
\end{aligned}
$$

Recalling the fact that $\omega_1(i), \omega_2(i)$, $i \in \mathcal{S}$ are given weights, we can introduce the following change of variables

$$
\begin{aligned}
\sigma_1(i) &= \theta_1(i) \ , \ \sigma_2(i) = \omega_1(i)\theta_2(i) \ , \ \sigma_3(i) = \omega_2(i)\theta_2(i) \\
\sigma_4(i) &= \sigma_6(i)\theta_3(i)\sigma_5(i) = \sigma_7(i)\theta_2(i) \\
\sigma_6(i) &= \omega_1^{-1}(i) \ , \ \sigma_7(i)\omega_2^{-1}(i) \qquad (3.96)
\end{aligned}
$$

It then follows from (3.90)-(3.96) that

$$
\Im_s^x[V(.,.)] \leq x^t(t)\Omega(X, \sigma, \omega, \tau, i)x(t) \qquad (3.97)
$$

with

$$
\begin{aligned}
\Omega(X,\sigma,\omega,\tau_{(i)},i) &= X(i)[A_o(i)+A_d(i)] + [A_o(i)+A_d(i)]^t X(i) \\
&+ X(i)\{\sum_{m=1}^{s} \alpha_{im} Y(m)\} X(i) + \sigma_1(i)[N_a^t(i)N_a(i) + N_d^t(i)N_d(i)] \\
&+ \sigma_1^{-1}(i)X(i)[M_a(i)M_a^t(i) + M_d(i)M_d^t(i)]X(i) \\
&+ \tau_{(i)}X(i)\left[\sigma_2^{-1}(i)M_d(i)M_d^t(i) + \sigma_3^{-1}(i)M_d(i)M_d^t(i)\right]X(i) \\
&+ \tau_{(i)}X(i)A_d(i)[\omega_1(i)I - \sigma_2(i)N_d^t(i)N_d(i)]^{-1}A_d^t(i)X(i) \\
&+ \tau_{(i)}X(i)A_d(i)[\omega_2(i)I - \sigma_3(i)N_d^t(i)N_d(i)]^{-1}A_d^t(i)X(i) \\
&+ \sigma_4^{-1}(i)\tau_{(i)}N_a^t(i)N_a(i) + \sigma_5^{-1}(i)\tau_{(i)}N_d^t(i)N_d(i) \\
&+ \tau_{(i)}A_o^t(i)[\sigma_6(i)I - \sigma_4(i)M_a(i)M_a^t(i)]^{-1}A_o(i) \\
&+ \tau_{(i)}A_d^t(i)[\sigma_7(i)I - \sigma_5(i)M_d(i)M_d^t(i)]^{-1}A_d(i) \qquad (3.98)
\end{aligned}
$$

where $\sigma(i)$ denotes $\{\sigma_1(i),\cdots,\sigma_7(i)\}$ and $\omega(i)$ denotes the combined weights $\{\omega_1(i),\omega_2(i)\}$. In view of the monotonic nondecreasing behavior of $\Omega(X,\theta,\omega,\tau_{(i)},i)$ and using the Schur complements, it can be readily verified that LMIs (3.81) ensure that $\Omega(X,\sigma,\omega,\tau_{(i)},i) < 0$. The remaining part of the proof is similar to that of **Theorem** 3.1. $\nabla\nabla\nabla$

Alternatively, by suppressing the uncertainties in system (Σ_{JS}), we obtain the nominal system

$$
\begin{aligned}
(\Sigma_{JS0}): \quad \dot{x}(t) &= [A_o(i)+A_d(i)]x(t) \\
&- A_d(t,i)\int_{-\tau_{\eta t}}^{0} A_o(i)x(t+\theta)d\theta \\
&+ A_d(t,i)\int_{-\tau_{\eta t}}^{0} A_d(i)x(t-\tau_{\eta_t}+\theta)d\theta \\
&- A_d(t,i)\int_{-\tau_{\eta t}}^{0} \Gamma(i)w(t+\theta)d\theta \\
x(t) &= \phi(t),\ t\in[-2\tau,0],\ \eta_o = i \qquad (3.99)
\end{aligned}
$$

Results on robust strong mode-dependent stability with $\mathcal{H}_\infty$ performance as well the corresponding results for the nominal system could be directly established from the last theorem and are stated below.

Theorem 3.8 *Consider the time-delay system* (Σ_{JS}). *Given scalars* $\gamma > 0,\ \omega_1(i) > 0,\ \omega_2(i) > 0,\ i \in \mathcal{S}$ *this system is* **RSSS with a disturbance attenuation** γ **for any mode-dependent time-delay** $\tau_{\eta_t} \equiv \tau_{(i)},\ i \in \mathcal{S}$ *and for all admissible uncertainties satisfying (3.52-3.53) if there exist matrices* $0 < X(i) = X^t(i) \in \Re^{n\times n},\ i \in \mathcal{S}$ *and scalars* $\beta_1(i) > 0,\ \beta_2(i) > 0,\ \beta_3(i) > 0,\ \beta_4(i) > 0,\ \beta_5(i) > 0,\ \beta_6(i) > 0,\ \beta_7(i) > 0,\ \upsilon(i) > 0,\ \ i \in \mathcal{S}$ *satisfying the system of LMIs for all* $i \in \mathcal{S}$

$$\begin{bmatrix} \Pi(i) & \Lambda_a(i,\tau_{(i)}) & \Lambda_b(i,\tau_{(i)}) & X(i)\Xi(i) \\ \Lambda_a^t(i,\tau_{(i)}) & -\Theta_a(i,\tau_{(i)})I & 0 & 0 \\ \Lambda_b^t(i,\tau_{(i)}) & 0 & -\Theta_b(i,\tau_{(i)})I & 0 \\ X(i)\Xi^t(i) & 0 & 0 & -\Theta_c(i) \end{bmatrix} < 0$$

$$\begin{bmatrix} -\omega_1(i)I & \sigma_2(i)N_d^t(i) \\ \sigma_2(i)N_d(i) & -\sigma_2(i)I \end{bmatrix} < 0\,, \begin{bmatrix} -\omega_2(i)I & \sigma_3(i)M_a(i) \\ \sigma_3(i)M_a(i) & -\sigma_3(i)I \end{bmatrix} < 0$$

$$\begin{bmatrix} -\omega_1^{-1}(i)I & \sigma_4(i)M_a(i) \\ \sigma_4(i)M_a^t(i) & -\sigma_4(i)I \end{bmatrix} < 0\,, \begin{bmatrix} -\omega_2^{-1}(i)I & \sigma_5(i)M_d(i) \\ \sigma_5(i)M_d^t(i) & -\sigma_5(i)I \end{bmatrix} < 0$$

$$\begin{bmatrix} X(i) & I \\ I & Y(i) \end{bmatrix} \geq 0$$

$$\begin{bmatrix} -\gamma^2 I + \Phi^t(i)\Phi(i) & \tau_{(i)}\upsilon(i)\Gamma^t(i) \\ \tau_{(i)}\upsilon(i)\Gamma(i) & -\tau_{(i)}\upsilon(i)I \end{bmatrix} < 0 \qquad (3.100)$$

for all admissible uncertainties satisfying (3.52-3.53) where

$$\begin{aligned} \Xi(i) &= [G^t(i) \ \ G^t(i)\Phi(i)] \\ \Theta_c(i) &= \begin{bmatrix} I & 0 \\ 0 & \gamma^2 I - \Phi^t(i)\Phi(i) - \tau_{(i)}\upsilon(i)\Gamma^t(i)\Gamma(i) \end{bmatrix} \end{aligned} \qquad (3.101)$$

Corollary 3.7 *Consider the jumping system* (Σ_{JSo}) *with* $(w(.) \equiv 0)$. *Given scalars* $\omega_1(i) > 0,\ \omega_2(i) > 0,\ i \in \mathcal{S}$ *this system is* **strong stochastically stable (SSS) for any mode-dependent time-delay** $\tau_{\eta_t} \equiv \tau_{(i)},\ i \in \mathcal{S}$ *if* *there exist matrices* $0 < X(i) = X^t(i) \in \Re^{n\times n}$, $i \in \mathcal{S}$ *satisfying the system of LMIs for all* $i \in \mathcal{S}$:

$$\begin{bmatrix} \Pi_o(i) & \Lambda_b(i,\tau_{(i)}) \\ \Lambda_b^t(i,\tau_{(i)}) & -\Theta_b(i,\tau_{(i)})I \end{bmatrix} < 0 \ , \ \begin{bmatrix} X(i) & I \\ I & Y(i) \end{bmatrix} \geq 0 \quad (3.102)$$

where

$$\begin{aligned} \Pi_o(i) &= X(i)[A_o(i)+A_d(i)] + [A_o(i)+A_d(i)]^t X(i) \\ &+ X(i)\{\sum_{m=1}^{s} \alpha_{im} Y(m)\} X(i) \end{aligned} \quad (3.103)$$

Corollary 3.8 *Consider the jumping system* (Σ_{JS0}). *Given scalars* $\gamma > 0$, $\omega_1(i) > 0$, $\omega_2(i) > 0$, $i \in \mathcal{S}$ *this system is* **SSS with a disturbance attenuation** γ **for any mode-dependent time-delay** $\tau_{\eta_t} \equiv \tau_{(i)}$, $i \in \mathcal{S}$ *and for all admissible uncertainties satisfying (3.52-3.53) if there exist matrices* $0 < X(i) = X^t(i) \in \Re^{n\times n}$, $i \in \mathcal{S}$ *satisfying the system of LMIs*

$$\begin{bmatrix} \Pi(i) & \Lambda_b(i,\tau_{(i)}) & X(i)\Xi(i) \\ \Lambda_b^t(i,\tau_{(i)}) & -\Theta_b(i,\tau_{(i)})I & 0 \\ X(i)\Xi^t(i) & 0 & -\Theta_c(i) \end{bmatrix} < 0$$

$$\begin{bmatrix} X(i) & I \\ I & Y(i) \end{bmatrix} \geq 0$$

$$\begin{bmatrix} -\gamma^2 I + \Phi^t(i)\Phi(i) & \tau_{(i)}\upsilon(i)\Gamma^t(i) \\ \tau_{(i)}\upsilon(i)\Gamma(i) & -\tau_{(i)}\upsilon(i)I \end{bmatrix} < 0 \quad (3.104)$$

3.3.4 Example 3.3

In order to illustrate the theoretical results of this section, we provide a numerical example. We consider a pilot-scale single-reach water quality system considered in **Example 2**. Focusing on functional delays, we use

$$W(1) = \begin{bmatrix} 2 & 0 \\ 0 & 2 \end{bmatrix} \ , \ W(2) = \begin{bmatrix} 1 & 0 \\ 0 & 1 \end{bmatrix} \ , \ W(3) = \begin{bmatrix} 2 & 0 \\ 0 & 2 \end{bmatrix}$$

and $\hat{\tau} = 0.8,\ \check{\tau} = 0.2$. The feasible solution of LMIs (3.58) is given by:

$$
\begin{aligned}
P(1) &= \begin{bmatrix} 2.9240 & 1.5361 \\ 1.5361 & 5.5422 \end{bmatrix}, \quad \epsilon(1) = 1.1254,\ \mu(1) = 1.1275 \\
P(2) &= \begin{bmatrix} 3.2054 & 2.1443 \\ 2.1443 & 6.7805 \end{bmatrix}, \quad \epsilon(2) = 2.7698,\ \mu(2) = 1.7894 \\
P(3) &= \begin{bmatrix} 2.8952 & 1.4293 \\ 1.4293 & 5.9865 \end{bmatrix}, \quad \epsilon(3) = 1.7134,\ \mu(3) = 1.8164
\end{aligned}
$$

which verifies **Theorem** 3.5. Note that the bounds on the functional delays less than the forgoing bounds which means less conservative stability results. For the strong functional delays, we solve LMIs (3.75) of **Theorem** 3.6 with $\omega_1(1) = 1,\ \omega_2(1) = 2,\ \omega_1(2) = 1,\ \omega_2(2) = 2,\ \omega_1(3) = 1,\ \omega_2(3) = 2$. The feasible results are:

$$
\begin{aligned}
X(1) &= \begin{bmatrix} 4.9377 & 2.2076 \\ 2.2076 & 2.1177 \end{bmatrix}, \quad \sigma_1(1) = 1.16529,\ \sigma_2(1) = 0.9825 \\
\sigma_3(1) &= 3.0178,\ \sigma_4(1) = 2.1114,\ \sigma_5(1) = 2.1456 \\
X(2) &= \begin{bmatrix} 5.5537 & 0.6743 \\ 0.6743 & 1.8289 \end{bmatrix}, \quad \sigma_1(2) = 1.1298,\ \sigma_2(2) = 0.8825 \\
\sigma_3(2) &= 2.7668,\ \sigma_4(2) = 2.4034,\ \sigma_5(2) = 2.0146 \\
X(3) &= \begin{bmatrix} 5.3457 & 0.7642 \\ 0.7642 & 1.7789 \end{bmatrix}, \quad \sigma_1(3) = 1.2918,\ \sigma_2(3) = 1.2576 \\
\sigma_3(3) &= 2.6658,\ \sigma_4(3) = 2.3344,\ \sigma_5(3) = 2.4126
\end{aligned}
$$

over the range $[0, 0.2421]$. This reads that the water quality system is stochastically stable for any functional delays $\tau_{(i)}$ satisfying $0 \leq \tau_{(i)} \leq 0.2421$. Next, we solve LMIs (3.100) with $\omega_1(1) = 1,\ \omega_2(1) = 2,\ \omega_1(2) = 1,\ \omega_2(2) = 2,\ \omega_1(3) = 1,\ \omega_2(3) = 2$. The feasible results are:

$$
\begin{aligned}
X(1) &= \begin{bmatrix} 4.7985 & 2.1054 \\ 2.1054 & 2.1168 \end{bmatrix} \\
\beta_1(1) &= 0.2987,\ \beta_2(1) = 1.0374,\ \beta_3(1) = 0.8117 \\
\beta_4(1) &= 1.1577,\ \beta_5(1) = 2.9487,\ \beta_6(1) = 5.2066 \\
\beta_7(1) &= 0.2987,\ \upsilon(1) = 11.0321
\end{aligned}
$$

$$
\begin{aligned}
X(2) &= \begin{bmatrix} 5.4325 & 0.5546 \\ 0.5546 & 1.9065 \end{bmatrix} \\
\beta_1(2) &= 0.3164, \ \beta_2(2) = 0.7374, \ \beta_3(2) = 1.2870 \\
\beta_4(2) &= 1.5063, \ \beta_5(2) = 6.0094, \ \beta_6(2) = 4.1286 \\
\beta_7(2) &= 0.3358, \ v(2) = 1.9987 \\
X(3) &= \begin{bmatrix} 5.2566 & 0.5534 \\ 0.5534 & 2.2375 \end{bmatrix} \\
\beta_1(3) &= 0.3364, \ \beta_2(3) = 0.7574, \ \beta_3(3) = 1.2970 \\
\beta_4(3) &= 1.5163, \ \beta_5(3) = 6.1094, \ \beta_6(3) = 4.2286 \\
\beta_7(3) &= 0.3458, \ v(3) = 3.8987, \quad \gamma = 4.2561
\end{aligned}
$$

over the range $[0, 0.3098]$. This again shows that the functional delays approach yields less conservative stability results.

3.4 Robust Stabilization

In this section, we consider the robust stabilization (closed-loop stability) problem using a feedback control law of the form

$$u(t) = -K(i)x(t) \ , \quad i \in \mathcal{S} \tag{3.105}$$

as applied to the uncertain JTD system (2.38). The results are split into two subsections: the first is for mode-independent and the second is for mode-dependent.

3.4.1 Mode-Independent Results

In this case, the uncertain closed-loop system is expressed for $\eta_t = i \in \mathcal{S}$ as:

$$
\begin{aligned}
(\Sigma_{\Delta c}): \ \dot{x}(t) &= A_{\Delta K}(t,i)x(t) + A_{\Delta d}(t,i)x(t-\tau) + \Gamma(t,i)w(t) \ , \quad t \geq 0 \\
x(s) &= \phi(s) \ , s \in [-\tau, 0] \ , \quad \eta_o = i
\end{aligned} \tag{3.106}
$$

$$
\begin{aligned}
z(t) &= G_o(i)x(t) + \Phi(i)w(t) && (3.107)\\
A_{\Delta K}(i) &= A_K(i) + M_a(i)\Delta(t,i)[N_a(i) + N_b(i)K(i)] && (3.108)
\end{aligned}
$$

We have the following results:

Theorem 3.9 *In the absence of input disturbance $w(t) \equiv 0$, controller (3.105) is a stabilizing controller for system (3.106)-(3.108) if, given matrix sequence $\bar{Q}(i) > 0$, $\hat{Q}(i) > 0$, there exist matrices $Y(i) = Y^t(i) > 0$, $Z(i)$, $L(i) = L^t(i) > 0$, $i \in \mathcal{S}$, satisfying the system of LMIs for all $i \in \mathcal{S}$*

$$
\begin{bmatrix} \bar{\Lambda}_t(i) & Y(i) & \Upsilon_a(i) & \Upsilon_b(i) \\ Y(i) & -L(i) & 0 & 0 \\ \Upsilon_a^t(i) & 0 & -\Upsilon_c(i) & 0 \\ \Upsilon_b^t(i) & 0 & 0 & -\Upsilon_d(i) \end{bmatrix} < 0
$$

$$
\begin{bmatrix} -L(i) & I \\ I & -\hat{Q}(i) \end{bmatrix} \geq 0 \qquad (3.109)
$$

for all admissible uncertainties satisfying (3.52)-(3.53) where

$$
\begin{aligned}
\bar{\Lambda}_t(i) &= Y(i)A_o^t(i) + A_o(i)Y(i) + B_o(i)Z(i) + Z^t(i)B_o^t(i)\\
&+ \alpha_{ii}Y(i) + [\varepsilon(i) + \varrho(i)]M_a(i)M_a^t(i)\\
\Upsilon_a(i) &= [Z^t(i)D_o^t(i) + Y(i)C_o^t(i) \;\; Z^t(i)N_b^t(i) + Y(i)N_a^t(i)]\\
\Upsilon_b(i) &= [\mathcal{R}(i) \;\; A_d(i)]\,,\; \Upsilon_c(i) = diag[I \;\; \varepsilon(i)I]\\
\Upsilon_d(i) &= diag[\mathcal{Y}(i) \;\; \bar{Q} - \varrho(i)N_d^t(i)N_d(i)] && (3.110)
\end{aligned}
$$

Proof: It follows from **Theorem** (3.1) and using the Schur complements that controller (3.105) is a stabilizing controller for system $(\Sigma_{\Delta c})$ if there exist matrices $P(i) = P^t(i) > 0$ satisfying the system of ARIs

$$
\begin{aligned}
A_{\Delta K}^t(i)P(i) + P(i)A_{\Delta K}(i) + \sum_{m=1}^{s} \alpha_{im}P(m) + \\
\hat{Q}(i) + P(i)A_{\Delta d}(i)\bar{Q}^{-1}(i)A_{\Delta d}^t(i)P(i) &< 0 && (3.111)
\end{aligned}
$$

By **Fact 1** and (3.108) we have

$$\begin{aligned}
&A^t_{\Delta K}(i)P(i) + P(i)A_{\Delta K}(i) \leq A^t_K(i)P(i) + P(i)A_K(i) + \\
&\varepsilon(i)P(i)M_a(i)M^t_a(i)P(i) + \\
&\varepsilon^{-1}(i)[N_a(i) + N_b(i)K(i)]^t[N_a(i) + N_b(i)K(i)]
\end{aligned} \tag{3.112}$$

for some scalars $\varepsilon(i) > 0,\ i \in \mathcal{S}$. Similarly, by **Fact 2** we get

$$\begin{aligned}
&P(i)A_{\Delta d}(i)\bar{Q}^{-1}(i)A^t_{\Delta d}(i)P(i) \leq \varrho(i)P(i)M_a(i)M^t_a(i)P(i) + \\
&P(i)A_d(i)[\bar{Q}(i) - \rho^{-1}(i)N^t_d(i)N_d(i)]^{-1}A^t_d(i)P(i)
\end{aligned} \tag{3.113}$$

for some scalars $\varrho(i) > 0,\ i \in \mathcal{S}$. Combining (3.112)-(3.113) into (3.111), it yields:

$$\begin{aligned}
&A^t_K(i)P(i) + P(i)A_K(i) + \sum_{m=1}^{s}\alpha_{im}P(m) + \hat{Q}(i) + \varepsilon(i)P(i)M_a(i)M^t_a(i)P(i) + \\
&\varrho(i)P(i)M_a(i)M^t_a(i)P(i) + \varepsilon^{-1}(i)[N_a(i) + N_b(i)K(i)]^t[N_a(i) + N_b(i)K(i)] + \\
&P(i)A_d(i)[\bar{Q}(i) - \rho^{-1}(i)N^t_d(i)N_d(i)]^{-1}A^t_d(i)P(i)\ <\ 0
\end{aligned} \tag{3.114}$$

Now substituting $Y(i) = P^{-1}(i),\ K(i)Y(i) = Z(i),$ and $L(i) = \hat{Q}^{-1}$ into (3.114) and manipulating using **Fact 3**, we obtain LMIs (3.109) subject to (3.110). $\nabla\nabla\nabla$

Theorem 3.10 *Given a prescribed constant $\gamma > 0$. The control law (3.105) is a weakly delay-dependent stabilizing controller for system (Σ_{nc}) with a disturbance attenuation γ with feedback gain given by $K(i) = Z(i)Y^{-1}(i),\ i \in \mathcal{S}$ if given matrix sequence $\bar{Q}(i) > 0,\ \hat{Q}(i) > 0,\ i \in \mathcal{S}$ there exist matrices $Y(i) = Y^t(i) >$*

0, $Z(i)$, $L(i) = L^t(i) > 0$, $i \in \mathcal{S}$, *satisfying the system of LMIs for all* $i \in \mathcal{S}$

$$\begin{bmatrix} \bar{\Lambda}_t(i) & Y(i) & \Omega^t(i) & \Upsilon_b(i) & \begin{matrix}\Gamma(i)+\\ Y(i)G_o^t(i)\Phi(i)\end{matrix} \\ Y(i) & -L(i) & 0 & 0 & 0 \\ \Omega^t(i) & 0 & -\Upsilon_c(i) & 0 & 0 \\ \Upsilon_b^t(i)(i) & 0 & 0 & -\Upsilon_d(i) & 0 \\ \begin{matrix}\Gamma^t(i)+\\ \Phi^t(i)G_o(i)Y(i)\end{matrix} & 0 & 0 & 0 & \begin{matrix}-\gamma^2 I+\\ \Phi^t(i)\Phi(i)\end{matrix} \end{bmatrix} < 0$$

$$\begin{bmatrix} -L(i) & I \\ I & -\hat{Q}(i) \end{bmatrix} \geq 0 \quad i \in \mathcal{S} \tag{3.115}$$

for all admissible uncertainties satisfying (3.52)-(3.53) and

$$\Omega(i) = [G_o(i)Y(i) \quad N_a(i)Y(i) + N_b(i)Z(i)] \tag{3.116}$$

Proof. It can be worked out by using the same technique as that used in **Theorem** 3.2. ∇∇∇

Theorem 3.11 *Given a prescribed constant* $\gamma > 0$. *The control law (3.105) is a strongly delay-dependent stabilizing controller for system* (Σ_{nc}) *with a disturbance attenuation* γ *with feedback gain given by* $K(i) = Z(i)Y^{-1}(i)$, $i \in \mathcal{S}$ *if there exist matrices* $0 < X(i) = X^t(i) \in \Re^{n \times n}$, $i \in \mathcal{S}$ *and scalars* $\sigma_1(i) > 0$, $\sigma_2(i) > 0$, $\sigma_3(i) > 0$, $\sigma_4(i) > 0$, $\sigma_5(i) > 0$, $\sigma_6(i) > 0$, $\sigma_7(i) > 0$, $i \in \mathcal{S}$ *satisfying the system of LMIs for all* $i \in \mathcal{S}$:

$$\begin{bmatrix} \Pi_{ac}(i) & \Lambda_{ac}(i,\tau) & \Lambda_{bc}(i,\tau) & \Lambda_n(i) \\ \Lambda_{ac}^t(i,\tau) & -\Theta_a(i,\tau)I & 0 & 0 \\ \Lambda_{bc}^t(i,\tau) & 0 & -\Theta_b(i,\tau)I & 0 \\ \Lambda_n^t(i) & 0 & 0 & -\Lambda_m(i) \end{bmatrix} < 0$$

$$\begin{bmatrix} -\omega_1(i)I & \sigma_2(i)N_d^t(i) \\ \sigma_2(i)N_d(i) & -\sigma_2(i)I \end{bmatrix} < 0 \,, \begin{bmatrix} -\omega_2(i)I & \sigma_3(i)M_a(i) \\ \sigma_3(i)M_a(i) & -\sigma_3(i)I \end{bmatrix} < 0$$

$$\begin{bmatrix} -\sigma_6(i)I & \sigma_4(i)M_a(i) \\ \sigma_4(i)M_a^t(i) & -\sigma_4(i)I \end{bmatrix} < 0 \,, \begin{bmatrix} -\sigma_7(i)I & \sigma_5(i)M_d(i) \\ \sigma_5(i)M_d^t(i) & -\sigma_5(i)I \end{bmatrix} < 0$$

$$\begin{bmatrix} X(i) & I \\ I & Y(i) \end{bmatrix} \geq 0 \,, \begin{bmatrix} \sigma_6(i) & I \\ I & \omega_1(i) \end{bmatrix} \geq 0$$

$$\begin{bmatrix} \sigma_7(i) & I \\ I & \omega_2(i) \end{bmatrix} \geq 0 \tag{3.117}$$

for all admissible uncertainties satisfying (3.52)-(3.53) where

$$\begin{aligned}
\Pi_{ac}(i) &= [A_o(i) + A_d(i)]Y(i) + Y(i)[A_o(i) + A_d(i)]^t \\
&+ \{\sum_{m=1}^{s} \alpha_{im} Y(m)\} \\
\Lambda_n(i) &= [\sigma_1(i)Y(i)N_a^t(i) \quad \sigma_1(i)Y(i)N_d^t(i)] \\
\Lambda_m(i) &= diag[\sigma(i)I \quad \sigma(i)I] \\
\Lambda_{ac}(i,\tau) &= [\Lambda_{ac1}(i,\tau) \quad \Lambda_{ac2}(i,\tau)] \\
\Lambda_{a1}(i,\tau) &= [M_a(i) \ M_d(i) \ \tau M_d(i)] \\
\Lambda_{a2}(i,\tau) &= [\tau M_d(i) \ \tau Y(i)N_a^t(i) \ \tau Y(i)N_d^t] \\
\Lambda_{bc}(i,\tau) &= [\tau A_d^t(i) \ \tau A_d^t(i) \ \tau Y(i)A_o^t(i) \ \tau Y(i)A_d^t(i)]
\end{aligned} \tag{3.118}$$

Proof. It can be worked out by using the same technique as that used in **Theorem** 3.3. ∇∇∇

Remark 3.4 *rem11 The corresponding results for the nominal case can be easily derived by supressing the uncertainties in* **Theorems** *3.9-3.11 and it is left to the reader as an excersize.*

3.4.2 Mode-Dependent Results

We now proceed to consider the problem of stabilizing the jumping system using a mode-dependent state feedback controller. Two distinct cases will be considered:

The first case is memoryless and the second is of delayed form.

3.4.3 Memoryless Feedback

The controller has the form

$$u(t) = K(\eta_t)\, x(t)\,, \quad \eta_t = i \in \mathcal{S} \tag{3.119}$$

which is designed to guarantee the stochastic stability of the resulting closed-loop system (3.47) with (3.119) for $\eta_t = i \in \mathcal{S}$:

$$\begin{aligned} (\Sigma_{\Delta CJ}): \quad \dot{x}(t) &= A_{\Delta K}(t,i)x(t) + A_{\Delta d}(t,i)x(t-\tau_{\eta_t}) + \Gamma(i)w(t), \\ x(t) &= \phi(t),\ t \in [-\tau, 0],\ \eta_o = i \end{aligned} \tag{3.120}$$

where

$$\begin{aligned} A_{\Delta K}(t,i) &= [A_o(i) + B_o(i)K(i)] + [\Delta A_o(t,i) + \Delta B_o(t,i)K(i)] \\ &= [A_o(i) + B_o(i)K(i)] + M_a(i)\Delta(t,i)[N_a(i) + N_b(i)K(i)] \\ &= A_{oK}(i) + M_a(i)\Delta(t,i)N_{aK}(i) \end{aligned} \tag{3.121}$$

and $K(i),\ i \in \mathcal{S}$ are mode-dependent constant gains to be designed. Thus we have the following result.

Theorem 3.12 *System $(\Sigma_{\Delta CJ})$ with $w(\cdot) \equiv 0$ is* **RSSFTD** *if there exist matrices $Y(i) = Y^t(i) > 0,\ Z(i),\ i \in \mathcal{S},\ L(i) > 0$ and scalars $\mu(i) > 0,\ \epsilon(i) > 0,\ i \in \mathcal{S}$ satisfying the LMIs for all $i \in \mathcal{S}$*

$$\begin{bmatrix} \Pi_t(i) & \Xi_a(i) & Y(i) \\ \Xi_a^t(i) & -\Xi_b(i) & 0 \\ Y(i) & 0 & -\hat{\tau}\beta^{-1}L(i) \end{bmatrix} < 0 \tag{3.122}$$

$$\begin{aligned} \Pi_t(i) &= Y(i)A_o^t(i) + A_o(i)Y(i) + B_o(i)Z(i) + Z^t(i)B_o^t(i) \\ &+ Y(i)\Big[\sum_{k=1}^{s} \alpha_{ik} Y^{-1}(k)\Big]Y(i) + \epsilon(i)M_a(i)M_a^t(i) \\ &+ \mu(i)M_d(i)M_d^t(i) + A_d(i)L(i)A_d^t(i) \\ \Xi_a(i) &= [Y(i)N_a^t(i) + Z(i)N_b^t(i) \quad A_d(i)L(i)N_d^t(i)] \\ \Xi_b(i) &= diag[\epsilon(i)I \quad \mu(i)I - N_d(i)L(i)N_d^t(i)] \end{aligned} \tag{3.123}$$

with the mode-dependent gain given by

$$K(i) \;=\; Z(i)Y^{-1}(i) \quad , \; i \in \mathcal{S}.$$

Proof: By **Theorem** 3.5 and **Fact 3**, it follows that system $(\Sigma_{\Delta CJ})$ is **RSS-FTD** if there exist matrices $P(i) = P^t(i) > 0, i \in \mathcal{S},\; W(i) > 0$ satisfying the system of ARIs

$$\begin{aligned}
& A_{oK}^t(i)P(i) + P(i)A_{oK}(i) + \sum_{k=1}^{s} \alpha_{ik} P(k) + \beta\, W + \\
& \epsilon^{-1}(i)N_{oK}^t(i)N_{aK}(i) + \epsilon(i)P(i)M_a(i)M_a^t(i)P(i) + \\
& P(i)A_d(i)[\hat{W} - \mu^{-1}(i)N_d^t(i)N_d(i)]^{-1}A_d^t(i)P(i) + \\
& \mu(i)P(i)M_d(i)M_d^t(i)P(i) \;\; < \; 0
\end{aligned} \tag{3.124}$$

Introducing $Y(i) = P^{-1}(i)$, $Z(i) = K(i)Y(i)$, $i \in \mathcal{S}$, $L(i) = \hat{W}^{-1}(i)$, pre- and post-multiplying (3.124) by $Y(i)$ and arranging using (3.123)-(3.124), we obtain:

$$\begin{aligned}
& Y(i)A_o^t(i) + A_o(i)Y(i) + B_o(i)Z(i) + Z^t(i)B_o^t(i) + \\
& Y(i)\Big[\sum_{k=1}^{s} \alpha_{ik} Y^{-1}(k)\Big]Y(i) + \epsilon(i)M_a(i)M_a^t(i) + \frac{\beta}{\hat{\tau}}Y(i)L^{-1}(i)Y(i) + \\
& \epsilon^{-1}(i)[Y(i)N_a^t(i) + Z(i)N_b^t(i)][N_a(i)Y(i) + N_b(i)Z(i)] + \mu(i)M_d(i)M_d^t(i) + \\
& A_d(i)\Big(L(i) + L(i)N_d^t(i)\Big[\mu(i)I - N_d(i)L(i)N_d^t(i)\Big]^{-1}N_d(i)L(i)\Big)A_d^t(i) \;\; = \\
& \Pi_t(i) + \frac{\beta}{\hat{\tau}}Y(i)L^{-1}(i)Y(i) + \epsilon^{-1}(i)[Y(i)N_a^t(i) + \\
& Z(i)N_b^t(i)][N_a(i)Y(i) + N_b(i)Z(i)] + \\
& A_d(i)L(i)N_d^t(i)\Big[\mu(i)I - N_d(i)L(i)N_d^t(i)\Big]^{-1}N_d(i)L(i)A_d^t(i) \;<\; 0
\end{aligned} \tag{3.125}$$

where the matrix inversion lemma [98] has been used in (3.125) for expanding the invertible term to avoid multiplication of variables in LMI setting. By the Schur complements, LMIs (3.122) immediately follows. $\nabla\nabla\nabla$

3.4.4 Delayed Feedback

In this case, the controller has the form

$$u(t) = K_d(\eta_t)\, x(t-\tau_{\eta_t})\,, \quad \eta_t = i \in \mathcal{S} \tag{3.126}$$

The resulting closed-loop system (3.47) with (3.126) is given by:

$$\begin{aligned}(\Sigma_{\Delta DJ}):\quad \dot{x}(t) &= A_\Delta(t,\eta_t)x(t) + A_{\Delta dK}(t,\eta_t)x(t-\tau_{\eta_t}) + \Gamma(\eta_t)w(t),\\ x(t) &= \phi(t),\ t\in[-\tau,0],\ \eta_o = i \end{aligned} \tag{3.127}$$

where for $\eta_t = i \in \mathcal{S}$

$$\begin{aligned} A_{\Delta dK}(t,i) &= [A_d(i) + B_o(i)K_d(i)] + [\Delta A_d(t,i) + \Delta B_o(t,i)K_d(i)]\\ &= [A_d(i) + B_o(i)K_d(i)] + M_a(i)\Delta(t,i)[N_d(i) + N_b(i)K_d(i)]\\ &= A_{dK}(i) + M_a(i)\Delta(t,i)N_{dK}(i) \end{aligned} \tag{3.128}$$

and $K_d(i),\ i \in \mathcal{S}$ are mode-dependent constant gains to be designed. It follows from **Theorem** 3.5 that the closed-loop system stability is guaranteed provided for some scalars $\epsilon(i) > 0,\ \mu(i) > 0,\ i \in \mathcal{S}$ the following inequality:

$$\begin{aligned} &A_o^t(i)P(i) + P(i)A_o(i) + \sum_{m=1}^{s} \alpha_{im}P(m) + \\ &\epsilon(i)N_a^t(i)N_a + \epsilon^{-1}(i)P(i)M_a(i)M_a^t(i)P(i) + \\ &\beta\, W + \mu^{-1}(i)P(i)M_d(i)M_d^t(i)P(i) + \\ &P(i)A_{dK}(i)[\hat{W}(i) - \mu(i)N_{dK}^t(i)N_{dK}(i)]^{-1}A_{dK}^t(i)P(i) \ < \ 0 \end{aligned} \tag{3.129}$$

holds. The following theorem summarizes the desired result.

Theorem 3.13 *System* $(\Sigma_{\Delta DJ})$ *with* $w(\cdot) \equiv 0$ *is* **RSSFTD** *if there exist matrices* $W^t(i) = W(i) > 0,\ Y(i) = Y^t(i) > 0,\ X(i) = X^t(i) > 0,\ Z(i),\ i \in \mathcal{S}$ *and* *scalars* $\mu(i) > 0,\ \epsilon(i) > 0,\ i \in \mathcal{S}$ *satisfying the LMIs*

$$\begin{bmatrix} \Pi_d(i) & M_a(i) & M_d(i) & \Omega(i) \\ M_a^t(i) & -\epsilon(i)I & 0 & 0 \\ M_d^t(i) & 0 & -\mu(i)I & 0 \\ \Omega^t(i) & 0 & 0 & \begin{matrix} -\hat{W}(i)+ \\ \sigma(i)\Psi^t(i)\Psi(i) \end{matrix} \end{bmatrix} < 0 \quad \forall i \in \mathcal{S}$$

$$\begin{bmatrix} \mu(i) & 1 \\ 1 & \sigma(i) \end{bmatrix} \geq 0 \ , \ \begin{bmatrix} Y(i) & I \\ I & X(i) \end{bmatrix} \geq 0 \tag{3.130}$$

$$\begin{aligned} \Pi_d(i) &= Y(i)A_o^t(i) + A_o(i)Y(i) + Y(i)\Big[\sum_{m=1}^{s} \alpha_{im} Y^{-1}(m)\Big]Y(i) \\ &+ \epsilon(i)N_a^t(i)N_a(i) + \beta Y(i)W(i)Y(i) \\ \Omega(i) &= A_d(i) + B_o(i)X(i)Z(i) \ , \quad \Psi(i) = N_d(i) + N_b(i)X(i)Z(i) \end{aligned} \tag{3.131}$$

with the mode-dependent gain given by

$$K_d(i) = Z(i)X(i) \quad , \ i \in \mathcal{S}$$

Proof: Introducing $Y(i) = P^{-1}(i),\ Z(i) = K_d(i)Y(i),\ i \in \mathcal{S}$, pre- and post-multiplying (3.129) by $Y(i)$ and arranging using (3.126)-(3.128), we obtain:

$$\begin{aligned} &Y(i)A_o^t(i) + A_o(i)Y(i) + Y(i)\Big[\sum_{m=1}^{s} \alpha_{im} Y^{-1}(m)\Big]Y(i) + \\ &\epsilon^{-1}(i)M_a(i)M_a^t(i) + \beta\, Y(i)W(i)Y(i) + \mu^{-1}(i)M_d(i)M_d^t(i) + \\ &\epsilon(i)N_a^t(i)N_a(i) + [A_d(i) + B_o(i)Y^{-1}(i)Z(i)] \\ &\Big[\hat{W} - \mu^{-1}[N_d(i) + N_b(i)Y^{-1}(i)Z(i)]^t[N_d(i) + N_b(i)Y^{-1}(i)Z(i)]\Big]^{-1} \\ &[A_d(i) + B_o(i)Y^{-1}(i)Z(i)]^t < 0 \end{aligned} \tag{3.132}$$

Rearranging (3.132) using (3.131) and the equality constraints $Y(i)X(i) = I\ ,\ \mu(i)\sigma(i) = 1$ and applying the Schur complements, LMIs (3.130) immediately follow. $\nabla\nabla\nabla$

To end this section and in line of **Theorem** 3.6, results on robust stabilization with $\mathcal{H}_\infty$ performance could be directly established. These are stated below.

Theorem 3.14 *System* $(\Sigma_{\Delta CJ})$ *is* **RSSFTD with a disturbance attenuation** γ *if there exist matrices* $Y(i) = Y^t(i) > 0,\ Z(i),\ i \in \mathcal{S},\ L(i) > 0$ *and scalars* $\mu(i) > 0,\ \epsilon(i) > 0,\ i \in \mathcal{S}$ *satisfying the LMIs for all* $i\ \in \mathcal{S}$

$$\begin{bmatrix} \Pi_t(i) & \Xi_a(i) & Y(i) & \begin{matrix} P(i)\Gamma(i)+ \\ G_o^t(i)\Phi(i) \end{matrix} \\ \Xi_a^t(i) & -\Xi_b(i) & 0 & 0 \\ Y(i) & 0 & -\hat{\tau}\beta^{-1}L(i) & 0 \\ \begin{matrix} \Gamma^t(i)P(i)+ \\ \Phi^t(i)G_o(i) \end{matrix} & 0 & 0 & \begin{matrix} -\gamma^2 I+ \\ \Phi^t(i)\Phi(i) \end{matrix} \end{bmatrix} < 0 \tag{3.133}$$

$$\begin{aligned} \Pi_{tw}(i) &= Y(i)A_o^t(i) + A_o(i)Y(i) + B_o(i)Z(i) + Z^t(i)B_o^t(i) \\ &+ Y(i)\Big[\sum_{k=1}^{s} \alpha_{ik} Y^{-1}(k)\Big]Y(i) + \epsilon(i)M_a(i)M_a^t(i) \\ &+ \mu(i)M_d(i)M_d^t(i) + A_d(i)L(i)A_d^t(i) + G_o^t(i)G_o(i) \end{aligned} \tag{3.134}$$

with the mode-dependent gain given by

$$K(i) = Z(i)Y^{-1}(i) \quad , \ i \in \mathcal{S}.$$

Theorem 3.15 *System* $(\Sigma_{\Delta DJ})$ *is* **RSSFTD with a disturbance attenuation** γ *if there exist matrices* $W^t(i) = W(i) > 0,\ Y(i) = Y^t(i) > 0,\ X(i) = X^t(i) > 0,\ Z(i),\ i \in \mathcal{S}$ *and scalars* $\mu(i) > 0,\ \epsilon(i) > 0,\ i \in \mathcal{S}$ *satisfying the LMIs for all* $i\ \in \mathcal{S}$

$$\begin{bmatrix} \Pi_{dw}(i) & M_a(i) & M_d(i) & \Omega(i) & \begin{matrix} P(i)\Gamma(i)+ \\ G_o^t(i)\Phi(i) \end{matrix} \\ M_a^t(i) & -\epsilon(i)I & 0 & 0 & 0 \\ M_d^t(i) & 0 & -\mu(i)I & 0 & 0 \\ \Omega^t(i) & 0 & 0 & \begin{matrix} -\hat{W}(i)+ \\ \sigma(i)\Psi^t(i)\Psi(i) \end{matrix} & 0 \\ \begin{matrix} \Gamma^t(i)P(i)+ \\ \Phi^t(i)G_o(i) \end{matrix} & 0 & 0 & 0 & \begin{matrix} -\gamma^2 I+ \\ \Phi^t(i)\Phi(i) \end{matrix} \end{bmatrix} < 0$$

$$\begin{bmatrix} \mu(i) & 1 \\ 1 & \sigma(i) \end{bmatrix} \geq 0 \ , \ \begin{bmatrix} Y(i) & I \\ I & X(i) \end{bmatrix} \geq 0 \tag{3.135}$$

$$\begin{aligned} \Pi_{dw}(i) &= Y(i)A_o^t(i) + A_o(i)Y(i) + Y(i)\Big[\sum_{m=1}^{s} \alpha_{im}Y^{-1}(m)\Big]Y(i) \\ &+ \epsilon(i)N_a^t(i)N_a(i) + \beta Y(i)W(i)Y(i) + G_o^t(i)G_o(i) \\ \Omega(i) &= A_d(i) + B_o(i)X(i)Z(i) \\ \Psi(i) &= N_d(i) + N_b(i)X(i)Z(i) \end{aligned} \tag{3.136}$$

with the mode-dependent gain given by

$$K_d(i) = Z(i)X(i) \quad , \ i \in \mathcal{S}$$

3.4.5 Example 3.4

Now, we provide a numerical example and consider the model of pilot-scale single-reach water quality system treated before with three-reaches and $\tau^* = 0.95$, $\tau^+ = 0.5$. An in **Example 2**, we let the Markov process governing the mode switching has generator

$$\Im = \begin{bmatrix} -6 & 4 & 2 \\ 2 & -5 & 3 \\ 3 & 1 & -4 \end{bmatrix}$$

The associated date for the three operating conditions (modes) are given there. Directing attention to the robust stabilization, we consider **Theorem** 3.9 for the memoryless feedback and solve LMIs (3.122) using

$$L(1) = \begin{bmatrix} 0.5 & 0 \\ 0 & 0.5 \end{bmatrix} \ , \ L(2) = \begin{bmatrix} 1 & 0 \\ 0 & 1 \end{bmatrix} \ , \ L(3) = \begin{bmatrix} 0.5 & 0 \\ 0 & 0.5 \end{bmatrix}$$

to get:

$$Y(1) = \begin{bmatrix} 12.2240 & 3.5321 \\ 3.5321 & 9.0422 \end{bmatrix} , \ Z(1) = \begin{bmatrix} 3.0124 & 0.5341 \\ -0.6125 & 1.0332 \end{bmatrix}$$

$$K(1) = \begin{bmatrix} 0.2585 & -0.0419 \\ -0.0937 & 0.1509 \end{bmatrix}, \ \epsilon(1) = 1.2154, \ \mu(1) = 1.2075$$

$$Y(2) = \begin{bmatrix} 9.2064 & 2.4423 \\ 2.4423 & 9.8605 \end{bmatrix}, \ Z(2) = \begin{bmatrix} 1.4564 & 0.6214 \\ -0.7250 & 1.2112 \end{bmatrix}$$

$$K(2) = \begin{bmatrix} 0.1514 & 0.0255 \\ -0.1192 & 0.1523 \end{bmatrix}, \ \epsilon(2) = 2.7698, \ \mu(2) = 1.7894$$

$$Y(3) = \begin{bmatrix} 10.8952 & 2.4293 \\ 2.4293 & 10.0065 \end{bmatrix}, \ Z(3) = \begin{bmatrix} 2.1004 & -0.2333 \\ 0.1925 & 1.2332 \end{bmatrix}$$

$$K(3) = \begin{bmatrix} 0.2093 & -0.0741 \\ -0.0104 & 0.1258 \end{bmatrix}, \ \epsilon(3) = 1.7134, \ \mu(3) = 1.8164$$

By considering **Theorem** 3.10, we solve LMIs (3.130) to obtain:

$$Y(1) = \begin{bmatrix} 11.2650 & 3.4520 \\ 3.3520 & 8.6422 \end{bmatrix}, \ Z(1) = \begin{bmatrix} 3.1004 & -0.3941 \\ -0.5825 & 1.1432 \end{bmatrix}$$

$$K_d(1) = \begin{bmatrix} 0.3295 & -0.1772 \\ -0.1051 & 0.1743 \end{bmatrix}, \ \epsilon(1) = 1.3154, \ \mu(1) = 1.3175$$

$$Y(2) = \begin{bmatrix} 8.2064 & 2.5423 \\ 2.5423 & 9.6605 \end{bmatrix}, \ Z(2) = \begin{bmatrix} 1.3564 & -0.5214 \\ -0.6250 & 1.4356 \end{bmatrix}$$

$$K_d(2) = \begin{bmatrix} 0.1982 & -0.1061 \\ -0.1331 & 0.1836 \end{bmatrix}, \ \epsilon(2) = 2.4469, \ \mu(2) = 1.8254$$

$$Y(3) = \begin{bmatrix} 11.8952 & 2.3943 \\ 2.3943 & 10.2165 \end{bmatrix}, \ Z(3) = \begin{bmatrix} 2.1214 & -0.2333 \\ -0.1925 & 1.3432 \end{bmatrix}$$

$$K_d(3) = \begin{bmatrix} 0.1922 & -0.0679 \\ -0.0448 & 0.1420 \end{bmatrix}, \ \epsilon(3) = 1.3498, \ \mu(3) = 1.8426$$

3.5 Notes and References

In this chapter, we have investigated the problems of stochastic stability and stabilization for a class of continuous-times JTD systems and have developed appropriate criteria based on LMIs. Our major concern has been on disclosing the inter playing effects among parametric uncertainties, patterns of time-delays and the jumping parameters. There are other research efforts along similar lines and interested reader can consult [19, 21, 22, 127, 128, 129, 131, 140] and their references. Development of strong delay-dependent stability criteria has been based on the Leibniz-Newton formula. Descriptor-type transformations have been proposed for nominal time-delay systems [53, 54] and their references. These transformations and others have been developed in [115, 116, 117, 118, 119, 120] and will be used in later chapters.

Chapter 4

Control System Design

4.1 Introduction

The problems of robust stability and performance for linear continuous-time JTD systems with parametric uncertainty have been fully analyzed in Chapter 3 where LMI-based criteria have been developed for different delay information patterns. This chapter contributes to the further development of methodologies for JTD systems by considering robust control design techniques. We will focus on robust guaranteed cost control approach and dynamic output-feedback techniques. In either case, we will pay closer attention to the role of delay factor and jumping parameters. Essentially, we will deal with of a class of linear continuous-time systems with real time-varying norm-bounded parametric uncertainties, unknown time-varying state-delay and Markovian jump parameters. In the sequel, the performance index considered is a quadratic cost function frequently used in linear quadratic regulators. In case of the robust performance analysis problem, we show that the addressed notions of stability (see Chapter 3) guarantee an upper bound on the expected value of a linear quadratic cost function. We then address the control synthesis problem and prove that a robust state-feedback controller can be constructed to render the closed-loop system robustly stochastically stable and guarantees an adequate level of performance. The feedback gain can be determined by solving a parameter-dependent algebraic Riccati (ARI) or linear matrix inequality (LMI). Next we deal with the design problem of dynamic output-feedback controllers for JTD systems for both weak- and strong-delay dependent schemes. Finally we consider controlling a class of JTD with state and input delays using a new state-transformation which readily exhibits the delay-dependent behavior.

4.2 Problem Description

We recall the probability space $(\Omega, \mathcal{F}, \mathbf{P})$ described in Chapter 1 and consider a class of continuous JTD systems for each possible value $\eta_t = i, \; i \in \mathcal{S}$, over that space by:

$$\begin{aligned} (\Sigma_J): \; \dot{x}(t) &= \Big[A_o(i) + \Delta A_o(t,i)\Big] x(t) + \Big[A_d(i) + \Delta A_d(t,i)\Big] x(t-\tau) \\ &+ \Big[B_o(i) + \Delta B_o(t,i)\Big] u(t) \\ &= A_{\Delta o}(t,i)x(t) + A_{\Delta d}(t,i)x(t-\tau) + B_{\Delta o}(t,i)u(t) \; , \\ & \quad x(t) = \phi(t), \; t \in [-\tau, 0], \; \eta_o = i, \; t \geq 0, \end{aligned} \tag{4.1}$$

where $x(t) \in \Re^n$ is the state vector and $u(t) \in \Re^m$ is the control input. Here, the unknown time-varying delay factor $\tau(t) \in (0, \tau^*]$ is such that $\dot{\tau}(t) \leq \tau^+ < 1$ with τ^* and τ^+ being finite known constants. In (4.1), for $\eta_t = i, \; i \in \mathcal{S}$: $A_o(i)$, $B_o(i)$ and $A_d(i)$ are known constant matrices of appropriate dimensions and $\Delta A_o(i)$, $\Delta B_o(i)$ and $\Delta A_d(i)$ are unknown matrices which represent time-varying parametric uncertainties and assumed to belong to certain bounded compact sets. The initial vector function is specified as $\xi_0 \equiv \langle x(0), x(s) \rangle = \langle x_o, \phi(s) \rangle$, where $\phi(\cdot) \in \mathcal{L}_2[-\tau, 0]$ and will be assumed, throughout this chapter, that it is independent of the process $\{\eta_t,\ t \in [0, \mathcal{T}]\}$. The admissible parameter uncertainties are assumed to be of the following forms

$$\begin{bmatrix} \Delta A_o(i) & \Delta B_o(i) & \Delta A_d(i) \end{bmatrix} = M_a(i)\Delta(i,t)\begin{bmatrix} N_a(i) & N_b(i) & N_d(i) \end{bmatrix} \quad (4.2)$$

where for any $\eta_t = i,\ i \in \mathcal{S}$, $M_a(i), N_a(i),\ N_b(i)$, and $N_d(i)$ are known constant matrices of appropriate dimensions and $\Delta(i,t)$ is an unknown time-varying matrix satisfying

$$\Delta^t(i,t)\Delta(i,t) \le I \quad (4.3)$$

Let $\mathsf{X}(\xi_0, \eta_0)$ denote the state trajectory in system (4.1) from the initial state (ξ_0, η_0).

4.3 Control Objective

Motivated by the well known linear quadratic control theory [36], we define the following cost function for JTD system (4.1) for $\eta_t = i \in \mathcal{S}$:

$$\begin{aligned} &\mathsf{J}\Big[\xi_0, \eta_0, \Delta A_o(i), \Delta B_o(i), \Delta A_d(i)\Big] \\ &= \mathbb{E}\Big\{\int_0^\infty \Big[x(s)^t Q(i) x(s) + u(s)^t R(i) u(s)\Big]\, ds \,\Big|\, \zeta_0, \eta_0\Big\} \end{aligned} \quad (4.4)$$

where $Q(i) > 0$ and $R(i) > 0$ are given state and control weighting matrices, respectively. In the sequel, we will design a memoryless state feedback control law

$$\mathcal{U}:\ u(i) \;=\; -K(i)x(t) \quad ,i \in \mathcal{S} \quad (4.5)$$

for the uncertain state-delay system (4.1) and cost function (4.4) such that the resulting closed-loop system is robustly stochastically stable with weak delay-dependence (see Chapter 3) and the corresponding cost function (4.4) satisfies

$$\mathsf{J}\Big[\zeta_0, \eta_0, \Delta A_o(i), \Delta B_o(i), \Delta A_d(i)\Big] \;\le\; \mathsf{J}^* \quad (4.6)$$

for all admissible parameter uncertainties satisfying (4.2)-(4.3), where J^* is a given number.

Towards our objective, we introduce the following definition.

Definition 4.1 *For uncertain state-delay system (4.1) with cost function (4.4), if there exists a control law $\mathcal{U}$ and a number $J^* > 0$ such that the resulting closed-loop system*

$$\begin{aligned} \dot{x}(t) &= \Big[A_o(i) + \Delta A_o(t,i) - [B_o(i) + \Delta B_o(i)]K(i)\Big]x(t) \\ &+ \Big[A_d(i) + \Delta A_d(t,i)\Big]x(t-\tau) \end{aligned} \tag{4.7}$$

is **RSSWDD** *and the corresponding cost function (4.4) satisfies (4.6) for all admissible parameter uncertainties satisfying (4.2)-(4.3), then we call $\mathcal{U}$ is a* **guaranteed cost control law** *and J^* is a* **guaranteed cost** *for system (4.1) and cost function (4.4).*

Recall from [98] that in the absence of parametric uncertainty in system (4.1) (that is $\Delta A_o(i,t) \equiv 0$, $\Delta B_o(i,t) \equiv 0$, $\Delta A_d(i,t) \equiv 0$), no time-delay ($\tau = 0$, $A_d = 0$) and there is only single system operating form (mode) ($\eta_t \equiv 1$) is concerned, the corresponding cost function of (4.4) for $u(t) \equiv 0$ is given by

$$\mathsf{J}(x_0) = x_0^t P x_0$$

where P is obtained by the algebraic Riccati equation

$$PA_o + A_o^t P + Q = 0$$

Here, P is called a **quadratic cost matrix (QCM)**.

4.4 Robust Performance Analysis

To examine the role of the time-delay factor on the performance of system (4.1), we divide our effort into two parts. The first part is concerned with weak delay-dependent stability in which the results depend only on the bound τ^+ of the derivative $\dot{\tau}$. The second part deals with strong delay-dependent stability in which the results depend on the bound τ^* of the instantaneous delay factor τ. The reader is advised to refer to Chapter 3 regarding stability issues.

4.4.1 Weak Delay Dependence

For this purpose, we introduce the following definition

Definition 4.2 *The uncertain time-delay system (4.1) with cost function (4.4) is said to be* **RSSWDD** *with a* **quadratic cost matrix (QCM)** , $0 < P(i) = P(i)^t \in \Re^{n\times n}, i \in \mathcal{S}$ *if there exist matrices* $0 < W(i) = W(i)^t \in \Re^{n\times n}, i \in \mathcal{S}$ *such that*

$$\begin{aligned} & A^t_{\Delta o}(i)P(i) + P(i)A_{\Delta o}i) + P(i)A_{\Delta d}(i)\bar{W}^{-1}A^t_{\Delta d}(i)P(i) + \\ & \sum_{m=1}^{s} \alpha_{im}P(m) + W(i) + Q(i) < 0 \\ & \qquad \forall\Delta(i) : \Delta^t(i)\ \Delta(i) \ \leq \ I \ \forall\, i \in \mathcal{S} \end{aligned}$$

The next theorem derives an upper bound on the cost function J.

Theorem 4.1 *Consider system (4.1) and cost function (4.4). If* $0 < P(i) = P^t(i) \in \Re^{n\times n}, i \ \in \ \mathcal{S}$ *is a* **QCM** *at mode* $\eta_t = i$*, then system (4.1) is* **RSSWDD** *and the cost function satisfies the bound*

$$\mathsf{J} \ \leq \ x_o^t P(i)x_o \ + \ \int_{-\tau}^{0} x^t(\alpha)W(i)x(\alpha)d\alpha \quad i \ \in \ \mathcal{S} \tag{4.8}$$

Conversely, if system (4.1) is **RSSWDD** *then there will be a* **QCM** *for this system and cost function (4.4).*

Proof:($\Longrightarrow$) Let $0 < P(i) = P^t(i) \in \Re^{n\times n}, i \ \in \ \mathcal{S}$ be a QCM for system (4.1) and cost function (4.4). It follows from **Definition** (4.2) that there exists a matrix $0 < W(i) = W^t(i) \in \Re^{n\times n}$, $i \in \mathcal{S}$ such that

$$X^t(i)\begin{bmatrix} \Pi_\Delta(i) & P(i)A_{\Delta d}(i) \\ A^t_{\Delta d}P(i) & -\bar{W}(i) \end{bmatrix} X(i) \quad < \quad 0 \; ,$$
$$\forall \Delta(i,t) : \;\; \Delta^t(i,t)\; \Delta(i,t) \;\; \leq \;\; I \; , \; i \;\in\; \mathcal{S} \; , \; X(i) \;\neq\; [0 \;\; 0] \qquad (4.9)$$

where

$$\begin{aligned} \Pi_\Delta(i) \quad &= \quad P(i)A_{\Delta o}(i) + A^t_{\Delta o}(i)P(i) \\ &+ \quad \sum_{m=1}^{s} \alpha_{im}P(m) + W(i) + Q(i) \\ X(t) \quad &= \quad [x^t(t) \;\; x^t(t \, - \tau)]^t \end{aligned} \qquad (4.10)$$

and $\bar{W}(i) = (1 - \tau^+)W(i)$. Therefore, system (4.1) is RSSWDD. Now by **Theorem** (3.1) and **Definition(4.2)**, the following bounds hold:

$$\begin{aligned} &\Im^x[V_w] \; \leq \; - \, x^t(t)Q(i)x(t) \;\; \Longrightarrow \;\; x^t(t)Q(i)x(t) \\ &\leq \;\; -\Im^x[V_w] \Longrightarrow \;\; \mathsf{J} \; \leq \; V_w(\xi, i) \end{aligned} \qquad (4.11)$$

In view of (4.4) and manipulating (4.11), it reduces to

$$\mathsf{J} \; \leq \; x^t_o P(i) x_o \; + \; \textstyle\int_{-\tau}^{0} x^t(\alpha)W(i)x(\alpha)d\alpha.$$

($\Longleftarrow$) Let system (4.1) be RSSWDD. It follows that there exist $0 < P(i) = P^t(i) \in \Re^{n\times n}$, $i \in \mathcal{S}$ and $0 < W(i) = W^t(i) \in \Re^{n\times n}, i \in \mathcal{S}$ such that

$$\begin{aligned} &A^t_{\Delta o}(i)P(i) + P(i)A_{\Delta o}(i) + P(i)A_{\Delta d}(i)\bar{W}^{-1}(i)A^t_{\Delta d}(i)P(i) + \\ &\sum_{m=1}^{s} \alpha_{im}P(m) + W(i) < 0 \\ &\forall \Delta(i) : \Delta(i)^t(i)\; \Delta(i) \;\; \leq \; I \; , \;\; \forall \, i \; \in \; \mathcal{S} \end{aligned} \qquad (4.12)$$

Hence, one can find some $\rho(i) \;>\; 0$, $i \in \mathcal{S}$ such that the following inequality holds:

$$\rho(i)^{-1}[A^t_{\Delta o}(i)P(i) + P(i)A_{\Delta o}(i) + P(i)A_{\Delta d}(i)\bar{W}^{-1}(i)A^t_{\Delta d}(i)P(i) +$$

$$
\begin{aligned}
&\sum_{m=1}^{s} \alpha_{im} P(m) + W(i) + Q(i) \ = \\
&[A^t_{\Delta o}(i)\breve{P}(i) + \breve{P}(i)A_{\Delta o}(i) + \breve{P}(i)A_{\Delta d}(i)\breve{W}^{-1}(i)A^t_{\Delta}(i)\breve{P}(i) + \\
&\sum_{m=1}^{s} \alpha_{im} P(m) + \breve{W}(i) + Q(i) \ < \ 0 \\
&\forall \Delta(i) : \Delta^t(i)\ \Delta(i) \ \leq \ I\ , \ \forall i \ \in \ \mathcal{S}
\end{aligned} \tag{4.13}
$$

The above inequality implies that there exists a matrix $\breve{W}(i) = \rho(i)^{-1}W(i)$ such that the matrix $\breve{P}(i) = \rho^{-1}(i)P(i)$ is a QCM for system (4.1). $\nabla\nabla\nabla$

By employing the uncertainty representation (4.2)-(4.3), the following theorem establishes a necessary and sufficient condition for the existence of a quadratic cost matrix associated with system (4.1):

Theorem 4.2 *A matrix $0 < P(i) = P^t(i) \in \Re^{n \times n}$, $i \in \mathcal{S}$ is a QCM for system (4.1) and cost function (4.4) if and only if there exist matrices $0 < W(i) = W^t(i) \in \Re^{n \times n}$, $i \in \mathcal{S}$ and scalars $\mu(i) > 0, i \in \mathcal{S}$ satisfying the LMIs for all $i \in \mathcal{S}$*

$$
\begin{aligned}
&\begin{bmatrix} \Pi_1(i) & P(i)M_a(i) & \Lambda_1(i) \\ M^t_a(i)P(i) & -\mu(i)I & 0 \\ \Lambda^t_1(i) & 0 & \begin{matrix} -W(i)+ \\ \mu N^t_d(i)N_d(i) \end{matrix} \end{bmatrix} < 0 \\
&\begin{bmatrix} -W(i) & \mu(i)N^t_d(i) \\ \mu(i)N_d(i) & \mu(i)I \end{bmatrix} < 0
\end{aligned} \tag{4.14}
$$

where

$$
\begin{aligned}
\Pi_1(i) &= P(i)A_o(i) + A^t_o(i)P(i) + \sum_{m=1}^{s} \alpha_{im} P(m) \\
&+ W(i) + Q(i) + \mu(i)N^t_a(i)N_a(i) \\
\Lambda_1(i) &= P(i)A_d(i) + \mu(i)N^t_a(i)N_d(i)
\end{aligned}
$$

Proof: By **Definition** (4.2) and **Fact 1** of the Appendix, system (4.1) with cost function (4.4) is **QCM** which implies that

$$\begin{bmatrix} \Pi_\Delta & P(i)A_{\Delta d}(i) \\ A^t_{\Delta d}(i)P(i) & -W(i) \end{bmatrix} =$$

$$\begin{bmatrix} \bar{\Pi}_\Delta & \bar{\Lambda}_1(i) \\ \bar{\Lambda}^t{}_1(i) & -W(i) \end{bmatrix} < 0 \tag{4.15}$$

where

$$\begin{aligned} \bar{\Pi}_\Delta(i) &= P(i)A_o(i) + A^t_o(i)P(i) \\ &+ \sum_{m=1}^{s} \alpha_{im} P(m) + W(i) + Q(i) \\ &+ P(i)M_a(i)\Delta(i)N_a(i) + N^t_a(i)\Delta^t(i)M^t_a(i)P(i) \\ \bar{\Lambda}_1(i) &= P(i)A_d(i) + P(i)M_a(i)\Delta(i)N_d(i) \end{aligned}$$

Inequality (4.15) holds if and only if

$$\begin{aligned} &\begin{bmatrix} \Pi_o(i) & P(i)A_d(i) \\ A^t_d(i)P(i) & -W(i) \end{bmatrix} \\ &+ \begin{bmatrix} P(i)M_a(i) \\ 0 \end{bmatrix} \Delta(i)[N_a(i) \quad N_d(i)] \\ &+ \begin{bmatrix} N^t_a(i) \\ N^t_d(i) \end{bmatrix} \Delta^t(i)[M^t_a(i)P(i) \quad 0] < 0 \\ &\forall \Delta(i) : \Delta^t(i)\, \Delta(i) \le I, \; \forall\, i \in \mathcal{S} \end{aligned} \tag{4.16}$$

where

$$\begin{aligned} \Pi_o(i) &= P(i)A_o(i) + A^t_o(i)P(i) \\ &+ \sum_{m=1}^{s} \alpha_{im} P(m) + W(i) + Q(i) \end{aligned}$$

By **Fact 4** of the Appendix, inequality (4.16) for some $\mu(i) > 0, \; i \in \mathcal{S}$ is equivalent to

$$\begin{bmatrix} \Pi_o(i) & P(i)A_d(i) \\ A^t_d(i)P(i) & -W(i) \end{bmatrix} +$$

$$\mu^{-1}(i)\begin{bmatrix} P(i)M_a(i) \\ 0 \end{bmatrix}[M_a^t(i)P(i) \quad 0] + \mu(i)\begin{bmatrix} N_a^t(i) \\ N_d^t(i) \end{bmatrix}[N_a(i) \quad N_d(i)] =$$

$$\begin{bmatrix} \Pi_1(i)+ \\ \mu^{-1}(i)P(i)M_a(i)M_a^t(i)P(i) & \Lambda_1(i) \\ \Lambda_1^t(i) & -W(i)+ \\ & \mu(i)N_d^t(i)N_d(i)] \end{bmatrix} < 0$$

$$\forall \Delta(i) : \Delta^t(i)\,\Delta(i) \leq I, \ \forall\, i \in \mathcal{S}. \tag{4.17}$$

Simple rearrangement of (4.17) yields LMIs (4.14). ∇∇∇

An alternative formulation of the existence of a quadratic cost matrix is provided by the following corollary:

Corollary 4.1 *A matrix $0 < P(i) = P^t(i) \in \Re^{n\times n}, i \in \mathcal{S}$ is a* **QCM** *for system (4.1) and cost function (4.4) if and only if there exist matrices $0 < W(i) = W^t(i) \in \Re^{n\times n}$, $i \in \mathcal{S}$ and scalars $\mu(i) > 0$, $i \in \mathcal{S}$ and satisfying the LMIs for all $i \in \mathcal{S}$*

$$\begin{bmatrix} \Pi_\mu(i) & P(i)M_a(i) & P(i)\Gamma(i) \\ M_a^t(i)P(i) & -\mu(i)I & 0 \\ \Gamma^t(i)P(i) & 0 & -I \end{bmatrix} < 0$$
$$\begin{bmatrix} -W(i) & \mu(i)A_d^t(i) \\ \mu(i)A_d(i) & -\mu(i)I \end{bmatrix} < 0 \tag{4.18}$$

where

$$\begin{aligned}
\Pi_\mu(i) &= P(i)\bar{A}(i) + \bar{A}^t(i)P(i) \\
&+ \sum_{m=1}^{s} \alpha_{im}P(m) + W(i) + Q(i) \\
&+ \mu(i)\Omega^t(i)\Omega(i) \\
\bar{W}_d(i) &= I + \mu^{-1}(i)N_d(i)W^{-1}(i)N_d^t(i) \\
W_d(i) &= W(i) - \mu(i)A_d^t(i)A_d(i) \\
\bar{A}(i) &= A_o(i) + \mu^{-1}(i)A_d(i)W_d^{-1}(i)N_d^t(i)N_a(i) \\
\Omega^t(i)\Omega(i) &= N_a^t(i)\bar{W}_d(i)N_a(i)
\end{aligned}$$

$$\Gamma(i)\Gamma^t(i) \;\; = \;\; A_d(i)W_d^{-1}(i)A_d^t(i) \tag{4.19}$$

Proof: Follows from direct application of the Schur complements to LMIs (4.15) using **Fact 3** of Appendix A and (4.19). $\nabla\nabla\nabla$

In the delayless case, the following corollary summarizes a bounded-real condition for the existence of a quadratic cost matrix

Corollary 4.2 *Consider system (4.1) with* $u(\cdot) \equiv 0$ *and* $\Delta A_d(i) \equiv 0$ *and let the matrix* $A_o(i)$, $\forall\, i \in \mathcal{S}$ *be Hurwitz. A matrix* $0 < P(i) = P^t(i) \in \Re^{n\times n}, i \in \mathcal{S}$ *is a* **QCM** *for this system with cost function (4.4) if and only if any one of the following equivalent conditions holds:*

(1) *For a given set of scalars* $\mu(i) > 0,\ i \in \mathcal{S}$, *there exists matrices* $0 < W(i) = W^t(i) \in \Re^{n\times n}\ , i \in \mathcal{S}$ *satisfying the LMIs for all* $i \in \mathcal{S}$

$$\begin{bmatrix} \Pi_1(i) & P(i)M_a(i) & P(i)A_d(i) \\ M_a^t(i)P(i) & -\mu(i)I & 0 \\ A_d^t(i)P(i) & 0 & -W(i) \end{bmatrix} < 0 \tag{4.20}$$

(2) *For a given set of scalars* $\mu(i) > 0,\ i \in \mathcal{S}$, *there exists matrices* $0 < W(i) = W^t(i) \in \Re^{n\times n}\ , i \in \mathcal{S}$ *satisfying the ARI*

$$\begin{aligned} & P(i)A_o(i) + A_o^t(i)P(i) + P(i)\mu^{-1}(i)M_a(i)M_a^t(i)P(i) + \\ & P(i)A_d(i)W^{-1}(i)A_d^t(i)P(i) + \sum_{m=1}^{s} \alpha_{im}P(m) + \\ & W(i) + Q(i) + \mu(i)N_a^t(i)N_a(i) \; < \; 0 \end{aligned} \tag{4.21}$$

(3) *For a given set of scalars* $\mu(i) > 0,\ i \in \mathcal{S}$, *t here exist matrices* $0 < W(i) = W^t(i) \in \Re^{n\times n}\ , i \in \mathcal{S}$ *satisfying the following* $H_\infty-$*norm bound*

$$\|\Xi(i)\ \ \Omega(i)\ \ \Gamma(i)\|_\infty < 1$$

where

$$\begin{aligned}
\Xi(i) &= \begin{bmatrix} \mu^{1/2}(i)N_a(i) \\ W^{1/2}(i) \\ Q^{1/2}(i) \\ \left(\sum_{m=1}^{s}\alpha_{im}P(m)\right)^{1/2} \end{bmatrix} \\
\Omega(i) &= \left[sI - A_o(i)\right]^{-1} \\
\Gamma(i) &= \left[\mu^{-1/2}(i)M_a(i) \;\; A_d(i)W^{-1/2}(i)\right] \qquad (4.22)
\end{aligned}$$

Proof: **(1)**$\Longleftrightarrow$**(2)** Follows easily from **Theorem** (4.2) and **corollary** (4.1) by setting $E_d(i) = 0\ , i\ \in\ \mathcal{S}$. **(1)**$\Longleftrightarrow$**(3)** can be easily derived by generalizing the results of [98] for single mode $\eta_t = 1$. $\nabla\nabla\nabla$

4.4.2 Strong Delay-Dependence

We now focus on a strong delay-dependent conditions that which are functions of the instantaneous value of the delay factor τ. Towards our goal, we consider system (4.1) and invoke the Libentiz-Netwon rule for some $\tau \in [0, \tau^*]$:

$$\begin{aligned}
x(t-\tau) &= x(t) - \int_{-\tau}^{0} \dot{x}(t+\theta)d\theta \\
&= x(t) - \int_{-\tau}^{0} A_{\Delta o}(i)x(t+\theta)d\theta \\
&- \int_{-\tau}^{0} A_{\Delta d}(i)x(t-\tau+\theta)d\theta \qquad (4.23)
\end{aligned}$$

Substituting (4.23) back into (4.1) we get:

$$\dot{x}(t) = [A_{\Delta o}(i) + A_{\Delta d}(i)]x(t) \; - \; A_{\Delta d}(i)\psi(x,i) \qquad (4.24)$$

where

$$\psi(x,i) = -\int_{-\tau}^{0} A_{\Delta o}(t+\theta, i)x(t+\theta)d\theta$$

$$- \int_{-\tau}^{0} A_{\Delta d}(t-\tau+\theta, i)x(t-\tau+\theta)d\theta \qquad (4.25)$$

which represents, in the terminology of time-delay systems [73], a functional differential equation with Markovian jump parameters and having initial conditions over the interval $[-2\tau^*, 0]$. It should be emphasized that the LMI-based sufficient stability criteria for system (4.24) has been established in Chapter 3.

Based on **Theorem** (3.3), we have the following definition:

Definition 4.3 *System (4.1) with cost function (4.4) is* **said to be RSSSDD for any constant time-delay** τ **satisfying** $0 \le \tau \le \tau^*$ *if there exist matrix* $0 < P(i) = P^t(i) \in \Re^{n \times n}$ *,* $i \in \mathcal{S}$ *and scalars* $r_1(i) > 0$ *,* $r_2(i) > 0$ *,* $i \in \mathcal{S}$ *satisfying the following ARIs for all* $i \in \mathcal{S}$

$$\begin{aligned}
& P(i)\left[A_{\Delta o}(i) + A_{\Delta d}(i)\right] + \left[A_{\Delta o}(i) + A_{\Delta d}(i)\right]^t P(i) + \\
& \tau^* P(i)A_{\Delta d}(i)A^t_{\Delta d}(i)P(i) + \sum_{m=1}^{s} \alpha_{im}P(m) + \\
& \tau^* r_1(i)A^t_{\Delta o}(i)A_{\Delta o}(i) + \tau^* r_2(i)A_{\Delta d}(i)A^t_{\Delta d}(i) + Q(i) \; < \; 0 \\
& \quad \forall \; \Delta(i,t): \;\; \Delta^t(i,t)\,\Delta(i,t) \; \le \; I \qquad (4.26)
\end{aligned}$$

By similarity to the weak-dependent case, the following theorem derives an upper bound on the cost function J under **RSSSDD** property.

Theorem 4.3 *Consider system (4.1) and cost function (4.4). Given scalars* $\tau^* > 0$, $\sigma(i) > 0, \mu(i) > 0, r_1(i) > 0, r_2(i) > 0$, $i \in \mathcal{S}$, *if* $0 < P(i) = P^t(i) \in \Re^{n \times n}$, $i \in \mathcal{S}$ *is a* **QCM** *then (4.1) is RSSSDD for any constant time-delay* $\tau \in [0, \tau^*]$ *and the cost function satisfies the bound*

$$\begin{aligned}
\mathsf{J} \;\; \le \;\; & x^t_o P(i)x_o \; + \int_{-\tau}^{0} r_1(i)x^t(\alpha)\sigma(i)N_a(i)^t N_a(i)x(\alpha)d\alpha \\
+ & \int_{-\tau}^{0} r_1(i)x^t(\alpha)A^t_o[I - \sigma(i)M_a(i)M^t_a(i)]A_o(i)x(\alpha)d\alpha
\end{aligned}$$

$$+ \int_{-\tau}^{0} r_2(i) x^t(\alpha) \mu(i) N_d^t(i) N_d(i) x(\alpha) d\alpha$$

$$+ \int_{-\tau}^{0} r_2(i) x^t(\alpha) A_d^t(i) [I - \mu(i) M_a(i) M_a^t(i)] A_d(i) x(\alpha) d\alpha \quad (4.27)$$

Conversely, if system (4.1) is **RSSSDD** *for any constant time-delay $\tau \in [0, \tau^*]$ then there will be a* **QCM** *for this system and cost function (4.4).*

Proof:($\Longrightarrow$) Let $0 < P(i) = P^t(i) \in \Re^{n \times n}$, $i \in \mathcal{S}$ be a QCM for system (4.1) and cost function (4.4). It follows from **Definition 4.3** that there exist scalars $\tau^* > 0, r_1(i) > 0, r_2(i) > 0$, $i \in \mathcal{S}$ such that

$$X^t(i) \begin{bmatrix} \Pi_{\Delta d} & \tau^* P(i) A_{\Delta d}(i) & \tau^* A^t_{\Delta o}(i) & \tau^* A^t_{\Delta d}(i) \\ \tau^* A^t_{\Delta d}(i) P(i) & -\tau^* I & 0 & 0 \\ \tau^* A^t_{\Delta o}(i) & 0 & -\tau^* r_1^{-1}(i) I & 0 \\ \tau^* A^t_{\Delta d}(i) & 0 & 0 & -\tau^* r_2^{-1}(i) I \end{bmatrix} X(i) < 0$$

$$\Pi_{\Delta d} = P(i)\left[A_{\Delta o}(i) + A_{\Delta d}(i)\right] + \left[A_{\Delta o}(i) + A_{\Delta d}(i)\right]^t P(i) +$$

$$Q(i) + \sum_{m=1}^{s} \alpha_{im} P(m)$$

$$\forall \Delta : \Delta^t \ \Delta \ \leq \ I \ , \ \ X(i) \neq [0 \ 0 \ 0 \ 0]$$

$$X(t) = [x^t(t) \ \ x^t(t-\tau) \ \ x^t(t) \ \ x^t(t-\tau)]^t \quad (4.28)$$

Note that the matrix in (4.28) is continuously dependent on τ^*. Therefore, system (4.1) is robustly stochastically stable for any constant time-delay τ satisfying $0 \leq \tau \leq \tau^*$. Now by **Theorem** (3.4) and **Definition 4.3**, the following bounds hold:

$$\Im^x[W_\Delta] \leq -x^t(t) Q(i) x(t) \Longrightarrow x^t(t) Q(i) x(t)$$

$$\leq -\Im^x[W_\Delta] \Longrightarrow \mathsf{J} \leq W_\Delta(\xi, i). \quad (4.29)$$

In view of (4.4) and manipulating (4.29) with the help of **Fact 2** of Appendix A for some scalars $\sigma(i) > 0$, $i \in \mathcal{S}$, $\mu(i) > 0$, $i \in \mathcal{S}$, it reduces to

$$\mathsf{J} \leq x_o^t P(i) x_o + \int_{-\tau}^{0} r_1(i) x^t(\alpha) \sigma(i) N_a(i)^t N_a(i) x(\alpha) d\alpha$$

$$+ \int_{-\tau}^{0} r_1(i) x^t(\alpha) A_o^t [I - \sigma(i) M_a(i) M_a^t(i)] A_o(i) x(\alpha) d\alpha$$
$$+ \int_{-\tau}^{0} r_2(i) x^t(\alpha) \mu(i) N_d^t(i) N_d(i) x(\alpha) d\alpha$$
$$+ \int_{-\tau}^{0} r_2(i) x^t(\alpha) A_d^t(i) [I - \mu(i) M_a(i) M_a^t(i)] A_d(i) x(\alpha) d\alpha$$

as desired.

($\Longleftarrow$) Let system (4.1) be robustly stochastically stable with strong-delay dependence for any constant time-delay τ satisfying $0 \le \tau \le \tau^*$. It follows that there exist matrix $0 < P(i) = P^t(i)$, $i \in \mathcal{S}$ and scalars $r_1(i) > 0, r_2(i) > 0$, $i \in \mathcal{S}$ such that

$$\begin{aligned}
& P(i)\big[A_{\Delta o}(i) + A_{\Delta d}(i)\big] + \big[A_{\Delta o}(i) + A_{\Delta d}(i)\big]^t P(i) + \\
& \tau^* P(i) A_{\Delta d}(i) A_{\Delta d}^t(i) P(i) + \sum_{m=1}^{s} \alpha_{im} P(m) + Q(i) + \\
& \tau^* r_1(i) A_{\Delta o}^t(i) A_{\Delta o}(i) + \tau^* r_2(i) A_{\Delta d}(i) A_{\Delta d}^t(i) \; < \; 0 \\
& \forall \; \Delta(i) : \;\; \Delta^t(i)\, \Delta(i) \; \le \; I \;\; , \;\; \forall i \, \in \, \mathcal{S}
\end{aligned} \tag{4.30}$$

Hence, one can find some $\rho(i) > 0$, $i \in \mathcal{S}$ such that the following inequality holds:

$$\begin{aligned}
& \vartheta^{-1}(i)\Big\{ P(i)\big[A_{\Delta o}(i) + A_{\Delta d}(i)\big] + \big[A_{\Delta o}(i) + A_{\Delta d}(i)\big]^t P(i) + \\
& \tau^* P(i) A_{\Delta d}(i) A_{\Delta d}^t(i) P(i) + \sum_{m=1}^{s} \alpha_{im} P(m) + \\
& \tau^* r_1(i) A_{\Delta o}^t(i) A_{\Delta o}(i) + \tau^* r_2(i) A_{\Delta d}(i) A_{\Delta d}^t(i) \Big\} + Q(i) \\
& = \Big\{ \breve{P}(i)\big[A_{\Delta o}(i) + A_{\Delta d}(i)\big] + \big[A_{\Delta o}(i) + A_{\Delta d}(i)\big]^t \breve{P}(i) + \\
& \tau^* \breve{P}(i) A_{\Delta d}(i) A_{\Delta d}^t(i) \breve{P}(i) + \sum_{m=1}^{s} \alpha_{im} \breve{P}(m) +
\end{aligned}$$

$$\tau^* r_1(i) A_{\Delta o}^t(i) A_{\Delta o}(i) + \tau^* r_2(i) A_{\Delta d}(i) A_{\Delta d}^t(i) \Big\} + Q(i) \ < \ 0 \qquad (4.31)$$

$$\forall \Delta : \ \Delta^t(t)\,\Delta(t) \ \leq \ I, \ \forall\, i \ \in \ \mathcal{S}.$$

It is redily seen that the above inequality implies that there exist scalars $\tau^o(i) = \vartheta^{-1}(i)\tau^* > 0, \breve{r}_1(i) = \vartheta^{-1}(i) r_1(i) > 0, \breve{r}_2(i) = \vartheta^{-1}(i) r_2(i) > 0$ such that the matrix $\breve{P}(i) = \vartheta^{-1}(i) P(i)$ is a QCM for system (4.4). $\nabla\nabla\nabla$

4.5 Guaranteed Cost Control

In this section, we focus attention on the problem of optimal guaranteed cost control based on state-feedback for the Markovian jump system (4.1) with uncertainties satisfying (4.2)-(4.3) and cost function as given by (4.4). To proceed further, we introduce the following definition:

Definition 4.4 *A state-feedback controller $u(t) = K_s(i)x(t)$ is said to define a* **quadratic guaranteed cost control (QGCC)** *associated with cost matrix $0 < P(i) = P^t(i) \in \Re^{n\times n}$, $i \in \mathcal{S}$ for system (4.1) and cost function (4.4), if there exists a matrix $0 < W(i) = W^t(i) \in \Re^{n\times n}$, $i \in \mathcal{S}$ such that*

$$\begin{aligned}
& P(i)\Big[A_c(i) + M_a(i)\Delta(i)N_c(i)\Big] + \Big[A_c(i) + M_a(i)\Delta(i)N_c(i)\Big]^t P(i) + \\
& W(i) + Q(i) + K_s^t(i)R(i)K_s(i) + \sum_{m=1}^{s} \alpha_{im} P(m) + \\
& P(i)\Big[A_d(i) + M_a(i)\Delta(i)N_d(i)\Big] W^{-1}(i) \\
& \cdot \Big[A_d(i) + M_a(i)\Delta(i)N_d(i)\Big]^t P(i) \ < \ 0 \qquad (4.32) \\
& \forall \Delta : \quad \Delta^t(t)\,\Delta(t) \ \leq \ I, \ \forall\, i \ \in \mathcal{S},
\end{aligned}$$

where

$$A_c(i) \ = \ A_o(i) + B(i)K_s(i), \quad N_c(i) \ = \ N_a(i) + N_b(i)K_s(i) \quad i \ \in \mathcal{S}. \qquad (4.33)$$

In line of our previous effort, the following two subsections contain the design of **QGCC** for both cases of weak-delay dependent and strong-delay dependent.

4.5.1 Weak-Delay Dependence

The following theorem represents one of the main results in this section. It establishes that the problem of determining a **QGCC** for system (4.1) and cost function (4.4) can be recast to a feasibility problem of an algebraic matrix inequality (AMI).

Theorem 4.4 *Suppose that there exist scalars* $\mu(i) > 0,\ i \in \mathcal{S}$ *and matrices* $0 < W(i) = W^t(i) \in \Re^{n\times n},\ i \in \mathcal{S}$ *such that the following ARE*

$$
\begin{aligned}
& P(i)\Big\{A_o(i) + \mu^{-1}(i)A_d(i)W_d^{-1}(i)N_d^t(i)N_a(i)\Big\} + \\
& \Big\{A_o(i) + \mu^{-1}(i)A_d(i)W_d^{-1}(i)N_d^t(i)N_a(i)\Big\}^t P(i) \\
& + \sum_{m=1}^{s} \alpha_{im}P(m) + W(i) + Q(i) + \mu^{-1}(i)N_a^t(i)\bar{W}_d^{-1}(i)E_o(i) \\
& + P(i)\Big[\mu(i)M_a(i)M_a^t(i) + A_d(i)W_d^{-1}(i)A_d^t(i)\Big]P(i) + \mu^{-1}(i)N_b^t(i)\bar{W}_d^{-1}(i)N_b \\
& -\Big\{P(i)B_o(i) + \mu^{-1}(i)\Big[N_a^t\bar{W}_d^{-1}(i) + P(i)A_d(i)W_d^{-1}(i)N_d^t(i)\Big]N_b\Big\}R(i)\Big\{B_o^t(i)P(i) \\
& \mu^{-1}(i)N_b^t(i)\Big[\bar{W}_d^{-1}(i)N_a(i) + N_d(i)W_d^{-1}(i)A_d^t(i)P(i)\Big]\Big\} \quad = \quad 0 \qquad (4
\end{aligned}
$$

has a stabilizing solution $0 < P(i) = P^t(i) \in \Re^{n\times n},\ i \in \mathcal{S}$. *In this case, the state-feedback controller*

$$
\begin{aligned}
u(t) &= K_s(i)x(t) \qquad (4.35) \\
K_s(i) &= -\bar{R}(i)\Big(B_o^t(i)P(i) \\
&+ \mu^{-1}(i)N_b^t(i)\bar{W}_d^{-1}(i)N_a(i)
\end{aligned}
$$

$$+ \quad \mu^{-1}(i)N_b^t(i)N_d(i)W_d^{-1}(i)A_d^t(i)P(i)\Big)$$

$$\bar{R} \quad = \quad \left[R(i) + \mu^{-1}(i)N_b^t(i)\bar{W}_d^{-1}(i)N_b(i)\right]^{-1} \tag{4.36}$$

is a QGCC for system (4.1) with cost matrix $\breve{P}(i)$, $\in \mathcal{S}$ *which satisfies* $P(i) < \breve{P}(i) < P(i) + \rho(i)I$, $i \in \mathcal{S}$ *for any* $\rho(i) > 0$, $i \in \mathcal{S}$.

Conversely given any QGCC with cost matrix $0 < \breve{P}(i) = \breve{P}^t(i)$, $i \in \mathcal{S}$, *there exists a scalar* $\mu(i) > 0$, $i \in \mathcal{S}$ *and a matrix* $0 < W(i) = W^t(i) \in \Re^{n \times n}$, $i \in \mathcal{S}$ *such that the ARE (4.34) has a stabilizing solution* $0 < P_*(i) = P_*^t(i)$, $i \in \mathcal{S}$ *where* $P_*(i) < \breve{P}(i)$, $i \in \mathcal{S}$.

Proof: ($\Longrightarrow$) Let the control law $u(t)$ be defined by (4.35). By substituting (4.33) and (4.36) into (4.34) and manipulating using **Fact 3** of the Appendix, it can be shown that (4.34) is equivalent to:

$$\begin{aligned} & P(i)A_c(i) + A_c^t(i)P(i) + \mu(i)P(i)M_a(i)M_a^t(i)P(i) + \\ & \mu^{-1}(i)N_c^t(i)N_c(i) + W(i) + Q(i) + \\ & K_s^t(i)R(i)K_s(i) + \sum_{m=1}^{s} \alpha_{im}P(m) + \\ & \left[P(i)A_d(i) + \mu^{-1}(i)N_c^t(i)N_d(i)\right]W_d^{-1}(i) \\ & \cdot\left[A_d^t(i)P(i) + \mu^{-1}(i)N_d^t(i)N_c(i)\right] \quad = \quad 0 \end{aligned} \tag{4.37}$$

By **Fact 4**, it follows that there exists matrices $0 < \breve{P}(i) = \breve{P}^t(i)$, $i \in \mathcal{S}$ such that

$$\begin{aligned} & \breve{P}(i)A_c(i) + A_c^t(i)\breve{P}(i) + \mu(i)\breve{P}(i)M_a(i)M_a^t(i)\breve{P}(i) + \\ & \mu^{-1}(i)N_c^t(i)N_c(i) + W(i) + Q(i) + \\ & K_s^t(i)R(i)K_s(i) + \sum_{m=1}^{s} \alpha_{im}P(m) + \end{aligned}$$

$$\left[\breve{P}(i)A_d(i) + \mu^{-1}(i)N_c^t(i)N_d(i)\right]W_d^{-1}(i)$$
$$\cdot\left[A_d^t(i)\breve{P}(i) + \mu^{-1}(i)N_d^t(i)N_c(i)\right] < 0 \tag{4.38}$$

which implies that there exist matrices $\Phi(i) > 0,\ i \in \mathcal{S}$, such that

$$\breve{P}(i)A_c(i) + A_c^t(i)\breve{P}(i) + \mu(i)\breve{P}(i)M_a(i)M_a^t(i)\breve{P}(i) +$$
$$\mu^{-1}(i)N_c^t(i)N_c(i) + W(i) + Q(i) +$$
$$K_s^t(i)R(i)K_s(i) + \sum_{m=1}^{s} \alpha_{im}P(m) +$$
$$\left[\breve{P}(i)A_d(i) + \mu^{-1}(i)N_c^t(i)N_d(i)\right]W_d^{-1}(i)$$
$$\cdot\left[A_d^t(i)\breve{P}(i) + \mu^{-1}(i)N_d^t(i)N_c(i)\right] + \Phi(i) = 0 \tag{4.39}$$

Given $\sigma(i) \in (0,1)$, $i \in \mathcal{S}$, it follows from (4.39) and the properties of the ARE that

$$\hat{P}(i)A_c(i) + A_c^t(i)\hat{P}(i) + \mu(i)\hat{P}(i)M_a(i)M_a^t(i)\hat{P}(i) +$$
$$\mu^{-1}(i)N_c^t(i)N_c(i) + W(i) + Q(i) +$$
$$K_s^t(i)R(i)K_s(i) + \sum_{m=1}^{s} \alpha_{im}P(m) +$$
$$\left[\hat{P}(i)A_d(i) + \mu^{-1}(i)N_c^t(i)N_d(i)\right]W_d^{-1}(i)$$
$$\cdot\left[A_d^t(i)\hat{P}(i) + \mu^{-1}(i)N_d^t(i)N_c(i)\right] + \Phi(i) = 0 \tag{4.40}$$

has a stabilizing solution $0 < \hat{P}(i) = \hat{P}^t(i),\ i \in \mathcal{S}$. In addition, $\hat{P}(i) > P(i),\ i \in \mathcal{S}$ and as $\sigma(i) \to 0,\ i \in \mathcal{S},\ \hat{P}(i) \to P(i),\ i \in \mathcal{S}$. Therefore, given any $\rho(i) > 0,\ i \in \mathcal{S}$, we can find a $\sigma(i) > 0,\ i \in \mathcal{S}$ such that $P(i) < \breve{P}(i) = \hat{P}(i) < P(i) + \rho(i)I$, $i \in \mathcal{S}$.

($\Longleftarrow$) Suppose that $u(t) = K_s(i)x(t)$, $i \in \mathcal{S}$ is a **QGCC** with a cost matrix $\breve{P}(i)$, $i \in \mathcal{S}$. By Theorem 4.2, it follows that there exist scalars $\mu(i) > 0$ and matrices $0 < W(i) = W^t(i), \forall i \in \mathcal{S}$, such that

$$P(i)A_c(i) + A_c^t(i)P(i) + \mu(i)P(i)M_a(i)M_a^t(i)P(i) +$$
$$\mu^{-1}(i)NE_c^t(i)N_c(i) + W(i) + Q(i) +$$
$$K_s^t(i)R(i)K_s(i) + \sum_{m=1}^{s} \alpha_{im}P(m)$$
$$+\Big[P(i)A_d(i) + \mu^{-1}(i)N_c^t(i)N_d(i)\Big]W_d^{-1}(i)$$
$$\cdot\Big[A_d^t(i)P(i) + \mu^{-1}(i)N_d^t(i)N_c(i)\Big] < 0 \tag{4.41}$$

In terms of (4.33), inequality (4.41) is equivalent to:

$$P(i)\big[A_o(i) + B_o(i)K_s(i)\big] + \big[A_o(i) + B_o(i)K_s(i)\big]^t P(i) +$$
$$\mu(i)P(i)M_a(i)M_a^t(i)P(i) + W(i) + Q(i) + \sum_{m=1}^{s} \alpha_{im}P(m) +$$
$$\mu^1(i)\big[N_a(i) + N_b(i)K_s(i)\big]^t\big[N_a(i) + N_b(i)K_s(i)\big]$$
$$+\Big\{P(i)A_d(i) + \mu^{-1}(i)\big[N_a(i) + N_b(i)K_s(i)\big]^t N_d(i)\Big\}W_d^{-1}(i)$$
$$\cdot\Big\{A_d^t(i)P(i) + \mu^{-1}(i)N_d^t(i)\big[N_a(i) + N_b(i)K_s(i)\big]\Big\} < 0, \ \forall i \in \mathcal{S} \tag{4.42}$$

For simplicity, we define

$$\bar{K}_s(i) = \mu^{-1/2}\,K_s(i)\ , \qquad \bar{R}(i) = \mu(i)\,R(i)$$
$$\bar{B}_o(i) = B_o(i) + \mu^{-1}(i)A_d(i)W_d^{-1}(i)N_d^t(i)N_b(i) \tag{4.43}$$

By substituting (4.19), (4.43) into (4.42), it follows that there exists $\bar{P}(i) > 0\ ,\ i \in \mathcal{S}$ satisfies

$$\bar{P}(i)\Big\{\bar{A}_o(i) + \mu^{1/2}(i)\bar{B}_o(i)\bar{K}_s(i)\Big\} +$$
$$\Big\{\bar{A}_o(i) + \mu^{1/2}(i)\bar{B}_o(i)\bar{K}_s(i)\Big\}^t \bar{P}(i) +$$

$$
\begin{aligned}
&\bar{P}(i)\left[\mu(i)H(i)H^t(i) + A_d(i)W_d^{-1}(i)A_d^t(i)\right]\bar{P}(i) + \\
&Q(i) + \bar{K}_s^t(i)\bar{R}(i)\bar{K}_s(i) + \sum_{m=1}^{s} \alpha_{im} P(m) + W(i) + \\
&\left\{\mu^{-1/2}(i)\bar{W}_d^{-1/2}(i)E_o(i) + \bar{W}_d^{-1/2}(i)E_b(i)\bar{K}_s(i)\right\}^t \\
&\cdot\left\{\mu^{-1/2}(i)\bar{W}_d^{-1/2}(i)E_o(i) + \bar{W}_d^{-1/2}(i)E_b(i)\bar{K}_s(i)\right\} < 0 \qquad (4.44)
\end{aligned}
$$

Now, for $i \in \mathcal{S}$ consider the state feedback $\mathcal{H}_\infty$ control problem of the following system [164]:

$$
\begin{aligned}
\dot{x}(t) &= \bar{A}(i) + \left[\mu^{1/2}(i)M_a(i) \quad A_d(i)W_d^{-1/2}(i)\right]w(t) \\
&+ \left[\mu^{1/2}(i)\bar{B}(i)\right]u(t) \\
z(t) &= \begin{bmatrix} \mu^{-1/2}(i)\bar{W}_d^{-1/2}(i)N_a(i) \\ 0 \\ \begin{bmatrix} W(i) + Q(i) \\ +\sum_{m=1}^{s} \alpha_{im}P(m) \end{bmatrix}^{1/2} \end{bmatrix} x(t) \\
&+ \begin{bmatrix} N_b(i)\bar{W}_d^{-1/2}(i) \\ \mu^{1/2}(i)R^{1/2}(i) \\ 0 \end{bmatrix} u(t) \qquad (4.45)
\end{aligned}
$$

It follows from [98, 164] that system (4.45) with the state feedback $u(t) = \bar{K}_s(i)x(t)$ has the $\mu(i)$-dependent ARE (4.34). Moreover, it has a stabilizing solution $P^*(i) \geq 0$, $i \in \mathcal{S}$ such that $P^*(i) < \bar{P}(i)$. Since $Q(i) > 0, W(i) > 0$, $i \in \mathcal{S}$, it follows that $P^*(i) > 0$, $i \in \mathcal{S}$. $\nabla\nabla\nabla$

4.5.2 Special Cases

From the foregoing theorem, the following corollaries can be readily obtained:

Corollary 4.3 *Consider system*

$$
\begin{aligned}
\dot{x}(t) &= [A_o(i) + \Delta A_o(i)]x(t) \\
&+ [B_o + \Delta B_o(i)]u(t) + A_d(i)x(t-\tau) \qquad (4.46)
\end{aligned}
$$

which is obtained from system (4.1) by setting $E_d(i) = 0$, $i \in \mathcal{S}$ *In this case, the ARE (4.33) and the controller (4.34)-(4.35) reduces to:*

$$P(i)A_o(i) + A_o^t(i)P(i) + P(i)\Big[\mu(i)M_a(i)M_a^t(i) + A_d(i)W^{-1}A_d^t(i)\Big]P(i)$$
$$+W(i) + Q(i) + \mu^{-1}(i)N_a^t(i)N_a(i) + \sum_{m=1}^{s}\alpha_{im}P(m)$$
$$-\Big\{P(i)B_o(i) + \mu^{-1}(i)N_a^t(i)N_b(i)\Big\}\hat{R}(i)$$
$$\cdot\Big\{P(i)B_o(i) + \mu^{-1}(i)N_a^t(i)N_b(i)\Big\} = 0 \tag{4.47}$$
$$K_s(i) = -\hat{R}(i)\Big\{P(i)B_o(i) + \mu^{-1}(i)N_a^t(i)N_b(i)\Big\} \tag{4.48}$$
$$\hat{R}(i) = \Big[R(i) + \mu^{-1}(i)N_b^t(i)N_b(i)\Big]^{-1} \tag{4.49}$$

This result recovers a general form of those produced in [98] for linear state-delay systems.

Corollary 4.4 *Consider system*

$$\dot{x}(t) = A_o(i)x(t) + B_o u(t) + A_d(i)x(t-\tau) \tag{4.50}$$

which is obtained from system (4.1) by setting $N_a = 0$, $N_d = 0$, $M_a = 0$, $N_b = 0$ *corresponding to the case of delay Markovian jump systems without uncertainties. In such case, the ARE (4.33) and the controller (4.34)-(4.35) reduce to:*

$$P(i)A_o(i) + A_o^t(i)P(i) +$$
$$P(i)\Big[A_d(i)W^{-1}A_d^t(i) - B_o(i)R^{-1}(i)B_o^t(i)\Big]P(i) = 0 \tag{4.51}$$
$$K_s(i) = -R^{-1}(i)B_o^t(i)P(i) \tag{4.52}$$

4.5.3 Strong Delay Dependence

Now, we direct attention on the problem of optimal guaranteed cost control based on state-feedback for system (4.1) with cost function given by (4.4) and adopting the notion of RSSSDD. The following definition is now given:

Definition 4.5 *A state-feedback controller $u(t) = K_s(i)x(t)$ is said to define a* **quadratic guaranteed cost control (QGCC)** *associated with cost matrix $0 < P(i) = P^t(i) \in \Re^{n\times n}$, $i \in \mathcal{S}$ for system (4.1) and cost function (4.4) for any constant time-delay τ satisfying $0 \le \tau \le \tau^*$ if, for a given $\tau^* > 0$, there exist matrices $0 < P(i) = P^t(i) \in \Re^{n\times n}$ and scalars $r_1(i) > 0$ and $r_2(i) > 0$, $i \in \mathcal{S}$ satisfying the ARI for all $i \in \mathcal{S}$*

$$\begin{aligned}
& P(i)A_{cd\Delta}(i) + A^t_{cd\Delta}(i)P(i) + \tau^* P(i)A_{\Delta d}(i)A^t_{\Delta d}(i)P(i) + \\
& K^t_s(i)R(i)K_s(i) + \sum_{m=1}^{s} \alpha_{im}P(m) + \\
& \tau^* r_1(i)A^t_{c\Delta}(i)A_{c\Delta}(i) + \tau^* r_2(i)A^t_{\Delta d}(i)A_{\Delta d}(i) < 0 \\
& \forall\, \Delta(i) : \Delta^t(i)\, \Delta(i) \le I
\end{aligned} \tag{4.53}$$

where

$$\begin{aligned}
A_{cd\Delta}(i) &= A_{cd}(i) + M_a(i)\Delta(i)N_{cd}(i) \\
&= \Big[A_o(i) + A_d(i) + B_o(i)K_s(i)\Big] \\
&+ M_a(i)\Delta(i)\Big[N_a(i) + N_d(i) + N_b(i)K_s(i)\Big] \\
A_{c\Delta}(i) &= A_c(i) + M_a(i)\,\Delta(i)\,N_c(i) \\
&= \Big[A_o(i) + B_o(i)K_s(i)\Big] \\
&+ M_a(i)\Delta(i)\Big[N_a(i) + N_b(i)K_s(i)\Big]
\end{aligned} \tag{4.54}$$

The following two theorems represents the main results in this subsection and complement those of the weak-delay dependent case. It is established that the problem of determining a QGCC for system (4.1) and cost function (4.4) having the RSSSDD property can be recast to an ARI or LMI-feasibility problem.

Theorem 4.5 *Given system (4.1) and cost function (4.4). Suppose that there exist matrices* $0 < P(i) = P^t(i) \in \Re^{n\times n}$, $i \in \mathcal{S}$ *and scalars* $r_1(i) > 0$, $r_2(i) > 0$, $\mu_1(i) > 0$, $\mu_2(i) > 0$, $\mu_3(i) > 0$, $\mu_4(i) > 0$, $i \in \mathcal{S}$ *and* $\tau^* > 0$ *such that for any constant time-delay* τ *such that* $0 \le \tau \le \tau^*$ *satisfying the ARIs for all* $i \in \mathcal{S}$

$$\begin{aligned}
&P(i)\big[A_o(i) + A_d(i)\big] + \big[A_o(i) + A_d(i)\big]^t P(i) + \sum_{m=1}^{s} \alpha_{im} P(m) + \\
&P(i)\big[\mu_1(i) M_a(i) M_a^t(i) + \tau^* M_1(i) M_1^t(i)\big] P(i) + \\
&\Big[\mu_1^{-1}(i)[N_a(i) + N_d(i)]^t [N_a(i) + N_d(i)] + \tau^* M_2^t(i) M_2(i)\Big] \\
&-\Big\{P(i)B_o(i) + \mu_1^{-1}(i)[N_a(i) + N_d(i)]^t N_b(i) + \tau^* N_1^t(i) N_1(i)\Big\} \mathsf{R}(i) \\
&\cdot\Big\{B_o^t(i)P(i) + \mu_1^{-1}(i) N_b^t(i)[N_a(i) + N_d(i)] + \tau^* N_1^t(i) N_1(i)\Big\}^t \\
&< 0
\end{aligned} \tag{4.55}$$

with

$$\begin{aligned}
\mathsf{R}(i) &= \Big\{R(i) + \mu_1^{-1}(i) N_b^t(i) N_b(i) + \tau^* N_2^t(i) N_2(i)\Big\}^{-1} \\
M_1(i) M_1^t(i) &= \mu_4^{-1}(i) M_a(i) M_a^t(i) + A_d(i)[I - \mu_4(i) N_d^t(i) N_d(i)]^{-1} A_d^t(i) \\
N_1^t(i) N_1(i) &= r_1(i)\Big(A_o^t[I - \mu_2(i) M_a(i) M_a^t(i)]^{-1} B_o(i) + \mu_2^{-1}(i) N_a^t(i) N_b(i)\Big) \\
N_2^t(i) N_2(i) &= r_1(i)\Big(\mu_2^{-1}(i) N_b^t(i) N_b(i) + B_o^t[I - \mu_2(i) M_a(i) M_a^t(i)]^{-1} B_o(i)\Big) \\
M_2^t(i) M_2(i) &= r_1(i)\mu_2^{-1}(i) N_a^t(i) N_a(i) + r_2(i)\mu_3^{-1}(i) N_d^t(i) N_d(i) \\
&+ r_1(i) A_o^t[I - \mu_2(i) M_a(i) M_a^t(i)]^{-1} A_o \\
&+ r_2(i) A_d^t(i)[I - \mu_3(i) M_a(i) M_a^t(i)]^{-1} A_d(i)
\end{aligned} \tag{4.56}$$

Then, the state-feedback controller

$$\begin{aligned} u(t) &= K_s(i)x(t) \\ K_s(i) &= -\mathsf{R}(i)\Big\{ B_o^t(i)P(i) + \mu_1^{-1}(i)N_b^t(i)[N_a(i) + N_d(i)] \\ &+ \tau^* N_1^t(i)N_1(i)\Big\} \end{aligned} \tag{4.57}$$

is a **QGCC** *for system (4.1) with cost matrix* $0 < P(i) = P^t(i) \in \Re^{n \times n}$, $i \in \mathcal{S}$.

Proof: From (4.54) using **Facts 2** and **3**, we get:

$$\begin{aligned} &P(i)A_{cd\Delta}(i) + A_{cd\Delta}^t(i)P(i) \\ &= P(i)\Big[A_o(i) + A_d(i) + B_o(i)K_s(i)\Big] + \\ &\Big[A_o(i) + A_d(i) + B_o(i)K_s(i)\Big]^t P(i) + \\ &P(i)M_a(i)\Delta(i)\Big[N_a(i) + N_d(i) + N_b(i)K_s(i)\Big] + \\ &\Big[N_a(i) + N_d(i) + N_b(i)K_s(i)\Big]^t \Delta^t(i)M_a^t(i)P(i) \\ &\le P(i)\Big[A_o(i) + A_d(i) + B_o(i)K_s(i)\Big] + \\ &\Big[A_o(i) + A_d(i) + B_o(i)K_s(i)\Big]^t P(i) + \\ &\mu_1(i)P(i)M_a(i)M_a^t(i)P(i) + \\ &\mu_1^{-1}(i)\Big[A_o(i) + A_d(i) + B_o(i)K_s(i)\Big]^t \\ &\cdot\Big[A_o(i) + A_d(i) + B_o(i)K_s(i)\Big] \end{aligned} \tag{4.58}$$

$$\begin{aligned} &A_{c\Delta}^t(i)A_{c\Delta}(i) \\ &= \Big[A_o(i) + B_o(i)K_s(i) + M_a(i)\Delta(i)\Big(N_a(i) + N_b(i)K_s(i)\Big)\Big]^t \end{aligned}$$

$$
\begin{aligned}
&\cdot\Big[A_o(i) + B_o(i)K_s(i) + M_a(i)\Delta(i)\Big(N_a(i) + N_b(i)K_s(i)\Big)\Big] \\
&\leq \mu_2^{-1}(i)\Big[N_a(i) + N_b(i)K_s(i)\Big]^t\Big[N_a(i) + N_b(i)K_s(i)\Big] \\
&+\Big[A_o(i) + B_o(i)K_s(i)\Big]^t[I - \mu_2(i)M_a(i)M_a^t(i)]^{-1} \\
&\Big[A_o(i) + B_o(i)K_s(i)\Big]
\end{aligned}
\tag{4.59}
$$

$$
\begin{aligned}
&\Big[A_d(i) + M_a(i)\Delta(i)N_d(i)\Big]^t\Big[A_d(i) + M_a(i)\Delta(i)N_d(i)\Big] \\
&\leq \mu_3^{-1}(i)N_d^t(i)N_d(i) + A_d^t(i)[I - \mu_3(i)M_a(i)M_a^t(i)]^{-1}A_d(i)
\end{aligned}
\tag{4.60}
$$

$$
\begin{aligned}
&\Big[A_d(i) + M_a(i)\Delta(i)N_d(i)\Big]\Big[A_d(i) + M_a(i)\Delta(i)N_d(i)\Big]^t \\
&\leq \mu_4^{-1}(i)M_a(i)M_a^t(i) + \\
&A_d(i)[I - \mu_4(i)N_d^t(i)E_d(i)]^{-1}A_d^t
\end{aligned}
\tag{4.61}
$$

for any scalars $\mu_1(i) > 0, ..., \mu_4(i) > 0,\ i\ \in\ \mathcal{S}$ satisfying

$$
\begin{aligned}
&[I - \mu_2(i)M_a(i)M_a^t(i)] > 0,\ [I - \mu_3(i)M_a(i)M_a^t(i)] > 0 \\
&[I - \mu_4(i)N_d^t(i)N_d(i)] > 0.
\end{aligned}
$$

It follows from **Definition** 4.5 by using (4.55) and (4.57)-(4.61) with some arrangement that

$$
\begin{aligned}
&P(i)\Big[A_o(i) + A_d(i) + B_o(i)K_s(i)\Big] + \\
&\Big[A_o(i) + A_d(i) + B_o(i)K_s(i)\Big]^t P(i) + \\
&P(i)\left[\mu_1(i)M_a(i)M_a^t(i) + \tau^* M_1(i)M_1^t(i)\right]P(i) + \sum_{m=1}^{s}\alpha_{im}P(m) +
\end{aligned}
$$

$$\Big[\mu_1^{-1}(i)[N_a(i)+N_d(i)]^t[N_a(i)+N_d(i)]+\tau M_2^t(i)M_2(i)\Big]+$$
$$K_s^t\Big\{B_o^t(i)P(i)+\mu_1^{-1}(i)N_b^t(i)[N_a(i)+N_d(i)]+\tau N_1^t(i)N_1(i)\Big\}$$
$$\cdot\Big\{P(i)B_o(i)+\mu_1^{-1}(i)[N_a(i)+N_d(i)]^tN_b(i)+\tau N_1^t(i)N_1(i)\Big\}K_s$$
$$+K_s^t\Big\{R(i)+\mu_1^{-1}(i)N_b^t(i)N_b(i)+\tau^*N_2^t(i)N_2(i)\Big\}K_s<0. \qquad (4.62)$$

Observe that (4.62) is continuously dependent on τ. On completing the squares in (4.51) with respect to $K_s(i),\ i\ \in\ \mathcal{S}$ and arranging terms, one obtains the control law (4.56) such that $0<P(i)=P^t(i),\ i\ \in\ \mathcal{S}$ satisfies ARI (4.54).$\nabla\nabla\nabla$

The next theorem provides a strong delay-dependent guaranteed cost controller and cost matrix for system (4.1) and cost function (4.4).

Theorem 4.6 *Consider system (4.1) and cost function (4.4). Suppose that there exist matrices* $0<Y(i)=Y^t(i),\ S(i)$ *and scalars* $r_1(i)>0,\ r_2(i)>0,\ \mu_1(i)>0,\ \mu_2(i)>0,\ \mu_3(i)>0,\ \mu_4(i)>0,\ i\in\mathcal{S}$ *and* $\tau^*>0$ *such that, for any constant time-delay* τ *being* $0\ \leq\ \tau\ \leq\ \tau^*$, *satisfying the LMIs for all* $i\in\mathcal{S}$

$$\begin{bmatrix}\Gamma(Y,\tau^*,\mu,i) & \tau^*M_1(i) & \tau^*Y(i)M_2^t(i) & \tau^*S(i)N_2^t(i)\\ \tau^*M_1^t(i) & -\tau^*I & 0 & 0\\ \tau^*M_2(i)Y(i) & 0 & -\tau^*I & 0\\ \tau^*N_2(i)S(i) & 0 & 0 & -\tau^*I\end{bmatrix}<0$$
$$\mu_2(i)M_a(i)M_a^t(i)-I<0\ ,\ \mu_4(i)E_d^t(i)E_d(i)-I<0$$
$$\mu_3(i)M_a(i)M_a^t(i)-I\ <\ 0 \qquad (4.63)$$

where

$$\begin{aligned}\Gamma(Y,\tau,\mu,i) &= [A_o(i)+A_d(i)]Y(i)+Y(i)[A_o(i)+A_d(i)]^t+\\ &+ \mu_1^{-1}(i)Y(i)[N_a(i)+N_d(i)]^t[N_a(i)+N_d(i)]Y(i)\\ &+ S(i)B_o^t+B_oS^t(i)+\mu_1(i)M_a(i)M_a^t(i)\\ &+ S(i)\Big[\mu_1^{-1}(i)N_b^t(i)[N_a(i)+N_d(i)]+\tau N_1^t(i)N_1(i)\Big]Y(i)\end{aligned}$$

$$+ \quad Y(i)\left[\mu_1^{-1}(i)[N_a(i) + N_d(i)]^t N_b + \tau N_1^t(i) N_1(i)\right] S(i)$$

$$+ \quad S(i)\left\{R(i) + \mu_1^{-1}(i) N_b^t(i) N_b(i)\right\} S(i) \tag{4.64}$$

Then, the state-feedback controller

$$u(t) = S(i) Y^{-1}(i) x(t) \tag{4.65}$$

is a **QGCC** *for system (4.1 and cost matrix* $0 < Y^{-1}(i) = Y^{-t}(i),\ i \in \mathcal{S}$

Proof: Starting from (4.62) , substituting $Y(i) = P^{-1}(i), K_s(i) = S(i)P(i)$ and manipulating using **Fact 1**, one immediately obtains (4.63). ∇∇∇

Remark 4.1 *It is important to observe the difference between* **Theorem** *4.5 and* **Theorem** *4.6. While the former gives closed-form solution for the feedback gain matrix after solving the ARIs (4.55) using numerical descent methods based on appropriate gridding [98], the latter provides a numerical value of the gain matrix based on the solution of LMIs (4.63) using the software package [57]. By and large, conversion of the ARIs (4.55) into an equivalent LMIs is not readily feasible.*

We end up this section by providing the following two corollaries which come quite naturally from Theorem 4.6.

Corollary 4.5 *Consider the time-delay system*

$$\dot{x}(t) = A_{\Delta o}(i) x(t) + A_d(i) x(t - \tau) \tag{4.66}$$

associated with cost function (4.4) and matrix $A_o(i),\ i \in \mathcal{S}$*, being Hurwitz. Suppose that there exist matrices* $0 < P(i) = P^t(i)$*, scalars* $r_1(i) > 0, r_2(i) > 0, \mu_1(i) > 0, \mu_2(i) > 0$ *,* $i \in \mathcal{S}$ *and* $\tau^* > 0$ *such that for any constant time-delay*

τ being $0 \leq \tau \leq \tau^*$ satisfying the ARI

$$
\begin{aligned}
&P(i)[A_o(i) + A_d(i)] + [A_o(i) + A_d(i)]^t P(i) + P(i)[\mu_1(i) M_a(i) M_a^t(i)] P(i) + \\
&\left[\mu_1^{-1}(i) N_a^t(i) N_a(i) + \tau^* \hat{M}_2^t(i) \hat{M}_2(i)\right] - \\
&\left\{P(i) B_o(i) + \mu_1^{-1} N_a^t(i) N_b(i) + \tau^* \hat{N}_1^t(i) \hat{N}_1(i)\right\} \\
&\cdot\left[R(i) + \mu_1^{-1}(i) N_b^t(i) N_b(i) + \tau^* \hat{N}_2^t(i) \hat{N}_2(i)\right]^{-1} \\
&\cdot\left\{B_o^t(i) P(i) + \mu_1^{-1}(i) N_b^t(i) N_a(i) + \tau^* \hat{N}_1^t(i) \hat{N}_1(i)\right\}^t < 0
\end{aligned}
\tag{4.67}
$$

with

$$
\begin{aligned}
\hat{M}_2^t(i) \hat{M}_2(i) &= r_1(i) \mu_2^{-1}(i) N_a^t(i) N_a(i) \\
&+ r_1(i) A_o^t(i)[I - \mu_2(i) M_a(i) M_a^t(i)]^{-1} A_o(i) \\
&+ r_2(i) A_d^t(i) A_d(i) \\
\hat{N}_1^t(i) \hat{N}_1(i) &= r_1(i) \Big(A_o^t(i)[I - \mu_2(i) M_a(i) M_a^t(i)]^{-1} B_o(i) \\
&+ \mu_2^{-1}(i) N_a^t(i) N_b(i) \Big) \\
\hat{N}_2^t(i) \hat{N}_2(i) &= r_1(i) \Big(\mu_2^{-1}(i) N_b^t(i) N_b(i) \\
&+ B_o^t(i)[I - \mu_2(i) M_a(i) M_a^t(i)]^{-1} B_o \Big)
\end{aligned}
\tag{4.68}
$$

Then, the state-feedback controller

$$
\begin{aligned}
u(t) &= K_s(i) x(t) \\
K_s &= -\bar{R}\left\{ B_o^t(i) P(i) + \mu_1^{-1}(i) N_b^t(i) N_a(i) + \tau^* \hat{N}_1^t(i) \hat{N}_1(i) \right\} \\
\bar{R} &= \left\{ R(i) + \mu_1^{-1}(i) N_b^t(i) N_b(i) + \tau^* \hat{N}_2^t(i) \hat{N}_2(i) \right\}^{-1}
\end{aligned}
\tag{4.69}
$$

is a **QGCC** *for system (4.1) with cost matrix* $0 < P(i) = P^t(i)$, $i \in \mathcal{S}$.

Corollary 4.6 *Consider the time-delay system*

$$\dot{x}(t) = A_{\Delta o}(i)x(t) + A_d(i)x(t-\tau) \tag{4.70}$$

associated with cost function (4.4) and matrix $A_o(i)$, $i \in \mathcal{S}$, *being Hurwitz. Suppose that there exist matrices* $0 < P(i) = P^t(i)$, *scalars* $r_1(i) > 0, r_2(i) > 0, \mu_1(i) > 0, \mu_2(i) > 0$, $i \in \mathcal{S}$ *and* $\tau^* > 0$ *such that, for any constant time-delay* τ *being* $0 \leq \tau \leq \tau^*$, *satisfying the LMIs*

$$\begin{bmatrix} \hat{\Gamma}(Y,\tau^*,\mu,i) & \tau^*\hat{M}_1(i) & \tau^*Y(i)\hat{M}_2^t(i) & \tau^*S(i)\hat{N}_2^t(i) \\ \tau^*\hat{M}_1^t(i) & -\tau^*I & 0 & 0 \\ \tau^*\hat{M}_2(i)Y(i) & 0 & -\tau^*I & 0 \\ \tau^*\hat{N}_2(i)S(i) & 0 & 0 & -\tau^*I \end{bmatrix} < 0$$

$$\mu_2(i)H(i)H^t(i) - I < 0 \tag{4.71}$$

where

$$\begin{aligned} \hat{\Gamma}(Y,\tau,\mu,i) &= [A_o(i) + A_d(i)]Y(i) + Y(i)[A_o(i) + A_d(i)]^t \\ &+ \mu_1(i)M_a(i)M_a^t(i) + \mu_1^{-1}(i)Y(i)N_a^t(i)N_a(i)Y(i) \\ &+ S(i)B_o^t(i) + B_o(i)S^t(i) \\ &+ S(i)\left[\mu_1^{-1}(i)N_b^t(i)N_a(i) + \tau N_1^t(i)N_1(i)\right]Y(i) \\ &+ Y(i)\left[\mu_1^{-1}(i)N_a^t(i)N_b(i) + \tau N_1^t(i)N_1(i)\right]S(i) \\ &+ S(i)[R(i) + \mu_1^{-1}(i)N_b^t(i)N_b(i)]S(i) \end{aligned} \tag{4.72}$$

Then, the state-feedback controller

$$u(t) = S(i)Y^{-1}(i)x(t) \tag{4.73}$$

is a **QGCC** *for system (4.1) with cost matrix* $0 < P(i) = P^t(i)$, $i \in \mathcal{S}$.

4.6 $\mathcal{H}_\infty$-State Feedback

In this section, we look into another problem in control system design by considering the synthesis of an $\mathcal{H}_\infty$-state feedback controller for the jumping system for $\eta_t = i \in \mathcal{S}$

$$\begin{aligned}(\Sigma_J): \quad \dot{x}(t) &= [A_o(i) + \Delta A_o(t,i)]x(t) + [A_d(i) + \Delta A_d(t,i)]x(t-\tau) \\ &+ [B_o(i) + \Delta B_o(t,i)]u(t) + \Gamma(i)w(t), \ t \geq 0 \\ &= A_{\Delta o}(t,i)x(t) + A_{\Delta d}(t,i)x(t-\tau) + B_{\Delta o}(t,i)u(t) \\ &+ \Gamma(i)w(t), \ x(t) = \phi(t), \ t \in [-\tau, 0], \ \eta_o = i, \ t \geq 0, \quad (4.74) \\ y(t) &= x(t), \quad (4.75) \\ z(t) &= G_o(i)x(t) + \Phi(i)w(t) + F_o(i)u(t), \quad (4.76)\end{aligned}$$

subject to (4.2)-(4.3) where $x(t) \in \Re^n$ is the state vector; $u(t) \in \Re^m$ is the control input; $w(t) \in \Re^q$ is the disturbance input which belongs to $\mathcal{L}_2[0, \mathcal{T}]$; $y(t) \in \Re^p$ is the measured output; $z(t) \in \Re^r$ is the controlled output which belongs to $\mathcal{L}_2\big[(\Omega, \mathcal{F}, \mathbf{P}), [0, \mathcal{T}]\big]$ and $\tau \in [0, \tau^*]$ is an unknown time-varying delay factor satisfying $0 \leq \tau \leq \tau^*$, $0 \leq \dot{\tau} \leq \tau^+$ where the bounds τ^* , $\tau^+ < 1$ are known constants.

Initially we focus on the nominal system for $\eta_t = i \in \mathcal{S}$.

$$\begin{aligned}(\Sigma_{Jo}): \quad \dot{x}(t) &= A_o(i)x(t) + A_d(i)x(t-\tau) + B_o(i)u(t) \\ &+ \Gamma(i)w(t), \ t \geq 0 \\ x(t) &= \phi(t), \ t \in [-\tau, 0], \ \eta_o = i \quad (4.77) \\ y(t) &= x(t) \quad (4.78) \\ z(t) &= G_o(i)x(t) + \Phi(i)w(t) + F_o(i)u(t) \quad (4.79)\end{aligned}$$

Observe that model (4.74)-(4.76) is more general than model (4.1). For simplicity in exposition, we define the following matrix expressions for $i \in \mathcal{S}$:

$$\begin{aligned}
\Lambda(i) &= \gamma^2 I - \Phi^t(i)\,\Phi(i)\ ,\ \ \Pi(i) = I - \gamma^2\,\Phi(i)\,\Phi^t(i) \\
\Xi(i) &= F_o^t(i)\Pi(i)F_o(t)\ ,\ \ \Omega(i) = P(i)\Gamma(i) + G_o^t(i)\Phi(i) \\
\Theta_o(i) &= P(i)A_o(i) + A_o^t(i)P(i) + \sum_{m=1}^{s} \alpha_{im} P(m) \\
&+ G_o^t(i)G_o(i) + \hat{Q}(i) \\
\Sigma(i) &= \big[P(i)B_o(i) + \Omega(i)\Lambda(i)\Phi^t(i)F_o(i) \\
&+ G_o^t(i)F_o(i)\big]\left\{F_o^t(i)\Pi(i)F_o(i)\right\}^{-1} \\
&\cdot \big[B_o^t(i)P(i) + F_o^t(i)G_o(i) + F_o^t(i)\Phi(i)\Lambda(i)\Omega^t(i)\big] \qquad (4.80)
\end{aligned}$$

The main result in this section is established by the following theorem.

Theorem 4.7 *Consider system (Σ_{Jo}). Then, for a given $\gamma > 0$, there exists a state-feedback controller $u(t)$ such that the closed-loop system is* **SSWDD** *and*

$$||z(t)||_{E_2} \ < \ \gamma\,||w(t)||_2$$

for all nonzero $w(t) \in \mathcal{L}_2[0, \infty)$, if for given matrix $Q(i) = Q^t(i) > 0,\ i \in \mathcal{S}$, there exist matrices $P(i) = P^t(i) > 0,\ i \in \mathcal{S}$, satisfying the system of ARIs for all $i \in \mathcal{S}$

$$\begin{aligned}
&P(i)A_o(i) + A_o^t(i)P(i) + \sum_{m=1}^{s} \alpha_{im} P(m) + \hat{Q}(i) + G^t(i)G(i) + \\
&P(i)A_d(i)\bar{Q}^{-1}(i)A_d^t(i)P(i) - \Sigma(i) < 0 \quad i \in \mathcal{S} \qquad (4.81)
\end{aligned}$$

Moreover, the feedback controller is $u(t) = K^*(i)x(t)$ has the gain:

$$\begin{aligned}
K^*(i) &= -R^{*-1}(i)\Big[B_o^t(i)P(i) + F_o^t(i)\Pi(i)G_o(i) \\
&+ F_o^t(i)\Phi(i)\Lambda(i)\Omega^t(i)\Big] \quad i \in \mathcal{S} \\
R^*(i) &= \Big\{F_o^t(i)\Pi(i)F_o(i)\Big\} \qquad (4.82)
\end{aligned}$$

Proof: Consider the Lyapunov functional $V(\cdot)$ as the form of (3.5). Evaluating the weak infinitesimal operator $\Im_f^x[\cdot]$ of the process $\{x(t), \eta_t, t \geq 0\}$ for system (4.77)-(4.79) at the point $\{t, x, \eta_t = i\}$, we get

$$\begin{aligned}
\Im_f^x[V] &= x^t(t)\Big\{A_o^t(i)P(i) + P(i)A_o(i) + \sum_{m=1}^{s} \alpha_{im}P(m) \\
&+ Q(i) + \sum_{m=1}^{s} \alpha_{im} \int_{-\tau}^{0} x^t(t+\theta)Q(i)x(t+\theta)d\theta\Big\}x(t) \\
&+ x^t(t-\tau)A_d^t(i)P(i)x(t) + x^t(t)P(i)A_d(i)x(t-\tau) \\
&- x^t(t-\tau)Q(i)x(t-\tau) + x^t(t)P(i)\Gamma(i)w(t) + w^t(t)\Gamma^t(i)P(i)x(t) \\
&+ x^t(t)P(i)B_o(i)u(t) + u^t(t)B_o^t(i)P(i)x(t) \\
&\leq x^t(t)\Big\{A_o^t(i)P(i) + P(i)A_o(i) + \sum_{m=1}^{s} \alpha_{im}P(m) + \hat{Q}(i)\Big\}x(t) \\
&+ x^t(t-\tau)A_d^t(i)P(i)x(t) + x^t(t)P(i)A_d(i)x(t-\tau) \\
&- x^t(t-\tau)\bar{Q}(i)x(t-\tau) + x^t(t)P(i)\Gamma(i)w(t) + w^t(t)\Gamma^t(i)P(i)x(t) \\
&+ x^t(t)P(i)B_o(i)u(t) + u^t(t)B_o^t(i)P(i)x(t)
\end{aligned} \tag{4.83}$$

Given the performance measure $\mathcal{J}$ with $\mathcal{T} \to \infty$. Algebraic manipulation in the mannar of the theorems of Chapter 3 but using (4.79)-(4.81) , (4.83) instead and the control law $u(i) = K^*(i)x(i)$ yields:

$$\mathcal{J}(x) \leq \mathbb{E}\int_0^\infty \chi^t(i)\Upsilon_f(i)\chi(i)dt \tag{4.84}$$

where

$$\begin{aligned}
\Upsilon_f(i) &= \begin{bmatrix} \Theta_o(i) + \Theta_k(i) & P(i)A_d(i) & \begin{matrix}\Omega(i)+ \\ K^{*t}(i)F_o^t(i)\Phi(i)\end{matrix} \\ A_d^t(i)P(i) & -\bar{Q}(i) & 0 \\ \begin{matrix}\Omega^t(i)+ \\ \Phi^t(i)F_o(i)K^*(i)\end{matrix} & 0 & -\Lambda(i) \end{bmatrix} \\
\chi(i) &= [x^t(i) \;\; x^t(t-\tau) \;\; w^t(i)] \\
\Theta_k(i) &= G_o^t(i)F_o(i)K^*(i) + K^{*t}(i)F_o^t(i)G_o(i)
\end{aligned}$$

$$
\begin{aligned}
&+ \quad K^{*t}(i)F_o^t(i)F_o(i)K^*(i) \\
&+ \quad K^{*t}(i)B_o^t(i)P(i) + P(i)B_o(i)K^*(i) \qquad (4.85)
\end{aligned}
$$

Application of **Fact 1** to (4.84)-(4.85) gives

$$
\begin{aligned}
\mathcal{J}(x) \quad &\leq \quad \mathbb{E}\Big\{ \int_0^\infty x^t(t) \Big[\Theta_o(i) + \Theta_k(i) \\
&+ \quad P(i)A_d(i)\bar{Q}^{-1}(i)A_d^t(i)P(i) \\
&+ \quad \Big(\Omega(i) + K^{*t}(i)F_o^t(i)\Phi(i) \Big) \Lambda^{-1}(i) \\
&\cdot \quad \Big(\Omega^t(i) + \Phi^t(i)F_o(i)K^*(i) \Big) \Big] x(i)dt \Big\} \qquad (4.86)
\end{aligned}
$$

Finally using the matrix inversion lemma [98] and the controller gain (4.82), it follows from (4.80), (4.81), (4.86) and the results of **Theorem** (3.1), that $\mathcal{J}(x) < 0$ and hence the resulting closed-loop system is **SSWDD** with disturbance attenuation γ, which completes the proof. $\nabla\nabla\nabla$

The next theorem provides an expression of $\mathcal{H}_\infty$ state feedback controller for system (Σ_J).

Theorem 4.8 *Consider system (Σ_J). Then, for a given $\gamma > 0$, there exists a state-feedback controller $u(t) = K^+(i)x(t)$ such that the closed-loop system is* **RSSWDD** *and*

$$
||z(t)||_{E_2} \quad < \quad \gamma \, ||w(t)||_2
$$

for all nonzero $w(t) \in \mathcal{L}_2[0,\infty)$, and all admissible parameter uncertainties satisfying (4.2)-(4.3), if given matrices $Q(i) = Q^t(i) > 0$, $i \in \mathcal{S}$, there exist matrices $P(i) = P^t(i) > 0$, $i \in \mathcal{S}$ and scalars $\epsilon_1(i) > 0$, $\epsilon_2(i) > 0$, $\mu(i) > 0, i \in \mathcal{S}$ satisfying the system of LMIs for all $i \in \mathcal{S}$

$$
\begin{bmatrix}
\begin{array}{c}\Theta_o(i) - \Sigma(i) + \\ \epsilon_1(i)N_a^t(i)N_a(i)\end{array} & \Lambda_{aa}(i) & P(i)A_d(i) \\
\Lambda_{aa}^t(i) & -\Lambda_{dd} & 0 \\
A_d^t(i)P(i) & 0 & \begin{array}{c}-\bar{Q}(i) + \\ \epsilon_2(i)N_d^t(i)N_d(i)\end{array}
\end{bmatrix} < 0
$$

$$\begin{bmatrix} -\bar{Q}(i) & N_d^t(i) \\ N_d(i) & -\epsilon_2(i)I \end{bmatrix} < 0 \tag{4.87}$$

where

$$\begin{aligned} \Lambda_{aa}(i) &= [P(i)M_a(i) \quad P(i)M_a(i) \quad P(i)M_a(i)] \\ \Lambda_{dd}(i) &= diag[\epsilon_1(i)I \quad \epsilon_2(i)I \quad \mu(i)I] \end{aligned} \tag{4.88}$$

Moreover, the feedback gain is given by:

$$\begin{aligned} K^+(i) &= -R_{++}^{-1}(i)\Big[B_o^t(i)P(i) + F_o^t(i)\Pi(i)G_o(i) \\ &+ F_o^t(i)\Phi(i)\Lambda(i)\Omega^t(i)\Big] \\ R_{++}(i) &= \Big\{F_o^t(i)\Pi(i)F_o(i) + \mu(i)N_b^t(i)N_b(i)\Big\} \end{aligned} \tag{4.89}$$

Proof: Consider the Lyapunov functional $V(\cdot)$ as the form of (3.5). Evaluating the weak infinitesimal operator $\Im_t^x[\cdot]$ of the process $\{x(t), \eta_t, t \geq 0\}$ for system (4.77)-(4.79) at the point $\{t, x, \eta_t = i\}$ using (4.83), we get

$$\begin{aligned} \Im_t^x[V] &\leq \Im_f^x[V] + x^t(t)\{\Delta A_o^t(i)P(i) + P(i)\Delta A_o(i)\}x(t) \\ &+ x^t(t-\tau)\Delta A_d^t(i)P(i)x(t) + x^t(t)P(i)\Delta A_d(i)x(t-\tau) \\ &+ u^t(t)\Delta B_o^t(i)P(i)x(i) + x^t(i)P(i)\Delta B_o(i)u(t) \\ &\leq \Im_f^x[V] + x^t(t)\left[\epsilon_1^{-1}(i) + \epsilon_2^{-1}(i) + \mu^{-1}(i)\right]P(i)M_a(i)M_a^t(i)P(i)x(i) \\ &+ \epsilon_1(i)x^t(t)N_a^t(i)N_a(i)x(t) + \epsilon_2(i)x^t(t-\tau)N_d^t(i)N_d(i)x(t-\tau) \\ &+ \mu(i)u^t(t)N_b^t(i)N_b(i)u(t) \end{aligned} \tag{4.90}$$

for some scalars $\epsilon_1(i) > 0$, $\epsilon_2 > 0$, $\mu(i) > 0$, $i \in \mathcal{S}$. On substituting $u(t) = K^+(i)x(i)$, we evaluate the performance measure $\mathcal{T} \to \infty$ in the manner of **Theorem** 4.7 to get:

$$\mathcal{J}(x) \leq \mathbb{E}\int_0^\infty \chi^t(i)\Upsilon_t(i)\chi(i)dt \tag{4.91}$$

where

$$\Upsilon_t(i) = \begin{bmatrix} \Theta_o(i)+\Theta_t(i) & P(i)A_d(i) & \begin{matrix}\Omega(i)+\\ K^{+^t}(i)F_o^t(i)\Phi(i)\end{matrix} \\ A_d^t(i)P(i) & \begin{matrix}-\bar{Q}(i)+\\ \epsilon_2(i)N_d^t(i)N_d(i)\end{matrix} & 0 \\ \begin{matrix}\Omega^t(i)+\\ \Phi^t(i)F_o(i)K^+(i)\end{matrix} & 0 & -\Lambda(i) \end{bmatrix}$$

$$\begin{aligned} \Theta_t(i) &= \left[\epsilon_1^{-1}(i)+\epsilon_2^{-1}(i)+\mu^{-1}(i)\right]P(i)M_a(i)M_a^t(i)P(i) \\ &+ G^t(i)F_o(i)K^+(i)+K^{+^t}(i)F_o^t(i)G(i) \\ &+ K^{+^t}(i)F_o^t(i)F_o(i)K^+(i)+K^{+^t}(i)B_o^t(i)P(i)+P(i)B_o(i)K^+(i) \\ &+ \epsilon_1(i)N_a^t(i)N_a(i)+\mu(i)K^{+^t}(i)N_b^t(i)N_b(i)K^+(i) \end{aligned} \tag{4.92}$$

Application of **Fact 1** to (4.92) using (4.80) gives

$$\begin{aligned} \mathcal{J}(x) &\le \mathbb{E}\Bigg\{\int_0^\infty x^t(t)\Big[\Theta_o(i)+\Theta_t(i) \\ &+ P(i)A_d(i)\Big[\bar{Q}(i)-\epsilon_2(i)N_d^t(i)N_d(i)\Big]^{-1}A_d^t(i)P(i) \\ &+ \Big(\Omega(i)+K^{+^t}(i)F_o^t(i)\Phi(i)\Big)\Lambda^{-1}(i) \\ &\cdot \Big(\Omega^t(i)+\Phi^t(i)F_o(i)K^+(i)\Big)\Big]x(i)dt\Bigg\} \end{aligned} \tag{4.93}$$

Substituting the controller gain (4.89) and arranging terms into the LMIs (4.87), it follows in the mannar of **Theorem** 4.7 that $\mathcal{J}(x) < 0$ and hence the resulting closed-loop system is **RSSWDD** with disturbance attenuation γ for all admissible parameter uncertainties, which completes the proof. ∇∇∇

Remark 4.2 *It should be emphasized that inequalities (4.87) are standard LMIs in the variables $P(i)$, $\mu(i)$, $\epsilon_1(i)$, $\epsilon_2(i)$, $i \in \mathcal{S}$ [25] which lend themselves to direct computation via the software environment [57].*

Remark 4.3 *Extension of the developed robustness results in this section can be made to the case where the jumping rates are subject to uncertainties. Specifically, we consider the transition probability from mode i at time t to mode j at time $t+\delta$, $i,j \in \mathcal{S}$ to be:*

$$\begin{aligned} p_{ij} &= Pr(\eta_{t+\delta} = j \mid \eta_t = i) \\ &= \begin{cases} (\alpha_{ij} + \Delta\alpha_{ij})\delta + o(\delta), & if\ i \neq j, \\ 1 + (\alpha_{ij} + \Delta\alpha_{ij})\delta + o(\delta), & if\ i = j \end{cases} \end{aligned} \tag{4.94}$$

with transition probability rates $(\alpha_{ij} + \Delta\alpha_{ij}) \geq 0$ for $i,j \in \mathcal{S}, i \neq j$ and

$$\alpha_{ii} + \Delta\alpha_{ii} = - \sum_{m=1, m\neq i}^{s} (\alpha_{im} + \Delta\alpha_{im}) \tag{4.95}$$

We assume that the uncertainties $\Delta\alpha_{ij}$ satisfies

$$||\Delta\alpha_{ij}|| \leq \beta_{ij} \ , \qquad \forall\, i,j \in \mathcal{S} \tag{4.96}$$

where β_{ij} are known scalars, $\forall\, i,j \in \mathcal{S}$.

In line of **Theorems** 4.7-4.8, we have the following robustness results:

Theorem 4.9 *Consider system (Σ_{Jo}) over the space $(\Omega, \mathcal{F}, \mathbf{P})$ where $\mathbf{P}$ is described by (4.94)-(4.96). Then, for a given $\gamma > 0$, there exists a state-feedback controller $u(t) = K^*(i)x(t)$ such that the closed-loop system is* **SSWDD** *and*

$$||z(t)||_{E_2} < \gamma\, ||w(t)||_2$$

for all nonzero $w(t) \in \mathcal{L}_2[0,\infty)$, if for given matrices $Q(i) = Q^t(i) > 0$, $i \in \mathcal{S}$, there exist matrices $P(i) = P^t(i) > 0$, satisfying the system of ARIs for all $i \in \mathcal{S}$

$$\begin{aligned} &P(i)A_o(i) + A_o^t(i)P(i) + \sum_{m=1}^{s} \Big(\alpha_{im} + \beta_{im}\Big) P(m) + \hat{Q}(i) + G^t(i)G(i) + \\ &P(i)A_d(i)\bar{Q}^{-1}(i)A_d^t(i)P(i) - \Sigma(i) < 0 \end{aligned} \tag{4.97}$$

Moreover, the feedback gain $K^*(i)$ is given by (4.82)

Theorem 4.10 *Consider system (Σ_J) over the space $(\Omega, \mathcal{F}, \mathbf{P})$ where $\mathbf{P}$ is described by (4.94)-(4.96). Then, for a given $\gamma > 0$, there exists a state-feedback controller $u(t) = K^+(i)x(t)$ such that the closed-loop system is* **RSSWDD** *and*

$$||z(t)||_{E_2} \; < \; \gamma \, ||w(t)||_2$$

for all nonzero $w(t) \in \mathcal{L}_2[0,\infty)$, and all admissible parameter uncertainties, if for given matrices $Q(i) = Q^t(i) > 0,\ i \in \mathcal{S}$, there exist matrices $P(i) = P^t(i) > 0,\ i \in \mathcal{S}$ and scalars $\epsilon_1(i) > 0,\ \epsilon_2(i) > 0,\ \mu(i) > 0, i \in \mathcal{S}$ satisfying the system of LMIs for all $i \in \mathcal{S}$

$$\begin{bmatrix} \begin{array}{c}\bar{\Theta}_o(i) - \Sigma(i) + \\ \epsilon_1(i) N_a^t(i) N_a(i)\end{array} & \Lambda_{aa}(i) & P(i)A_d(i) \\ \Lambda_{aa}^t(i) & -\Lambda_{dd}(i) & 0 \\ A_d^t(i)P(i) & 0 & \begin{array}{c}-\bar{Q}(i) + \\ \epsilon_2(i) N_d^t(i) N_d(i)\end{array} \end{bmatrix} < 0$$

$$\begin{bmatrix} -\bar{Q}(i) & N_d^t(i) \\ N_d(i) & -\epsilon_2(i) I \end{bmatrix} < 0 \qquad (4.98)$$

The feedback gain $K^+(i)$ is given by (4.89) and

$$\begin{aligned} \bar{\Theta}_o(i) &= P(i)A_o(i) + A_o^t(i)P(i) + \sum_{m=1}^{s} \Big(\alpha_{im} + \beta_{im} \Big) P(m) \\ &+ G^t(i)G(i) + \hat{Q}(i) \end{aligned}$$

4.7 Numerical Examples

In this section, we provide three examples to illustrate the developed theories.

4.7.1 Example 4.1

Consider a two-mode JTD system with mode switching generator

$$\Im = \begin{bmatrix} -2 & 2 \\ 3 & -3 \end{bmatrix}$$

and having the dynamics

Mode 1:

$$\begin{aligned}
\dot{x}(t) &= \left(\begin{bmatrix} 2 & 1 \\ 0 & 1 \end{bmatrix} + \begin{bmatrix} 0.1 \\ 0.2 \end{bmatrix} \; sin\ 0.5\pi\ t\ [0\ \ 1]\right) x(t) \\
&+ \left(\begin{bmatrix} 0.1 & 0.3 \\ 0 & 0.2 \end{bmatrix} + \begin{bmatrix} 0.1 \\ 0.2 \end{bmatrix} \; sin\ 0.5\pi\ t\ [1\ \ 1]\right) x(t-\tau) \\
&+ \left(\begin{bmatrix} 0.5 \\ 2 \end{bmatrix} + \begin{bmatrix} 0.1 \\ 0.2 \end{bmatrix} \; sin\ 0.5\pi\ t\ 0.4\right) u(t) + \begin{bmatrix} 0.1 \\ 0.1 \end{bmatrix} w(t) \\
z(t) &= [1\ \ 1]x(t) + 0.1w(t) + u(t)
\end{aligned}$$

Mode 2:

$$\begin{aligned}
\dot{x}(t) &= \left(\begin{bmatrix} 1 & 0.7 \\ 0 & 2 \end{bmatrix} + \begin{bmatrix} 0.2 \\ 0.1 \end{bmatrix} \; cos\ 0.5\pi\ t\ [1\ \ 0]\right) x(t) \\
&+ \left(\begin{bmatrix} 0.2 & 0 \\ 0.4 & 0.1 \end{bmatrix} + \begin{bmatrix} 0.2 \\ 0.1 \end{bmatrix} \; cos\ 0.5\pi\ t\ [1\ \ 1]\right) x(t-\tau) \\
&+ \left(\begin{bmatrix} 0.4 \\ 1 \end{bmatrix} + \begin{bmatrix} 0.2 \\ 0.1 \end{bmatrix} \; cos\ 0.5\pi\ t\ 0.6\right) u(t) + \begin{bmatrix} 0.2 \\ 0.2 \end{bmatrix} w(t) \\
z(t) &= [1\ \ 0]x(t) + 0.2w(t) + 2\ u(t)
\end{aligned}$$

The result of solving ARIs (4.34) of **Theorem** 4.4 by descent methods [98] using

$$\begin{aligned}
W(1) &= \begin{bmatrix} 8 & 2 \\ 2 & 12 \end{bmatrix}, \quad \mu(1) = 0.2\ , \quad \tau^* = 2 \\
W(2) &= \begin{bmatrix} 5 & 1 \\ 1 & 6 \end{bmatrix}, \quad \mu(2) = 0.35\ , \quad \tau^+ = 0.8
\end{aligned}$$

yields the stabilizing matrices

$$P(1) = \begin{bmatrix} 25.884 & -8.740 \\ -8.740 & 4.269 \end{bmatrix}, \quad P(2) = \begin{bmatrix} 17.645 & 1.124 \\ 1.124 & 6.836 \end{bmatrix}$$

The associated feedback gains for the guaranteed cost control are given by

$$\begin{aligned}
K_s(1) &= [-16.3114 \quad 3.6044] \\
K_s(2) &= [12.6743 \quad -4.2015]
\end{aligned}$$

Turning to **Theorem** 4.6, we solve LMIs (4.63 with

$$r_1(1) = 2.5 \quad , \quad r_2(1) = 1.7 \quad , \quad r_1(2) = 3.2 \quad , \quad r_2(2) = 1.4$$

We get the feasible solutions

$$\begin{aligned}
Y(1) &= \begin{bmatrix} 8.212 & -2.254 \\ -2.254 & 4.228 \end{bmatrix} , \quad S(1) = [5.278 \quad -16.349] \\
\mu_1(1) &= 1.9872 \ , \ \mu_2(1) = 2.4562 \\
\mu_3(1) &= 0.9843 \ , \ \mu_4(1) = 5.1207 \\
Y(2) &= \begin{bmatrix} 5.365 & 1.015 \\ 1.015 & 3.369 \end{bmatrix} , \quad S(2) = [8.322 \quad -10.451] \\
\mu_1(2) &= 0.8792 \ , \ \mu_2(2) = 3.2456 \\
\mu_3(2) &= 1.4983 \ , \ \mu_4(2) = 2.7102
\end{aligned}$$

for $\tau \leq 0.95$. The gains of the corresponding strong delay-dependent guaranteed cost controllers are given by

$$Gain(1) = [-0.4904 \quad -4.1283] \ , \ Gain(2) = [2.2673 \quad -3.7852]$$

4.7.2 Example 4.2

Using the JTD system of **Example 1**, we proceed to solve LMIs (4.87) of **Theorem** 4.8 to obtain an $\mathcal{H}_\infty$ state-feedback controller. The feasible solution is given by

$$\begin{aligned}
P(1) &= \begin{bmatrix} 7.334 & -1.234 \\ -1.234 & 2.805 \end{bmatrix} , \quad \epsilon_1(1) = 2.456 \ , \ \epsilon_2(1) = 3.124 \ , \ \mu(1) = 0.257 \\
K^+(i) &= [-0.0950 \quad -5.1848] \\
P(2) &= \begin{bmatrix} 2.654 & 0.722 \\ 0.722 & 2.975 \end{bmatrix} , \quad \epsilon_1(2) = 1.564 \ , \ \epsilon_2(2) = 1.412 \ , \ \mu(2) = 0.128 \\
K^+(i) &= [4.7934 \quad -2.3401]
\end{aligned}$$

4.7.3 Example 4.3

Consider a continuous-time JTD system of the type (4.1) with $\mathcal{S} = \{1, 2\}$:

Mode 1:

$$\begin{aligned}
\dot{x}(t) &= \left[\begin{array}{cc} -5 & 3 \\ -2 & -1 \end{array}\right] + \left[\begin{array}{c} 0 \\ 0.1 \end{array}\right] \; sin\; 0.25\pi\; t\; [0 \;\; 1] \Big) x(t) \\
&+ \left(\left[\begin{array}{cc} 2 & -1 \\ 1 & -0.5 \end{array}\right] + \left[\begin{array}{c} 0 \\ 0.1 \end{array}\right] \; sin\; 0.5\pi\; t\; [0 \;\; 1] \right) x(t-\tau) \\
&+ \left(\left[\begin{array}{c} 0 \\ 0.1 \end{array}\right] + \left[\begin{array}{c} 0 \\ 0.1 \end{array}\right] \; sin\; 0.5\pi\; t\; 0.5 \right) u(t) \\
z(t) &= [1 \;\; 1] x(t) + 2\, u(t)
\end{aligned}$$

Mode 2:

$$\begin{aligned}
\dot{x}(t) &= \left(\left[\begin{array}{cc} -3 & 1 \\ -3 & -1 \end{array}\right] + \left[\begin{array}{c} 0.2 \\ 0.1 \end{array}\right] \; cos\; 0.5\pi\; t\; [1 \;\; 0] \right) x(t) \\
&+ \left(\left[\begin{array}{cc} 2 & -1 \\ 1 & -1 \end{array}\right] + \left[\begin{array}{c} 0.2 \\ 0.1 \end{array}\right] \; cos\; 0.5\pi\; t\; [0.5 \;\; 1] \right) x(t-\tau) \\
&+ \left(\left[\begin{array}{c} 0.1 \\ 0 \end{array}\right] + \left[\begin{array}{c} 0.2 \\ 0.1 \end{array}\right] \; cos\; 0.5\pi\; t\; 0.6 \right) u(t) \\
z(t) &= [1 \;\; 0] x(t) + u(t)
\end{aligned}$$

With the Markovian process η_t be described by

$$\Im = \left[\begin{array}{cc} -4 & 4 \\ 3 & -3 \end{array}\right]$$

Following the procedure of **Example 1**using

$$\begin{aligned}
W(1) &= \left[\begin{array}{cc} 10 & 0 \\ 0 & 10 \end{array}\right] \;, \quad \mu(1) = 2.5 \;, \quad \tau^* = 1.2 \\
W(2) &= \left[\begin{array}{cc} 5 & 0 \\ 0 & 5 \end{array}\right] \;, \quad \mu(2) = 3.5 \;, \quad \tau^+ = 0.8
\end{aligned}$$

We get the stabilizing matrices

$$P(1) = \left[\begin{array}{cc} 0.3304 & -0.1401 \\ -0.1401 & 0.6439 \end{array}\right] \;, \quad P(2) = \left[\begin{array}{cc} 0.5163 & -0.276 \\ -0.276 & 0.676 \end{array}\right]$$

The associated feedback gains for the guaranteed cost control are given by

$$\begin{aligned} K_s(1) &= [0.071 \quad -0.322] \\ K_s(2) &= [0.138 \quad -0.338] \end{aligned}$$

The closed-loop state and control trajectories are plotted in Figs 4.1-4.2, from which it is demonstrated that the derived controller assures the desired robust stability.

4.8 $\mathcal{H}_\infty$-Output Feedback

In this section, we move one step ahead to complement the $\mathcal{H}_\infty$ control synthesis examined in the foregoing section. Here we consider the design of an $\mathcal{H}_\infty$-output feedback controller for system (Σ_{JO}) for $i \in \mathcal{S}$:

$$\begin{aligned} (\Sigma_{JO}): \quad \dot{x}(t) &= A_o(i)x(t) + A_d(i)x(t-\tau(t)) \\ &+ B_o(i)u(t) + \Gamma(i)w(t), \quad t \geq 0 \\ x(t) &= \phi(t) \quad , t \in [-\tau^*, 0]\,, \; \eta_o = i, \qquad (4.99) \\ y(t) &= C_o(i)x(t) + D_o(i)w(t) \qquad (4.100) \\ z(t) &= G_o(i)x(t) + F_o(i)u(t) + \Phi(i)w(t) \qquad (4.101) \end{aligned}$$

4.8.1 Weak Delay-Dependence

We consider a dynamic output feedback controller for $i \in \mathcal{S}$, as given by:

$$\begin{aligned} (\Sigma_C): \quad \dot{\zeta}(t) &= A_C(i)\zeta(t) + B_C(i)y(t) \\ u(t) &= C_C(i)\zeta(t) + D_C(i)y(t) \qquad (4.102) \end{aligned}$$

where $\zeta(t) \in \Re^{nc}$ is the state of the controller and the gain matrices $A_C(i) \in \Re^{nc \times nc}$, $B_C(i) \in \Re^{nc \times p}$, $C_C(i) \in \Re^{m \times nc}$, $D_C(i) \in \Re^{m \times p}$ are controller matrices to be determined. Combining (4.99)-(4.101) and (4.102) for $i \in \mathcal{S}$, we obtain the jumping closed-loop system where

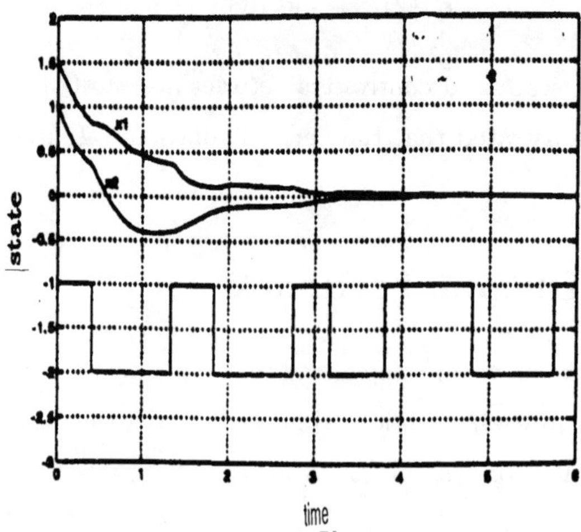

Fig 4.1: Closed-loop State Trajectories

$$
\begin{aligned}
\xi(t) &= \begin{bmatrix} x(t) \\ \zeta(t) \end{bmatrix}, \quad A_{JCd}(i) = \begin{bmatrix} A_d(i) & 0 \\ 0 & 0 \end{bmatrix} \\
A_{JC}(i) &= \begin{bmatrix} A_o(i) + B_o(i)D_C(i)C(i) & B_o(i)C_C(i) \\ B_C(i)C_o(i) & A_C(i) \end{bmatrix} \\
\Gamma_{JC}(i) &= \begin{bmatrix} \Gamma(i) + B_o(i)D_C(i)D_o(i) \\ B_C(i)D_o(i) \end{bmatrix} \\
G_{JC}(i) &= [G_o(i) + F_o(i)D_C(i)C_o(i) \quad F_o(i)C_C(i)] \\
\Phi_{JC}(i) &= \Phi(i) + F_o(i)D_C(i)D_o(i)
\end{aligned} \tag{4.103}
$$

$$
\begin{aligned}
(\Sigma_{JC}): \quad \dot{\xi}(t) &= A_{JC}(i)\xi(t) + A_{JCd}(i)\xi(t - \tau(t)) + \Gamma_{JC}(i)w(t) \\
\xi(t) &= \phi_{JC}(t) \quad ,t \in [-\tau^*, 0]
\end{aligned}
$$

$$z(t) = G_{JC}(i)\xi(t) + \Phi_{JC}(i)w(t) \tag{4.104}$$

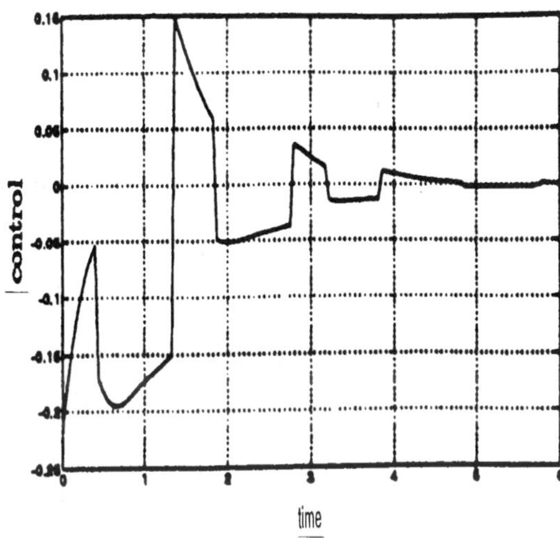

Fig 4.2: Closed-loop Control Trajectories

To facilitate further development, we group the available system information into the following matrices:

$$\begin{aligned}
\mathcal{K}(i) &= \begin{bmatrix} D_C(i) & C_C(i) \\ B_C(i) & A_C(i) \end{bmatrix} \\
\bar{A}(i) &= \begin{bmatrix} A_o(i) & 0 \\ 0 & 0 \end{bmatrix}, \ \bar{A}_d(i) = \begin{bmatrix} A_d(i) \\ 0 \end{bmatrix} \\
\bar{\Gamma}(i) &= \begin{bmatrix} \Gamma(i) \\ 0 \end{bmatrix}, \ \bar{I} = [I \quad 0] \\
\bar{B}(i) &= \begin{bmatrix} B_o(i) & 0 \\ 0 & I \end{bmatrix}, \ \bar{C}(i) = \begin{bmatrix} C_o(i) & 0 \\ 0 & I \end{bmatrix} \\
\bar{G}(i) &= [G_o(i) \quad 0] \\
\bar{D}(i) &= \begin{bmatrix} D_o(i) \\ 0 \end{bmatrix}, \ \bar{F}(i) = [F_o(i) \quad 0]
\end{aligned} \tag{4.105}$$

This enables us to cast the matrices of the jumping closed-loop system (4.104) into the affine form:

$$
\begin{aligned}
A_{JC}(i) &= \bar{A}(i) + \bar{B}(i)\mathcal{K}\bar{C}(i) \ , \ \ \Gamma_{JC}(i) = \bar{\Gamma}(i) + \bar{B}(i)\mathcal{K}\bar{D}(i) \\
G_{JC}(i) &= \bar{G}(i) + \bar{F}(i)\mathcal{K}\bar{C}(i) \ , \ \ \Phi_{JC}(i) = \Phi(i) + \bar{F}(i)\mathcal{K}\bar{D}(i) \\
A_{JCd}(i) &= \bar{A}_d(i)\bar{I}.
\end{aligned}
\qquad (4.106)
$$

Based on **Theorems** (3.1)-(3.2), immediate results follow for the closed-loop system (Σ_{JC}) and are summarized, without proof, by the next lemmas.

Lemma 4.1 *Consider the closed-loop system* (Σ_{JC}). *If for any given matrix* $Q(i) = Q^t(i) \in \Re^{n\times n} > 0,\ i \in \mathcal{S}$, *there exist matrices* $\mathcal{P}(i) = \mathcal{P}^t(i) \in \Re^{(n+n_C)\times(n+n_C)} > 0,\ i \in \mathcal{S}$, *satisfying the system of LMIs for all* $i \in \mathcal{S}$

$$
\begin{bmatrix} \Pi_K(i) & \mathcal{P}(i)A_{JCd}(i) \\ A^t_{JCd}(i)\mathcal{P}(i) & -(1-\tau^*)Q(i) \end{bmatrix} < 0
$$

where

$$
\begin{aligned}
\Pi_K(i) &= \mathcal{P}(i)A_{JC}(i) + A^t_{JC}(i)\mathcal{P}(i) \\
&+ \sum_{m=1}^{s} \alpha_{im}\mathcal{P}(m) + \bar{I}^tQ(i)\bar{I}
\end{aligned}
$$

then system (Σ_{JC}) *is* **SSWDD**.

Lemma 4.2 *Consider the jumping closed-loop system* (Σ_{JC}) . *If for any given matrices* $Q(i) = Q^t(i) \in \Re^{n\times n} > 0,\ i \in \mathcal{S}$ *and a scalar* $\gamma > 0$, *there exist matrices* $\mathcal{P}(i) = \mathcal{P}^t(i) \in \Re^{(n+n_C)\times(n+n_C)} > 0,\ i \in \mathcal{S}$ *satisfying the system of LMIs for all* $i \in \mathcal{S}$

$$
\begin{bmatrix} \Pi_K(i) & \mathcal{P}(i)\Gamma_{JC}(i) & G^t_{JC}(i) & \mathcal{P}A_{JCd}(i) \\ \Gamma^t_{JC}(i)\mathcal{P}(i) & -I & \gamma^{-1}\bar{\Phi}^t(i) & 0 \\ G_{JC}(i) & \gamma^{-1}\bar{\Phi}(i) & -I & 0 \\ A^t_{JCd}(i)\mathcal{P}(i) & 0 & 0 & -(1-\tau^*)Q(i) \end{bmatrix} < 0 \qquad (4.107)
$$

then system (Σ_{JC}) *is* **SSWDD with a disturbance attenuation** γ.

Remark 4.4 *It should be observed that* **Lemmas** *(4.1)-(4.2) and together provide LMI-based sufficient stochastic stability criteria with weakly delay-dependence of the closed-loop system for a given controller matrix* $\mathcal{K}$.

Interestingly enough, we learn from **Lemma** (4.2), that system (4.104) is stochastically stable and weakly delay-dependent with a disturbance attenuation γ if there exist matrices $0 < \mathcal{P}(i) = \mathcal{P}^t(i) \in \Re^{n+n_C \times n+n_C}$ such that for all $i \in \mathcal{S}$, LMIs (4.107) holds. Our immediate goal is to compute the controller matrix $\mathcal{K}$. Towards our goal, we cast the LMIs (4.107) for all $i \in \mathcal{S}$ into the following affine form:

$$\Xi(i) + \Lambda(i)\Upsilon(i)\mathcal{K}(i)\Omega^t(i) + \Omega(i)\mathcal{K}^t(i)\Upsilon^t(i)\Lambda^t(i) < 0 \tag{4.108}$$

where

$$\begin{aligned}
\Lambda(i) &= diag[\mathcal{P}(i) \quad I \quad I \quad I] \\
\Upsilon(i) &= [\bar{B}(i) \quad 0 \quad \bar{F}^t(i) \quad 0]^t \\
\Omega(i) &= [\bar{C}(i) \quad \bar{D}(i) \quad 0 \quad 0]^t \\
\Xi(i) &= \begin{bmatrix} \Pi_O & \mathcal{P}(i)\bar{\Gamma}(i) & \bar{G}^t(i) & \mathcal{P}\bar{A}_d(i) \\ \bar{\Gamma}^t(i)\mathcal{P}(i) & -I & \gamma^{-1}\Phi^t(i) & 0 \\ \bar{G}(i) & \gamma^{-1}\Phi(i) & -I & \\ \bar{A}_d^t(i)\mathcal{P}(i) & 0 & 0 & -(1-\tau^*)Q(i) \end{bmatrix} \\
\Pi_O &= \mathcal{P}(i)\bar{A}(i) + \bar{A}^t(i)\mathcal{P}(i) + \sum_{m=1}^{s} \alpha_{im}\mathcal{P}(m) + \bar{I}^t Q(i)\bar{I}
\end{aligned}$$

It follows from [7, 56] for some gain matrix $\mathcal{K}$ and for $i \in \mathcal{S}$ that inequality (4.108) holds if and only if the following inequalities hold for all $i \in \mathcal{S}$:

$$\Upsilon_*^t(i)\,\Lambda^{-1}(i)\,\Xi(i)\,\Lambda^{-1}(i) < 0 \quad , \quad \Omega_*^t(i)\,\Xi(i)\,\Omega(i) < 0 \tag{4.109}$$

where $\Upsilon_*(i)$ and $\Omega_*(i)$ are the orthogonal complements of $\Upsilon(i)$ and $\Omega(i)$ for $i \in \mathcal{S}$, respectively.

By rewriting

$$\begin{aligned} \mathcal{P}(i) &= \begin{bmatrix} \mathcal{X}(i) & \mathcal{Z}(i) \\ \mathcal{Z}^t(i) & \mathcal{W}_x(i) \end{bmatrix} \\ \mathcal{P}^{-1}(i) &= \begin{bmatrix} \mathcal{Y}(i) & \mathcal{T}(i) \\ \mathcal{T}^t(i) & \mathcal{W}_y(i) \end{bmatrix} \quad i \in \mathcal{S} \end{aligned} \tag{4.110}$$

where $\mathcal{X}(i) \in \Re^{n \times n}$, $\mathcal{Y}(i) \in \Re^{n \times n}$ and $\mathcal{Z}(i) \in \Re^{n \times n_C}$, $\mathcal{T}(i) \in \Re^{n \times n_C}$, for all $i \in \mathcal{S}$ and choosing

$$\begin{bmatrix} B(i) \\ F(i) \end{bmatrix}_* = \begin{bmatrix} S_1(i) \\ S_2(i) \end{bmatrix} , \quad \begin{bmatrix} C^t(i) \\ D^t(i) \end{bmatrix}_* = \begin{bmatrix} S_3(i) \\ S_4(i) \end{bmatrix}$$

it can be easily shown with the aid of (4.107) that

$$\Upsilon_*(i) = \begin{bmatrix} S_1(i) & 0 & 0 \\ 0 & 0 & 0 \\ 0 & I & 0 \\ S_2(i) & 0 & 0 \\ 0 & 0 & I \end{bmatrix} , \quad \Omega_*(i) = \begin{bmatrix} R_1(i) & 0 & 0 \\ 0 & 0 & 0 \\ R_2(i) & 0 & 0 \\ 0 & I & 0 \\ 0 & 0 & I \end{bmatrix} .$$

Now we are in a position to give the general solvability conditions for a dynamic output-feedback controller of the type (4.102) guaranteeing that system (Σ_{JO}) is stochastically stable and weakly delay-dependent with disturbance attenuation γ. This is summarized by the following theorem

Theorem 4.11 *Consider the closed-loop system (Σ_{JC}) with matrices described in (4.103)-(4.106). Given constant matrices $Q(i) = Q^t(i) \in \Re^{n \times n} > 0$, $i \in \mathcal{S}$ and a scalar $\gamma > 0$, there exists a dynamic output feedback controller of the type (4.102) such that the jumping closed-loop system (Σ_{JC}) is* **SSWDD with a disturbance attenuation** *γ if there exist matrices $0 < \mathcal{X}(i) = \mathcal{X}^t(i) \in \Re^{n \times n}$, $0 < \mathcal{Y}(i) = \mathcal{Y}^t(i) \in \Re^{n \times n}$, $i \in \mathcal{S}$ such that*

$$\begin{bmatrix} \mathcal{X} & I \\ I & \mathcal{Y} \end{bmatrix} \geq 0 \tag{4.111}$$

and satisfying the system of LMIs for all $i \in \mathcal{S}$

$$\Pi_x^t(i)\begin{bmatrix} \Xi_{xx}(i) & \mathcal{X}G^t(i) & \Gamma(i) & A_d(i) \\ G(i)\mathcal{X} & -I & \gamma^{-1}\Phi^t(i) & 0 \\ \Gamma^t(i) & \gamma^{-1}\Phi(i) & -I & 0 \\ A_d^t(i) & 0 & 0 & -(1-\tau^*)Q(i) \end{bmatrix}\Pi_x(i) < 0 \qquad (4.112)$$

$$\Pi_y^t(i)\begin{bmatrix} \Xi_{yy}(i) & \mathcal{Y}\Gamma(i) & G^t(i) & \mathcal{Y}A_d(i) \\ \Gamma^t\mathcal{Y}(i) & -I & \gamma^{-1}\Phi^t(i) & 0 \\ G(i) & \gamma^{-1}\Phi(i) & -I & 0 \\ A_d^t(i)\mathcal{Y} & 0 & 0 & -(1-\tau^*)Q(i) \end{bmatrix}\Pi_y^t(i) < 0 \qquad (4.113)$$

where

$$\begin{aligned} \Xi_{xx}(i) &= \mathcal{X}(i)A_o^t(i) + A_o(i)\mathcal{X}(i) \\ &+ \sum_{m=1}^{s}\alpha_{im}\mathcal{X}(m) + \mathcal{X}Q(i)\mathcal{X} \\ \Xi_{yy}(i) &= \mathcal{Y}(i)A_o(i) + A_o^t(i)\mathcal{Y}(i) \\ &+ \sum_{m=1}^{s}\alpha_{im}\mathcal{Y}(m) + Q(i) \\ \Pi_x(i) &= \begin{bmatrix} S_1(i) & 0 & 0 \\ S_2(i) & 0 & 0 \\ 0 & I & 0 \\ 0 & 0 & I \end{bmatrix}, \quad \Pi_y(i) = \begin{bmatrix} R_1(i) & 0 & 0 \\ R_2(i) & 0 & 0 \\ 0 & I & 0 \\ 0 & 0 & I \end{bmatrix} \end{aligned}$$

Proof: It is straightforward to see that inequality (4.111) for $i \in \mathcal{S}$ holds if and only if there exist matrices $0 < \mathcal{P}(i) = \mathcal{P}^t(i) \in \Re^{n+n_C \times n+n_C}$ satisfying (4.110). Standard matrix manipulations yields inequalities (4.112)-(4.113) and the proof is completed. ∇∇∇

Remark 4.5 *It is interesting to note that (4.112)-(4.113) are basic LMI convex feasibility problems which can be solved quite effectively by the* **MATLAB LMI** *Control Toolbox [57]. It essentially provides existence conditions of* $\gamma-$*suboptimal* $\mathcal{H}_\infty$ *controllers of arbitrary order by parameterizing the* $\gamma-$*suboptimal* $\mathcal{H}_\infty$ *controllers in terms of positive-definite solutions of LMI's. However, it does not address the explicit computation of the dynamic controller itself.*

Next, we consider the computation of the $\mathcal{H}_\infty$ controller structure $\mathcal{K}$. Given the solutions $\mathcal{X}(i)$ and $\mathcal{Y}(i)$ of (4.112)-(4.113), respectively, for $i \in \mathcal{S}$, compute full-column-rank matrices $\mathcal{T}(i) \in \Re^{n \times n_C}$ and $\mathcal{Z}(i) \in \Re^{n \times n_C}$ for $i \in \mathcal{S}$ such that:

$$\mathcal{T}(i)\, \mathcal{Z}^t(i) \;=\; I \,-\, \mathcal{X}(i)\, \mathcal{Y}(i).$$

It follows from [7, 56] that the unique solution $\mathcal{P}(i)$, $i \in \mathcal{S}$ is obtained from the following equation

$$\begin{bmatrix} \mathcal{Y}(i) & I \\ \mathcal{Z}^t(i) & 0 \end{bmatrix} \;=\; \mathcal{P}(i) \begin{bmatrix} I & \mathcal{X}(i) \\ 0 & \mathcal{T}^t(i) \end{bmatrix}. \tag{4.114}$$

The solution of (4.114) always exists since $\mathcal{Y}(i)$ and $\Omega_y(i)$ has full-column rank. The following theorem summarizes the main result.

Theorem 4.12 *Consider the closed-loop system* (Σ_{JC}) *with matrices described in (4.103)-(4.106). Suppose there exist matrices* $0 < \mathcal{X}(i) = \mathcal{X}^t(i) \in \Re^{n \times n}$, $0 < \mathcal{Y}(i) = \mathcal{Y}^t(i) \in \Re^{n \times n}$, $i \in \mathcal{S}$ *satisfying the LMIs (4.112)-(4.113) and given* $\mathcal{P}(i)$, $i \in \mathcal{S}$ *solving (4.114). If*

$$Rank\ [I \,-\, \mathcal{X}(i)\, \mathcal{Y}(i)] \;=\; n_c\, n\,, \qquad i \in \mathcal{S} \tag{4.115}$$

then there exist $\gamma-$*suboptimal* $\mathcal{H}_\infty$ *controller structure* $\mathcal{K}$ *satisfying inequality (4.108).*

4.8.2 Strong Delay-Dependence

Direct application of the foregoing results to the case with strong delay-dependence and with reference **Theorems** (4.11)-(4.12), we anticipate technical difficulties due to the presence of the product terms like $G_o^t(i)\Phi(i)$, or $G_o^t(i)G_o(i)$. This will yield quadratic functional dependence on the controller matrices $\mathcal{K}$ and hence the LMI-based stability condition in $\mathcal{P}$ can not converted into the linear affine form (4.108) in $\mathcal{K}$. Based thereon, we follow here a basically different route and consider an observer-based output feedback control scheme for $i \in \mathcal{S}$, given by the following form:

$$
\begin{aligned}
(\Sigma_O): \ \dot{\beta}(t) &= A_O(i)\beta(t) + B_O(i)[y(t) - C_o(i)\beta(t)] \\
u(t) &= C_O(i)\beta(t)
\end{aligned}
\tag{4.116}
$$

where $\beta(t) \in \Re^n$ is the state of the controller and the matrices $A_O(i) \in \Re^{n\times n}$, $B_O(i) \in \Re^{n\times p}$, $C_O(i) \in \Re^{m\times n}$, are gain matrices to be selected. Combining (4.99)-(4.101) and (4.116) for $i \in \mathcal{S}$, we obtain the following closed-loop system

$$
\begin{aligned}
(\Sigma_{JO}): \ \dot{\omega}(t) &= A_{JO}(i)\omega(t) + A_{JOd}(i)\xi(t-\tau) \\
&+ \Gamma_{JO}(i)w(t) \\
\xi(t) &= \phi_{JO}(t) \quad ,t \in [-\tau^*, 0] \\
z(t) &= G_{JO}(i)\xi(t) + \Phi_{JO}(i)w(t)
\end{aligned}
\tag{4.117}
$$

where

$$
\begin{aligned}
\omega(t) &= \begin{bmatrix} x(t) \\ \beta(t) \end{bmatrix} \\
A_{JO}(i) &= \begin{bmatrix} A_o(i) & B_o(i)C_O(i) \\ B_O(i)C_o(i) & A_O(i) - B_O(i)C_o(i) + B_O(i)F_o(i)C_O(i) \end{bmatrix} \\
A_{JOd}(i) &= \begin{bmatrix} A_d(i) & 0 \\ 0 & 0 \end{bmatrix}, \ \Gamma_{JO}(i) = \begin{bmatrix} \Gamma(i) \\ B_O(i)D(i) \end{bmatrix} \\
G_{JO}(i) &= [G_o(i) \ \ F_o(i)C_O(i)], \ \Phi_{JC}(i) = \Phi(i)
\end{aligned}
\tag{4.118}
$$

Based on **Theorem** (3.5) and **Theorem** (3.6), immediate results follow for the jumping closed-loop system (Σ_{JO}) and are summarized by the next lemmas.

Lemma 4.3 *Consider the jumping closed-loop system* (Σ_{JO}) *and given a scalar* $\tau^* > 0$. *If there exist matrices* $\mathcal{P}(i) = \mathcal{P}^t(i) \in \Re^{2n\times 2n} > 0$, $i \in \mathcal{S}$ *and scalars*

$\varepsilon(i) > 0, \mu(i) > 0$, $i \in \mathcal{S}$ *satisfying the system of LMIs for all* $i \in \mathcal{S}$*:*

$$\begin{bmatrix} \Xi_{JO} & \tau^* A^t_{JO}(i) & \tau^* A^t_{JOd}(i) \\ \tau^* A_{JO}(i) & -\tau^* \varepsilon(i) I & 0 \\ \tau^* A_{JOd}(i) & 0 & -\tau^* \mu(i) I \end{bmatrix} < 0$$

where

$$\begin{aligned} \Xi_{JO} &= \mathcal{P}(i)[A_{JO}(i) + A_{JOd}(i)] + [A_{JO}(i) + A_{JOd}(i)]^t \mathcal{P}(i) \\ &+ \sum_{m=1}^{s} \alpha_{im} \mathcal{P}(m) + \tau^*(\varepsilon(i)\mu(i)) \mathcal{P}(i) A_{JOd}(i) A^t_{JOd}(i) \mathcal{P}(i) \end{aligned}$$

then system (Σ_{JO}) *is* **SSSDD for any time-delay** τ **satisfying** $0 < \tau \leq \tau^*$ **and** $\dot{\tau} \leq \tau^+ < 1$.

Proof: It can be readily obtained from **Theorem** (3.5) and taking into account the matrices of (4.118). $\nabla\nabla\nabla$

Lemma 4.4 *Consider the jumping closed-loop system* (Σ_{JO}) *and given a scalar* $\tau^* > 0$. *If there exist scalars* $\sigma(i) > 0$, $i \in \mathcal{S}$ *and matrices* $\mathcal{P}(i) = \mathcal{P}^t(i) \in \Re^{2n \times 2n} > 0$, $i \in \mathcal{S}$ *and scalars* $\varepsilon(i) > 0, \mu(i) > 0$, $i \in \mathcal{S}$ *satisfying the system of LMIs for all* $i \in \mathcal{S}$*:*

$$\begin{bmatrix} \Xi_{JD}(i) & \tau^* A^t_{JO}(i) & \tau^* A^t_{JOd}(i) & G^t_{JO}(i)\Phi_{JO}(i) \\ \tau^* A_{JO}(i) & -\tau^* \varepsilon(i) I & 0 & 0 \\ \tau^* A_{JOd}(i) & 0 & -\tau^* \mu(i) I & 0 \\ \Phi^t_{JO}(i) G_{JO}(i) & 0 & 0 & \begin{array}{c} -\gamma^2 I \\ +\Phi^t(i)\Phi(i) \\ +\tau^* \sigma(i) \Gamma^t_{JO}(i) \Gamma_{JO}(i) \end{array} \end{bmatrix} < 0 \tag{4.119}$$

$$\begin{bmatrix} -\gamma^2 I + \Phi^t(i)\Phi(i) & \tau^* \sigma(i) \Gamma^t_{JO}(i) \\ \tau^* \sigma(i) \Gamma_{JO}(i) & -\tau^* \sigma(i) I \end{bmatrix} < 0$$

where

$$\Xi_{JD}(i) = \mathcal{P}(i)[A_{JO}(i) + A_{JOd}(i)] + [A_{JO}(i) + A_{JOd}(i)]^t \mathcal{P}(i)$$

$$+ \quad \sum_{m=1}^{s} \alpha_{im}\mathcal{P}(m) + G_{JO}^t(i)G_{JO}(i)$$
$$+ \quad \tau^*(\varepsilon(i) + \mu(i))\mathcal{P}(i)A_{JOd}(i)A_{JOd}^t(i)\mathcal{P}(i)$$

then system (Σ_{JO}) *is* **SSSDD with a disturbance attenuation** γ **for any time-delay** τ **satisfying** $0 < \tau \leq \tau^*$ *and* $\dot{\tau} \leq \tau^+ < 1$.

Proof: It can be derived along the same line as that of **Theorem** (3.4). ∇∇∇

In a similar way, we observe that **Lemmas** (4.3)-(4.4) establish LMI-based sufficient stochastic stability criteria with strongly delay-dependence for the closed-loop system (Σ_{JO}). Our next objective is to develop conditions that can be used for computing the gains of the observer-based output feedback controller.

Now by applying **Fact 1** to (4.119) we obtain the following algebraic matrix inequality (AMI):

$$\begin{aligned}
&\mathcal{P}(i)[A_{JO}(i) + A_{JOd}(i)] + [A_{JO}(i) + A_{JOd}(i)]^t\mathcal{P}(i) + \\
&\sum_{m=1}^{s} \alpha_{im}\mathcal{P}(m) + G_{JO}^t(i)G_{JO}(i) + \\
&\tau^*\varepsilon^{-1}(i)A_{JO}^t(i)A_{JO}(i) + \tau^*\mu^{-1}(i)A_{JOd}^t(i)A_{JOd}(i) + \\
&\tau^*(\varepsilon(i)\mu(i))\mathcal{P}(i)A_{JOd}(i)A_{JOd}^t(i)\mathcal{P}(i) + \\
&\gamma^{-2}G_{JO}^t(i)\Phi(i)\hat{\mathcal{R}}^{-1}\Phi^t(i)G_{JO}(i) \quad < \quad 0
\end{aligned} \tag{4.120}$$

where

$$\hat{\mathcal{R}} \triangleq I - \gamma^{-2}\Phi^t(i)\Phi(i) - \tau^*\sigma(i)\Gamma_{JO}^t(i)\Gamma_{JO}(i)$$

and other matrices are given by (4.118). In order to develop our last result, we need the following assumption.

Assumption 4.1: For all $i \in \mathcal{S}$ the matrix $C(i)$ has a full rank.

The following theorem summarizes the main solvability conditions for a dynamic output-feedback controller of the type (4.116) guaranteeing that system (Σ_{JO}) is **SSSDD with disturbance attenuation** γ.

Theorem 4.13 *Consider the jumping closed-loop system (Σ_{JO}) with matrices described in (4.118). Given $\gamma > 0$, $\tau^* > 0$, $\varepsilon(i) > 0$, $\mu(i) > 0$, $\sigma(i)$, $i \in \mathcal{S}$, there exists an observer-based output feedback controller of the type (4.116) such that this system* **SSSDD with a disturbance attenuation** γ *if there exist matrices $0 < \mathcal{M}(i) = \mathcal{M}^t(i) \in \Re^{n\times n}$, $0 < \mathcal{N}(i) = \mathcal{N}^t(i) \in \Re^{n\times n}$, $i \in \mathcal{S}$ such that*

$$\begin{aligned}\mathcal{R}_O(i) \;&\triangleq\; \Big\{\gamma^2 I \;-\; \Phi^t(i)\Phi(i) \\ &-\; \tau^*\sigma(i)[\Gamma^t(i)\Gamma(i) + D^t(i)\mathcal{B}^t\mathcal{B}(i)D(i)\Big\} \;>\; 0\end{aligned}$$

and satisfying the system of simultaneous LMIs for all $i \in \mathcal{S}$

$$\begin{bmatrix} \Xi_M(i) & \tau^* A^t(i) & \tau^* A_d^t(i) & G^t(i)\Phi(i) & \begin{matrix}(\varepsilon(i)/\tau^*)\mathcal{N} \\ +A^t(i)\end{matrix} \\ \tau^* A(i) & -\tau^*\varepsilon(i)I & 0 & 0 & 0 \\ \tau^* A_d(i) & 0 & -\tau^*\mu(i)I & 0 & 0 \\ \Phi^t(i)G(i) & 0 & 0 & -\mathcal{R}_O(i) & 0 \\ \begin{matrix}(\varepsilon(i)/\tau^*)\mathcal{N} \\ +A(i)\end{matrix} & 0 & 0 & 0 & -I \end{bmatrix} < 0 \tag{4.121}$$

with the ARIs for all $i \in \mathcal{S}$

$$\begin{aligned}&\mathcal{N}(i)A(i) + A^t(i)\mathcal{N}(i) + \sum_{m=1}^{s} \alpha_{im}\mathcal{N}(m) - \\ &(\tau^*/\varepsilon(i))[(\varepsilon(i)/\tau^*)\mathcal{N} + A^t(i)][(\varepsilon(i)/\tau^*)\mathcal{N} + A(i)] + \\ &\mathcal{N}[(\varepsilon(i)/\tau^*)\mathcal{N} + A(i)]\Delta_*(i)\Pi^t(i)\Pi(i)[(\varepsilon(i)/\tau^*)\mathcal{N} + \\ &A^t(i)]\Delta_*^t(i) \;<\; 0\end{aligned} \tag{4.122}$$

where

$$
\begin{aligned}
\Xi_M(i) &= \mathcal{M}(i)[A(i) + A_d(i)] + [A(i) + A_d(i)]^t \mathcal{M}(i) \\
&+ \sum_{m=1}^{s} \alpha_{im} \mathcal{M}(m) + \tau^* \varepsilon(i) \mathcal{M}(i) A_d(i) A_d^t(i) \mathcal{M}(i) \\
&+ G^t(i) G(i) + \tau^* \mu(i) \mathcal{M}(i) A_d(i) A_d^t(i) \mathcal{M}(i) \\
\mathcal{B}^t \mathcal{B}(i) &= C_*^t(i)[(\varepsilon(i)/\tau^*)\mathcal{N} + A^t(i)][(\varepsilon(i)/\tau^*)\mathcal{N} + A(i)] C_*(i) \\
\Pi^t(i)\Pi(i) &= (\tau^*/\varepsilon(i)) B^t(i) B(i) \; + \; F^t(i)\{I + \Phi^t(i) \mathcal{R}_O^{-1}(i) \Phi(i)\} F(t) \\
\Delta(i) &= \mathcal{M}B(i) \; + \; G^t(i)\{F(i) + \Phi^t(i) \mathcal{R}_O^{-1}(i) \Phi(i)\} \\
&+ (\tau^*/\varepsilon(i)) A^t(i) B(i)
\end{aligned}
\tag{4.123}
$$

and the associated controller matrices are given by:

$$
\begin{aligned}
A_O(i) &= A(i) \\
B_O(i) &= C_*^t(i)[(\varepsilon(i)/\tau^*)\mathcal{N} + A(i)] \\
C_O(i) &= \Delta_*(i)[(\varepsilon(i)/\tau^*)\mathcal{N} + A^t(i)]\mathcal{N}
\end{aligned}
\tag{4.124}
$$

Proof: Start from (4.120) for $i \in \mathcal{S}$ and let

$$
\mathcal{P} = \begin{bmatrix} \mathcal{M} & 0 \\ 0 & \mathcal{N} \end{bmatrix}.
$$

Standard algebraic manipulations using (4.118) subject to **Assumption 2** yield inequalities (4.121)-(4.122), and the desired controller matrices (4.124). $\nabla\nabla\nabla$

Remark 4.6 *A simple comparison of* **Theorem** *(4.11) and* **Theorem** *(4.13) reveals that the controller (4.102) is of arbitrary order and the solvability conditions are LMI-based yielding stochastic stability with weak delay-dependence. The computations of the gain matrices are implicit. On the other hand, the controller (4.116) is full-order the solvability conditions are AMI-based yielding stochastic stability with strong delay-dependence. The gain matrices have explicit expressions.*

4.8.3 Example 4.4

In this section, we demonstrate the application of the results in **Theorems** (4.11)-(4.13) to robust stabilization and control of combustion in rocket motor chambers. In a typical model [30, 48], a liquid mono-propellant rocket motor with a pressure feeding system is considered. Assuming non-steady flow and tacking non0uniform lag into account, a linearized model of the feeding system and the combustion chamber equations has been developed. This model is of the form (4.99)-(4.101) with the state variables being x_1 the instantaneous combustion chamber pressure, x_2 the instantaneous mass flow upstream of the capacitance, x_3 the instantaneous mass rate of the injected propellant from their steady values and x_4 is the ratio between the deviation of the instantaneous pressure in the feeding line from steady state value and twice the injector pressure drop in steady operation. We consider two operating modes $\{1, 2\}$ with the following data:

$$
\begin{aligned}
A_o(1) &= \begin{bmatrix} -0.15 & 0 & 0 & 0 \\ 0 & 0 & 0 & -5 \\ -0.6 & 0 & -0.6 & 0.6 \\ 0 & 1 & -1 & 0 \end{bmatrix}, \; A_d(1) = \begin{bmatrix} -0.85 & 0 & 1 & 0 \\ 0 & 0 & 0 & 0 \\ 0 & 0 & 0 & 0 \\ 0 & 0 & 0 & 0 \end{bmatrix} \\
B_o(1) &= \begin{bmatrix} 0 \\ 0 \\ 1 \\ 0 \end{bmatrix}, \; \Gamma(1) = \begin{bmatrix} 0 \\ 5 \\ 0 \\ 0 \end{bmatrix}, \; C_o(1) = [1 \;\; 0 \;\; 0 \;\; 0] \; D_o(1) = 1 \\
G_o(1) &= [1 \;\; 0 \;\; 0 \;\; 0] \, , \; F_o(1) = 1 \, , \; \Phi(1) = 0.01 \\
A_o(2) &= \begin{bmatrix} 0.15 & 0 & 0 & 0 \\ 0 & 0 & 0 & -5 \\ -0.6 & 0 & -0.6 & 0.6 \\ 0 & 1 & -1 & 0 \end{bmatrix}, \; A_d(2) = \begin{bmatrix} -1.15 & 0 & 1 & 0 \\ 0 & 0 & 0 & 0 \\ 0 & 0 & 0 & 0 \\ 0 & 0 & 0 & 0 \end{bmatrix} \\
B_o(2) &= \begin{bmatrix} 0 \\ 0 \\ 1 \\ 0 \end{bmatrix}, \; \Gamma(1) = \begin{bmatrix} 0 \\ 5 \\ 0 \\ 0 \end{bmatrix}, \; C_o(2) = [1 \;\; 0 \;\; 0 \;\; 0] \; D_o(2) = 1
\end{aligned}
$$

$$G_o(2) = [1 \ \ 0 \ \ 0 \ \ 0] \ , \ F_o(2) = 1 \ , \ \Phi(2) = 0.02$$
$$\tau^* = 1.5 \ , \ \tau^+ = 0.7$$

and mode switching generator

$$\Im = \begin{bmatrix} -3 & 3 \\ 2 & -2 \end{bmatrix}$$

Implementation of the **LMI** toolbox [57] shows that the rocket motor chamber system is stochastically stable with weak delay dependence.

For the strong stability results, it has been found from the ensuing feasible solution that the chamber system is indeed stochastically stable for $\tau \leq 0.95$. For robust performance, it is found that the minimum value of γ which admits a feasible solution is $\gamma_{min} = 8$.

On the part of control design, we consider the dynamic controller (4.102) with $n_C = 2$. By applying **Theorem** (4.12), the feasible solutions yields controller matrices:

$$A_C(1) = \begin{bmatrix} -0.2 & 0 \\ 0.7 & -0.4 \end{bmatrix} , \ D_C(1) = 2.5$$
$$C_C(1) = [-3.6547 \ \ -6.1985] \ , \ B_C(1) = \begin{bmatrix} 1.73 \\ 1.45 \end{bmatrix}$$
$$A_C(2) = \begin{bmatrix} -0.6 & 0 \\ 1.2 & -0.5 \end{bmatrix} , \ D_C(2) = 3.3$$
$$C_C(2) = [-7.4236 \ \ -9.6017] \ , \ B_C(2) = \begin{bmatrix} 1.34 \\ 1.29 \end{bmatrix}$$
$$\gamma = 1.16$$

Typical state and control response are displayed in Fig 4.3.

On the other hand, by applying **Theorem** (4.13) we obtain the following matrices of the controller (4.116)

$$B_O(1) = \begin{bmatrix} 1.1 \\ 0.7 \\ -1.5 \\ 2.3 \end{bmatrix} , \ B_O(2) = \begin{bmatrix} -2.2 \\ -0.3 \\ 2.4 \\ 1.35 \end{bmatrix}$$

$$C_O(1) = [2.3476 \quad -6.0150 \quad -4.9005 \quad -11.2356]$$

$$C_O(2) = [-10.0506 \quad -5.1523 \quad -3.7105 \quad 3.4116] \quad , \quad \gamma = 1.10$$

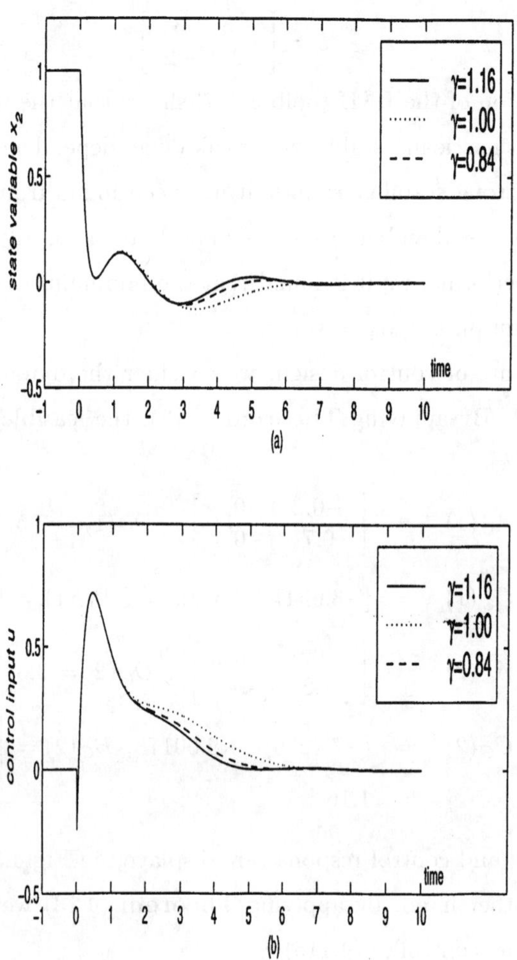

Fig 4.3: State and Control Input Response: WDD Case

In this case, typical state and control response are displayed in Fig 4.4.

4.9 Transformation Method

So far we have learned that strong delay-dependent approach to JTD systems yield, by and large, less conservative results. The purpose of this section is to extend further the results developed in this chapter by developing $\mathcal{H}_\infty$-control for a class of JLS with delays in the state and the input. To tackle this problem in the proper way, a new transformation method exhibiting delay-dependent behavior is developed to solve the problem of $\mathcal{H}_\infty$ control for a class of uncertain systems with Markovian jump parameters as well as state and input delays. The parametric uncertainties are assumed to be real, time-varying and norm-bounded that appear in the state, input and delayed-state matrices and the time-delay factors are known. Complete results for instantaneous state feedback control designs are established which guarantee the strong-delay dependent stochastic stability with a prescribed $\mathcal{H}_\infty$-performance. The solutions are provided in terms of a finite set of coupled LMIs.

4.9.1 Problem Description

Given the probability space $(\Omega, \mathcal{F}, \mathbf{P})$ where Ω is the sample space, $\mathcal{F}$ is the algebra of events and $\mathbf{P}$ is the probability measure defined on $\mathcal{F}$. Let the random form process $\{\eta_t, t \in [0, \mathcal{T}]\}$ be as defined in Chapter 1.

We consider a class of uncertain systems with Markovian jump parameters with state and input delays described over the space $(\Omega, \mathcal{F}, \mathbf{P})$ for $i \in \mathcal{S}$ by:

$$
\begin{aligned}
(\Sigma_{J_{SI}}): \ \dot{x}(t) &= [A_o(i) + E_o(i)\Delta(t,i)N(i)]x(t) \\
&+ [A_d(i) + E_d(i)\Delta(t,i)N(i)]x(t-\tau) \\
&+ [B_o(i) + E_b(i)\Delta(t,\eta_t)N(i)]u(t) \\
&+ [B_d(i) + E_d(i)\Delta(t,i)M(i)]u(t-\psi) \\
&+ \Gamma(i)w(t), \ t \geq 0\,, x(t) = 0, \ t \in [-\max\{\tau, \psi\}, 0] \qquad (4.125)
\end{aligned}
$$

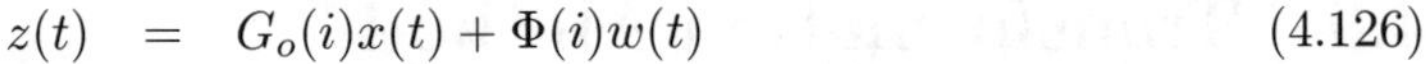

$$z(t) \quad = \quad G_o(i)x(t) + \Phi(i)w(t) \tag{4.126}$$

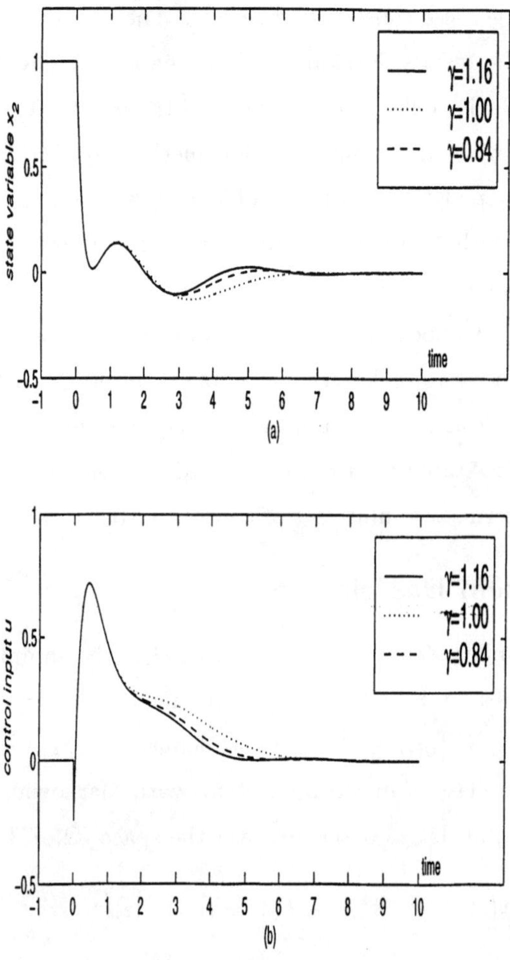

Fig 4.4: State and Control Input Response: SDD Case

where $x(t) \in \Re^n$ is the state vector; $u(t) \in \Re^m$ is the control input; $w(t) \in \Re^q$ is the disturbance input which belongs to $\mathcal{L}_2[0, \mathcal{T}]$; $z(t) \in \Re^r$ is the controlled output which belongs to $\mathcal{L}_2\big[(\Omega,\mathcal{F},\mathbf{P}),[0,\mathcal{T}]\big]$ to be attenuated. The time delays $\tau,\ \psi$ are known constants with $\hat{d} = \max\{\tau,\ \psi\}$ and where the matrices $A_o(i),\ B_o,\ A_d(i),\ B_d, \Gamma(i), G_o(i), G_n(i), F_o$ and $\Phi(i)$ are known real constants of appropriate dimensions which describe the nominal system of Σ_J. Also, $E_o(i) \in \Re^{n\times\alpha}$, $E_d(i) \in \Re^{n\times\alpha}$, $E_b(i) \in \Re^{n\times\alpha}$, $M(i) \in \Re^{\beta\times m}$ and $N(i) \in \Re^{\beta\times n}$ are known real constant matrices where the elements of $\Delta(t,i)$ are Lebesgue measurable for any $i \in \mathcal{S}$ such that:

$$||\Delta(t,i)||_2 \ \leq\ 1 \tag{4.127}$$

For a prescribed $\gamma > 0$, we introduce the following measure

$$\mathcal{J} = \int_0^\infty \ \mathbb{E}[z^t(t)z(t) - \gamma^2\ w^t(t)w(t)]dt \tag{4.128}$$

to aid in assessing the performance of system $\Sigma_{J_{SI}}$.

Our purpose hereafter is to design an $\mathcal{H}_\infty$ controller of the jumping system $\Sigma_{J_{SI}}$ which will guarantee desirable dynamical behavior. Here we use the instantaneous state feedback control law of the form:

$$u(t) \ =\ K(i)x(t) \quad ,\ i \in \mathcal{S} \tag{4.129}$$

The application of (4.129) to (4.125)-(4.126) yields the closed-loop system for $\eta_t = i\ \in \mathcal{S}$

$$\begin{aligned}
(\Sigma_{J_{SK}}):\quad \dot{x}(t) &= A_{\Delta K}(i)x(t) + A_{\Delta d}(i)x(t-\tau) \\
&+ B_{\Delta d}(i)K(i)x(t-\psi) + \Gamma(\eta_t)w(t),\ t \geq 0\ , \\
x(t) &= 0,\ t \in [-\max\{\tau,\psi\},0] \qquad (4.130) \\
z(t) &= G_o(i)x(t) + F_o(i)u(t) + \Phi(i)w(t) \qquad (4.131)
\end{aligned}$$

where

$$A_{\Delta o}(i) \ =\ A_o(i) + E_o(i)\Delta(t,i)N(i)$$

$$
\begin{aligned}
B_{\Delta o}(i) &= B_o(i) + E_b(i)\Delta(t,i)N(i) \\
A_{\Delta K}(i) &= A_K(i) + E_K(i)\Delta(t,i)M_K(i) \\
A_{\Delta d}(i) &= A_d(i) + E_d(i)\Delta(t,i)N(i) \\
A_K(i) &= A_o(i) + B_o(i)K(i) \\
B_{\Delta d}(i) &= [B_d(i) + E_d(i)\Delta(t,i)M(i)] \\
E_K(i) &= \begin{bmatrix} E_o(i) \\ E_b(i) \end{bmatrix}, \quad M_K(i) = [N(i) \quad M(i)K(i)]
\end{aligned} \tag{4.132}
$$

4.9.2 Model Transformation

In order to exhibit the delay-dependence behavior, we introduce for each possible value $\eta_t = i, \ i \in \mathcal{S},$ the following state transformation

$$
\begin{aligned}
\sigma(t) &= x(t) + \int_{t-\tau}^{t} A_{\Delta d}(t,i)\, x(s)ds \\
&+ \int_{t-\psi}^{t} B_{\Delta d}(t,i)K(i)\, x(r)dr
\end{aligned} \tag{4.133}
$$

into (4.130) to yield

$$
\begin{aligned}
\dot{\sigma}(t) &= \Big([A_{\Delta o}(t,i) + A_{\Delta d}(t,i) + [B_{\Delta o}(t,i) + B_{\Delta d}(t,i)]K(i)\Big)\, x(t) \\
&+ \Gamma(i)w(t) \\
&= A_{\Delta C}(t,K,i)\, x(t) + \Gamma(i)w(t)
\end{aligned} \tag{4.134}
$$

Define the augmented state-vector

$$
\zeta(t) = \begin{bmatrix} \sigma(t) \\ x(t) \end{bmatrix} \in \Re^{2n} \tag{4.135}
$$

By combining (4.126) and (4.133)-(4.134), we obtain the transformed system

$$
\begin{aligned}
(\Sigma_T): \quad \dot{\zeta}(t) &= \Lambda_{\Delta}(K,i)\zeta(t) + \int_{t-\tau}^{t} \Upsilon_{\Delta a}(i)\, \zeta(s)ds \\
&+ \int_{t-\psi}^{t} \Upsilon_{\Delta b}(i)\, \zeta(r)dr
\end{aligned}
$$

$$
\begin{aligned}
& + \quad \bar{\Gamma}(i)w(t) \\
\zeta(t) &= \bar{\phi}(t), \; t \in [-2\tau, 0], \; \eta_o = i, \; t \geq 0 \\
z(t) &= \bar{G}(i)\zeta(t) + \Phi(i)w(t), \qquad (4.136)
\end{aligned}
$$

where

$$
\begin{aligned}
\Lambda_\Delta(K,i) &= \begin{bmatrix} 0 & A_{\Delta C}(t,K,i) \\ -I & I \end{bmatrix} \\
&\triangleq \Lambda_o(K,i) + \hat{E}_K(i)\Delta(t,i)\hat{M}_K(i) \\
\Upsilon_{\Delta a}(i) &= \begin{bmatrix} 0 & 0 \\ 0 & A_{\Delta d}(t,i) \end{bmatrix} \\
&\triangleq \Upsilon_a(i) + \bar{E}_d(i)\Delta(t,i)\bar{N}(i) \\
\Upsilon_{\Delta b}(i) &= \begin{bmatrix} 0 & 0 \\ 0 & B_{\Delta d}(t,i)K(i) \end{bmatrix} \\
&\triangleq \Upsilon_b(i) + \bar{E}_d(i)\Delta(t,i)\bar{M}(i)K(i) \\
\bar{\Gamma}(i) &= \begin{bmatrix} \Gamma(i) \\ 0 \end{bmatrix}, \; \bar{G}(i) = [0 \quad G(i)] \\
\hat{M}_K(i) &= [0 \quad M_K(i)] \;, \; \bar{M}(i) = [0 \quad M(i)] \qquad (4.137)
\end{aligned}
$$

and

$$
\begin{aligned}
\Lambda_o(K,i) &= \begin{bmatrix} 0 & A_K(i) \\ -I & I \end{bmatrix}, \; \Upsilon_a(i) \begin{bmatrix} 0 & 0 \\ 0 & A_d(i) \end{bmatrix} \\
\Upsilon_b(i) &= \begin{bmatrix} 0 & 0 \\ 0 & B_d(i)K(i) \end{bmatrix}, \; \hat{E}_K(i) = \begin{bmatrix} E_K(i) \\ 0 \end{bmatrix} \\
\bar{E}_d(i) &= \begin{bmatrix} 0 \\ E_d(i) \end{bmatrix}, \; \bar{N}(i) = [0 \quad N(i)] \;, \; E_1 = \begin{bmatrix} I \\ 0 \end{bmatrix} \\
\bar{P}(i) &= U\mathbb{P}(i) \; ; \; U = \begin{bmatrix} I & 0 \\ 0 & 0 \end{bmatrix} \\
\mathbb{P}(i) &= \begin{bmatrix} P_\sigma(i) & 0 \\ P_d(i) & P_x(i) \end{bmatrix}, \; E_2 = \begin{bmatrix} 0 \\ I \end{bmatrix} \qquad (4.138)
\end{aligned}
$$

Invoking the stability definitions of Chapter 3, the following corresponding definitions are provided for system (4.136)-(4.138):

Definition 4.6 *System Σ_T is said to be* **robustly stochastically stable with strong-delay dependence (RSSSDD)** *for any time-delay $d \in [-\hat{d} = -\max\{\tau, \psi$ if for zero initial vector function $\phi \equiv 0$ defined on the interval $[-\hat{d}, 0]$ and initial mode $\eta_o \in \mathcal{S}$*

$$\lim_{\mathcal{T}\to\infty} \left\{ \int_0^{\mathcal{T}} \mathbb{E}\{||\zeta(t,\phi)||^2\}\, dt \right\} < +\infty$$

for all admissible un certainties satisfying (4.127)

Definition 4.7 *System Σ_T is said to be* **RSSSDD with a disturbance attenuation** *$\gamma > 0$ for any time-delay $d \in [-\hat{d} = -\max\{\tau, \psi\}, 0]$ if for zero initial vector function $\phi \equiv 0$ defined on the interval $[-\hat{d}, 0]$ and initial mode $\eta_o = i \in \mathcal{S}$* **Definition** *(4.6) is met and $\mathcal{J} < 0$ for all admissible uncertainties satisfying (4.127)*

Theorem 4.14 *System Σ_T with $w(.) \equiv 0$ is said to be* **RSSSDD** *if given gain matrices $K(i)$ $i \in \mathcal{S}$, and weighting matrices $Q_x(i) = Q_x^t(i) > 0$, $R_x(i) = R_x^t(i) > 0$ such that $Q(i) = Q_x(i)(1 - \hat{\alpha}\tau) > 0$, $R(i) = R_x(i)(1 - \hat{\alpha}\psi) > 0$ there exist scalars $\epsilon(i) > 0$, $\varrho(i) > 0$, $\mu(i) > 0$, $i \in \mathcal{S}$, and matrices $0 < \mathbb{P}(i) = \mathbb{P}^t(i)$ $i \in \mathcal{S}$, such that the following LMIs hold for all $i \in \mathcal{S}$:*

$$\Xi_n(i) = \begin{bmatrix} \Pi(i) & \Omega_a(i) & \Omega_b & \begin{matrix}\bar{G}^t(i)\Phi(i)+ \\ \mathbb{P}^t(i)\bar{\Gamma}(i)\end{matrix} \\ \Omega_a^t(i) & -\Theta_a(i) & 0 & 0 \\ \Omega_b^t(i) & 0 & -\Theta_b(i) & 0 \\ \begin{matrix}\Phi^t(i)\bar{G}(i)+ \\ \bar{\Gamma}^t(i)\mathbb{P}(i)\end{matrix} & 0 & 0 & \begin{matrix}-\gamma^2 I+ \\ \Phi^t(i)\Phi(i)\end{matrix} \end{bmatrix} < 0 \tag{4.139}$$

$$\begin{aligned} \Pi(i) &= \mathbb{P}^t(i)\Lambda_o(K,i) + \Lambda_o^t(K,i)\mathbb{P}(i) \\ &+ \sum_{m=1}^{s} \alpha_{im}\bar{P}(m) + \varepsilon(i)\hat{M}_K^t(i)\hat{M}_K(i) + \bar{G}^t(i)\bar{G}(i) \\ &+ \tau(1-\hat{\alpha}\tau)^{-1}E_2 Q(i) E_2^t + \psi(1-\hat{\alpha}\psi)^{-1}E_2 K^t(i) R(i) K(i) E_2^t \end{aligned}$$

$$
\begin{aligned}
\Omega_a(i) &= [I\!P^t(i)\hat{E}_K(i) \quad \tau I\!P^t(i)E_2E_d(i) \quad I\!P^t(i)E_2E_d(i)] \\
\Omega_b(i) &= [\tau I\!P^t(i)E_2A_d(i) \quad \psi I\!P^t(i)E_2B_d(i)] \\
\Theta_a(i) &= diag[\varepsilon(i)I \quad \tau\varrho(i)I \quad \mu(i)I] \\
\Theta_b(i) &= diag[\tau[Q(i) - \varrho(i)N^t(i)N(i)] \quad \psi[R(i) - \mu(i)M^t(i)M(i)]]
\end{aligned}
$$

Proof: Let $\mathbf{x}_s(t) \stackrel{\Delta}{=} \mathbf{x}(s+t),\ t-\tau \leq s \leq t$ and define the process $\{(\mathbf{x}(t), \eta_t),\ t \geq 0\}$ over the state space $\bar{\mathcal{C}}$. It should be observed that $\{(\mathbf{x}(t), \eta_t),\ t \geq 0\}$ is strong Markovian [78] and so is $\{(\zeta(t), \eta_t),\ t \geq 0\}$. Now for $\eta_t = i \in \mathcal{S}$, and given $Q_x(i) = Q_x^t(i) > 0,\ R_x(i) = R_x^t(i) > 0$, let the Lyapunov functional $V(\cdot) : \Re^{2n} \times \Re_+ \times \mathcal{S} \to \Re_+$ of the transformed system be selected as

$$
\begin{aligned}
V(t,\zeta,i) &= V_p(t,\zeta,i) + V_q(t,\zeta,i) + V_r(t,\zeta,i) \\
V_p(t,\zeta,i) &= \zeta^t(t)\bar{P}(i)\zeta(t) \\
V_q(t,\zeta,i) &= \int_{-\tau}^{0}\int_{t+\beta}^{t} \zeta^t(s)E_2Q_x(i)E_2^t\zeta(s)\, ds d\beta \\
V_r(t,\zeta,i) &= \int_{-\psi}^{0}\int_{t+\beta}^{t} \zeta^t(r)E_2K^t(i)R_x(i)K(i)E_2^t\zeta(r)\, dr d\beta \qquad (4.140)
\end{aligned}
$$

The weak infinitesimal operator $\Im^\zeta[\cdot]$ of the process $\{\zeta(t), i, t \geq 0\}$ for system (4.136) at the point $\{t, \zeta, i\}$ is given by [78, 45]:

$$
\begin{aligned}
\Im^\zeta[V_p] &= \partial V_p/\partial t + \partial V_p/\partial\zeta\ \dot{\zeta}(t)\ |_{\eta_t = i} \\
&+ \sum_{m=1}^{s} \alpha_{im}V_p(t,\zeta,i,m) \qquad (4.141)
\end{aligned}
$$

Using (4.136)-(4.138) we get:

$$
\begin{aligned}
\partial V_p/\partial\zeta\ \dot{\zeta}(t) &= 2\zeta^t(t)\ U\mathbb{P}^t(i)\ \dot{\zeta}(t) = 2\sigma^t(t)\ P_\sigma^t(i)\ \dot{\sigma}(t) \\
&= 2\zeta^t(t)\mathbb{P}^t(i)\begin{bmatrix} \dot{\sigma}(t) \\ 0 \end{bmatrix}
\end{aligned}
$$

$$
\begin{aligned}
&= 2\zeta^t(t)\mathbb{P}^t(i)\begin{bmatrix} A_{\Delta C}(t,K,i)x(t)+\Gamma(i)w(t) \\ \\ -\sigma(t)+x(t)+ \\ \int_{t-\tau}^{t} A_{\Delta d}(t,i)\; x(s)ds+ \\ \int_{t-\psi}^{t} B_{\Delta d}(t,i)K(i)\; x(r)dr \end{bmatrix} \\
&= 2\zeta^t(t)\mathbb{P}^t(i)\Lambda_\Delta(K,i)\zeta(t)+2\zeta^t(t)\mathbb{P}^t(i)\bar{\Gamma}(i)w(t) \\
&+ 2\int_{t-\tau}^{t}\zeta^t(t)\mathbb{P}^t(i)\Upsilon_{\Delta a}(i)\zeta(s)\; ds \\
&+ 2\int_{t-\psi}^{t}\zeta^t(t)\mathbb{P}^t(i)\Upsilon_{\Delta b}(i)\zeta(s)\; ds
\end{aligned}
$$

$$
\sum_{m=1}^{s}\alpha_{im}V_p(t,\zeta,i,m) = \zeta^t(t)\sum_{m=1}^{s}\alpha_{im}\bar{P}\zeta(t) \tag{4.142}
$$

By **Fact 1** from the Appendix and (4.138), we have

$$
\begin{aligned}
& 2\zeta^t(t)\mathbb{P}^t(i)\Lambda_\Delta(K,i)\zeta(t) \\
=\;& \zeta^t(t)\mathbb{P}^t(i)\Big(\Lambda_o(K,i)+\hat{E}_K(i)\Delta(t,i)\hat{M}_K(i)\Big)\zeta(t) \\
+\;& \zeta^t(t)\Big(\Lambda_o^t(K,i)+\hat{M}_K^t(i)\Delta^t(t,i)\hat{E}_K^t(i)\Big)\mathbb{P}(i)\zeta(t) \\
\le\;& \zeta^t(t)\Big(\mathbb{P}^t(i)\Lambda_o(K,i)+\Lambda_o^t(K,i)\mathbb{P}(i) \\
+\;& \varepsilon^{-1}(i)\mathbb{P}^t(i)\hat{E}_K(i)\hat{E}_K^t(i)\mathbb{P}(i)+\varepsilon(i)\hat{M}_K^t(i)\hat{M}_K(i)\Big)\zeta(t)
\end{aligned} \tag{4.143}
$$

$$
\begin{aligned}
& 2\int_{t-\tau}^{t}\zeta^t(t)\mathbb{P}^t(i)\Upsilon_{\Delta a}(i)\zeta(s)\; ds \\
=\;& 2\int_{t-\tau}^{t}\zeta^t(t)\mathbb{P}^t(i)E_2A_{\Delta d}(t,i)x(s)\; ds \\
\le\;& \tau\zeta^t(t)\mathbb{P}^t(i)E_2A_{\Delta d}(t,i)Q^{-1}(i)A_{\Delta d}^t(t,i)E_2^t\mathbb{P}^t(i)\zeta(t) \\
+\;& \int_{t-\tau}^{t}x^t(s)Q(i)x(s)\; ds \\
=\;& \tau\zeta^t(t)\mathbb{P}^t(i)E_2A_{\Delta d}(t,i)Q^{-1}(i)A_{\Delta d}^t(t,i)E_2^t\mathbb{P}(i)\zeta(t) \\
+\;& \int_{t-\tau}^{t}\zeta^t(s)E_2Q(i)E_2^t\zeta(s)\; ds
\end{aligned}
$$

$$
\begin{aligned}
&\leq \tau\zeta^t(t)\bigg(\mathbb{P}^t(i)E_2A_d(i)[Q(i)-\varrho(i)N^t(i)N(i)]^{-1}A_d^tE_2^t\mathbb{P}(i) \\
&+ \varrho^{-1}(i)\mathbb{P}^t(i)E_2E_d(i)E_d^t(i)E_2^t\mathbb{P}(i)\bigg)\zeta(t) \\
&+ \int_{t-\tau}^t \zeta^t(s)E_2Q(i)E_2^t\zeta(s)\,ds \qquad (4.144)
\end{aligned}
$$

$$
\begin{aligned}
&\int_{t-\psi}^t \zeta^t(t)\mathbb{P}^t(i)\Upsilon_{\Delta b}(i)\zeta(s)\,ds \\
&= 2\int_{t-\psi}^t \zeta^t(t)\mathbb{P}^t(i)E_2B_{\Delta d}(t,i)K(i)x(r)\,dr \\
&\leq \psi\zeta^t(t)\mathbb{P}^t(i)E_2B_{\Delta d}(t,i)R^{-1}(i)A_{\Delta d}^t(t,i)E_2^t\mathbb{P}^t(i)\zeta(t) \\
&+ \int_{t-\psi}^t x^t(r)K^t(i)R(i)K(i)x(r)\,dr \\
&= \psi\zeta^t(t)\mathbb{P}^t(i)E_2B_{\Delta d}(t,i)R^{-1}(i)B_{\Delta d}^t(t,i)E_2^t\mathbb{P}(i)\zeta(t) \\
&+ \int_{t-\psi}^t \zeta^t(r)E_2K^t(i)R(i)K(i)E_2^t\zeta(r)\,dr \\
&\leq \psi\zeta^t(t)\bigg(\mathbb{P}^t(i)E_2B_d(i)[R(i)-\mu(i)M^t(i)M(i)]^{-1}B_d^tE_2^t\mathbb{P}(i) \\
&+ \mu^{-1}(i)\mathbb{P}^t(i)E_2E_d(i)E_d^t(i)E_2^t\mathbb{P}(i)\bigg)\zeta(t) \\
&+ \int_{t-\psi}^t \zeta^t(r)E_2K^t(i)R(i)K(i)E_2^t\zeta(r)\,dr \qquad (4.145)
\end{aligned}
$$

for some $\varepsilon(i) > 0$, $\varrho(i) > 0$, $\mu(i) > 0$, $i \in \mathcal{S}$. In a similar way, it can be established that

$$
\begin{aligned}
\partial V_q/\partial\zeta\;\dot{\zeta}(t) &= \tau\zeta^t(t)E_2Q_x(i)E_2^t\zeta(t) - \int_{-\tau}^0 \zeta^t(t+\beta)E_2Q_x(i)E_2^t\zeta(t+\beta)d\beta \\
&+ \sum_{m=1}^{s}\alpha_{im}\int_{-\tau}^0\int_{t+\beta}^t \zeta^t(s)E_2Q_x(m)E_2^t\zeta(s)dsd\beta \\
&\leq \tau\zeta^t(t)E_2Q_x(i)E_2^t\zeta(t) - \int_{-\tau}^0 \zeta^t(t+\beta)E_2Q_x(i)E_2^t\zeta(t+\beta)d\beta \\
&+ \hat{\alpha}\tau\int_{t-\tau}^t \zeta^t(s)E_2Q_x(i)E_2^t\zeta(s)ds \qquad (4.146)
\end{aligned}
$$

$$
\begin{aligned}
\partial V_r/\partial\zeta\ \dot{\zeta}(t) &= \psi\zeta^t(t)E_2K^t(i)R_x(i)K(i)E_2^t\zeta(t) \\
&- \int_{-\psi}^{0}\zeta^t(t+\beta)E_2K^t(i)R_x(i)K(i)E_2^t\zeta(t+\beta)d\beta \\
&+ \sum_{m=1}^{s}\alpha_{im}\int_{-\psi}^{0}\int_{t+\beta}^{t}\zeta^t(r)E_2K^t(i)R_x(m)K(i)E_2^t\zeta(r)drd\beta \\
&\le \psi\zeta^t(t)E_2K^t(i)R_x(i)K(i)E_2^t\zeta(t) \\
&- \int_{-\psi}^{0}\zeta^t(t+\beta)E_2K^t(i)R_x(i)K(i)E_2^t\zeta(t+\beta)d\beta \\
&+ \hat{\alpha}\psi\int_{t-\tau}^{t}\zeta^t(s)E_2K^t(i)R_x(i)K(i)E_2^tds \qquad (4.147)
\end{aligned}
$$

By selecting $Q(i) = Q_x(i)(1-\hat{\alpha}\tau),\ i \in \mathcal{S}$ and $R(i) = R_x(i)(1-\hat{\alpha}\psi),\ i \in \mathcal{S}$, we employ standard matrix manipulations of (4.140)-(4.147) using (4.136) and arranging terms to yield:

$$
\begin{aligned}
\Im^{\zeta}[V] &= \zeta^t(t)\Big\{\mathbb{P}^t(i)\Lambda_o(K,i) + \Lambda_o^t(K,i)\mathbb{P}(i) + \sum_{m=1}^{s}\alpha_{im}\bar{P}(m) \\
&+ \varepsilon^{-1}(i)\mathbb{P}^t(i)\hat{E}_K(i)\hat{E}_K^t(i)\mathbb{P}(i) + \varepsilon(i)\hat{M}_K^t(i)\hat{M}_K(i) \\
&+ \tau\mathbb{P}^t(i)E_2A_d(i)[Q(i) - \varrho(i)N^t(i)N(i)]^{-1}A_d^tE_2^t\mathbb{P}(i) \\
&+ \tau\varrho^{-1}(i)\mathbb{P}^t(i)E_2E_d(i)E_d^t(i)E_2^t\mathbb{P}(i) \\
&+ \psi\mathbb{P}^t(i)E_2B_d(i)[R(i) - \mu(i)M^t(i)M(i)]^{-1}B_d^tE_2^t\mathbb{P}(i) \\
&+ \psi\mu^{-1}(i)\mathbb{P}^t(i)E_2E_d(i)E_d^t(i)E_2^t\mathbb{P}(i) + \tau(1-\hat{\alpha}\tau)^{-1}E_2Q(i)E_2^t \\
&+ \psi(1-\hat{\alpha}\psi)^{-1}E_2K^t(i)R(i)K(i)E_2^t\Big\}\zeta(t) + 2\zeta^t\mathbb{P}^t(i)\bar{\Gamma}(i)w(t) \\
&= \zeta^t(t)\Xi(K,i)\zeta(t) + 2\zeta^t\mathbb{P}^t(i)\bar{\Gamma}(i)w(t) \qquad (4.148)
\end{aligned}
$$

for some scalars $\epsilon(i) > 0$, $\sigma(i) > 0$, $i \in \mathcal{S}$.

By taking $w(t) \equiv 0, \bar{\Gamma}(i) \equiv 0, \Phi(i) \equiv 0$, the robust stability of system (4.130) readily follows from (4.139) and **Fact 3** for all $i \in \mathcal{S}$. Thus we conclude that $\Im^{\zeta}[V] < 0$ for all $\zeta \neq 0$ and $\Im^{\zeta}[V] \le 0$ for all ζ.

Since $||\zeta(t+\beta)|| \leq \varphi||\zeta(t)||$, $\forall \beta \in [-\max\{\tau, \psi\}, 0]$ and some $\varphi > 0$ [85], it follows from (4.140) that $V(\zeta, i) \leq \zeta^t(t)\bar{P}(i)\zeta(t) + \sigma||\zeta||^2$ where $\sigma = \big(\tau \max_i \lambda_M[Q_x(i)] + \psi \max_i \lambda_M[K^t(i)R(i)K(i)\big)$. Therefore, for all $\zeta \neq 0$, we have

$$\begin{aligned} \frac{\Im^{\zeta}[V]}{V(\zeta, i)} &\leq \frac{\zeta^t \Xi(i) \zeta^t}{\zeta^t \bar{P}(i)\zeta + \sigma||\zeta||^2} \\ &\leq -\xi \triangleq -\min_{i \in \mathcal{S}} \left\{ \frac{\lambda_m[-\Xi(i)]}{\lambda_M[\bar{P}] + \sigma} \right\} \end{aligned} \tag{4.149}$$

It is readily seen from (4.149) that $\xi > 0$ and hence we get

$$\Im^{\zeta}[V] \leq -\xi\, V(t, \zeta, i)$$

It follows from [78] by using the Gronwall-Bellman lemma [98] and letting

$$\zeta(t = 0, \phi, \eta_o) = \zeta_o$$

one has

$$\mathbb{E}[V(\zeta, i)|\phi, \eta_o] \leq e^{-\xi\, t}\, V(\zeta_o, i) \tag{4.150}$$

Since

$$\mathbb{E}\Big\{ \int_{-\tau}^{0} \zeta^t(t+\theta) E_2 Q(i) E_2^t \zeta(t+\theta) d\theta | \phi, \eta_o \Big\} \geq 0$$

and

$$\mathbb{E}\Big\{ \int_{-\psi}^{0} \zeta^t(t+\theta) E_2 K^t(i) R(i) K(i) E_2^t \zeta(t+\theta) d\theta | \phi, \eta_o \Big\} \geq 0$$

it is easy to see from (4.140) that

$$\begin{aligned} \mathbb{E}\Big\{ \zeta^t(t)\bar{P}(i)\zeta(t) | \phi, \eta_o \Big\} &\leq e^{-\xi\, t} V(\zeta_o, i) \\ &\Longrightarrow \mathbb{E}\Big\{ \int_0^{\mathcal{T}} \zeta^t(t)\bar{P}(i)\zeta(t) dt | \phi, \eta_o = i \Big\} \\ &\leq \Big[\int_0^{\mathcal{T}} e^{-\xi\, t} dt \Big] V(\zeta_o, i) = \frac{1}{\xi}[e^{-\xi\, \mathcal{T}} - 1]\, V(\zeta_o, i) \end{aligned}$$

$$
\begin{aligned}
\Longrightarrow & \quad \lim_{\mathcal{T}\to\infty} \mathbb{E}\Big\{ \int_0^{\mathcal{T}} \zeta^t(t)\bar{P}(i)\zeta^t(t)dt|\phi, \eta_o = i \Big\} \\
\leq & \quad \frac{1}{xi}\zeta_o^t \bar{P}(\eta_o)\zeta_o \\
+ & \quad \frac{\tau[Q(\eta_o)] + \psi K^t(\eta_o)[R(\eta_o)]K(\eta_o)}{\xi} ||\zeta(t+\theta)||_*^2 \\
& \quad \forall \theta \in [-\max\{\tau, \psi\}, 0] \qquad (4.151)
\end{aligned}
$$

where

$$
||x(t+\theta)||_*^2 \stackrel{\Delta}{=} \sup_{\theta\in[-\max\{\tau,\psi\},0]} ||x(t+\theta)||_2^2
$$

Letting

$$
\hat{P}(i) = \max_{i\in\mathcal{S}} \left\{ \frac{\bar{P}(\eta_o)||\zeta_o||^2 + \tau[Q(\eta_o)] + \psi K^t(\eta_o)[R(\eta_o)]K(\eta_o)||x(t+\theta)||_*^2}{\zeta[\bar{P}(\eta_o)]||\zeta_o||^2} \right\}
$$

it then follows for $i \in \mathcal{S}$ that

$$
\begin{aligned}
& \lim_{\mathcal{T}\to\infty} \mathbb{E}\Big\{ \int_0^{\mathcal{T}} \zeta^t(t)\zeta(t)dt|\phi, \eta_o = i \Big\} \\
& \leq \zeta_o^t \lambda_M(\hat{P}(i))\zeta_o < +\infty
\end{aligned}
$$

which, in the light of **Definition** 4.6, shows that system (Σ_T) is **RSSSDD**.

Now we proceed to show the robust performance. With some algebraic manipulations using (4.131) and (4.148), we obtain:

$$
\begin{aligned}
\mathcal{J}(\zeta) &= \mathbb{E}\Big\{ \int_0^{\infty} [z^t(t)z(t) - \gamma^2 w^t(t)w(t) \\
&+ \Im^{\zeta}[V] - \Im^{\zeta}[V]]dt \Big\} \\
&\leq \mathbb{E}\Big\{ \int_0^{\infty} [\zeta^t(t)\bar{G}^t(i)\bar{G}(i)\zeta(t) \\
&+ \zeta^t(t)\bar{G}^t(i)\Phi(i)w(t) + w^t(t)\Phi^t(i)\bar{G}(i)\zeta(t) \\
&- w^t(t)[\gamma^2 I - \Phi^t(i)\Phi(i)]w(t) \\
&+ \zeta^t(t)\Xi(K,i)\zeta(t) + 2\zeta^t \mathbb{P}^t(i)\bar{\Gamma}(i)w(t)]dt \Big\} \\
&= \mathbb{E}\Big\{ \int_0^{\infty} \chi^t(i)\Xi_w(i)\chi(i)dt \Big\} \qquad (4.152)
\end{aligned}
$$

It follows from inequality (4.139) that $\mathcal{J}(\zeta) < 0$ and hence system (4.130)-(4.131) is **RSSSDD with disturbance attenuation** $\gamma > 0$. $\nabla\nabla\nabla$

Now consider the following block matrices

$$\mathcal{X}(i) \triangleq \begin{bmatrix} \mathcal{X}_\sigma(i) & 0 \\ \mathcal{X}_d(i) & \mathcal{X}_x(i) \end{bmatrix} = \begin{bmatrix} \mathbb{P}_\sigma^{-1}(i) & 0 \\ -\mathcal{X}_x(i)\mathbb{P}_d\mathcal{X}_x(i) & \mathbb{P}_x^{-1}(i) \end{bmatrix} \quad (4.153)$$

we now establish the following analytical result

Theorem 4.15 *Given a scalar* $\gamma > 0$ *and weighting matrices* $Q_x(i) = Q_x^t(i) > 0$, $R_x(i) = R_x^t(i) > 0$ *such that* $Q(i) = Q_x(i)(1 - \hat{\alpha}\tau) > 0$, $R(i) = R_x(i)(1 - \hat{\alpha}\psi) > 0$, *system* Σ_T *is* **RSSSDD with disturbance attenuation** γ *by the controller (4.129) if there exist scalars* $\epsilon(i) > 0$, $\varrho(i) > 0$, $\mu(i) > 0$, $i \in \mathcal{S}$, *and matrices* $\mathcal{X}_\sigma(i) > 0$, $\mathcal{X}_d(i) > 0$, $\mathcal{X}_x(i) > 0$, $Y_d(i)$, $Y_x(i)$, $i \in \mathcal{S}$, *such that the following LMIs have a feasible solution for all* $i \in \mathcal{S}$

$$\begin{bmatrix} \Pi_d(i) & \Upsilon_{xy}(i) & \Gamma(i) \\ \Upsilon_{xy}^t(i) & -\Upsilon_{xyd}(i) & 0 \\ \Gamma^t(i) & 0 & -\gamma^2 I + \Phi^t(i)\Phi(i) \end{bmatrix} < 0 \quad (4.154)$$

$$\begin{bmatrix} \mathcal{X}_x(i) + \mathcal{X}_x^t(i) & \Upsilon_{eb}(i) & \mathcal{X}_x^t(i)G^t(i)\Phi(i) \\ \Upsilon_{eb}^t & -\Upsilon_{ebd}(i) & 0 \\ \Phi^t(i)G(i)\mathcal{X}_x(i) & 0 & -\gamma^2 I + \Phi^t(i)\Phi(i) \end{bmatrix} < 0 \quad (4.155)$$

$$\begin{aligned} \mathcal{X}_\sigma(i) &= \mathcal{X}_d(i) + \mathcal{X}_x^t(i)A_o^t(i) + Y_x^t(i)B_o^t(i) + \\ &\Gamma(i)[\gamma^2 I - \Phi^t(i)\Phi(i)]\Phi^t(i)G(i)\mathcal{X}_x(i) + \\ &\varepsilon^{-1}(i)E_b(i)E_a^t(i) + (1 - \hat{\alpha}\tau)\mathcal{X}_d^t(i)Q(i)\mathcal{X}_x(i) + \\ &(1 - \hat{\alpha}\psi)Y_d^t(i)R(i)Y_x(i) \end{aligned} \quad (4.156)$$

where

$$\Upsilon_{xy} = [E_a(i) \quad \tau\mathcal{X}_d(i)Q(i) \quad \psi Y_d^t(i)R(i)]$$

$$
\begin{aligned}
\Upsilon_{xyd} &= diag[\varepsilon(i)I \quad \tau(1-\hat{\alpha}\tau)Q(i) \quad \psi(1-\hat{\alpha}\psi)R(i)] \\
\Upsilon_{eb}(i) &= [E_b(i) \quad \bar{E}_d(i) \quad \tau A_{dq} \quad \psi B_{dr}] \\
\Upsilon_{ebd}(i) &= diag[\varepsilon(i)I \quad \varrho_{\mu}(i)I \quad \tau\hat{Q}(i) \quad \psi\hat{R}^t(i)] \\
\Pi_d(i) &= A_o(i)\mathcal{X}_d(i) + B_o(i)Y_d(i) + \mathcal{X}_x^t(i)A_o^t(i) + Y_d^t(i)B_o^t(i) \\
\bar{E}_d(i) &= [\tau E_d(i) \quad \psi E_d(i)] \ , \ A_{dq}(i) = [A_d(i) \quad \mathcal{X}_x^t(i)Q(i)] \\
\psi B_{dr}(i) &= [B_d(i) \quad Y_x^t(i)R(i)] \ , \ \varrho_{\mu}(i) = diag[\tau\varrho(i) \quad \psi\mu(i)] \\
\hat{Q}(i) &= diag[Q(i) - \varrho(i)N^t(i)N(i) \quad (1-\hat{\alpha}\tau)Q(i)] \\
\hat{R} &= diag[R(i) - \mu(i)M^t(i)M(i) \quad (1-\hat{\alpha}\psi)R(i)] \qquad (4.157)
\end{aligned}
$$

Moreover the state-feedback matrix gain is $K(i) = Y_x(i)\mathcal{X}_x^{-1}(i),\ i \in \mathcal{S}$

Proof: Premultiplying (4.139) by $\mathbb{P}^{-t}$, post multiplying the result by $\mathbb{P}^{-1}$, using (4.138) and (4.153) along with $K(i) = Y_x(i)\mathcal{X}_x^{-1}(i),\ i \in \mathcal{S}$ and arranging terms, we obtain the LMIs (4.154)- (4.155) corresponding to the blocks $\Xi_w(1,1)(i)$ and $\Xi_w(2,2)(i)$ respectively. Equation (4.156) ensures that the block $\Xi_w(1,2)(i) \equiv 0$ which completes the proof. $\nabla\nabla\nabla$

4.9.3 Example 4.5

We consider a two-mode system of the type (4.125)-(4.126) with Markov process generator

$$
\Im = \begin{bmatrix} -3 & 3 \\ 2 & -2 \end{bmatrix}
$$

and time-delays $\tau = 0.5,\ \psi = 0.4$. The system dynamics are given by:

Mode 1:

$$
\begin{aligned}
A_o(1) &= \begin{bmatrix} -2 & 0 \\ 0 & -0.9 \end{bmatrix}, \ A_d(1) = \begin{bmatrix} -1 & 0 \\ -1 & 1 \end{bmatrix}, \ B_o(1) = \begin{bmatrix} 1 & 0 \\ 0 & 0.5 \end{bmatrix} \\
E_o(1) &= \begin{bmatrix} 0.1 & 0 \\ 0 & 0.1 \end{bmatrix}, \ N(1) = \begin{bmatrix} 1 & 0 \\ 0 & 1 \end{bmatrix}, \ E_d(1) = \begin{bmatrix} 0.1 & 0 \\ 0 & 0.1 \end{bmatrix}
\end{aligned}
$$

$$
\begin{aligned}
E_b(1) &= \begin{bmatrix} 0.1 & 0 \\ 0 & 0.1 \end{bmatrix}, \; B_d(1) = \begin{bmatrix} 0.1 & 0 \\ 0 & 0.1 \end{bmatrix}, \; \Gamma(1) = \begin{bmatrix} 0.5 & 0 \\ 0 & 0.5 \end{bmatrix} \\
G_o(1) &= [1 \quad 0.4], \; \Phi(1) = [0.9 \quad 1]
\end{aligned}
$$

Mode 2:

$$
\begin{aligned}
A_o(2) &= \begin{bmatrix} -2 & 0 \\ 0 & -0.95 \end{bmatrix}, \; A_d(2) = \begin{bmatrix} -1 & 0 \\ -1 & -1 \end{bmatrix}, \; B_o(2) = \begin{bmatrix} 0.8 & 0 \\ 0 & 1 \end{bmatrix} \\
E_o(2) &= \begin{bmatrix} 0.1 & 0 \\ 0 & 0.1 \end{bmatrix}, \; N(2) = \begin{bmatrix} 1 & 0 \\ 0 & 1 \end{bmatrix}, \; E_d(2) = \begin{bmatrix} 0.1 & 0 \\ 0 & 0.1 \end{bmatrix} \\
E_b(2) &= \begin{bmatrix} 0.1 & 0 \\ 0 & 0.1 \end{bmatrix}, \; B_d(2) = \begin{bmatrix} 0.1 & 0 \\ 0 & 0.1 \end{bmatrix}, \; \Gamma(2) = \begin{bmatrix} 0.4 & 0 \\ 0 & 0.4 \end{bmatrix} \\
G_o(2) &= [0.5 \quad 1], \; \Phi(2) = [0.4 \quad 1]
\end{aligned}
$$

Choosing the disturbance attenuation level $\gamma = 1.5$ and invoking the **LMI**-toolbox [57], the feasible solution of (4.154)- (4.156) yields the matrix gains as

$$
K(1) = \begin{bmatrix} 1.002 & 0.391 \\ -2.785 & -0.531 \end{bmatrix}, \; K(2) = \begin{bmatrix} -0.241 & 0.312 \\ 0.501 & 1.104 \end{bmatrix}
$$

4.10 Notes and References

We have addressed the basic problems of control design using different approaches for a class of linear ime-delay systems with Markovian jump parameters and norm-bounded uncertainties. Throughout the chapter, systematic development of concepts have been adhered. The main emphasize, beside the rigorous analysis, has been on delay-dependent patterns. Basically, we have designed the following:

1) State-feedback guaranteed cost controller 2) $\mathcal{H}_\infty$ state-feedback controller 3) $\mathcal{H}_\infty$ dynamic output-feedback controller 4) $\mathcal{H}_\infty$ state-feedback controller using a new model transformation

We have established that the robust control design problem for the JTD system under consideration with uncertain parameters can be essentially solved in terms of the solutions of a finite set of coupled ARIs or LMIs. In all cases the closed-loop stochastic stability is guaranteed. Numerical examples are provided.

The methods developed thus far are amenable to various extensions including multi-state delays, multi-input delays and/or output-feedback guaranteed cost controllers.

Chapter 5

Simultaneous $\mathcal{H}_2/\mathcal{H}_\infty$ Control

5.1 Introduction

From the previous chapters, it should become increasingly apparent that Jumping Time-Delay systems (JTDS) are important classes of dynamical systems whose structures vary in response to random changes and possessing inherent state and/or input delays. By focusing on the structural variations that could result from abrupt phenomena such as parameter shifting, component and interconnection failures, it has been generally recognized that a tractable model would be in the form of hybrid continuous and discrete states [42, 47]. For a recent account of the subject, see [24]. Without jumping parameters, time-delay dynamical systems have been the subject of extensive research during the part two-decades; for good coverage on the available results the reader is referred to [87, 89, 95, 98] and recent results can be found in [53].

In Chapter 4, control analysis and design for classes of JTDS with emphasis on delay-dependence patterns and robustness [36] are presented. Regarding robust control system design, the simultaneous (mixed) $\mathcal{H}_2/\mathcal{H}_\infty$ control is of particular interest in this chapter for which some of the known results are found in [14, 69, 72, 145, 146, 163]. Control design results of JTDs based on separate $\mathcal{H}_2$ and $\mathcal{H}_\infty$ specifications can be found [96, 105, 165] and their references.

One popular methodology to simultaneous $\mathcal{H}_2/\mathcal{H}_\infty$ control design consists of determining feedback controller that certain performance measure ($\mathcal{H}_2$ specifications) subject to robustness constraints ($\mathcal{H}_\infty$ specifications). In this light, the material covered in this chapter presents new control design approaches pertaining to the class of linear uncertain systems with Markovian jump parameters. We provide three different methods and give a self-contained detailed analysis for each. The approaches are:

1) Direct Approach

2) State Transformation Approach

3) Descriptor Approach

It should be noted that in presenting the three approaches we opt to preserve the identity of each approach and make the analytical development complete and self-contained. Indeed, this will on the expense of repeating similar steps.

5.2 Problem Statement

We consider a class of stochastic systems with Markovian jump parameters and functional state-delay described over the space $(\Omega, \mathcal{F}, \mathbf{P})$ by:

$$
\begin{aligned}
(\Sigma_{Jo}): \quad \dot{x}(t) &= A_o(\eta_t)x(t) + A_d(\eta_t)x(t-\tau_{\eta_t}) \\
&+ B_o(\eta_t)u(t) + \Gamma(\eta_t)w(t) \\
x(s) &= \phi(s), \; s \in [-\tau, 0], \; \eta_o = i \qquad (5.1) \\
y(t) &= C_o(\eta_t)x(t) + D_o(\eta_t)u(t) \\
&+ \Psi(\eta_t)w(t), \qquad (5.2)
\end{aligned}
$$

$$\begin{aligned} z(t) &= G_o(\eta_t)x(t) + F_o(\eta_t)u(t) \\ &+ \Phi(\eta_t)w(t), \end{aligned} \tag{5.3}$$

where $x(t) \in \Re^n$ is the state vector; $u(t) \in \Re^m$ is the control input; $w(t) \in \Re^q$ is the disturbance input which belongs to $\mathcal{L}_2[0,\infty]$; $z(t) \in \Re^r$ is the controlled output which belongs to $\mathcal{L}^2\left[(\Omega,\mathcal{F},\mathbf{P}),[0,\infty]\right]$ and τ_{η_t} denotes the time-delay in the jumping system when the mode is η_t with

$$\hat{\tau} \triangleq \max_j\{\tau_j, j \in \mathcal{S}\} \ , \ \check{\tau} \triangleq \min_j\{\tau_j, j \in \mathcal{S}\}$$

Note that system (5.1)-(5.3) have the following features:

1) The jumping parameters are modeled as a continuous-time, discrete-state Markov process and the uncertainties are norm-bounded.

2) An integral part of the system dynamics is a delayed state in which the time-delays are mode-dependent.

3) In general, the functional relationship could be expressed analytically or presented in tabular form.

For simplicity in exposition, the matrices associated with the i-th mode will be denoted in the sequel by

$$\begin{aligned} A_o(\eta_t) &\triangleq A_o(i) \ , \ B_o(\eta_t) \triangleq B_o(i) \ , \ \Gamma(\eta_t) \triangleq \Gamma(i) \\ G_o(\eta_t) &\triangleq G_o(i) \ , \ C_o(\eta_t) \triangleq C_o(i) \ , \ F_o(\eta_t) \triangleq F_o(i) \\ A_d(\eta_t) &\triangleq A_d(i) \ , \ \Phi(\eta_t) \triangleq \Phi(i) \ , \ D_o(\eta_t) \triangleq D_o(i) \\ \Psi(\eta_t) &\triangleq \Psi(i) \end{aligned} \tag{5.4}$$

where $A_o(i), B_o(i), A_d(i), G_o(i), C_o(i), D_o(i), F_o(i), \Gamma(i), \Psi(i)$ and $\Phi(i)$ are known real constant matrices of appropriate dimensions which describe the nominal model (5.1)-(5.3). The initial condition is specified as $\langle x(0), x(s)\rangle = \langle x_o, \phi(s)\rangle$, where $\phi(.) \in \mathcal{L}_2[-\tau, 0]$.

5.3 Direct Approach

The purpose of this section is to extend the results of [22, 27, 72, 96, 105, 114, 165] further by studying a general class of JTDS with mode-dependent state-delays. Here as usual, the jumping parameters are treated as continuous-time, discrete-state Markov process. The parametric uncertainties are assumed to be real, time-varying and norm-bounded and the time-delay factor depends on the operating mode. Unlike [22, 27, 72], we follow [114] and employ a Lyapunov functional candidate which exhibits the interplay between the time-delay and jumping parameters and consider functional (mode-dependent) delays. In particular the objective is to develop methods for determining a stabilizing controller with mode-dependent delay (functional time-delay) which minimizes the upper bound of an $\mathcal{H}_2$ performance measure while guaranteeing that a prescribed upper bound on an $\mathcal{H}_\infty$ performance is attained for all possible $w(t) \in \mathcal{L}_2[0, \infty]$. It will be shown at the outest that this problem is to the existence of a positive definite solution of a family of linear matrix inequalities (LMIs).

Throughout this chapter, for a prescribed $\gamma > 0$, we introduce the performance measures:

$\mathcal{H}_2$ performance measure

$$\mathcal{J}_2 \triangleq \mathbb{E}\Big[\int_0^\infty y^t(s)y(s)\, ds\Big] \tag{5.5}$$

$\mathcal{H}_\infty$ performance measure

$$\mathcal{J}_\infty \triangleq \mathbb{E}\Big[\int_0^\infty \{z^t(s)z(s) - \gamma^2 w^t(s)w(s)\}\, ds\Big] \tag{5.6}$$

The problem of simultaneous $\mathcal{H}_2/\mathcal{H}_\infty$ state-feedback control could be phrazed as follows:

Given system (5.1)-(5.3) determine a linear control law

$$u(t) = K(i)\,x(t) \quad i \in \mathcal{S} \tag{5.7}$$

which achieves the minimal value of $\mathcal{H}_2$ performance measure while guaranteeing that $\mathcal{H}_\infty$ performance measure is bounded by γ for all $w(t) \in \mathcal{L}_2[0,\infty]$.

Remark 5.1 *As demonstrated in the previous chapters, the concept of functional time-delay arises from the fact that in jumping systems there are real situations reflecting the dependence between time-delay and mode of operation [114]. It should be observed that the objective of the mixed $\mathcal{H}_2/\mathcal{H}_\infty$ feedback control under consideration is to minimize the energy of the output $y(t)$ while simultaneously satisfying the prescribed $\mathcal{H}_\infty$-norm bound of the controlled system $w(t) \to z(t)$.*

For simplicity in exposition, we divide our effort in this section into two parts: The first part deals with the nominal model and the second part treats the uncertain model.

5.3.1 Mode-Dependent Nominal Model

The theorems established in the sequel show that designing a simultaneous $\mathcal{H}_2/\mathcal{H}_\infty$ controller for system (Σ_{Jo}) is essentially related to the existence of a positive definite solution of a family of linear matrix inequalities (LMIs). We direct attention to the case in which the time-delay varies with the mode of operation and this is referred to as functional time-delay. Throughout this work, we consider the case where the bounds of the functional dependence (maximum and minimum) are the only available information.

Considering system (5.1-5.3) for $\eta_t = i \in \mathcal{S}$ and under the feedback law (5.7), it takes the form:

$$(\Sigma_{JKo}) \;\; \dot{x}(t) = A_K(i)x(t) + A_d(i)x(t-\tau_i)$$

$$
\begin{aligned}
&+ \quad \Gamma(i)w(t)\ ,\ x(s) = \phi(s),\ s \in [\max -\tau, 0] && (5.8) \\
y(t) &= C_K(i)x(t) + \Phi(i)w(t) && (5.9) \\
z(t) &= G_K(i)x(t) + \Phi(i)w(t) && (5.10) \\
A_K(i) &= A_o(i) + B_o(i)K(i) \\
C_K(i) &= C_o(i) + D_o(i)K(i) \\
G_K(i) &= G_o(i) + F_o(i)K(i) && (5.11)
\end{aligned}
$$

5.3.2 $\mathcal{H}_2$ Performance

Definition 5.1 *System* (Σ_{Jo}) *is said to be* **stochastically stable with functional time-delays (SSFTD)** *if given matrices* $W^t(i) = W(i) > 0$ *and mode-dependent delays* $(\tau_{\eta_i} = \tau_i,\ i \in \mathcal{S})$, *there exist matrices* $P(i) = P^t(i) > 0,\ i \in \mathcal{S}$ *such that the LMIs for all* $i \in \mathcal{S}$

$$
\begin{aligned}
\Upsilon(t,i) &\triangleq \begin{bmatrix} \Pi_t(t,i) & P(i)A_d(i) \\ A_d^t(i)P(i) & -\hat{W}(i) \end{bmatrix} < 0 && (5.12) \\
\Pi_t(t,i) &= P(i)A_o(i) + A_o^t(i)P(i) \\
&+ \sum_{k=1}^{s} \alpha_{ik}P(k) + \beta\, W(i) \\
\beta &= \hat{\tau} + \hat{\alpha}\big(\hat{\tau} - \check{\tau}\big)\big(\hat{\tau} + \check{\tau}\big) \\
\hat{W}(i) &= \hat{\tau}\, W(i) && (5.13)
\end{aligned}
$$

The following theorem is established

Theorem 5.1 *In the absence of input disturbance* $w(t) \equiv 0$, *controller (5.7) is an* $\mathcal{H}_2$*-optimal controller for system* (Σ_{Jo}) *minimizing the* $\mathcal{H}_2$*-performance measure (5.5) if given matrices* $W^t(i) = W(i) > 0$ *and mode-dependent delays* $(\tau_{\eta_i} = \tau_i,\ i \in \mathcal{S})$, *there exist matrices* $P(i) = P^t(i) > 0,\ i \in \mathcal{S}$ *satisfying the LMIs for all* $i \in \mathcal{S}$

$$
\Pi_t(i) = \begin{bmatrix} \Xi_{Ko}(i) & P(i)A_d(i) \\ A_d^t(i)P(i) & -\hat{W}(i) \end{bmatrix} < 0 \qquad (5.14)
$$

where

$$\Xi_{Ko}(i) = A_K^t(i)P(i) + P(i)A_K(i) + \sum_{m=1}^{s} \alpha_{im}P(m) + \beta\, W(i) + C_K^t(i)C_K(i) \tag{5.15}$$

An upper bound on the $\mathcal{H}_2$ performance measure is given by

$$\mathcal{J}_2 \leq \mathsf{J}^+ \triangleq \Big[x^t(0)P(\eta_o)x(0) + \tau_o \int_{-\tau_o}^{0} x^t(s)W(\eta_o)x(s)ds \Big]^{1/2} \tag{5.16}$$

Proof: Let $\mathbf{x}_s(t) \triangleq x(s+t),\ t-\tau_{\eta_t} \leq s \leq t$ and define the process $\{(\mathbf{x}(t),\eta_t),\ t \geq 0\}$ over the state space $\bar{\mathcal{C}}$. It should be observed that $\{(\mathbf{x}(t),\eta_t),\ t \geq 0\}$ is strong Markovian [78]. Now introduce the following Lyapunov-Krasovskii functional:

$$\begin{aligned}
V_t(\mathbf{x}(t),\eta_t) &= V_n(\mathbf{x}(t),\eta_t) + V_d(\mathbf{x}(t),\eta_t) \\
&+ V_f(\mathbf{x}(t),\eta_t) + V_h(\mathbf{x}(t),\eta_t) \\
V_d(\mathbf{x}(t),\eta_t) &= \int_{-\tau_{\eta_t}}^{0}\int_{t+\beta}^{t} x^t(r)W(\eta_t)x(r)\ dr\ d\beta \\
V_f(\mathbf{x}(t),\eta_t) &= \hat{\alpha}\Big(\hat{\tau}-\check{\tau}\Big)\int_{-\check{\tau}}^{0}\int_{t+\beta}^{t} x^t(r)W(\eta_t)x(r)\ dr\ d\beta \\
V_h(\mathbf{x}(t),\eta_t) &= \hat{\alpha}\int_{-\hat{\tau}}^{-\check{\tau}}\int_{t+\beta}^{t} x(r-t+\hat{\tau})W(\eta_t)x(r-t+\hat{\tau})\ dr\ d\beta \\
V_n(\mathbf{x}(t),\eta_t) &= x^t(t)P(\eta_t)x(t)
\end{aligned} \tag{5.17}$$

The weak infinitesimal operator $\Im_t^x[\cdot]$ of the process $\{(\mathbf{x}(t),\eta_t),\ t \geq 0\}$ for system (Σ_{Jo}) at the point $\{t,x,\eta_t\}$ is given by [?, 78]:

$$\begin{aligned}
\Im_k^x[V_k] &= \frac{\partial V_k}{\partial t} + \dot{x}^t(t)\frac{\partial V_k}{\partial x}\,|_{\eta_t=i} + \sum_{m=1}^{s}\alpha_{im}V_k(t,x,i,m) \\
\Im_t^x[V_t] &= \sum_{k=n,d,f,h} \Im_m^x[V_m]
\end{aligned} \tag{5.18}$$

Select $\eta_t = i \in \mathcal{S}$, $\mathbf{x} \in \mathcal{C}[-\tau_{(i)}, 0]$. Upon applying (5.17)-(5.18) to system (5.9)-(5.11) it yields:

$$\begin{aligned} \Im_n^x[V_n] &= x^t(t)\Big\{ P(i)A_K(i) + A_K^t(i)P(i) + \sum_{k=1}^{s} \alpha_{\eta_t k} P(k) \Big\} x(t) \\ &+ 2x^t(t)P(i)A_d(i)x(t-\tau_{\eta_t}) \end{aligned} \tag{5.19}$$

$$\begin{aligned} \Im_d^x[V_d] &= \tau_{\eta_t}\Big[x^t(t)W(i)x(t) \; - \; x^t(t-\tau_{\eta_t})W(i)x(t-\tau_{\eta_t}) \Big] \\ &+ \sum_{m=1}^{s} \alpha_{im} \int_{-\tau_{\eta_t}}^{0} \int_{t+\beta}^{t} x^t(r)W(i)x(r) dr \; d\beta \end{aligned} \tag{5.20}$$

Standard manipulations lead to

$$\begin{aligned} & \sum_{m=1}^{s} \alpha_{im} \int_{-\tau_{\eta_t}}^{0} \int_{t+\beta}^{t} x^t(r)W(i)x(r) \; dr \; d\beta \\ = \;& \alpha_{ii} \int_{-\tau_{\eta_t}}^{0} \int_{t+\beta}^{t} x^t(r)W(i)x(r) \; dr \; d\beta \\ + \;& \sum_{m=1, m\neq i}^{s} \alpha_{im} \int_{-\tau_{\eta_t}}^{0} \int_{t+\beta}^{t} x^t(r)Wx(r) \; dr \; d\beta \\ \leq \;& -|\alpha_{ii}| \int_{-\check{\tau}}^{0} \int_{t+\beta}^{t} x^t(r)W(i)x(r) \; dr \; d\beta \\ + \;& \sum_{m=1, m\neq i}^{s} \alpha_{im} \int_{-\check{\tau}}^{0} \int_{t+\beta}^{t} x^t(r)W(i)x(r) \; dr \; d\beta \\ = \;& |\alpha_{ii}| \int_{-\check{\tau}}^{-\hat{\tau}} \int_{t+\beta}^{t} x^t(r)W(i)x(r) \; dr \; d\beta \\ \leq \;& \hat{\alpha} \int_{-\check{\tau}}^{-\hat{\tau}} \int_{t+\beta}^{t} x^t(r)W(i)x(r) \; dr \; d\beta \\ = \;& \hat{\alpha}\Big(\hat{\tau} - \check{\tau} \Big) \int_{t-\check{\tau}}^{t} x^t(r)W(i)x(r) \; dr \\ + \;& \hat{\alpha} \int_{t-\check{\tau}}^{t-\hat{\tau}} x^t(t+\beta)W(i)x(t+\beta) \; dr \end{aligned} \tag{5.21}$$

Therefore

$$
\begin{aligned}
\Im_d^x[V_d] \;\leq\; & \tau_{\eta_t}\Big[x^t(t)W(i)x(t) \;-\; x^t(t-\tau_{\eta_t})W(i)x(t-\tau_{\eta_t})\Big] \\
+ \; & \hat{\alpha}\Big(\hat{\tau}-\check{\tau}\Big)\int_{t-\check{\tau}}^{t} x^t(r)W(i)x(r)\,dr \\
+ \; & \hat{\alpha}\int_{t-\check{\tau}}^{t-\hat{\tau}} x^t(r+\hat{\tau})W(i)x(r+\hat{\tau})\,dr
\end{aligned} \tag{5.22}
$$

In a similar way, it is easy to show that:

$$
\begin{aligned}
\Im_f^x[V_f] \;=\; & \hat{\alpha}\check{\tau}\Big(\hat{\tau}-\check{\tau}\Big)x^t(t)W(i)x(t) \\
- \; & \hat{\alpha}\Big(\hat{\tau}-\check{\tau}\Big)\int_{t-\check{\tau}}^{t} x^t(\beta)W(i)x(\beta)\,d\beta \\
\Im_h^x[V_h] \;=\; & \hat{\alpha}\hat{\tau}\Big(\hat{\tau}-\check{\tau}\Big)x^t(t)W(i)x(t) \\
- \; & \hat{\alpha}\int_{t-\check{\tau}}^{t-\hat{\tau}} x^t(r+\hat{\tau})W(i)x(r+\hat{\tau})\,dr
\end{aligned} \tag{5.23}
$$

Now by combining (5.19) through (5.23) and arranging terms, we obtain

$$
\begin{aligned}
\Im_t^x[V_t] \;\leq\; & x^t(t)\Big\{P(i)A_K(i) + A_K^t(i)P(i \\
+ \; & \sum_{k=1}^{s}\alpha_{\eta_t k}P(k) + \beta\,W(i)\Big\}x(t) \\
+ \; & 2x^t(t)P(i)A_d(i)x(t-\tau_{\eta_t}) \\
- \; & x^t(t-\tau_{\eta_t})\hat{W}(i)x(t-\tau_{\eta_t})
\end{aligned} \tag{5.24}
$$

which is negative from (5.12)-(5.13). By **Fact 3** of Appendix A, it follows that

$$
\begin{aligned}
& [x^t(t) \quad x^t(t-\tau_{\eta_t})]\Pi_t(i)\begin{bmatrix} x(t) \\ x(t-\tau_{\eta_t}) \end{bmatrix} = \\
& x^t(t)\Big\{P(i)A_K(i) + A_K^t(i)P(i) + \sum_{m=1}^{s}\alpha_{im}P(m) + \\
& \beta\,W(i) + P(i)A_d(i)\hat{W}^{-1}(i)A_d^t(i)P(i)\Big\}x(t) \\
\triangleq \; & x^t(t)\Lambda(i)x(t) < 0\,, \qquad \forall\; i \in \mathcal{S}.
\end{aligned} \tag{5.25}
$$

This is equivalent to the LMIs (5.14). Hence, we conclude that $\Im_t^x[V_t] < 0$ for all $x \neq 0$ and $\Im_t^x[V_t] \leq 0$ for all x.

Since $||x(t+\upsilon)|| \leq \varphi||x(t)||$, $\forall \upsilon \in [-\tau_\eta, 0]$ and some $\varphi > 0$ [85], it follows from (5.25) that

$$V_t(x,i) \leq x^t(t)P(i)x(t) + \mu||x||^2$$

where

$$\mu = \varphi\kappa(\max_i \lambda_M[W(i)])$$

$$\kappa = \hat{\tau} + \hat{\alpha}[\hat{\tau}^2 + \hat{\tau} - \check{\tau}(1+\hat{\tau})]$$

Therefore, for all $x \neq 0$, we have

$$\begin{aligned} \frac{\Im_t^x[V_t]}{V_t(x,i)} &\leq \frac{x^t\Lambda(i)x}{x^tP(i)x + \mu||x||^2} \leq -\zeta \\ &\triangleq -\min_{i\in\mathcal{S}}\left\{\frac{\lambda_m[-\Lambda(i)]}{\lambda_M[P(i)] + \mu}\right\} \end{aligned} \tag{5.26}$$

It is readily seen from (5.26) that $\zeta > 0$ and hence we get

$$\Im_t^x[V_t] \leq -\zeta\, V_t(x,i)$$

It follows from [78] by using the Gronwall-Bellman lemma [98] and letting $x(t = 0, \phi, \eta_o) = x_o$, one has

$$\mathbb{E}[V_t(x,i)|\phi,\eta_o] \leq e^{-\zeta\, t}\, V_t(x_o,i) \tag{5.27}$$

Since

$$\mathbb{E}\left\{\int_{-\tau_\eta}^{0} x^t(t+\theta)W(i)x(t+\theta)d\theta|\phi,\eta_o\right\} \geq 0$$

it is easy to see from (5.27) that

$$\begin{aligned} &\mathbb{E}\left\{x^t(t)P(i)x(t)|\phi,\eta_o\right\} \leq e^{-\zeta\, t}V_t(x_o,i) \Longrightarrow \\ &\mathbb{E}\left\{\int_0^{\mathcal{T}} x^t(t)P(i)x(t)dt|\phi,\eta_o = i\right\} \end{aligned}$$

$$
\leq \left[\int_0^{\mathcal{T}} e^{-\zeta\, t} dt \right] V_t(x_o, i) = \frac{1}{\zeta}[e^{-\zeta\, \mathcal{T}} \; - \; 1] \; V_t(x_o, i) \Longrightarrow
$$

$$
\lim_{\mathcal{T}\to\infty} \mathbb{E}\left\{ \int_0^{\mathcal{T}} x^t(t)P(i)x(t)dt|\phi, \eta_o = i \right\}
$$

$$
\leq \; \frac{1}{\zeta} x_o^t P(\eta_o) x_o + \frac{\kappa}{\zeta}[W(\eta_o)]||x(t+\theta)||_*^2, \quad \forall\theta \in [-\tau_\eta, 0] \tag{5.28}
$$

where

$$
||x(t+\theta)||_*^2 \stackrel{\Delta}{=} \sup_{\theta\in[-\tau,0]} ||x(t+\theta)||_2^2
$$

Let

$$
\bar{P}(i) \;=\; \max_{i\in\mathcal{S}} \left\{ \frac{P(\eta_o)||x_o||^2 + \kappa[W(\eta_o)]||x(t+\theta)||_*^2}{\zeta[P(\eta_o)]||x_o||^2} \right\}
$$

it follows from (5.28) for $i \in \mathcal{S}$ that

$$
\lim_{\mathcal{T}\to\infty} \mathbb{E}\left\{ \int_0^{\mathcal{T}} x^t(t)x(t)dt|\phi, \eta_o = i \right\}
$$

$$
\leq x_o^t \lambda_M(\bar{P}(i))x_o < +\infty
$$

which, in view of Chapter 3, shows that system (Σ_{JKo}) is **SSFTD** under the control law (5.7).

Now by the Dynkin's formula and (5.14)

$$
\begin{aligned}
&\mathbb{E}[V(x(t), \eta_t)] - V(x(0), \eta_o) = \\
&\mathbb{E}\left[\int_0^t \Im_t^x[V(x(v), \eta_v)]dv \right] \Longrightarrow \\
&\mathbb{E}[V(x(t_f), \eta_{t_f})] - V(x(0), \eta_o) \\
&\leq \mathbb{E}[\int_0^{t_f} \chi^t(v)\Pi_t(\eta_v)\chi(v)dv]
\end{aligned} \tag{5.29}
$$

where

$$
\chi^t(v) = [x^t(v) \;\; x^t(t - \tau(v)]
$$

Letting $t_f \to \infty$ and in view the system stability, it follows that

$$
\mathbb{E}\left[\int_0^{\infty} y^t(r)y(r)dr \right]
$$

$$
\begin{aligned}
&\leq V(x(0), \eta_o)] = \\
&x^t(0)P(\eta_o)x(0) + \\
&\tau_o \int_{-\tau_o}^{0} x^t(s)W(\eta_o)x(s)ds \Longrightarrow \\
&||y|| \leq \Bigg[x^t(0)P(\eta_o)x(0) + \\
&\tau_o \int_{-\tau_o}^{0} x^t(s)W(\eta_o)x(s)ds\Bigg]^{1/2} = \mathsf{J}^+
\end{aligned} \tag{5.30}
$$

which completes the proof. ∇∇∇

The following theorem provides an LMI-based method for computing the feedback gains.

Theorem 5.2 *The feedback gain associated with the $\mathcal{H}_2$-optimal controller for system (Σ_{Jo}) is given by $K(i) = Z(i)Y^{-1}(i)$, $i \in \mathcal{S}$ where the matrices $Y(i) = Y^t(i) > 0$, $Z(i)$, $L(i) = L^t(i) > 0$, $i \in \mathcal{S}$, satisfy the system of LMIs for all $i \in \mathcal{S}$*

$$
\begin{bmatrix} \Lambda_t(i) & \Upsilon_r(i) & \mathcal{R}(i) \\ \Upsilon_r^t(i) & -\Upsilon_d(i) & 0 \\ \mathcal{R}^t(i) & 0 & -\mathcal{Y}(i) \end{bmatrix} < 0
$$

$$
\begin{bmatrix} -L(i) & I \\ I & -\beta W(i) \end{bmatrix} \geq 0 \tag{5.31}
$$

where

$$
\begin{aligned}
\Upsilon_r(i) &= [Y(i) \quad Z^t(i)D_o^t(i) + Y(i)C_o^t(i) \quad A_d(i)] \\
\Upsilon_d(i) &= diag[L(i) \quad I \quad \hat{W}(i)] \\
\Lambda_t(i) &= Y(i)A_o^t(i) + A_o(i)Y(i) + B_o(i)Z(i) \\
&+ Z^t(i)B_o^t(i) + \alpha_{ii}Y(i) \\
\mathcal{Y}(i) &= diag\Big[Y(1)...Y(i-1) \quad Y(i+1)...Y(s)\Big] \\
\mathcal{R}(i) &= \Big[\sqrt{\alpha_{i1}}Y(i).....\sqrt{\alpha_{is}}Y(i)\Big]
\end{aligned} \tag{5.32}
$$

Proof: By **Fact 3** of Appendix A, LMI (5.14) is equivalent to

$$A_K^t(i)P(i) + P(i)A_K(i) + \sum_{m=1}^{s} \alpha_{im}P(m) + \beta W(i) +$$
$$C_K^t(i)C_K(i) + P(i)A_d(i)\hat{W}^{-1}(i)A_d^t(i)P(i) < 0 \tag{5.33}$$

Using (5.11), premultiplying (5.33) by $Y(i) = P^{-1}(i)$ and postmultiplying the result by $Y(i)$ with

$$L(i) = \hat{Q}^{-1}(i) \ , \ K(i)Y(i) = Z(i)$$

we obtain:

$$\begin{aligned} & Y(i)A_o^t(i) + A_o(i)Y(i) + B_o(i)Z(i) + Z^t(i)B_o^t(i) \\ + \ & A_d(i)\hat{W}^{-1}(i)A_d^t(i) + Y(i)L^{-1}(i)Y(i) \\ + \ & [Z^t(i)D_o^t(i) + Y(i)C_o^t(i)][C_o(i)Y(i) + D_o(i)Z(i)] \\ + \ & Y(i)\Big[\sum_{m=1}^{s} \alpha_{im}Y^{-1}(m)\Big]Y(i) < 0 \end{aligned} \tag{5.34}$$

With reference to (5.32), inequality (5.34) becomes

$$\begin{aligned} & \Lambda_t(i) + \mathcal{R}(i)\mathcal{Y}^{-1}\mathcal{R}^t(i) + Y(i)L^{-1}(i)Y(i) \\ + \ & [Z^t(i)D_o^t(i) + Y(i)C_o^t(i)][C_o(i)Y(i) + D_o(i)Z(i)] \\ + \ & A_d(i)\hat{W}^{-1}(i)A_d^t(i) < 0 \end{aligned} \tag{5.35}$$

By **Fact 3**, LMIs (5.31) follows from (5.35). ∇∇∇

5.3.3 $\mathcal{H}_\infty$ Performance

With reference to the developed results of Chapter 4, we summarize the main results of the direct approach by the next two theorems

Theorem 5.3 *Given a prescribed constant* $\gamma > 0$. *Controller (5.7) renders system* (Σ_{JKo}) **SSFTD** *with a disturbance attenuation level* γ *for all* $w(t) \in \mathcal{L}_2[0,\infty$ *if given matrices* $W^t(i) = W(i) > 0$ *and mode-dependent delays* $(\tau_{\eta_i} = \tau_i,\ i \in \mathcal{S})$, *there exist matrices* $P(i) = P^t(i) > 0,\ i \in \mathcal{S}$ *satisfying the LMIs for all* $i \in \mathcal{S}$

$$\Pi_r(i) \triangleq \begin{bmatrix} \Xi_{KK} & P(i)A_d(i) & \begin{array}{c} P(i)\Gamma(i)+ \\ G_o^t(i)\Phi(i) \end{array} \\ A_d^t(i)P(i) & -\hat{W}(i) & 0 \\ \begin{array}{c} \Gamma^t(i)P(i)+ \\ \Phi^t(i)G_o(i) \end{array} & 0 & \begin{array}{c} -\gamma^2 I+ \\ \Phi^t(i)\Phi(i) \end{array} \end{bmatrix} < 0 \tag{5.36}$$

where

$$\begin{aligned} \Xi_{KK} &= A_K^t(i)P(i) + P(i)A_K(i) + \sum_{m=1}^{s} \alpha_{im} P(m) \\ &+ \beta W(i) + G_K^t(i)G_K(i) \end{aligned}$$

Moreover,

$$\begin{aligned} ||z(t)|| &= \bigg[\gamma^2 ||w(t)||^2 + x^t(0)P(\eta_o)x(0) \\ &+ \tau_o \int_{-\tau_o}^{0} x^t(s)W(\eta_o)x(s)ds\bigg]^{1/2} \end{aligned} \tag{5.37}$$

Proof: The stochastic stability follows as a result of **Theorem** 5.1. To show that system (Σ_{JKo}) has a disturbance attenuation γ, we let the Lyapunov functional $V(t,x,i)$ be given by (5.17). By evaluating the weak infinitesimal operator $\Im_s^x[\cdot]$ of the process $\{x(t), \eta_t = i, t \geq 0\}$ for system (5.8)-(5.10) at the point $\{t, x, \eta_t = i\}$ using (5.11) and manipulating we get

$$\begin{aligned} \Im_s^x[V] &\leq \Im_t^x[V] + x^t(t)P(i)\Gamma(i)w(t) + w^t(t)\Gamma^t(i)P(i)x(t) \\ &\triangleq \Im_+^x[V] \end{aligned} \tag{5.38}$$

By Dynkin's formula, one has

$$\mathbb{E}\bigg\{\int_0^T \Im_+^x[V(.)]dt\bigg\} = \mathbb{E}\bigg\{V(x(t),\eta_t)|_{t=T}\bigg\} - V(x(t),\eta_t)|_{t=0}$$

With standard matrix manipulations using (5.8)-(5.11) and (5.36), it follows from (5.38) that:

$$z^t(t)z(t) - \gamma^2\ w^t(t)w(t) + \Im_+^x[V] \leq \chi_w^t(t)\Pi_r(i)\chi_w(t) \tag{5.39}$$

where

$$\chi_w^t(t) = [x^t(t)\ \ x^t(t-\tau)\ \ w^t(t)]$$

Therefore from (5.6), we have:

$$\begin{aligned} \mathcal{J}_T &\leq \mathbb{E}\{\int_0^T \chi_w^t(t)\Pi_r(i)\chi_w(t)dt\} \\ &- \mathbb{E}\{V(x(t),\eta_t)|_{t=T} + V(x(t),\eta_t)|_{t=0}\} \end{aligned} \tag{5.40}$$

In view of (5.36) and the fact that

$$\mathbb{E}\left\{V(x(t),\eta_t)|_{t=T} \geq 0\right\}$$

then (5.40) leads to

$$\begin{aligned} \mathcal{J}_T &\leq V(x(0),\eta_0) \Longrightarrow \\ \mathcal{J}_\infty &\leq V(x(0),\eta_0) \Longrightarrow \\ ||z(t)||^2 - \gamma^2\ ||w(t)||^2 &\leq x_0^t P(\eta_o)x_0 \\ &+ \tau_o \int_{-\tau_o}^0 \phi^t(\beta)W(\eta_o)\phi(\beta)d\beta \Longrightarrow \\ ||z(t)||^2 &\leq \gamma^2\ ||w(t)||^2 + x_0^t P(\eta_o)x_0 \\ &+ \tau_o \int_{-\tau_o}^0 \phi^t(\beta)W(\eta_o)\phi(\beta)d\beta \end{aligned} \tag{5.41}$$

which completes the proof. $\nabla\nabla\nabla$

Theorem 5.4 *Given a prescribed constant* $\gamma > 0$. *The feedback gain associated with the* $\mathcal{H}_\infty$*-controller for system* (Σ_{Jo}) *is given by* $K(i) = Z(i)Y^{-1}(i),\ i \in \mathcal{S}$

where matrices $Y(i) = Y^t(i) > 0,\ Z(i),\ L(i) = L^t(i) > 0,\ i \in \mathcal{S}$, *satisfy the system of LMIs for all* $i \in \mathcal{S}$

$$\begin{bmatrix} \Lambda_t(i) & \Upsilon_r(i) & \mathcal{R}(i) & \begin{array}{c}\Gamma(i)+\\ Y(i)G_o^t(i)\Phi(i)\end{array} \\ \Upsilon_r(i) & -\Upsilon_d(i) & 0 & 0 \\ \mathcal{R}^t(i) & 0 & -\mathcal{Y}(i) & 0 \\ \begin{array}{c}\Gamma^t(i)+\\ \Phi^t(i)G_o(i)Y(i)\end{array} & 0 & 0 & \begin{array}{c}-\gamma^2 I+\\ \Phi^t(i)\Phi(i)\end{array} \end{bmatrix} < 0$$

$$\begin{bmatrix} -L(i) & I \\ I & -\beta W(i) \end{bmatrix} \geq 0 \qquad (5.42)$$

Proof: Follows from parallel development to **Theorem** 5.2. $\nabla\nabla\nabla$

5.3.4 Mixed Performance

With the foregoing results at hand, the solution to the simultaneous $\mathcal{H}_2/\mathcal{H}_\infty$ control problem posed earlier is readily founded by the following theorem:

Theorem 5.5 *Given a prescribed constant* $\gamma > 0$. *The feedback gain* $K(i) = Z(i)Y^{-1}(i),\ i \in \mathcal{S}$ *is a* **simultaneous** $\mathcal{H}_2/\mathcal{H}_\infty$ **controller** *satisfying the performance measure (5.6) for system (??-53.3) if there exist matrices* $Y(i) = Y^t(i) > 0,\ Z(i),\ L(i) = L^t(i) > 0,\ \Upsilon = \Upsilon^t > 0,\ i \in \mathcal{S}$, *such the system of generalized eigenvalue problems*

$$\min \Big[\lambda + Tr(\Upsilon)\Big]$$
$$\textit{subject to LMIs } (5.31)\,,\ (5.42) \textit{ and}$$

$$\begin{bmatrix} -\lambda & \phi^t(0) \\ \phi(0) & -Y(i) \end{bmatrix} < 0\,,\ \begin{bmatrix} -\Upsilon & X^t \\ X & -L(i) \end{bmatrix} < 0 \qquad (5.43)$$

has a feasible solution for all $i \in \mathcal{S}$

Proof: On observing that

$$x^t(0)P(\eta_t = i)x(0) \quad \leq \quad \max \Big[x^t(0)P(\eta_t = 1)x(0), ..., x^t(0)P(\eta_t = s)x(0)\Big]$$

$$
\begin{aligned}
&\triangleq \lambda \Longrightarrow \\
&- \lambda + \phi^t(0) Y^{-1}(i) \phi(0) \; < \; 0 \qquad (5.44)
\end{aligned}
$$

and in similar way using the cyclic properties of matric trace [25]

$$
\begin{aligned}
\int_{-\tau(0)}^{0} x^t(s) W(\eta_o) x(s) ds &= \\
\int_{-\tau(0)}^{0} Tr\Big[x^t(s) W(\eta_o) x(s)\Big] ds &= \\
Tr\Big[X X^t L^{-1}(i)\Big] &= Tr\Big[X^t L^{-1}(i) X\Big] < Tr(\Upsilon) \Longrightarrow \\
&- \Upsilon + X^t L^{-1}(i) X < 0 \qquad (5.45)
\end{aligned}
$$

where $XX^t = \int_{-\tau(0)}^{0} x^t(s)x(s)ds$. By **Fact 3**, (5.44)-(5.45) leads to LMIs (5.43) and achieving the simultaneous $\mathcal{H}_2/\mathcal{H}_\infty$ goal leads to the above minimization subject to LMIs (5.43). $\nabla\nabla\nabla$

Remark 5.2 *It should be noted that* **Theorem** *5.5 presents a design procedure to compute the simultaneous $\mathcal{H}_2/\mathcal{H}_\infty$ controller as the solution of a convex minimization problem which can conveniently solved by the MATLAB software system [57].*

5.3.5 Example 5.1

In order to illustrate the foregoing results, we provide a numerical example. We consider a pilot-scale single-reach water quality system which can fall into the type (5.1)-(5.3) with mode-dependent delays. Let the Markov process governing the mode switching has generator

$$
\Im = \begin{bmatrix} -6 & 4 & 2 \\ 2 & -5 & 3 \\ 3 & 1 & -4 \end{bmatrix}
$$

For the three operating conditions (modes), the associated date are:

Mode 1:

$$\begin{aligned}
A_o(1) &= \begin{bmatrix} -3 & -2 \\ 1 & 0 \end{bmatrix}, \; A_d(1) = \begin{bmatrix} 0 & 0.3 \\ -0.3 & -0.2 \end{bmatrix} \\
B_o(1) &= \begin{bmatrix} 1 & 0 \\ 0 & 2 \end{bmatrix}, \; \Gamma(1) = \begin{bmatrix} 0.1 \\ 0.2 \end{bmatrix}, \; C_o(1) = \begin{bmatrix} 1 & 0 \\ 0 & 1 \end{bmatrix} \\
\Phi(1) &= \begin{bmatrix} 0.2 \\ 0.1 \end{bmatrix}, \; G_o(1) = [0.1 \;\; 0.1], \; \tau(1) = 0.15 \\
D_o(1) &= \begin{bmatrix} 0.1 & 0 \\ 0 & 0.1 \end{bmatrix}, \; F_o(1) = [0.2 \;\; 0.2], \; \Psi = 0.25
\end{aligned}$$

Mode 2:

$$\begin{aligned}
A_o(2) &= \begin{bmatrix} -1 & 0 \\ 2 & -2 \end{bmatrix}, \; A_d(2) = \begin{bmatrix} -0.5 & -0.6 \\ -0.2 & -0.1 \end{bmatrix} \\
B_o(2) &= \begin{bmatrix} 1 & 0 \\ 0 & 1 \end{bmatrix}, \; \Gamma(2) = \begin{bmatrix} 0.2 \\ 0.1 \end{bmatrix}, \; C_o(2) = \begin{bmatrix} 0.5 & 0 \\ 0 & 0.5 \end{bmatrix} \\
\Phi(2) &= \begin{bmatrix} 0.1 \\ 0.2 \end{bmatrix}, \; G_o(2) = [0.2 \;\; 0.1], \; \tau(1) = 0.85 \\
D_o(2) &= \begin{bmatrix} 0.2 & 0 \\ 0 & 0.2 \end{bmatrix}, \; F_o(1) = [0.2 \;\; 0.2], \; \Psi(1) = 0.15
\end{aligned}$$

Mode 3:

$$\begin{aligned}
A_o(3) &= \begin{bmatrix} -4 & -3 \\ 1 & 0 \end{bmatrix}, \; A_d(3) = \begin{bmatrix} 0.1 & 0.2 \\ -0.4 & 0 \end{bmatrix} \\
B_o(3) &= \begin{bmatrix} 1 & 0 \\ 0 & 1 \end{bmatrix}, \; \Gamma(3) = \begin{bmatrix} 0.1 \\ 0.1 \end{bmatrix}, \; C_o(3) = \begin{bmatrix} 0.4 & 0 \\ 0 & 0.6 \end{bmatrix} \\
\Phi(3) &= \begin{bmatrix} 0.1 \\ 0.2 \end{bmatrix}, \; G_o(3) = [0.1 \;\; 0.2], \; \tau(1) = 0.55 \\
D_o(2) &= \begin{bmatrix} 0.15 & 0 \\ 0 & 0.15 \end{bmatrix}, \; F_o(1) = [0.2 \;\; 0.1], \; \Psi(1) = 0.2
\end{aligned}$$

Invoking the software environment [57] and using the data

$$\begin{aligned}
W(1) &= \begin{bmatrix} 2 & 0 \\ 0 & 2 \end{bmatrix}, \; W(2) = \begin{bmatrix} 1 & 0 \\ 0 & 1 \end{bmatrix} \\
W(3) &= \begin{bmatrix} 2 & 0 \\ 0 & 2 \end{bmatrix}, \; \phi(t) = \begin{bmatrix} e^{0.5t} \\ e^{0.2t} \end{bmatrix}
\end{aligned}$$

the feasible solution of the LMIs (5.45) is given by:

$$\begin{aligned} Y(1) &= \begin{bmatrix} 12.2240 & 3.5321 \\ 3.5321 & 9.0422 \end{bmatrix}, \; Z(1) = \begin{bmatrix} 3.0124 & 0.5341 \\ -0.6125 & 1.0332 \end{bmatrix} \\ Y(2) &= \begin{bmatrix} 9.2064 & 2.4423 \\ 2.4423 & 9.8605 \end{bmatrix}, \; Z(2) = \begin{bmatrix} 1.4564 & 0.6214 \\ -0.7250 & 1.2112 \end{bmatrix} \\ Y(3) &= \begin{bmatrix} 10.8952 & 2.4293 \\ 2.4293 & 10.0065 \end{bmatrix}, \; Z(3) = \begin{bmatrix} 2.1004 & -0.2333 \\ 0.1925 & 1.2332 \end{bmatrix} \\ \Upsilon &= \begin{bmatrix} 4.2054 & 2.1443 \\ 2.1443 & 6.7805 \end{bmatrix}, \; \lambda = 5.2154 \end{aligned}$$

The feedback gains and the associated performance bound are

$$\begin{aligned} K(1) &= \begin{bmatrix} 0.2585 & -0.0419 \\ -0.0937 & 0.1509 \end{bmatrix}, \; K(2) = \begin{bmatrix} 0.1514 & 0.0255 \\ -0.1192 & 0.1523 \end{bmatrix} \\ K(3) &= \begin{bmatrix} 0.2093 & -0.0741 \\ -0.0104 & 0.1258 \end{bmatrix}, \; \mathcal{J} = 7.0814 \end{aligned}$$

5.3.6 Uncertain Model Design

In this section, we develop the robust counterparts of **Theorems** 5.1-5.5 for the uncertain system with $\eta_t = i \in \mathcal{S}$

$$\begin{aligned} (\Sigma_J): \; \dot{x}(t) &= [A_o(i) + \Delta A_o(t,i)]x(t) + [A_d(i) + \Delta A_d(t,i)]x(t - \tau_{\eta_t}) \\ &+ [B_o(i) + \Delta B_o(t,i)]u(t) + \Gamma(i)w(t) \\ &= A_{\Delta o}(t,i)x(t) + A_{\Delta d}(t,i)x(t - \tau_i) + B_{\Delta o}(t,i)u(t) \\ &+ \Gamma(i)w(t), \; x(t) = \phi(t), \; t \in [-\tau, 0] \end{aligned} \tag{5.46}$$

$$y(t) = C_o(i)x(t) + D_o(i)u(t) + \Psi(i)w(t) \tag{5.47}$$

$$z(t) = G_o(i)x(t) + F_o(i)u(t) + \Phi(i)w(t) \tag{5.48}$$

where the nominal matrices are described by (5.4). The matrices $\Delta A_o(t,i)$, $\Delta A_d(t,i)$ and $\Delta B_o(t,i)$ are real, time-varying matrix functions representing the norm-bounded uncertainties. For $\eta_t = i$, the admissible uncertainties are assumed to be modeled in the form:

$$[\Delta A_o(t,i)\ \ \Delta A_d(t,i)\ \ \Delta B_o(t,i)] = M_a(i)\Delta(t,i)[N_a(i)\ \ N_d(i)\ \ N_b(i)] \quad (5.49)$$

where $M_a(i)$, $M_g(i)$, $N_a(i)$, $N_b(i)$ and $N_d(i)$ are known real constant matrices, with appropriate dimensions, and $\Delta(t,i)$ being unknown time-varying matrix function satisfying

$$||\Delta(t,i)||_2 \le 1 \quad (5.50)$$

where the elements of $\Delta(t,i)$ are Lebesgue measurable for any $i \in \mathcal{S}$.

Now under the feedback control (5.7), the uncertain closed loop system is expressed for $\eta_t = i \in \mathcal{S}$ as:

$$\begin{aligned}
(\Sigma_{JK}):\ \dot{x}(t) &= A_{\Delta K}(t,i)x(t) + A_{\Delta d}(t,i)x(t-\tau_i) \\
&+ \Gamma(t,i)w(t) \\
x(s) &= \phi(s)\ ,\ s\ \in\ \max[-\tau,0] \quad (5.51) \\
y(t) &= C_K(i)x(t) + \Phi(i)w(t) \\
z(t) &= G_K(i)x(t) + \Phi(i)w(t) \quad (5.52) \\
A_{\Delta K}(i) &= A_K(i) + M_a(i)\Delta(t,i)N_{aK}(i) \\
N_{aK}(i) &= N_a(i) + N_b(i)K(i) \quad (5.53)
\end{aligned}$$

5.3.7 $\mathcal{H}_2$-Performance

The following results are developed.

Theorem 5.6 *In the absence of input disturbance $w(t) \equiv 0$, controller (5.7) is an $\mathcal{H}_2$-optimal controller for system (5.51)-(5.53) minimizing the $\mathcal{H}_2$-performance measure (5.5) if given matrices $W^t(i) = W(i) > 0$ and mode-dependent delays $(\tau_{\eta_i} = \tau_i,\ i \in \mathcal{S})$, there exist matrices $P(i) = P^t(i) > 0,\ i \in \mathcal{S}$ and scalars*

$\epsilon(i) > 0,\ \mu(i) > 0,\ i \in \mathcal{S}$, satisfying the system of LMIs for all $i \in \mathcal{S}$

$$\begin{bmatrix} \Pi_{sK}(i) & P(i)M_a(i) & P(i)M_d(i) & P(i)A_d(i) \\ M_a^t(i)P(i) & -\epsilon(i)I & 0 & 0 \\ M_d^t(i)P(i) & 0 & -\mu(i)I & 0 \\ A_d^t(i)P(i) & 0 & 0 & \begin{matrix} -\hat{W}(i)+ \\ \mu(i)N_d^t(i)N_d(i) \end{matrix} \end{bmatrix} < 0 \tag{5.54}$$

$$\Pi_{sK}(i) = A_K^t(i)P(i) + P(i)A_K(i) + \sum_{k=1}^{s} \alpha_{ik}P(k) + \beta\, W(i)$$
$$+\epsilon(i)N_{aK}^t(i)N_{aK}(i) + \bar{C}_K^t\bar{C}_K \tag{5.55}$$

for all admissible uncertainties satisfying (5.49)-(5.50)

Proof: It follows from **Theorem** (5.1) that controller (5.7) is $\mathcal{H}_\infty$-optimal controller for system (Σ_{JK}) if there exist matrices $P(i) = P^t(i) > 0$ satisfying the system of ARIs

$$A_{\Delta K}^t(i)P(i) + P(i)A_{\Delta K}(i) + \sum_{m=1}^{s} \alpha_{im}P(m) + \beta W(i)$$
$$+ \quad P(i)A_{\Delta d}(i)\hat{W}^{-1}(i)A_{\Delta d}^t(i)P(i) + \bar{C}_K^t\bar{C}_K \ < \ 0 \tag{5.56}$$

By **Fact 1** and (5.53) we have

$$\begin{aligned} A_{\Delta K}^t(i)P(i) + P(i)A_{\Delta K}(i) \ \le\ & A_K^t(i)P(i) + P(i)A_K(i) \\ +\ & \varepsilon(i)[N_a(i) + N_b(i)K(i)]^t[N_a(i) + N_b(i)K(i)] \\ +\ & \varepsilon^{-1}(i)P(i)M_a(i)M_a^t(i)P(i) \end{aligned} \tag{5.57}$$

for some scalars $\varepsilon(i) > 0,\ i \in \mathcal{S}$. Similarly, by **Fact 2** we get

$$P(i)A_{\Delta d}(i)\hat{W}^{-1}(i)A_{\Delta d}^t(i)P(i) \le \mu^{-1}(i)P(i)M_a(i)M_a^t(i)P(i) +$$
$$A_d(i)[\hat{W}(i) - \mu(i)N_d^t(i)N_d(i)]^{-1}A_d^t(i)P(i) \tag{5.58}$$

for some scalars $\mu(i) > 0,\ i \in \mathcal{S}$. Combining (5.57)-(5.58) into (5.56), it yields the ARIs:

$$
\begin{aligned}
& A_K^t(i)P(i) + P(i)A_K(i) + \sum_{m=1}^{s} \alpha_{im} P(m) + \beta W(i) \\
+ \;& \varepsilon^{-1}(i)P(i)M_a(i)M_a^t(i)P(i) + \mu^{-1}(i)P(i)M_a(i)M_a^t(i)P(i) \\
+ \;& \varepsilon(i)[N_a(i) + N_b(i)K(i)]^t[N_a(i) + N_b(i)K(i)] \\
+ \;& P(i)A_d(i)[\hat{W}(i) - \mu(i)N_d^t(i)N_d(i)]^{-1}A_d^t(i)P(i) \\
+ \;& [C_o(i) + D_o(i)K(i)]^t[C_o(i) + D_o(i)K(i)] \; < \; 0
\end{aligned} \tag{5.59}
$$

By **Fact 3**, ARIs (5.59) are equivalent to LMIs (5.54). $\nabla\nabla\nabla$

A method for computing the feedback gains is provided by the following theorem.

Theorem 5.7 *The feedback gain associated with the $\mathcal{H}_2$-optimal controller for system (5.51-5.53) is given by $K(i) = Z(i)Y^{-1}(i),\ i \in \mathcal{S}$ if there exist matrices $Y(i) = Y^t(i) > 0,\ Z(i),\ L(i) = L^t(i) > 0,\ i \in \mathcal{S}$ and scalars $\varepsilon(i) > 0,\ \mu(i) > 0,\ i \in \mathcal{S}$ satisfying the system of LMIs for all $i \in \mathcal{S}$*

$$
\begin{aligned}
& \begin{bmatrix} \bar{\Lambda}_s(i) & \Upsilon_{ur}(i) & \mathcal{R}(i) \\ \Upsilon_{ur}^t(i) & -\Upsilon_{ud}(i) & 0 \\ \mathcal{R}^t(i) & 0 & -\mathcal{Y} \end{bmatrix} < 0 \\
& \begin{bmatrix} -L(i) & I \\ I & -\beta W(i) \end{bmatrix} \geq 0
\end{aligned} \tag{5.60}
$$

where

$$
\begin{aligned}
\bar{\Lambda}_s(i) &= Y(i)A_o^t(i) + A_o(i)Y(i) + B_o(i)Z(i) + Z^t(i)B_o^t(i) \\
&+ \alpha_{ii}Y(i) + [\varepsilon^{-1}(i) + \mu^{-1}(i)]M_a(i)M_a^t(i) \\
\Upsilon_{ur}(i) &= [Y(i) \quad Z^t(i)D_o^t(i) + Y(i)C_o^t(i) \quad Z^t(i)N_b^t(i) + Y(i)N_a^t(i) \quad A_d(i)] \\
\Upsilon_{ud}(i) &= diag[L(i) \quad I \quad \varepsilon(i)I \quad \hat{W} - \mu(i)N_d^t(i)N_d(i)]
\end{aligned} \tag{5.61}
$$

The feedback gain associated with the $\mathcal{H}_2$-optimal controller has $K(i) = Z(i)Y^{-1}(i),\ i \in \mathcal{S}$ and an upper bound on the $\mathcal{H}_2$ performance measure is given by

$$\begin{aligned} \mathcal{J}_2 &\leq \mathsf{J}^+ \triangleq \Big[x^t(0)Y^{-1}(\eta_o)x(0) \\ &+ \tau_o \int_{-\tau_o}^{0} x^t(s)W(\eta_o)x(s)ds \Big]^{1/2} \end{aligned} \tag{5.62}$$

Proof: It follows from (5.59) directly by substituting

$$Y(i) = P^{-1}(i) \ , \ K(i)Y(i) = Z(i) \ , \ L(i) = (\beta W)^{-1}$$

and manipulating the result using **Fact 3**, we obtain LMIs (5.60) subject to (5.61). The bound (5.62) follows immediately. $\nabla\nabla\nabla$

5.3.8 $\mathcal{H}_\infty$-Performance

Building on **Theorems** (5.6-5.7), it is an appealing task to consider the $\mathcal{H}_\infty$-performance. The next two theorems summarize the main results pertaining to this task.

Theorem 5.8 *Given a prescribed constant $\gamma > 0$. Controller (5.7) renders system (Σ_{JK})* **SSFTD** *with a disturbance attenuation level γ for all $w(t) \in \mathcal{L}_2[0, \infty f$ given matrices $W^t(i) = W(i) > 0$ and mode-dependent delays $(\tau_{\eta_i} = \tau_i,\ i \in \mathcal{S})$, there exist matrices $P(i) = P^t(i) > 0,\ i \in \mathcal{S}$ and scalars $\epsilon(i) > 0,\ \mu(i) > 0,\ i \in \mathcal{S}$, satisfying the LMIs for all $i \in \mathcal{S}$*

$$\Pi_{rK}(i) \triangleq \begin{bmatrix} \bar{\Pi}_{sK} & \Upsilon_{mu}(i) & \begin{array}{c} P(i)\Gamma(i)+ \\ G_o^t(i)\Phi(i) \end{array} \\ \Upsilon_{mu}^t(i) & -\Upsilon_{du}^t(i)I & 0 \\ \begin{array}{c} \Gamma^t(i)P(i)+ \\ \Phi^t(i)G_o(i) \end{array} & 0 & \begin{array}{c} -\gamma^2 I+ \\ \Phi^t(i)\Phi(i) \end{array} \end{bmatrix} < 0 \tag{5.63}$$

$$\bar{\Pi}_{sK}(i) = A_K^t(i)P(i) + P(i)A_K(i) + \sum_{k=1}^{s} \alpha_{ik}P(k) + \beta\, W(i)$$

$$
\begin{aligned}
&+ \quad \epsilon(i)N^t_{aK}(i)N_{aK} + \bar{G}^t_K\bar{G}_K \\
\Upsilon_{mu}(i) &= [P(i)M_a(i) \quad P(i)M_d(i) \quad P(i)A_d(i)] \\
\Upsilon_{du}(i) &= diag[\varepsilon(i)I \quad \mu(i)I \quad \hat{W}(i)]
\end{aligned} \tag{5.64}
$$

for all admissible uncertainties satisfying (5.49)-(5.50). Moreover,

$$
\begin{aligned}
||z(t)|| &= \Bigg[\gamma^2||w(t)||^2 + x^t(0)P(\eta_o)x(0) \\
&+ \tau_o \int_{-\tau_o}^{0} x^t(s)W(\eta_o)x(s)ds\Bigg]^{1/2}
\end{aligned} \tag{5.65}
$$

Theorem 5.9 *Given a prescribed constant $\gamma > 0$. The feedback gain associated with the $\mathcal{H}_\infty$-controller for system (Σ_{JK}) is given by $K(i) = Z(i)Y^{-1}(i)$, $i \in \mathcal{S}$ if given $W^t(i) = W(i) > 0$, there exist matrices $Y(i) = Y^t(i) > 0$, $Z(i)$, $L(i) = L^t(i) > 0$, $i \in \mathcal{S}$ and scalars $\epsilon(i) > 0$, $\mu(i) > 0$, $i \in \mathcal{S}$ satisfying the system of LMIs for all $i \in \mathcal{S}$*

$$
\begin{aligned}
&\begin{bmatrix}
\bar{\Lambda}_t(i) & \Upsilon_{ur}(i) & \mathcal{R}(i) & \begin{array}{c}\Gamma(i)+ \\ Y(i)G^t_o(i)\Phi(i)\end{array} \\
\Upsilon^t_{ur}(i) & -\Upsilon_{ud}(i) & 0 & 0 \\
\mathcal{R}^t(i) & 0 & -\mathcal{Y}(i) & 0 \\
\begin{array}{c}\Gamma^t(i)+ \\ \Phi^t(i)G_o(i)Y(i)\end{array} & 0 & 0 & \begin{array}{c}-\gamma^2 I+ \\ \Phi^t(i)\Phi(i)\end{array}
\end{bmatrix} < 0 \\
&\begin{bmatrix}
-L(i) & I \\
I & -\beta W(i)
\end{bmatrix} \geq 0
\end{aligned} \tag{5.66}
$$

for all admissible uncertainties satisfying (5.49)-(5.50). Moreover

$$
\begin{aligned}
||z(t)|| &= \Bigg[\gamma^2||w(t)||^2 + x^t(0)Y^{-1}(\eta_o)x(0) \\
&+ \tau_o \int_{-\tau_o}^{0} x^t(s)W(\eta_o)x(s)ds\Bigg]^{1/2}
\end{aligned} \tag{5.67}
$$

Proof. It can be worked out by using the same technique as that used in **Theorem** 5.6. ∇∇∇

5.3.9 Mixed Performance

The simultaneous $\mathcal{H}_2/\mathcal{H}_\infty$ control for system (5.51-5.53) is readily phrazed by the following theorem:

Theorem 5.10 *Given a prescribed constant* $\gamma > 0$. *The feedback gain* $K(i) = Z(i)Y^{-1}(i),\ i \in \mathcal{S}$ *is a* **simultaneous $\mathcal{H}_2/\mathcal{H}_\infty$ controller** *satisfying the performance measure (5.6) for system* (Σ_{JK}) *if there exist matrices* $Y(i) = Y^t(i) > 0,\ Z(i),\ L(i) = L^t(i) > 0,\ \Upsilon = \Upsilon^t > 0,\ i \in \mathcal{S}$, *and scalars* $\epsilon(i) > 0,\ \mu(i) > 0,\ i \in \mathcal{S}$ *such the system of generalized eigenvalue problems for all* $i \in \mathcal{S}$

$$\min \left[\lambda + Tr(\Upsilon)\right]$$
$$\textit{subject to LMIs } (5.60)\,,\ (5.66) \ \textit{ and}$$
$$\begin{bmatrix} -\lambda & \phi^t(0) \\ \phi(0) & -Y(i) \end{bmatrix} < 0\,, \begin{bmatrix} -\Upsilon & X^t \\ X & -L(i) \end{bmatrix} \geq 0 \qquad (5.68)$$

has a feasible solution for all admissible uncertainties satisfying (5.49)-(5.50)

5.3.10 Example 5.2

Here we utilize the basic data of **Example 1** in addition to

Mode 1:

$$M_a(1) = \begin{bmatrix} 0.2 \\ 0.1 \end{bmatrix},\ N_b(1) = [0.3\ \ 0.3]\,,\ N_a(1) = [0.2\ \ 0.4]$$
$$N_d(1) = [0.1\ \ 0.3]$$

Mode 2:

$$M_a(2) = \begin{bmatrix} 0.1 \\ 0.2 \end{bmatrix},\ N_b(2) = [0.1\ \ 0.1]\,,\ N_a(2) = [0.4\ \ 0.2]$$
$$N_d(2) = [0.3\ \ 0.1]$$

Mode 3:

$$M_a(3) = \begin{bmatrix} 0.1 \\ 0.1 \end{bmatrix},\ N_b(3) = [0.2\ \ 0.3]\,,\ N_a(3) = [0.2\ \ 0.3]$$
$$N_d(3) = [0.1\ \ 0.3]$$

Using the software package [57] the feasible results are:

$$\begin{aligned}
Y(1) &= \begin{bmatrix} 11.2650 & 3.4520 \\ 3.3520 & 8.6422 \end{bmatrix}, \; Z(1) = \begin{bmatrix} 3.1004 & -0.3941 \\ -0.5825 & 1.1432 \end{bmatrix} \\
K(1) &= \begin{bmatrix} 1.2385 & -0.1672 \\ -0.1151 & 2.1463 \end{bmatrix}, \; \epsilon(1) = 1.3154, \; \mu(1) = 1.3175 \\
Y(2) &= \begin{bmatrix} 8.2064 & 2.5423 \\ 2.5423 & 9.6605 \end{bmatrix}, \; Z(2) = \begin{bmatrix} 1.3564 & -0.5214 \\ -0.6250 & 1.3546 \end{bmatrix} \\
K(2) &= \begin{bmatrix} 1.1892 & -0.1061 \\ -0.1331 & 1.1836 \end{bmatrix}, \; \epsilon(2) = 2.4469, \; \mu(2) = 1.8254 \\
Y(3) &= \begin{bmatrix} 11.8952 & 2.3943 \\ 2.3943 & 10.2165 \end{bmatrix}, \; Z(3) = \begin{bmatrix} 2.1214 & -0.2333 \\ -0.1925 & 1.3432 \end{bmatrix} \\
K(3) &= \begin{bmatrix} 1.1722 & -0.0679 \\ -0.0448 & 1.1220 \end{bmatrix}, \; \epsilon(3) = 1.3498 \; , \; \mu(3) = 1.8426 \\
\lambda &= 3.7542 \; , \quad \mathcal{J} = 5.7783
\end{aligned}$$

5.3.11 Extensions

The foregoing developed robustness results can naturally extended to the case where the jumping rates are subject to uncertainties. Specifically, we consider the transition probability from mode i at time t to mode j at time $t+\delta$, $i, j \in \mathcal{S}$ to be:

$$\begin{aligned}
p_{ij} &= Pr(\eta_{t+\delta} = j \mid \eta_t = i) \\
&= \begin{cases} (\alpha_{ij} + \Delta\alpha_{ij})\delta + o(\delta), & if\ i \neq j, \\ 1 + (\alpha_{ij} + \Delta\alpha_{ij})\delta + o(\delta), & if\ i = j \end{cases}
\end{aligned} \tag{5.69}$$

with transition probability rates $(\alpha_{ij} + \Delta\alpha_{ij}) \geq 0$ for $i, j \in \mathcal{S}, i \neq j$ and

$$\alpha_{ii} + \Delta\alpha_{ii} = - \sum_{m=1, m\neq i}^{s} (\alpha_{im} + \Delta\alpha_{im}) \tag{5.70}$$

We assume that the uncertainties $\Delta\alpha_{ij}$ satisfies

$$||\Delta\alpha_{ij}|| \leq \psi_{ij} \; , \qquad \forall\, i, j \in \mathcal{S} \tag{5.71}$$

where ψ_{ij} are known scalars, $\forall\, i, j \in \mathcal{S}$.

Extending on **Theorems** 5.5 and 5.10 , we have the following results:

Theorem 5.11 *Consider system* (Σ_J) *over the space* $(\Omega, \mathcal{F}, \mathbf{P})$ *where* $\mathbf{P}$ *is described by (5.69)-(5.71). Given a prescribed constant* $\gamma > 0$. *The feedback gain* $K(i) = Z(i)Y^{-1}(i),\ i \in \mathcal{S}$ *is a* **simultaneous** $\mathcal{H}_2/\mathcal{H}_\infty$ **controller** *satisfying the performance measure (5.5) for system (5.1)-(52.4) if there exist matrices* $Y(i) = Y^t(i) > 0,\ Z(i),\ L(i) = L^t(i) > 0,\ \Upsilon = \Upsilon^t > 0,\ i \in \mathcal{S}$, *such the system of generalized eigenvalue problems has a feasible solution*

$$\min \Big[\lambda \,+\, Tr(\Upsilon)\Big] \qquad subject \quad to \tag{5.72}$$

$$\begin{bmatrix} \Lambda_{tt}(i) & \Upsilon_r(i) & \bar{\mathcal{R}}(i) \\ \Upsilon_r^t(i) & -\Upsilon_d(i) & 0 \\ \mathcal{R}^t(i) & 0 & -\mathcal{Y}(i) \end{bmatrix} < 0 \tag{5.73}$$

$$\Lambda_{tt}(i) = Y(i)A_o^t(i) + A_o(i)Y(i) + B_o(i)Z(i) +$$
$$Z^t(i)B_o^t(i) + \Big(\alpha_{ii} + \psi_{ii}\Big)Y(i)$$

$$\bar{\mathcal{R}}(i) = \Big[\sqrt{\alpha_{i1} + \psi_{i1}}Y(i) \ldots\ldots \sqrt{\alpha_{is} + \psi_{is}}Y(i)\Big]$$

$$\begin{bmatrix} -\lambda & \phi^t(0) \\ \phi(0) & -Y(i) \end{bmatrix} < 0\,, \begin{bmatrix} -\Upsilon & X^t \\ X & -L(i) \end{bmatrix} < 0$$

$$\begin{bmatrix} -L(i) & I \\ I & -\beta W(i) \end{bmatrix} \geq 0 \tag{5.74}$$

$$\begin{bmatrix} \Lambda_{tt}(i) & \Upsilon_r(i) & \bar{\mathcal{R}}(i) & \begin{matrix}\Gamma(i)+ \\ Y(i)G_o^t(i)\Phi(i)\end{matrix} \\ \Upsilon_r^t(i) & -\Upsilon_d(i) & 0 & 0 \\ \bar{\mathcal{R}}^t(i) & 0 & -\mathcal{Y}(i) & 0 \\ \begin{matrix}\Gamma^t(i)+ \\ \Phi^t(i)G_o(i)Y(i)\end{matrix} & 0 & 0 & \begin{matrix}-\gamma^2 I+ \\ \Phi^t(i)\Phi(i)\end{matrix} \end{bmatrix} < 0 \tag{5.75}$$

Theorem 5.12 *Consider system* (Σ_J) *over the space* $(\Omega, \mathcal{F}, \mathbf{P})$ *where* $\mathbf{P}$ *is described by (5.69)-(5.71). Given a prescribed constant* $\gamma > 0$. *The feedback gain* $K(i) = Z(i)Y^{-1}(i),\ i \in \mathcal{S}$ *is a* **simultaneous** $\mathcal{H}_2/\mathcal{H}_\infty$ **controller** *sat-*

isfying the performance measure (5.5) for system (Σ_{JK}) *if there exist matrices* $Y(i) = Y^t(i) > 0,\ Z(i),\ L(i) = L^t(i) > 0,\ \Upsilon = \Upsilon^t > 0,\ i \in \mathcal{S}$, *and scalars* $\epsilon(i) > 0,\ \mu(i) > 0,\ i \in \mathcal{S}$ *such the system of generalized eigenvalue problems*

$$\min \left[\lambda + Tr(\Upsilon)\right] \quad subject\ \ to$$

$$\begin{bmatrix} \bar{\Lambda}_{ss}(i) & \Upsilon_{ur}(i) & \bar{\mathcal{R}}(i) \\ \Upsilon^t_{ur}(i) & -\Upsilon_{ud}(i)Y(i) & 0 \\ \bar{\mathcal{R}}^t(i) & 0 & -\mathcal{Y}(i) \end{bmatrix} < 0 \tag{5.76}$$

$$\begin{bmatrix} \bar{\Lambda}_{ss}(i) & \Upsilon_{ur}(i) & \bar{\mathcal{R}}(i) & \Gamma(i) + Y(i)G^t_o(i)\Phi(i) \\ \Upsilon_{ur}(i) & -\Upsilon_{ud}(i) & 0 & 0 \\ \bar{\mathcal{R}}^t(i) & 0 & -\mathcal{Y}(i) & 0 \\ \Gamma^t(i) + \Phi^t(i)G_o(i)Y(i) & 0 & 0 & -\gamma^2 I + \Phi^t(i)\Phi(i) \end{bmatrix} < 0$$

$$\begin{bmatrix} -L(i) & I \\ I & -\beta W(i) \end{bmatrix} \geq 0\,, \begin{bmatrix} -\lambda & \phi^t(0) \\ \phi(0) & -Y(i) \end{bmatrix} < 0$$

$$\begin{bmatrix} -\Upsilon & X^t \\ X & -L(i) \end{bmatrix} < 0 \tag{5.77}$$

where

$$\begin{aligned} \bar{\Lambda}_{ss}(i) &= Y(i)A^t_o(i) + A_o(i)Y(i) + B_o(i)Z(i) + Z^t(i)B^t_o(i) \\ &+ \left(\alpha_{ii} + \psi_{ii}\right)Y(i) + [\varepsilon(i) + \mu(i)]M_a(i)M^t_a(i) \end{aligned} \tag{5.78}$$

has a feasible solution for all admissible uncertainties satisfying (5.49)-(5.50)

This concludes the part on the direct approach to simultaneous $\mathcal{H}_2/\mathcal{H}_\infty$ control design for a class of uncertain JTDS with mode-dependent delays.

5.4 State Transformation Approach

This section introduces a new model transformation method for the mixed $\mathcal{H}_2/\mathcal{H}_\infty$ control of a class of uncertain JTDS. Through this method, the delay-dependence dynamics is naturally brought up in the design procedure. A state-feedback control is derived for both the nominal and uncertain systems such that the $\mathcal{H}_2$-performance measure is minimized while guaranteeing a prescribed $\mathcal{H}_\infty$-norm bound on the controlled system.

In the sequel, reference is made for a class of stochastic uncertain time-delay systems with Markovian jump parameters (Σ_J) described over the space $(\Omega, \mathcal{F}, \mathbf{P})$ by (5.46)-(5.48).

5.4.1 State Transformation

In order to exhibit the delay-dependence of the system dynamics for each possible value $\eta_t = i,\ i \in \mathcal{S},$ we introduce the following state transformation

$$\sigma(t) \ = \ x(t) \ + \int_{t-\tau}^{t} A_{\Delta d}(t,i)\, x(s) ds \tag{5.79}$$

into (5.46) to yield

$$\begin{aligned} \dot{\sigma}(t) \quad &= \quad [A_{\Delta o}(t,i) + A_{\Delta d}(t,i)]\, x(t) \\ &+ \quad B_{\Delta o}(t,\eta_t) u(t) + \Gamma(i) w(t) \end{aligned} \tag{5.80}$$

Define the augmented state-vector

$$\varsigma(t) = \begin{bmatrix} \sigma(t) \\ x(t) \end{bmatrix} \quad \in \Re^{2n} \tag{5.81}$$

By combining (5.46) and (5.79)-(5.81), we obtain the transformed system

$$\begin{aligned} (\Sigma_T): \quad \dot{\varsigma}(t) \quad &= \quad \Lambda_{\Delta}(i)\varsigma(t) \ + \int_{t-\tau}^{t} \Upsilon(i)\ \varsigma(s) ds \\ &+ \quad \bar{B}_{\Delta o}(t,i) u(t) + \bar{\Gamma}(i) w(t) \\ \varsigma(t) \quad &= \quad \bar{\phi}(t),\ t \in [-2\tau, 0] \end{aligned} \tag{5.82}$$

$$y(t) \quad = \quad \bar{C}(i)\varsigma(t) + D_o(i) u(t) + \Phi(i) w(t) \tag{5.83}$$

$$z(t) \quad = \quad \bar{G}(i)\varsigma(t) + F_o(i) u(t) + \Psi(i) w(t) \tag{5.84}$$

where

$$
\begin{aligned}
\bar{\Gamma}(i) &= \begin{bmatrix} \Gamma(i) \\ 0 \end{bmatrix}, \quad \bar{B}_{\Delta o}(i) = \begin{bmatrix} B_{\Delta o}(i) \\ 0 \end{bmatrix} \\
\bar{G}(i) &= \begin{bmatrix} 0 & A_{\Delta o}(t,i) + A_{\Delta d}(t,i) \\ -I & I \end{bmatrix}, \quad \Upsilon(i) = \begin{bmatrix} 0 & 0 \\ 0 & A_{\Delta d}(t,i) \end{bmatrix} \\
\Lambda_{\Delta}(i) &= [0 \quad G_o(i)], \; \bar{C}(i) = [0 \quad C_o(i)]
\end{aligned}
\tag{5.85}
$$

5.4.2 Nominal Design

In the absence of uncertainties ($\Delta_a \equiv 0$, $\Delta_d \equiv 0$), we extract from system (Σ_T) for $\eta_t = i \in \mathcal{S}$ the nominal jump system which under the feedback law

$$
\begin{aligned}
u(t) = K(i)x(t) &= [0 \quad K(i)]\zeta(t) \\
&\stackrel{\Delta}{=} \bar{K}(i)\zeta(t)
\end{aligned}
\tag{5.86}
$$

takes the form:

$$
\begin{aligned}
(\Sigma_{To}): \; \dot{\zeta}(t) &= \Lambda_{Ko}(i)\zeta(t) + \int_{t-\tau}^{t} \Upsilon_o(i)\, \zeta(s) ds \\
&+ \bar{\Gamma}(i) w(t) \\
\zeta(t) &= \bar{\phi}(t), \; t \in [-2\tau, 0]
\end{aligned}
\tag{5.87}
$$

$$
y(t) = \bar{C}_K(i)\zeta(t) + \Phi(i)w(t) \tag{5.88}
$$

$$
z(t) = \bar{G}_K(i)\zeta(t) + \Psi(i)w(t) \tag{5.89}
$$

where

$$
\begin{aligned}
A_{Kd}(i) &= A_{od}(i) + B_o(i)K(i), \quad A_{od}(i) = A_o(i) + A_d(i) \\
\Lambda_{Ko}(i) &= \begin{bmatrix} 0 & A_{Kd}(i) \\ -I & I \end{bmatrix}, \quad \Upsilon_o(i) = \begin{bmatrix} 0 & 0 \\ 0 & A_d(i) \end{bmatrix} \\
\bar{\Gamma}(i) &= \begin{bmatrix} \Gamma(i) \\ 0 \end{bmatrix}, \quad \bar{C}_K(i) = [0 \quad C_o(i) + D_o(i)K(i)] \\
\bar{G}_K(i) &= [0 \quad G_o(i) + F_o(i)K(i)]
\end{aligned}
\tag{5.90}
$$

For convenience, we introduce the matrices for $i \in \mathcal{S}$

$$
\begin{aligned}
\bar{P}(i) &= U\, \mathbb{P}(i) \ ; \ U = \begin{bmatrix} I & 0 \\ 0 & 0 \end{bmatrix} \\
\mathbb{P}(i) &= \begin{bmatrix} P_\sigma(i) & 0 \\ P_d(i) & P_x(i) \end{bmatrix}, \ E_1 = \begin{bmatrix} I \\ 0 \end{bmatrix}, \ E_2 = \begin{bmatrix} 0 \\ I \end{bmatrix} \\
Y(i) &= \mathbb{P}^{-1}(i) = \begin{bmatrix} Y_\sigma(i) & 0 \\ Y_d(i) & Y_x(i) \end{bmatrix} \\
Y_\sigma &= P_\sigma^{-1}(i), \ Y_x = P_x^{-1}(i), \ Y_d = -Y_x(i) P_d(i) Y_\sigma(i)
\end{aligned} \tag{5.91}
$$

5.4.3 $\mathcal{H}_2$-Performance

The analytical result is established by the following theorem:

Theorem 5.13 *In the absence of input disturbance $w(t) \equiv 0$, controller (5.86) is an $\mathcal{H}_2$-optimal controller for system (Σ_{To}) minimizing the $\mathcal{H}_2$-performance measure (5.5) if, given matrix sequence $Q_x(i) = Q_x^t(i) > 0$, $i \in \mathcal{S}$ there exist matrices $\mathbb{P}(i) = \mathbb{P}^t(i) > 0$, $i \in \mathcal{S}$, satisfying the system of LMIs for all $i \in \mathcal{S}$*

$$
\Pi_t(i) = \begin{bmatrix} \Xi_{Kno}(i) & \tau \mathbb{P}^t(i) E_2 A_d(i) & \tau E_2 Q_x(i) \\ \tau A_d^t(i) E_2^t \mathbb{P}(i) & -\tau Q_x(i) & 0 \\ \tau Q_x(i) E_2^t & 0 & -\tau Q_x(i) \end{bmatrix} < 0 \tag{5.92}
$$

where

$$
\begin{aligned}
\Xi_{Kno}(i) &= \Lambda_{Ko}^t(i) \mathbb{P}(i) + \mathbb{P}(i) \Lambda_{Ko}(i) + \sum_{m=1}^{s} \alpha_{im} \mathbb{P}(m) \\
&+ \bar{C}_K^t(i) \bar{C}_K(i) + \rho(i) \tau^2 E_2 \sum_{m=1}^{s} \alpha_{im} Q_x(m) E_2^t
\end{aligned}
$$

An upper bound on the $\mathcal{H}_2$ performance measure is given by

$$
\begin{aligned}
\mathcal{J}_2 &\leq \mathsf{J}^+ \triangleq \Big[x^t(0) \bar{P}(\eta_o) x(0) \\
&+ \tau(0) \int_{-\tau(0)}^{0} \zeta^t(s) E_2 Q_x(\eta_o) E_2^t \zeta(s)\, ds \Big]^{1/2}
\end{aligned} \tag{5.93}
$$

Proof: Let $\mathbf{x}_s(t) \triangleq x(s+t),\ t-\tau \leq s \leq t$ and define the process $\{(\mathbf{x}(t), \eta_t),\ t \geq 0\}$ over the state space $\bar{\mathcal{C}}$. It should be observed that $\{(\mathbf{x}(t), \eta_t),\ t \geq 0\}$ is strong Markovian [78] so is the process $\{(\zeta(t), \eta_t),\ t \geq 0\}$. Now for $\eta_t = i \in \mathcal{S}$, and given $Q_x(i) = Q_x^t(i) > 0$, let the Lyapunov functional $V(\cdot) : \Re^n \times \Re_+ \times \mathcal{S} \to \Re_+$ of the transformed system be selected as

$$
\begin{aligned}
V(t, \zeta, i) &= \zeta^t(t) \bar{P}(i) \zeta(t) \\
&+ \int_{t-\tau}^{t} \int_{t+\theta}^{t} \zeta^t(s) E_2 Q_x(i) E_2^t \zeta(s)\, ds d\theta
\end{aligned} \tag{5.94}
$$

The weak infinitesimal operator $\Im_1^\zeta[\cdot]$ of the process $\{\zeta(t), i, t \geq 0\}$ for system (5.87)-(5.88) at the point $\{t, x, i\}$ is given by [45, 78]:

$$
\Im_1^\zeta[V] = \partial V/\partial t + \partial V/\partial \zeta\ \dot{\zeta}(t)\ |_{\eta_t = i} + \sum_{m=1}^{s} \alpha_{im} V(t, \zeta, i, m) \tag{5.95}
$$

Using (5.90)-(5.91) and (5.87), we get:

$$
\begin{aligned}
\partial V/\partial \zeta\ \dot{\zeta}(t) &= 2\zeta^t(t)\, U \mathbb{P}^t(i)\ \dot{\zeta}(t) = 2\sigma^t(t)\ P_\sigma^t(i)\ \dot{\sigma}(t) \\
&= 2\zeta^t(t) \mathbb{P}^t(i) \begin{bmatrix} \dot{\sigma}(t) \\ 0 \end{bmatrix} \\
&= 2\zeta^t(t) \mathbb{P}^t(i) \begin{bmatrix} A_{Kd}(i) x(t) \\ -\sigma(t) + x(t) + \int_{t-\tau}^{t} A_d(i)\ x(s) ds \end{bmatrix} \\
&= 2\zeta^t(t) \mathbb{P}^t(i) \Lambda_{Ko}(i) \zeta(t) \\
&+ 2 \int_{t-\tau}^{t} \zeta^t(t) \mathbb{P}^t(i) \Upsilon_o(i) \zeta(\theta)\, d\theta
\end{aligned} \tag{5.96}
$$

Hence, it follows from (5.94)-(5.96) that

$$
\begin{aligned}
\Im_1^\zeta[V] &= \zeta^t(t) \big[\Lambda_{Ko}^t(i) \mathbb{P}(i) + \mathbb{P}^t(i) \Lambda_{Ko}(i) + \sum_{m=1}^{s} \alpha_{im} \mathbb{P}(m)\big] \zeta(t) \\
&+ 2 \int_{t-\tau}^{t} \zeta^t(t) \mathbb{P}^t(i) \Upsilon_o(i) \zeta(\theta)\, d\theta \\
&+ \sum_{m=1}^{s} \alpha_{im} \int_{t-\tau}^{t} \int_{\theta}^{t} \zeta^t(s) E_2 Q_x(m) E_2^t \zeta(s)\, ds d\theta
\end{aligned}
$$

$$
\begin{aligned}
&- \int_{t-\tau}^{t} \zeta^t(\theta) E_2 Q_x(i) E_2^t \zeta(\theta)\, d\theta \\
&+ \int_{t-\tau}^{t} \zeta^t(t) E_2 Q_x(i) E_2^t \zeta(t)\, d\theta
\end{aligned}
\tag{5.97}
$$

Since for some $\rho(i) > 0$, $i \in \mathcal{S}$

$$
\begin{aligned}
&\sum_{m=1}^{s} \alpha_{im} \int_{t-\tau}^{t} \int_{t+\theta}^{t} \zeta^t(s) E_2 Q_x(m) E_2^t \zeta(s)\, ds d\theta \le \\
&\tau^2\, \rho(i) \zeta^t(t) E_2 \sum_{m=1}^{s} \alpha_{im} Q_x(m) E_2^t \zeta(t)
\end{aligned}
\tag{5.98}
$$

and by **Fact 1**, we have

$$
\begin{aligned}
& 2 \int_{t-\tau}^{t} \zeta^t(t) \mathbb{P}^t(i) \Upsilon_o(i) \zeta(\theta)\, d\theta \\
= \;& 2 \int_{t-\tau}^{t} \zeta^t(t) \mathbb{P}^t(i) E_2 A_d(t,i) x(\theta)\, d\theta \\
\le \;& \tau \zeta^t(t) \mathbb{P}^t(i) E_2 A_d(i) Q_x^{-1}(i) A_d^t(i) E_2^t \mathbb{P}^t(i) \zeta(\theta) \\
+ \;& \int_{t-\tau}^{t} x^t(s) Q_x(i) x(s)\, ds \\
+ \;& \tau \zeta^t(t) \mathbb{P}^t(i) E_2 A_d(i) Q_x^{-1}(i) A_d^t(i) E_2^t \mathbb{P}(i) \zeta(t) \\
+ \;& \int_{t-\tau}^{t} \zeta^t(\theta) E_2 Q_x(i) E_2^t \zeta(\theta)\, d\theta
\end{aligned}
\tag{5.99}
$$

Now, it follows from (5.97)-(5.99) that

$$
\begin{aligned}
& \Im_1^{\zeta}[V] \\
\le \;& \zeta^t(t) \Big[\Lambda_{Ko}^t(i) \mathbb{P}(i) + \mathbb{P}^t(i) \Lambda_{Ko}(i) \\
+ \;& \sum_{m=1}^{s} \alpha_{im} \mathbb{P}(m) + \rho(i) \tau^2 E_2 \sum_{m=1}^{s} \alpha_{im} Q_x(m) E_2^t \\
+ \;& \tau \mathbb{P}^t(i) E_2 A_d(i) Q_x^{-1}(i) A_d^t(i) E_2^t \mathbb{P}(i) + \tau E_2 Q_x(i) E_2^t \Big] \zeta(t) \\
\triangleq \;& \zeta^t(t) \Pi_o(i) \zeta(t)
\end{aligned}
\tag{5.100}
$$

In view of (5.92) and the fact that

$$\bar{C}_K^t(i)\bar{C}_K(i) > 0$$

it follows that

$$\Pi_o(i) < 0$$

is guaranteed for all $i \in \mathcal{S}$ and we conclude that

$$\Im_1^\zeta[V] \; < \; 0 \; \forall \; \zeta \neq 0 \quad , \quad \Im_1^\zeta[V] \; \leq \; 0 \, \forall \; \zeta$$

Since $||\zeta(t+\beta)|| \; \leq \varphi||\zeta(t)|, \; \forall \beta \in [-\tau, 0]$ and some $\varphi > 0$ [85], it follows from (5.94) that

$$V(\zeta, i) \leq \zeta^t(t)\bar{P}(i)\zeta(t) + \mu||\zeta||^2$$

where

$$\mu = \varphi \, \tau \big(\max_i \lambda_M[\bar{P}(i)] + \tau \, \lambda_M[E_2 Q_x(i) E_2^t] \big)$$

Therefore, for all $\zeta \neq 0$, we have

$$\begin{aligned} \frac{\Im_1^\zeta[V]}{V(\zeta, i)} &\leq \frac{\zeta^t \Pi_o(i)\zeta}{\zeta^t \bar{P}(i)\zeta + \mu||\zeta||^2} \\ &\leq -\xi \\ &\triangleq -\min_{i \in \mathcal{S}} \left\{ \frac{\lambda_m[-\Pi_o(i)]}{\lambda_M[\bar{P}(i)] + \mu} \right\} \end{aligned} \tag{5.101}$$

It is readily seen from (5.101) that $\xi > 0$ and hence we get

$$\Im_1^\zeta[V] \; \leq -\xi \, V(t, \zeta, i)$$

It follows from [78] by using the Gronwall-Bellman lemma [98] and letting $\zeta(t = 0, \phi, \eta_o) = \zeta_o$, one has

$$\mathbb{E}[V(\zeta, i) \big| \phi, i] \; \leq \; e^{-\xi \, t} \, V(\zeta_o, i) \tag{5.102}$$

Since

$$\mathbb{E}\left\{\int_{t-\tau}^{t}\int_{t+\theta}^{t}\zeta^t(s)E_2Q_x(i)E_2^t\zeta(s)\;dsd\theta\middle|\phi,\eta_o\right\}\geq 0$$

it is easy to see from (5.102) that

$$\begin{aligned}
&\mathbb{E}\left\{\zeta^t(t)\bar{P}(i)\zeta(t)\middle|\phi,i\right\}\leq e^{-\zeta\,t}V(x_o,i)\Longrightarrow\\
&\mathbb{E}\left\{\int_0^{\mathcal{T}}\zeta^t(t)\bar{P}(i)\zeta(t)dt\middle|\phi,\eta_o=i\right\}\leq\\
&\left[\int_0^{\mathcal{T}}e^{-\xi\,t}dt\right]V(\zeta_o,i)=\frac{1}{\zeta}[e^{-\xi\,\mathcal{T}}\;-\;1]\;V(\zeta_o,i)\Longrightarrow\\
&\lim_{\mathcal{T}\to\infty}\;\mathbb{E}\left\{\int_0^{\mathcal{T}}\zeta^t(t)\bar{P}(i)\zeta(t)dt\middle|\phi,\eta_o-i\right\}\leq\\
&\frac{1}{\xi}\zeta_o^t\bar{P}(\eta_o)\zeta_o+\frac{\tau^+}{\xi}[E_2Q_x((\eta_o))E_2^t]||\zeta(t+\theta)||_*^2,\quad\forall\theta\in[-\tau,0]
\end{aligned}\tag{5.103}$$

where

$$||\zeta(t+\theta)||_*^2\triangleq\sup_{\theta\in[-\tau,0]}\;||\zeta(t+\theta)||_2^2$$

Let

$$\Xi_P(i)\;=\;\max_{i\in\mathcal{S}}\left\{\frac{\bar{P}(\eta_o)||\zeta_o||^2+\tau^+[E_2Q_x(\eta_o)E_2^t]||x(t+\theta)||_*^2}{\zeta[\bar{P}(\eta_o)]||\zeta_o||^2}\right\}$$

it follows from (5.103) for $i\in\mathcal{S}$ that

$$\begin{aligned}
&\mathbb{E}\left\{\int_0^{\infty}\zeta^t(t)\zeta(t)dt|\phi,\eta_o=i\right\}\\
\leq\;&\zeta_o^t\lambda_M[\Xi_P(i)]\zeta_o<+\infty
\end{aligned}$$

which, in the light of Chapter 3, shows that system (Σ_T) is **SSSDD** under the control law (5.86). Now by the Dynkin's formula and (5.92)

$$\begin{aligned}
&\mathbb{E}[V(\zeta(t),\eta_t)]-V(\zeta(0),\eta_o)=\\
&\mathbb{E}\left[\int_0^t\Im_1^\zeta[V(\zeta(v),\eta_v)]dv\right]\Longrightarrow\\
&\mathbb{E}[V(\zeta(t_f),\eta_{t_f})]-V(\zeta(0),\eta_o)\leq\\
&\mathbb{E}\left[\int_0^{t_f}\chi^t\;\Pi_t(\eta_v)\;\chi dv\right]
\end{aligned}\tag{5.104}$$

By letting $t_f \to \infty$ and in view the system stability, it follows that

$$\begin{aligned}
& \mathbb{E}\left[\int_0^\infty y^t(r)y(r)dr\right] \leq V[\zeta(0), \eta_o)] \\
= & \ \zeta^t(0)\bar{P}(\eta_o)\zeta(0) \\
+ & \int_{-\tau(0)}^{0}\int_{t+\theta}^{t} \zeta^t(s)E_2Q_x(\eta_o)E_2^t\zeta(s)\, dsd\theta \Longrightarrow \\
||y|| \leq & \Bigg[\zeta^t(0)\bar{P}(\eta_o)\zeta(0) \\
+ & \ \tau(0)\int_{-\tau(0)}^{0} \zeta^t(s)E_2Q_x(\eta_o)E_2^t\zeta(s)\, ds\Bigg]^{1/2} \\
= & \ \mathsf{J}^+
\end{aligned} \tag{5.105}$$

which completes the proof. ∇∇∇

Remark 5.3 *It should be noted that* **Theorem** *5.13 establishes an LMI-based sufficient condition for the existence of state-feedback controller (5.86) and hence it depends on* $K(i),\ i \in \mathcal{S}$.

The following theorem provides a method for computing the feedback gains

Theorem 5.14 *The feedback gain associated with the* $\mathcal{H}_2$*-optimal controller for system* (Σ_{To}) *is given by* $K(i) = Z(i)Y^{-1}(i),\ i \in \mathcal{S}$ *where the matrices* $Y_d(i) = Y_d^t(i) > 0,\ Y_x(i) = Y_x^t(i) > 0,\ Y_\sigma(i) = Y_\sigma^t(i) > 0,\ Z(i),\ L(i),\ i \in \mathcal{S}$*, satisfy the system of LMIs and AIs for all* $i \in \mathcal{S}$

$$\begin{aligned}
& \begin{bmatrix} \Lambda_d(i) & \Omega_{nn}(i) & \mathcal{R}_\sigma(i) \\ \Omega_{nn}^t(i) & -\Omega_{mm}(i) & 0 \\ \mathcal{R}_\sigma^t(i) & 0 & -\mathcal{Y}(i) \end{bmatrix} < 0 \\
& \begin{bmatrix} Y_x(i) + Y_x^t(i) & \Omega_{rr}(i) & \mathcal{R}_\sigma(i) \\ \Omega_{rr}^t(i) & -\Omega_{qq}(i) & 0 \\ \mathcal{R}_\sigma^t(i) & 0 & -\mathcal{Y}(i) \end{bmatrix} < 0 \\
& -Y_\sigma(i) + Y_d(i) + Y_x^t(i)A_{od}^t(i) + L^t(i)B_o^t(i) + \\
& \mathcal{R}_d(i)\mathcal{Y}^{-1}(i)\mathcal{R}_d(i) + Y_x^t(i)Q_x(i)Y_d(i)
\end{aligned} \tag{5.106}$$

$$Y_x(i)[C_o^t(i) + L^t(i)D_o^t(i)][C_o(i) + D_o(i)L(i)]Y_x^t(i) +$$
$$\rho(i)\tau^2 Y_x^t(i)\mathcal{Q}_x(i)Y_d(i) \geq 0$$
$$Z(i)\, Y_d^{-1}(i) \;-\; L(i)\, Y_x^{-1}(i) \;\geq\; 0 \tag{5.107}$$

where

$$\begin{aligned}
\Lambda_d(i) &= Y_d(i)A_{od}^t(i) + A_{od}(i)Y_d(i) + B_o(i)Z(i) \\
&+ Z^t(i)B_o^t(i) + \alpha_{ii}Y(i) \\
\Omega_{nn}(i) &= [Y_d^t(i)C_o^t(i) + Y_d^t(i)L^t(i)D_o^t(i) \;\; \Omega_{mm}(i)] \\
\Omega_{mm}(i) &= [\tau Y_d^t(i) \;\; \rho(i)\tau Y_d^t(i)] \\
\Omega_{dd}(i) &= diag[I \;\; \tau Q_x(i) \;\; \rho(i)\tau\mathcal{Q}_x(i)] \\
\Omega_{rr}(i) &= [Y_x^t(i)C_o^t(i) + Y_x^t(i)L^t(i)D_o^t(i) \;\; \Omega_{qq}(i)] \\
\Omega_{qq}(i) &= [\tau Y_x^t(i) \;\; \rho(i)\tau Y_x^t(i) \;\; \tau A_d(i)] \\
\Omega_{hh}(i) &= diag[I \;\; \tau Q_x(i) \;\; \rho(i)\tau\mathcal{Q}_x(i) \;\; \tau Q_x(i)] \\
\mathcal{Y}(i) &= diag\left[Y(1)...Y(i-1) \;\; Y(i+1)...Y(s)\right] \\
\mathcal{Q}_x(i) &= \sum_{m=1}^{s} \alpha_{im}\, Q_x(m) \\
\mathcal{R}(i) &= \left[\sqrt{\alpha_{i1}}Y(i).....\sqrt{\alpha_{is}}Y(i)\right] \\
&= \begin{bmatrix} \mathcal{R}_\sigma(i) & 0 \\ \mathcal{R}_d(i) & \mathcal{R}_d(i) \end{bmatrix}
\end{aligned} \tag{5.108}$$

Proof: By **Fact 4**, condition (5.92) is equivalent to

$$\begin{aligned}
& \Lambda_{Ko}^t(i)\mathbb{P}(i) + \mathbb{P}^t(i)\Lambda_{Ko}(i) \\
+ & \sum_{m=1}^{s} \alpha_{im}\mathbb{P}(m) + \bar{C}_K^t(i)\bar{C}_K(i) + \\
+ & \rho(i)\tau^2 E_2 \sum_{m=1}^{s} \alpha_{im}Q_x(m)E_2^t + \tau E_2 Q_x(i)E_2^t \\
+ & \tau\mathbb{P}^t(i)E_2 A_d(i)Q_x^{-1}(i)A_d^t(i)E_2^t\mathbb{P}(i) < 0
\end{aligned} \tag{5.109}$$

Using (5.90) and (5.108), premultiplying (5.109) by $Y^t(i) = \mathbb{P}^{-t}(i)$ and postmultiplying the result by $Y(i)$, we obtain:

$$
\begin{aligned}
& Y^t(i)\Lambda_{Ko}^t(i) + \Lambda_{Ko}(i)Y(i) + \alpha_{ii}Y(i) + \mathcal{R}(i)\mathcal{Y}^{-1}(i)\mathcal{R}^t(i) \\
+ \quad & \rho(i)\tau^2 Y^t(i)E_2 \sum_{m=1}^{s} \alpha_{im} Q_x(m)E_2^t Y(i) + \tau E_2 A_d(i) Q_x^{-1}(i) A_d^t(i) E_2^t \\
+ \quad & \tau Y^t(i) E_2 Q_x(i) E_2^t Y(i) + Y^t(i)\bar{C}_K^t(i)\bar{C}_K(i)Y(i) < 0
\end{aligned} \qquad (5.110)
$$

Letting

$$
K(i)Y_d(i) = Z(i), \; K(i)Y_x(i) = L(i)
$$

inequality (5.110) can be expanded into blocks and upon using **Fact 3** LMIs (5.106) subject to AIs (5.107) follow immediately. $\nabla\nabla\nabla$

5.4.4 $\mathcal{H}_\infty$ Performance

Turning to the $\mathcal{H}_\infty$ performance, the following theorems summarize the main results:

Theorem 5.15 *Given a prescribed constant $\gamma > 0$, controller (5.86) renders system (Σ_{To})* **RSSSDD** *with a disturbance attenuation level γ for all $w(t) \in \mathcal{L}_2[0,\infty)$ if given matrix sequence $Q_x(i) = Q_x^t(i) > 0$, $i \in \mathcal{S}$ there exist matrices $I\!P(i) = I\!P^t(i) > 0$, $i \in \mathcal{S}$, satisfying the system of LMIs $i \in \mathcal{S}$*

$$
\Pi_t(i) = \begin{bmatrix}
\Xi_{Kco}(i) & \Omega_{ff}(i) & \begin{matrix} I\!P^t(i)\bar{\Gamma}(i) + \\ \bar{G}_K^t(i)\Phi(i) \end{matrix} \\
\Omega_{ff}^t & -\Omega_{pp}(i) & 0 \\
\begin{matrix} \bar{\Gamma}^t(i) I\!P(i) + \\ \Phi^t(i)\bar{G}_K(i) \end{matrix} & 0 & \begin{matrix} -\gamma^2 I + \\ \Phi^t(i)\Phi(i) \end{matrix}
\end{bmatrix} < 0 \qquad (5.111)
$$

where

$$
\Xi_{Kco}(i) = \Lambda_{Ko}^t(i) I\!P(i) + I\!P(i)\Lambda_{Ko}(i) + \sum_{m=1}^{s} \alpha_{im} I\!P(m)
$$

$$
\begin{aligned}
&+ \quad \bar{G}_K^t(i)\bar{G}_K(i) + \rho(i)\tau^2 E_2 \sum_{m=1}^{s} \alpha_{im} Q_x(m) E_2^t \\
\Omega_{ff}(i) &= [\tau I\!P^t(i) E_2 A_d(i) \quad \tau E_2 Q_x(i)] \\
\Omega_{pp}(i) &= diag[\tau Q_x(i) - \tau\mu(i)\bar{N}_d^t(i)\bar{N}_d(i) \quad \tau Q_x(i)]
\end{aligned}
$$

Moreover

$$
\begin{aligned}
||z(t)|| &= \Big[\gamma^2 ||w(t)||^2 + \zeta^t(0)\bar{P}(\eta_o)\zeta(0) \\
&+ \tau(0) \int_{-\tau(0)}^{0} \zeta^t(s) E_2 Q_x(\eta_o) E_2^t \zeta(s)\, ds\Big]^{1/2}
\end{aligned} \tag{5.112}
$$

Proof: The stochastic stability follows as a result of **Theorem** 5.13. To show that system (Σ_{To}) has a disturbance attenuation γ, we let the Lyapunov functional $V(t, \zeta, i)$ be given by (5.94). By evaluating the weak infinitesimal operator $\Im_2^x[\cdot]$ of the process $\{\zeta(t), \eta_t = i, t \geq 0\}$ for system (5.87)-(5.89) at the point $\{t, \zeta, \eta_t = i\}$ we get

$$
\begin{aligned}
\Im_2^\zeta[V] &\leq \Im_1^\zeta[V] + \zeta^t(t)P(i)\bar{\Gamma}(i)w(t) + w^t(t)\bar{\Gamma}^t(i)P(i)\zeta(t) \\
&\triangleq \Im_+^\zeta[V]
\end{aligned} \tag{5.113}
$$

By Dynkin's formula, one has

$$
I\!E\Big\{ \int_0^T \Im_+^\zeta[V(.)]dt \Big\} = I\!E\Big\{ V(\zeta(t), \eta_t)|_{t=T} \Big\} - V(\zeta(t), \eta_t)|_{t=0}
$$

By standard matrix manipulations using (5.89)-(5.90) and (5.100), it follows from (5.113) that:

$$
z^t(t)z(t) - \gamma^2 w^t(t)w(t) + \Im_+^\zeta[V] \leq \begin{bmatrix} \zeta(t) \\ \zeta(t-\tau) \\ w(t) \end{bmatrix}^t \Pi_r(i) \begin{bmatrix} \zeta(t) \\ \zeta(t-\tau) \\ w(t) \end{bmatrix} \tag{5.114}
$$

Therefore we have:

$$
\mathcal{J}_T \leq I\!E\Big\{ \int_0^T \begin{bmatrix} \zeta(t) \\ \zeta(t-\tau) \\ w(t) \end{bmatrix}^t \Pi_r(i) \begin{bmatrix} \zeta(t) \\ \zeta(t-\tau) \\ w(t) \end{bmatrix} dt \Big\}
$$

$$- \quad \mathbb{E}\Big\{ V(\zeta(t), \eta_t)\big|_{t=T} + V(\zeta(t), \eta_t)\big|_{t=0} \tag{5.115}$$

In view of (5.111) and the fact that

$$\mathbb{E}\{V(\zeta(t), \eta_t)|_{t=T} \geq 0\}$$

then (5.115) leads to

$$\begin{aligned}
\mathcal{J}_T &\leq V(\zeta(0), \eta_0) \Longrightarrow \mathcal{J}_\infty \leq V(\zeta(0), \eta_0) \Longrightarrow \\
||z(t)||^2 - \gamma^2 \, ||w(t)||^2 &\leq x_0^t P(\eta_o) x_0 + \tau(0) \int_{-\tau(0)}^{0} \phi^t(\beta) Q(\eta_o) \phi(\beta) d\beta \Longrightarrow \\
||z(t)||^2 &\leq \gamma^2 \, ||w(t)||^2 + x_0^t P(\eta_o) x_0 \\
&+ \tau(0) \int_{-\tau(0)}^{0} \phi^t(\beta) Q(\eta_o) \phi(\beta) d\beta
\end{aligned} \tag{5.116}$$

which completes the proof. ∇∇∇

Theorem 5.16 *Given a prescribed constant $\gamma > 0$. The feedback gain associated with the $\mathcal{H}_\infty$-controller for system (Σ_{To}) is given by $K(i) = Z(i)Y_d^{-1}(i) = L(i)Y_x^{-1}(i),\ i \in \mathcal{S}$ where the matrices $Y_d(i) = Y_d^t(i) > 0,\ Y_x(i) = Y_x^t(i) > 0,\ Y_\sigma(i) = Y_\sigma^t(i) > 0,\ Z(i),\ L(i),\ i \in \mathcal{S}$, satisfy the system of LMIs*

$$\begin{bmatrix} \Lambda_d(i) & \Omega_{ss}(i) & \mathcal{R}_\sigma(i) \\ \Omega_{ss}^t(i) & -\Omega_{zz}(i) & 0 \\ \mathcal{R}_\sigma^t(i) & 0 & -\mathcal{Y}(i) \end{bmatrix} < 0$$

$$\begin{bmatrix} Y_x(i) + Y_x^t(i) & \Omega_{vv}(i) & \mathcal{R}_x(i) \\ \Omega_{vv}^t(i) & -\Omega_{ww}(i) & 0 \\ \mathcal{R}_x^t(i) & 0 & -\mathcal{Y}(i) \end{bmatrix} < 0 \tag{5.117}$$

$$\begin{aligned}
&- \quad Y_\sigma(i) + Y_d(i) + Y_x^t(i) A_{od}^t(i) + L^t(i) B_o^t(i) \\
&+ \quad \mathcal{R}_d(i) \mathcal{Y}^{-1}(i) \mathcal{R}_d(i) + Y_x^t(i) Q_x(i) Y_d(i) + \rho(i)\tau^2 Y_x^t(i) \mathcal{Q}_x(i) Y_d(i) \\
&+ \quad Y_x(i)[C_o^t(i) + L^t(i) D_o^t(t)][C_o(i) + D_o(t) L(i)] Y_x^t(i) \\
&+ \quad [Y_x^t(i) G_o^t(i) + L^t(i) F_o^t(i)] \Phi(i) \big(\gamma^2 I - \Phi^t(i) \Phi(i)\big)
\end{aligned}$$

$$\cdot \quad [\Gamma^t(i) + \Phi^t(i)\big(G_o(i)Y_d(i) + F_o(i)Z(i)\big)] \geq 0$$

$$Z(i)\, Y_d^{-1}(i) \;-\; L(i)\, Y_x^{-1}(i) \geq 0 \tag{5.118}$$

where

$$\begin{aligned}
\Omega_{ss}(i) &= [\Omega_{nn}(i) \quad Y_d^t(i)C_o^t(i)\Phi(i) + Z^t(i)D_o^t(i)\Phi(i)] \\
\Omega_{zz}(i) &= diag[\Omega_{mm}(i) \quad \gamma^2 I - \Phi^t(i)\Phi(i)] \\
\Omega_{vv}(i) &= [\Omega_{vw}(i) \quad \Omega_{wv}(i)] \\
\Omega_{vw}(i) &= [G_o(i)Y_d(i) + D_o(t)L(i) \quad Y_x^t(i)G_o^t(i)\Phi(i) + L^t(i)F_o^t(i)\Phi(i)] \\
\Omega_{wv}(i) &= [\tau Y_x^t(i) \quad \rho(i)\tau Y_x^t(i) \quad \tau A_d(i)] \\
\Omega_{ww}(i) &= diag[\tau Q_x(i) \quad \rho(i)\tau Q_x(i) \quad \tau Q_x(i)]
\end{aligned} \tag{5.119}$$

Proof: Follows from parallel development to **Theorem** 5.14. ∇∇∇

5.4.5 Mixed $\mathcal{H}_2/\mathcal{H}_\infty$ Control

Now we are ready to provide the solution to the mixed $\mathcal{H}_2/\mathcal{H}_\infty$ control problem. It is summarized by the following theorem:

Theorem 5.17 *Given a prescribed constant $\gamma > 0$. The feedback gain $K(i) = Z(i)Y_d^{-1}(i) = L(i)Y_x^{-1}(i)$, $i \in \mathcal{S}$ is a* **mixed $\mathcal{H}_2/\mathcal{H}_\infty$ controller** *satisfying the performance measure (5.6) for system (Σ_{To}) if there exist matrices $Y_d(i) = Y_d^t(i) > 0$, $Y_x(i) = Y_x^t(i) > 0$, $Y_\sigma(i) = Y_\sigma^t(i) > 0$, $Z(i)$, $L(i)$, $W = W^t > 0$, $i \in \mathcal{S}$, such the system of generalized eigenvalue problems*

$$\min \Big[\lambda \;+\; Tr(W)\Big]$$

$$subject \;\; to \;\; LMIs \;\; (5.106), (5.107), (5.117), (5.118) \;\; and$$

$$\begin{bmatrix} -W & X^t(i) \\ X(i) & M_x(i) \end{bmatrix} < 0 \;, \; \begin{bmatrix} -\lambda & \bar{\phi}^t(0)E_1 \\ E_1^t\bar{\phi}(0) & Y_x(i) \end{bmatrix} < 0 \tag{5.120}$$

has a feasible solution

Proof: On observing that

$$\begin{aligned}
\zeta^t(0)\bar{P}(\eta_t = i)\zeta(0) &\leq \max\left[\zeta^t(0)\bar{P}(\eta_t = 1)\zeta(0), \ldots\ldots, \zeta^t(0)\bar{P}(\eta_t = s)\zeta(0)\right] \\
&\triangleq \lambda \Longrightarrow \\
&- \lambda + \bar{\phi}^t(0)E_1 Y_\sigma^{-1}(i)E_1^t\bar{\phi}(0) < 0 \qquad (5.121)
\end{aligned}$$

and in similar way using the cyclic properties of matric trace [25]

$$\begin{aligned}
&\int_{-\tau(0)}^{0} \zeta^t(s)E_2 Q_x(\eta_o)E_2^t\zeta(s)ds = \\
&= \int_{-\tau(0)}^{0} Tr\left[x^t(s)Q_x(\eta_o)x(s)\right]ds \\
&= Tr\left[XX^tQ_x(i)\right] \\
&= Tr\left[X^tM_x^{-1}(i)X\right] \\
&< Tr(W) \Longrightarrow \\
&- W + X^tM_x^{-1}(i)X < 0 \qquad (5.122)
\end{aligned}$$

where

$$XX^t = \int_{-\tau(0)}^{0} x^t(s)x(s)ds$$

Utilizing the results of **Theorems** 5.14 and 5.16 and achieving the objective of mixed $\mathcal{H}_2/\mathcal{H}_\infty$ control leads to the minimization of

$$[\lambda + Tr(W)]$$

subject to LMIs (5.106), (5.107), (5.117), (5.118). Relations (5.121) and (5.122) are expressed by LMI (5.120), which completes the proof. $\nabla\nabla\nabla$

5.4.6 Design with Uncertainties

In this section, we consider the robust counterparts of **Theorems** 5.13-5.17 for system (Σ_J) under the feedback control (5.7). Using the transformation (5.79) and under the feedback law

$$\begin{aligned} u(t) &= K(i)x(t) \\ &= [0 \quad K(i)]\zeta(t) \\ &\triangleq \bar{K}(i)\zeta(t) \end{aligned}$$

the uncertain closed loop system is expressed for $\eta_t = i \in \mathcal{S}$ as:

$$\begin{aligned} (\Sigma_{T\Delta}): \quad \dot{\zeta}(t) &= \Lambda_{K\Delta}(i)\zeta(t) \\ &+ \int_{t-\tau}^{t} \Upsilon_{\Delta}(i)\, \zeta(s)ds + \bar{\Gamma}(i)w(t) \\ \zeta(t) &= \bar{\phi}(t), \; t \in [-2\tau, 0] \qquad (5.123) \\ y(t) &= \bar{C}_K(i)\zeta(t) + \Phi(i)w(t) \qquad (5.124) \\ z(t) &= \bar{G}_K(i)\zeta(t) + \Psi(i)w(t) \qquad (5.125) \end{aligned}$$

where

$$\begin{aligned} N_{aK}(i) &= N_a(i) + N_b(i)K(i)\ , \ \bar{N}_d(i) = [0 \quad N_d(i)] \\ \bar{N}_{aK}(i) &= [0 \quad N_{aK}(i)] \\ A_{\Delta Kd}(i) &= A_{\Delta o}(i) + A_{\Delta d}(i) + B_{\Delta o}(i)K(i) \\ &= A_{Kd}(i) + M_a(i)\Delta_a(t,i)N_{aK}(i) + M_a(i)\Delta_d(t,i)N_d(i) \\ \Lambda_{\Delta K}(i) &= \begin{bmatrix} 0 & A_{\Delta Kd}(i) \\ -I & I \end{bmatrix} = \begin{bmatrix} 0 & A_{Kd}(i) \\ -I & I \end{bmatrix} \\ &+ \begin{bmatrix} M_a(i) \\ 0 \end{bmatrix} \Delta_a(t,i)\bar{N}_{aK} + \begin{bmatrix} M_a(i) \\ 0 \end{bmatrix} \Delta_d(t,i)\bar{N}_d(i) \\ &= \Lambda_{oK}(i) + \bar{M}_a(i)\Delta_a(t,i)\bar{N}_{aK} + \bar{M}_a(i)\Delta_d(t,i)\bar{N}_d(i) \\ \Upsilon_{\Delta}(i) &= \begin{bmatrix} 0 & 0 \\ 0 & A_{\Delta d}(i) \end{bmatrix} \\ &= \begin{bmatrix} 0 & 0 \\ 0 & A_d(i) \end{bmatrix} + \begin{bmatrix} 0 \\ M_a(i) \end{bmatrix} \Delta_d(t,i)\bar{N}_d(i) \\ &= \Upsilon_o(i) + \hat{M}_a(i)\Delta_d(t,i)\bar{N}_d(i) \qquad (5.126) \end{aligned}$$

5.4.7 $\mathcal{H}_2$ Performance

For $\mathcal{H}_2$ performance, we have the following results:

Theorem 5.18 *In the absence of input disturbance $w(t) \equiv 0$, controller (5.7) is an $\mathcal{H}_2$-optimal controller for system $(\Sigma_{T\Delta})$ minimizing the $\mathcal{H}_2$-performance measure (5.5) if, given matrix sequence $Q_x(i) = Q_x^t(i) > 0,\ i \in \mathcal{S}$ there exist matrices $I\!P(i) = I\!P^t(i) > 0,\ i \in \mathcal{S}$, and scalars $\varepsilon(i) > 0,\ \nu(i) > 0,\ \mu(i) > 0,\ i \in \mathcal{S}$ satisfying the system of LMIs $i \in \mathcal{S}$*

$$\Pi_t(i) \triangleq \begin{bmatrix} \Xi_t(i) & \Xi_{m_1}(i) & \Xi_{m_2}(i) \\ \Xi_{m_1}^t(i) & -\Xi_{d_1}(i) & 0 \\ \Xi_{m_2}^t(i) & 0 & -\Xi_{d_2}(i) \end{bmatrix} < 0 \tag{5.127}$$

where

$$\begin{aligned} \Xi_t(i) &= \Lambda_{Ko}^t(i) I\!P(i) + I\!P(i)\Lambda_{Ko}(i) + \sum_{m=1}^{s} \alpha_{im} I\!P(m) \\ &+ \bar{C}_K^t(i)\bar{C}_K(i) + \rho(i)\tau^2 E_2 \sum_{m=1}^{s} \alpha_{im} Q_x(m) E_2^t \\ &+ \varepsilon(i)\bar{N}_{aK}^t(i)\bar{N}_{aK}(i) + \mu(i) N_d^t(i) N_d(i) \\ \Xi_{m_1}(i) &= [\tau I\!P^t(i) E_2 A_d(i) \quad \tau E_2 Q_x(i)] \\ \Xi_{d_1}(i) &= diag[\tau Q_x(i) - \tau\mu(i)\bar{N}_d^t(i)\bar{N}_d(i) \quad \tau Q_x(i)] \\ \Xi_{m_2}(i) &= [I\!P^t(i)\bar{M}_a(i) \quad I\!P^t(i)\bar{M}_a(i) \quad \tau I\!P^t(i)\hat{M}_a(i)] \\ \Xi_{d_2}(i) &= diag[\varepsilon(i) I \quad \nu(i) I \quad \tau\mu(i) I] \end{aligned} \tag{5.128}$$

for all admissible uncertainties satisfying (5.49)-(5.50). An upper bound on the $\mathcal{H}_2$ performance measure is given by

$$\begin{aligned} \mathcal{J}_2 &\le \mathsf{J}^+ \triangleq \Big[x^t(0)\bar{P}(\eta_o)x(0) \\ &+ \tau(0)\int_{-\tau(0)}^{0} \zeta^t(s) E_2 Q_x(\eta_o) E_2^t \zeta(s)\, ds \Big]^{1/2} \end{aligned} \tag{5.129}$$

Proof: It follows from **Theorems** (5.13) and (5.14) that controller (5.7) is $\mathcal{H}_2$-optimal controller for system $(\Sigma_{T\Delta})$ if there exist matrices $P(i) = P^t(i) > 0$ satisfying the system of ARIs

$$\begin{aligned} & \Lambda^t_{\Delta K}(i)\mathbb{P}(i) + \mathbb{P}^t(i)\Lambda_{\Delta K}(i) + \sum_{m=1}^{s} \alpha_{im}\bar{P}(m) \\ + \;& \bar{C}^t_K(i)\bar{C}^t_K(i) + \rho(i)\tau^2 E_2 \sum_{m=1}^{s} \alpha_{im} Q_x(m) E^t_2 \\ + \;& \tau\mathbb{P}^t(i)E_2 A_{\Delta d}(i) Q_x^{-1}(i) A^t_{\Delta d}(i) E^t_2 \mathbb{P}(i) + \tau E_2 Q_x(i) E^t_2 \;<\; 0 \end{aligned} \quad (5.130)$$

By **Fact 2** and (5.126) we have

$$\begin{aligned} & \Lambda^t_{\Delta K}(i)\mathbb{P}(i) + \mathbb{P}^t(i)\Lambda_{\Delta K}(i) \le \Lambda^t_{oK}(i)\mathbb{P}(i) + \mathbb{P}^t(i)\Lambda_{oK}(i) \\ + \;& \varepsilon^{-1}(i)\mathbb{P}^t(i)\bar{M}_a(i)\bar{M}^t_a(i)\mathbb{P}(i) + \nu^{-1}(i)\mathbb{P}^t(i)\bar{M}_a(i)\bar{M}^t_a(i)\mathbb{P}(i) \\ + \;& \varepsilon(i)\bar{N}^t_{aK}(i)\bar{N}_{aK}(i) + \nu(i)\bar{N}_d(i)\bar{N}^t_d(i) \end{aligned} \quad (5.131)$$

for some scalars $\varepsilon(i) > 0,\ \nu(i) > 0,\ i \in \mathcal{S}$.

Similarly, by **Fact 3** we get

$$\begin{aligned} & \mathbb{P}^t(i)E_2 A_{\Delta d}(i) Q_x^{-1}(i) A^t_{\Delta d}(i) E^t_2 \mathbb{P}(i) \le \mu^{-1}(i)\mathbb{P}^t(i)\hat{M}_a(i)\hat{M}^t_a(i)\mathbb{P}(i) \\ + \;& \mathbb{P}^t(i)E_2 A_d(i)[Q_x(i) - \mu(i)N_d(i)N^t_d(i)]^{-1} A^t_d(i) E^t_2 \mathbb{P}(i) \end{aligned} \quad (5.132)$$

for some scalars $\varrho(i) > 0,\ i \in \mathcal{S}$. Combining (5.131)-(5.132) into (5.130), it yields:

$$\begin{aligned} & \Lambda^t_{oK}(i)\mathbb{P}(i) + \mathbb{P}^t(i)\Lambda_{oK}(i) + \sum_{m=1}^{s} \alpha_{im}\bar{P}(m) + \bar{C}^t_K(i)\bar{C}^t_K(i) \\ + \;& \varepsilon^{-1}(i)\mathbb{P}^t(i)\bar{M}_a(i)\bar{M}^t_a(i)\mathbb{P}(i) + \nu^{-1}(i)\mathbb{P}^t(i)\bar{M}_a(i)\bar{M}^t_a(i)\mathbb{P}(i) \\ + \;& \varepsilon(i)\bar{N}^t_{aK}(i)\bar{N}_{aK}(i) + \tau\mu^{-1}(i)\mathbb{P}^t(i)\hat{M}_a(i)\hat{M}^t_a(i)\mathbb{P}(i) \\ + \;& \tau\mathbb{P}^t(i)E_2 A_d(i)[Q_x(i) - \mu(i)\bar{N}_d(i)\bar{N}^t_d(i)]^{-1} A^t_d(i) E^t_2 \mathbb{P}(i) \\ + \;& \rho(i)\tau^2 E_2 \sum_{m=1}^{s} \alpha_{im} Q_x(m) E^t_2 \end{aligned}$$

$$+ \quad \tau E_2 Q_x(i) E_2^t + \nu(i) \bar{N}_d(i) \bar{N}_d^t(i) \ < \ 0 \tag{5.133}$$

Using **Fact 3**, we obtain LMIs (5.127) and the performance bound follows immediately. $\nabla\nabla\nabla$

A method for computing the feedback gain matrix is provided by the next theorem.

Theorem 5.19 *The feedback gain associated with the $\mathcal{H}_2$-optimal controller for system $(\Sigma_{T\Delta})$ is given by $K(i) = L(i) Y_x^{-1}(i),\ i \in \mathcal{S}$ where the matrices $Y_d(i) = Y_d^t(i) > 0,\ Y_x(i) = Y_x^t(i) > 0,\ Y_\sigma(i) = Y_\sigma^t(i) > 0,\ Z(i),\ L(i),\ i \in \mathcal{S}$ and scalars $\varepsilon(i) > 0,\ \nu(i) > 0,\ \mu(i) > 0,\ i \in \mathcal{S}$ satisfying the system of LMIs and AIs for all $\ i \in \mathcal{S}$*

$$\begin{bmatrix} \hat{\Lambda}_d(i) & \Omega_{nn}(i) & \mathcal{R}_\sigma(i) & \bar{\Xi}_m(i) \\ \Omega_{nn}^t(i) & -\Omega_{mm}^t(i) & 0 & 0 \\ \mathcal{R}_\sigma^t(i) & 0 & -\mathcal{Y}(i) & 0 \\ \bar{\Xi}_m(i) & 0 & 0 & -\Xi_d(i) \end{bmatrix} < 0$$

$$\begin{bmatrix} \bar{\Xi}_t(i) & \Omega_{rr}(i) & \mathcal{R}_x(i) \\ \Omega_{rr}^t(i) & -\Omega_{qq}^t(i) & 0 \\ \mathcal{R}_\sigma^t(i) & 0 & -\mathcal{Y}(i) \end{bmatrix} < 0$$

$$\begin{aligned} - \quad & Y_\sigma(i) + Y_d(i) + Y_x^t(i) A_{od}^t(i) + L^t(i) B_o^t(i) \\ + \quad & \mathcal{R}_d(i) \mathcal{Y}^{-1}(i) \mathcal{R}_d(i) + Y_x^t(i) Q_x(i) Y_d(i) \\ + \quad & Y_x(i) [C^t(i) + L^t(i) D^t(t)][C(i) + D(t) L(i)] Y_x^t(i) \\ + \quad & \varrho(i) \tau^2 Y_x^t(i) \mathcal{Q}_x(i) Y_d(i) \geq 0 \end{aligned} \tag{5.134}$$

$$Z(i)\, Y_d^{-1}(i) \ - \ L(i)\, Y_x^{-1}(i) \ \geq \ 0 \tag{5.135}$$

for all admissible uncertainties satisfying (5.49)-(5.50) where

$$\begin{aligned} \hat{\Lambda}_d(i) \quad = \quad & Y_d(i) A_{od}^t(i) + A_{od}(i) Y_d(i) + B_o(i) Z(i) + Z^t(i) B_o^t(i) + \alpha_{ii} Y(i) \\ + \quad & \varepsilon(i) [Y_d^t(i) N_a^t(i) + Z^t(i) N_b^t(i)][N_a(i) Y_d(i) + N_b(i) Z(i)] \\ + \quad & \nu(i) Y_d^t(i) N_d^t(i) N_d(i) Y_d(i) \end{aligned}$$

$$
\begin{aligned}
\bar{\Xi}_m(i) &= [\bar{M}_a(i) \;\; \bar{M}_a(i) \;\; \tau \hat{M}_a(i)] \\
\bar{\Xi}_t(i) &= Y_x(i) + Y_x^t(i) + \varepsilon(i)[Y_d^t(i)N_a^t(i) + L^t(i)N_b^t(i)][N_a(i)Y_x(i) + N_b(i)L(i)] \\
&+ \nu(i)Y_x^t(i)N_d^t(i)N_d(i)Y_x(i) \qquad (5.136)
\end{aligned}
$$

Proof: By substituting $Y(i) = P^{-1}(i),\ K(i)Y_d(i) = Z(i),\ K(i)Y_x(i) = L(i)$, into (5.133) and manipulating we obtain LMIs (5.134) subject to ARIs (5.135). The performance bound (5.126) follows immediately. ∇∇∇

5.4.8 $\mathcal{H}_\infty$ Performance

With regards to the $\mathcal{H}_\infty$ performance, the following theorems summarize the main results:

Theorem 5.20 *Given a prescribed constant $\gamma > 0$, controller (5.7) renders system $(\Sigma_{T\Delta})$* **RSSSDD** *with a disturbance attenuation level γ for all $w(t) \in \mathcal{L}_2[0,\infty)$ if given matrix sequence $Q_x(i) = Q_x^t(i) > 0,\ i \in \mathcal{S}$ there exist matrices $I\!P(i) = I\!P^t(i) > 0,\ i \in \mathcal{S}$ and scalars $\varepsilon(i) > 0,\ \nu(i) > 0,\ \mu(i) > 0,\ i \in \mathcal{S}$ satisfying the system of LMIs $i \in \mathcal{S}$*

$$
\Pi_{tt}(i) =
\begin{bmatrix}
\Xi_{tt}(i) & \Omega_{ff}(i) & \begin{matrix} I\!P^t(i)\bar{\Gamma}(i) + \\ \bar{G}_K^t(i)\Phi(i) \end{matrix} & \Xi_m(i) \\
\Omega_{ff}^t(i) & -\Omega_{pp}(i) & 0 & 0 \\
\begin{matrix} \bar{\Gamma}^t(i)I\!P(i) + \\ \Phi^t(i)\bar{G}_K(i) \end{matrix} & 0 & \begin{matrix} -\gamma^2 I + \\ \Phi^t(i)\Phi(i) \end{matrix} & 0 \\
\Xi_m^t(i) & 0 & 0 & -\Xi_d(i)
\end{bmatrix} < 0 \quad (5.137)
$$

for all admissible uncertainties satisfying (5.49)-(5.50) where

$$
\begin{aligned}
\Xi_{tt}(i) &= \Lambda_{Ko}^t(i)I\!P(i) + I\!P(i)\Lambda_{Ko}(i) + \sum_{m=1}^{s} \alpha_{im} I\!P(m) + \bar{G}_K^t(i)\bar{G}_K(i) \\
&+ \rho(i)\tau^2 E_2 \sum_{m=1}^{s} \alpha_{im} Q_x(m) E_2^t + \varepsilon(i)\bar{N}_{aK}^t(i)\bar{N}_{aK}(i) + \mu(i)N_d^t(i)N_d(i)
\end{aligned}
$$

$$+ \quad \rho(i)\tau^2 E_2 \sum_{m=1}^{s} \alpha_{im} Q_x(m) E_2^t \tag{5.138}$$

Moreover

$$\begin{aligned} ||z(t)|| &= \Big[\gamma^2 ||w(t)||^2 + \zeta^t(0)\bar{P}(\eta_o)\zeta(0) \\ &+ \tau(0) \int_{-\tau(0)}^{0} \zeta^t(s) E_2 Q_x(\eta_o) E_2^t \zeta(s)\, ds\Big]^{1/2} \end{aligned} \tag{5.139}$$

Proof: Follows by parallel development to **Theorem** 5.15. $\nabla\nabla\nabla$

The following theorem gives a procedure to compute the feedback gain.

Theorem 5.21 *Given a prescribed constant $\gamma > 0$. The feedback gain associated with the $\mathcal{H}_\infty$-controller for system $(\Sigma_{T\Delta})$ is is given by $K(i) = L(i)Y_x^{-1}(i),\ i \in \mathcal{S}$ if there exist matrices $Y_d(i) = Y_d^t(i) > 0,\ Y_x(i) = Y_x^t(i) > 0,\ Y_\sigma(i) = Y_\sigma^t(i) > 0,\ Z(i),\ L(i),\ i \in \mathcal{S}$ and scalars $\varepsilon(i) > 0,\ \nu(i) > 0,\ \mu(i) > 0,\ i \in \mathcal{S}$ satisfying the system of LMIs and algebraic inequalities (AIs) for all $i \in \mathcal{S}$*

$$\begin{bmatrix} \hat{\Lambda}_d(i) & \Omega_{nn}(i) & \mathcal{R}_\sigma(i) & \bar{\Xi}_m(i) \\ \Omega_{nn}^t(i) & -\Omega_{mm}(i) & 0 & 0 \\ \mathcal{R}_\sigma^t(i) & 0 & -\mathcal{Y}(i) & 0 \\ \bar{\Xi}_m(i) & 0 & 0 & -\Xi_d(i) \end{bmatrix} < 0$$

$$\begin{bmatrix} \bar{\Xi}_t(i) & \Omega_{rr}(i) & \mathcal{R}_x(i) \\ \Omega_{rr}^t(i) & -\Omega_{qq}(i) & 0 \\ \mathcal{R}_\sigma^t(i) & 0 & -\mathcal{Y}(i) \end{bmatrix} < 0$$

$$\begin{aligned} &- \quad Y_\sigma(i) + Y_d(i) + Y_x^t(i) A_{od}^t(i) + L^t(i) B_o^t(i) \\ &+ \quad \mathcal{R}_d(i)\mathcal{Y}^{-1}(i)\mathcal{R}_d(i) + Y_x^t(i) Q_x(i) Y_d(i) \\ &+ \quad Y_x(i)[G^t(i) + L^t(i)F^t(t)][G(i) + F(t)L(i)]Y^t(i) \\ &+ \quad \varrho(i)\tau^2 Y_x^t(i) \mathcal{Q}_x(i) Y_d(i) \geq 0 \end{aligned} \tag{5.140}$$

$$Z(i)\, Y_d^{-1}(i) \;-\; L(i)\, Y_x^{-1}(i) \;\geq\; 0 \tag{5.141}$$

for all admissible uncertainties satisfying (5.49)-(5.50). Moreover,

$$||z(t)|| \quad = \quad \Big[\gamma^2 ||w(t)||^2 + x^t(0) Y^{-1}(\eta_o) x(0)$$

$$+ \int_{-\tau(0)}^{0} x^t(s)Q(\eta_o)x(s)ds\Bigg]^{1/2} \tag{5.142}$$

Proof. It can be worked out by using the same technique as that used in **Theorem** 5.19. ∇∇∇

5.4.9 Mixed Performance

Finally, the solution to the mixed $\mathcal{H}_2/\mathcal{H}_\infty$ control problem for the uncertain jumping system $(\Sigma_{\Delta c})$ is contained in the following theorem:

Theorem 5.22 *Given a prescribed constant $\gamma > 0$. The feedback gain $K(i) = Z(i)Y_x^{-1}(i),\ i \in \mathcal{S}$ is a* **mixed $\mathcal{H}_2/\mathcal{H}_\infty$ controller** *satisfying the performance measure (5.6) for system $(\Sigma_{T\Delta})$ if there exist matrices $Y_d(i) = Y_d^t(i) > 0,\ Y_x(i) = Y_x^t(i) > 0,\ Y_\sigma(i) = Y_\sigma^t(i) > 0,\ Z(i),\ L(i),\ i \in \mathcal{S}$ and scalars $\varepsilon(i) > 0,\ \nu(i) > 0,\ \mu(i) > 0,\ i \in \mathcal{S}$ such that the system of generalized eigenvalue problems*

$$\begin{array}{c} \min \Big[\lambda + Tr(W)\Big] \\ subject \;\; to \;\; LMIs \;\; 5.134, 5.135, 5.140, 5.141 \quad and \\ \begin{bmatrix} -\lambda & \phi^t(0) \\ \phi(0) & -Y(i) \end{bmatrix} < 0\,, \begin{bmatrix} -W & X^t \\ X & -L(i) \end{bmatrix} < 0 \end{array} \tag{5.143}$$

has a feasible solution for all admissible uncertainties satisfying (5.49)-(5.50).

5.5 Examples

Two examples will be presented regarding the mixed $\mathcal{H}_2/\mathcal{H}_\infty$ controller design for the nominal and the uncertain models, respectively.

5.5.1 Example 5.3

In order to illustrate **Theorem** 5.17, we consider a pilot-scale multi-reach water quality system which can fall into the type (5.1)-(5.3). Let the Markov process governing the mode switching has generator

$$\Im = \begin{bmatrix} -7 & 3 & 4 \\ 2 & -5 & 3 \\ 5 & 4 & -9 \end{bmatrix}$$

For the three operating conditions (modes), the associated date are:

Mode 1:

$$\begin{aligned}
A_o(1) &= \begin{bmatrix} -0.2 & 0 \\ 0 & -0.1 \end{bmatrix}, \; A_d(1) = \begin{bmatrix} -0.1 & 0 \\ -0.1 & -0.1 \end{bmatrix} \\
\Gamma(1) &= \begin{bmatrix} 2 \\ 1 \end{bmatrix}, \; B_o(1) = \begin{bmatrix} 1 & 0 \\ 0 & 1 \end{bmatrix}, \; C(1) = \begin{bmatrix} 2 & 0 \\ 0 & 2 \end{bmatrix} \\
D(1) &= \begin{bmatrix} 1 & 0 \\ 0 & 1 \end{bmatrix}, \; G(1) = [0.1 \;\; 0.1], \; \Psi(1) = \begin{bmatrix} 0.5 \\ 0.4 \end{bmatrix} \\
\Phi(1) &= 0.4, \; F(1) = [1 \;\; 2]
\end{aligned}$$

Mode 2:

$$\begin{aligned}
A_o(2) &= \begin{bmatrix} -2 & -1 \\ 0 & -2 \end{bmatrix}, \; A_d(2) = \begin{bmatrix} 0 & 1 \\ 1 & 0 \end{bmatrix} \\
\Gamma(2) &= \begin{bmatrix} 1 \\ 2 \end{bmatrix}, \; B_o(2) = \begin{bmatrix} 1 & 0 \\ 0 & 1 \end{bmatrix} \\
C(2) &= \begin{bmatrix} 1 & 0 \\ 0 & 2 \end{bmatrix}, \; D(2) = \begin{bmatrix} 1 & 0 \\ 0 & 1 \end{bmatrix} \\
G(2) &= [0.2 \;\; 0.1], \; \Psi(2) = \begin{bmatrix} 0.4 \\ 0.3 \end{bmatrix} \\
\Phi(2) &= 0.5, \; F(2) = [2 \;\; 1]
\end{aligned}$$

Mode 3:

$$\begin{aligned}
A_o(3) &= \begin{bmatrix} -1.9 & 0 \\ 0 & -1 \end{bmatrix}, \; A_d(3) = \begin{bmatrix} -0.9 & 0 \\ -1 & -1.1 \end{bmatrix} \\
\Gamma(3) &= \begin{bmatrix} 1 \\ 1 \end{bmatrix}, \; B_o(3) = \begin{bmatrix} 2 & 0 \\ 0 & 1 \end{bmatrix} \\
C(3) &= \begin{bmatrix} 2 & 0 \\ 0 & 2 \end{bmatrix}, \; D(3) = \begin{bmatrix} 1 & 0 \\ 0 & 1 \end{bmatrix}
\end{aligned}$$

$$
\begin{aligned}
G(3) &= [0.2 \quad 0.2]\,,\ \Psi(3) = \begin{bmatrix} 0.3 \\ 0.2 \end{bmatrix} \\
\Phi(3) &= 0.3\,,\ F(2) = [1 \quad 1]
\end{aligned}
$$

Invoking the software environment [57], we solve the system of LMIs (5.120) using

$$
\begin{aligned}
Q_\sigma(1) &= \begin{bmatrix} 2 & 0 \\ 0 & 1 \end{bmatrix}, \quad Q_\sigma(2) = \begin{bmatrix} 1 & 0 \\ 0 & 1 \end{bmatrix} \\
Q_\sigma(3) &= \begin{bmatrix} 2 & 0 \\ 0 & 2 \end{bmatrix}, \ \phi = \begin{bmatrix} e^{0.3\,t} \\ e^{0.3\,t} \end{bmatrix}, \quad \tau = 0.6
\end{aligned}
$$

The feasible solution is given by:

$$
\begin{aligned}
Y_x(1) &= \begin{bmatrix} 3.2354 & 1.1532 \\ 1.1532 & 3.0422 \end{bmatrix}, \ Y_\sigma(1) = \begin{bmatrix} 0.9765 & 0.2240 \\ 0.2240 & 3.1367 \end{bmatrix} \\
Y_d(1) &= \begin{bmatrix} 2.0065 & 1.1213 \\ 1.1213 & 3.2524 \end{bmatrix} \\
Y_x(2) &= \begin{bmatrix} 4.2064 & 1.4423 \\ 1.4423 & 4.8605 \end{bmatrix}, \ Y_\sigma(2) = \begin{bmatrix} 1.3264 & 0.7443 \\ 0.7443 & 2.8987 \end{bmatrix} \\
Y_d(2) &= \begin{bmatrix} 1.6554 & 1.4142 \\ 1.4142 & 6.8055 \end{bmatrix} \\
Y_x(3) &= \begin{bmatrix} 6.8095 & 2.4729 \\ 2.4729 & 7.1265 \end{bmatrix}, \ Y_\sigma(3) = \begin{bmatrix} 2.1134 & 1.7944 \\ 1.7944 & 5.2634 \end{bmatrix} \\
Y_d(3) &= \begin{bmatrix} 1.7791 & 0.7988 \\ 0.7988 & 2.275 \end{bmatrix} \\
Z(1) &= \begin{bmatrix} 1.0124 & -0.2341 \\ -0.4125 & 1.0332 \end{bmatrix}, \ Z(2) = \begin{bmatrix} 1.4564 & -0.6214 \\ -0.7250 & 1.2112 \end{bmatrix} \\
Z(3) &= \begin{bmatrix} 2.1004 & -0.8333 \\ -0.2975 & 1.2332 \end{bmatrix} \\
W &= \begin{bmatrix} 5.2054 & 2.1443 \\ 2.1443 & 4.7805 \end{bmatrix}, \ \lambda = 1.4152\,, \ \gamma = 2.2245
\end{aligned}
$$

The feedback gains and the associated performance bound are

$$
\begin{aligned}
K(1) &= \begin{bmatrix} 0.5661 & -0.2194 \\ -0.3710 & 0.3614 \end{bmatrix}, \ K(2) = \begin{bmatrix} 0.3618 & -0.2961 \\ -0.2533 & 0.4080 \end{bmatrix} \\
K(3) &= \begin{bmatrix} 0.3577 & -0.2804 \\ -0.1168 & 0.2484 \end{bmatrix}, \quad \mathcal{J} = 11.2013
\end{aligned}
$$

This show that the water quality model is **SSSDD** with a disturbance attenuation level of $\gamma = 2.2245$.

5.5.2 Example 5.4

In order to illustrate **Theorem** (5.22), we use the data of Example 1 in addition to

$$\begin{aligned}
M_a(1) &= \begin{bmatrix} 0.2 \\ 0.1 \end{bmatrix}, \ N_a(1) = [0.4 \quad 0.2] \\
N_d(1) &= [0.3 \quad 0.1], \ N_b(1) = [0.1 \quad 0.3] \\
M_a(2) &= \begin{bmatrix} 0.1 \\ 0.1 \end{bmatrix}, \ N_a(2) = [0.2 \quad 0.2] \\
N_d(2) &= [0.2 \quad 0.1], \ N_b(2) = [0.2 \quad 0.2] \\
M_a(3) &= \begin{bmatrix} 0.1 \\ 0.2 \end{bmatrix}, \ N_a(3) = [0.3 \quad 0.3] \\
N_d(3) &= [0.1 \quad 0.2], \ N_b(3) = [0.1 \quad 0.3]
\end{aligned}$$

Using the software LMILab [57], the feasible solution is summarized by

$$\begin{aligned}
K(1) &= \begin{bmatrix} 0.8532 & 0.9260 \\ -1.4317 & -1.2628 \end{bmatrix}, \ K(2) = \begin{bmatrix} 0.9145 & -0.6128 \\ 0.5844 & 1.9912 \end{bmatrix} \\
K(3) &= \begin{bmatrix} 1.1425 & 0.6603 \\ -0.3123 & 0.4912 \end{bmatrix}
\end{aligned}$$

for

$$\varepsilon(1) = 1.3345 \ , \ \nu(1) = 0.9144 \ , \ \mu(1) = 2.4367$$

$$\varepsilon(2) = 2.3567 \ , \ \nu(2) = 2.5433 \ , \ \mu(2) = 1.5321$$

$$\varepsilon(3) = 5.2355 \ , \ \nu(3) = 0.6673 \ , \ \mu(3) = 2.3226$$

and

$$\begin{aligned}
W &= \begin{bmatrix} 4.9154 & 1.3553 \\ 1.3553 & 4.1105 \end{bmatrix}, \ \lambda = 2.7451 \\
\gamma &= 1.3504 \ , \ \mathcal{J} = 11.7413
\end{aligned}$$

5.6 Descriptor Approach

In this section, we develop a descriptor approach to simultaneous $\mathcal{H}_2/\mathcal{H}_\infty$ control design for a class of uncertain JTDS. The rationale behind this approach is to exhibit the delay-dependence dynamics in the design procedure. Also this approach shares the same objective and features like the model transformation approach, the idea and analytical development are different. The time-delay factor is treated as a constant within a prespecified range. The main analytical tool is the constructive use of Lyapunov-Krasovskii functional with mode-dependent weighting matrices to disclose the interplay between the time-delay dynamics and the jumping behavior.

Throughout this section, we will refer to the class of stochastic time-delay systems (Σ_{Jo}) with Markovian jump parameters described over the space $(\Omega, \mathcal{F}, \mathbf{P})$ and we seek to determine a strong delay dependent stabilizing controller which minimizes the upper bound of an $\mathcal{H}_2$ performance measure while guaranteeing that a prescribed upper bound on an $\mathcal{H}_\infty$ performance is attained for all possible $w(t) \in \mathcal{L}_2[0, \infty]$.

Note that system Σ_{Jo} is jumping system with discrete delay. In the sequel, the main thrust is to transform this system to an appropriate form in order to exhibit its delay dependence behavior. We will accomplish this by the following method.

5.6.1 Descriptor Transformation

We employ the descriptor system approach [53] and thus rewrite system (Σ_{Jo}) into the descriptor form for every mode $i \in \mathcal{S}$:

$$\begin{aligned}
(\Sigma_{Do}): \quad \dot{x}(t) &= \sigma(t) \\
0 &= -\sigma(t) + [A_o(i) + A_d(i)]x(t) + B_o(i)u(t)
\end{aligned}$$

$$
\begin{aligned}
& - & A_d(i)\int_{t-\tau(t)}^{t}\sigma(s)ds + \Gamma(i)w(t) \qquad (5.144)\\
y(t) &= & C(i)x(t) + D(i)u(t) + \Psi(i)w(t) \qquad (5.145)\\
z(t) &= & G(i)x(t) + F(i)u(t) + \Phi(i)w(t) \qquad (5.146)
\end{aligned}
$$

Under the control law (5.7), system (Σ_{Do}) becomes:

$$
\begin{aligned}
(\Sigma_{DKo}): \ \dot{x}(t) &= \sigma(t)\\
0 &= -\sigma(t) + A_{Kd}(i)x(t)\\
&- A_d(i)\int_{t-\tau(t)}^{t}\sigma(s)ds + \Gamma(i)w(t) \qquad (5.147)\\
y(t) &= C_K(i)x(t) + \Psi(i)w(t) \qquad (5.148)\\
z(t) &= G_K(i)x(t) + \Phi(i)w(t) \qquad (5.149)
\end{aligned}
$$

where

$$
\begin{aligned}
C_K(i) &= C(i) + D(i)K(i)\ ,\ G_K(i) = G(i) + F(i)K(i)\\
A_{od}(i) &= A_o(i) + A_d(i)\\
A_{Kd}(i) &= A_o(i) + B_o(i)K(i) + A_d(i)\\
&= A_{od}(i) + B_o(i)K(i) \qquad (5.150)
\end{aligned}
$$

5.6.2 Simultaneous Nominal Design

The theorems established in the sequel show that designing a simultaneous $\mathcal{H}_2/\mathcal{H}_\infty$ controller for system (Σ_{DKo}) is essentially related to the existence of a positive definite solution of a family of linear matrix inequalities (LMIs).

For convenience, we introduce the matrices for $i \in \mathcal{S}$

$$
\begin{aligned}
\bar{P}(i) &= U\,\mathbb{P}(i)\ ;\ U = \begin{bmatrix} I & 0\\ 0 & 0 \end{bmatrix}\ ;\ \mathbb{P}(i) = \begin{bmatrix} P_x(i) & 0\\ P_d(i) & P_\sigma(i) \end{bmatrix}\\
E_1 &= \begin{bmatrix} I\\ 0 \end{bmatrix}\ ,\ E_2 = \begin{bmatrix} 0\\ I \end{bmatrix} \qquad (5.151)
\end{aligned}
$$

$$
\begin{aligned}
Y(i) &= \mathbb{P}^{-1}(i) = \begin{bmatrix} Y_x(i) & 0 \\ Y_d(i) & Y_\sigma(i) \end{bmatrix} \\
Y_x &= P_x^{-1}(i),\ Y_\sigma = P_\sigma^{-1}(i),\ Y_d = -Y_x(i)P_d(i)Y_\sigma(i) \\
\mathcal{A}_{oKd}(i) &= \begin{bmatrix} 0 & I \\ A_{Kd}(i) & -I \end{bmatrix},\ \bar{A}_d(i) = \begin{bmatrix} 0 \\ A_d(i) \end{bmatrix} \\
\tau_s &= \tau[1+2\hat{\alpha}\tau],\ \bar{C}_K(i) = [C_K(i) \quad 0]
\end{aligned} \tag{5.152}
$$

Theorem 5.23 *In the absence of input disturbance $w(t) \equiv 0$, controller (5.7) is an $\mathcal{H}_2$-optimal controller for system (Σ_{DKo}) minimizing the $\mathcal{H}_2$-performance measure (8.6) if, given matrix sequence $Q_\sigma(i) = Q_\sigma^t(i) > 0,\ i \in \mathcal{S}$ there exist matrices $P_x(i) = P_x^t(i) > 0,\ P_\sigma(i) = P_\sigma^t(i) > 0,\ P_d(i) = P_d^t(i) > 0,\ i \in \mathcal{S}$, satisfying the system of LMIs for all $i \in \mathcal{S}$*

$$
\Pi_{td}(i) = \begin{bmatrix} \Omega_1(i) + C_K^t(i)C_K(i) & \Omega_2(i) & \tau P_d^t(i)A_d(i) \\ \Omega_2^t(i) & -\Omega_3(i) & \tau P_\sigma^t(i)A_d(i) \\ \tau A_d^t(i)P_d(i) & \tau A_d^t(i)P_d(i) & -Q_\sigma(i) \end{bmatrix} < 0 \tag{5.153}
$$

where

$$
\begin{aligned}
\Omega_1(i) &= P_d^t(i)A_{Kd}(i) + A_{Kd}^t(i)P_d(i) + \sum_{m=1}^{s} \alpha_{im}P_x(m) \\
\Omega_2(i) &= P_x^t - P_d^t(i) + A_{Kd}^t(i)P_\sigma(i) \\
\Omega_3(i) &= P_\sigma^t(i) + P_\sigma(i) - \tau_s Q_\sigma(i)
\end{aligned} \tag{5.154}
$$

An upper bound on the $\mathcal{H}_2$ performance measure is given by

$$
\begin{aligned}
\mathcal{J}_2 &\le \mathsf{J}^+ \triangleq \Big[x^t(0)P_x(\eta_o)x(0) \\
&+ \tau(0)\int_{-\tau(0)}^{0} \sigma^t(s)Q_\sigma(\eta_o)\sigma(s)\,ds\Big]^{1/2}
\end{aligned} \tag{5.155}
$$

Proof: Let $\mathbf{x}_s(t) \triangleq x(s+t),\ t-\tau \le s \le t$ and define the process $\{(\mathbf{x}(t),\eta_t),\ t \ge 0\}$ over the state space $\bar{\mathcal{C}}$. It should be observed that

$\{(\mathbf{x}(t), \eta_t),\ t \geq 0\}$ is strong Markovian [78]. Now for $\eta_t = i \in \mathcal{S}$, and given $Q_\sigma(i) = Q^t_\sigma(i) > 0$, let the Lyapunov-Krasovskii functional $V(\cdot) : \Re^n \times \Re_+ \times \mathcal{S} \rightarrow \Re_+$ of the transformed system (Σ_{Do}) be selected as:

$$\begin{aligned} V(t,i) &\triangleq V_x(t,i) + V_\sigma(t,i) + V_a(t,i) + V_c(t,i) \\ V_\sigma(t,i) &= \int_{-\tau(t)}^{0} \int_{t+\theta}^{t} \sigma^t(s)\, Q_\sigma(i)\, \sigma(s)\, ds\, d\,\theta \\ V_a(t,i) &= \hat{\alpha} \int_{-\tau(t)}^{0} \int_{t+\theta}^{t} \sigma^t(s)\, Q_\sigma(i)\, \sigma(s)\, (s-t+\tau) ds\, d\,\theta \\ V_c(t,i) &= \hat{\alpha}\, \tau \int_{-\tau(t)}^{0} \int_{t+\theta}^{t} \sigma^t(s)\, Q_\sigma(i)\, \sigma(s)\, ds\, d\,\theta \\ V_x(t,i) &= x^t(t) P_x(i) x(t) \end{aligned} \tag{5.156}$$

The weak infinitesimal operator $\Im_x[\cdot]$ of the process $\{x(t), i, t \geq 0\}$ for system (5.147)-(5.149) at the point $\{t, x, i\}$ is given by [78]:

$$\Im_x[V] = \partial V_x/\partial t + \partial V_x/\partial x\, \dot{x}(t) \mid_{\eta_t = i} + \sum_{m=1}^{s} \alpha_{im} V_x(t,i,m) \tag{5.157}$$

Using (5.156)-(5.157) and (5.148), we get:

$$\begin{aligned} \partial V_x/\partial x \dot{x}(t) &= 2x^t(t) P^t_x(i)\, \dot{x}(t) \\ &= 2[x^t(t) \quad \sigma^t(t)] \mathbb{P}^t(i) \begin{bmatrix} \dot{x}(t) \\ 0 \end{bmatrix} \\ &= 2[x^t(t) \quad \sigma^t(t)]\, \mathbb{P}^t(i) \Bigg\{ \begin{bmatrix} \sigma(t) \\ -\sigma(t) + A_{oKd}(i) x(t) \end{bmatrix} \\ &+ \begin{bmatrix} 0 \\ -A_d(i) \int_{t-\tau}^{t} \sigma(s) ds \end{bmatrix} \Bigg\} \end{aligned} \tag{5.158}$$

Simple manipulations using (5.147) yield:

$$\begin{aligned} &2[x^t(t) \quad \sigma^t(t)]\, \mathbb{P}^t(i) \begin{bmatrix} \sigma(t) \\ -\sigma(t) + A_{oKd}(i) x(t) \end{bmatrix} = \\ &[x^t(t) \quad \sigma^t(t)] \left\{ \mathbb{P}^t(i) \mathcal{A}_{oKd}(i) + \mathcal{A}^t_{oKd}(i) \mathbb{P}(i) \right\} \begin{bmatrix} x(t) \\ \sigma(t) \end{bmatrix} \end{aligned} \tag{5.159}$$

Using **Fact 3**, it follows that

$$
\begin{aligned}
& 2[x^t(t) \quad \sigma^t(t)]\, \mathbb{P}^t(i) \left[\begin{array}{c} 0 \\ -A_d(i) \int_{t-\tau}^{t} \sigma(s)ds \end{array} \right] \\
= & \; 2 \int_{t-\tau}^{t} [x^t(t) \quad \sigma^t(t)]\, \mathbb{P}^t(i) \bar{A}_d(i) \sigma(s) ds \\
\leq & \; \tau \left[\begin{array}{c} x(t) \\ \sigma(t) \end{array} \right]^t \mathbb{P}^t(i) \bar{A}_d(i) Q_\sigma^{-1}(i) \bar{A}_d^t(i) \mathbb{P}(i) \left[\begin{array}{c} x(t) \\ \sigma(t) \end{array} \right] \\
+ & \int_{t-\tau}^{t} \sigma^t(s) Q_\sigma(i) \sigma(s)\; ds
\end{aligned}
\tag{5.160}
$$

Therefore from (5.157)-(5.160) for all $i \in \mathcal{S}$ we get:

$$
\begin{aligned}
\Im_x[V] \;\; \leq \;\; & \left[\begin{array}{c} x(t) \\ \sigma(t) \end{array} \right]^t \Big\{ \mathbb{P}^t(i) \mathcal{A}_{oKd}(i) + \mathcal{A}_{oKd}^t(i) \mathbb{P}(i) \\
+ & \; \tau\, \mathbb{P}^t(i) \bar{A}_d(i) Q_\sigma^{-1}(i) \bar{A}_d^t(i) \mathbb{P}(i) \Big\} \left[\begin{array}{c} x(t) \\ \sigma(t) \end{array} \right] \\
+ & \; x^t(t) \Big[\sum_{m=1}^{s} \alpha_{im} P_x(m) \Big] x(t) \\
+ & \int_{t-\tau}^{t} \sigma^t(s) Q_\sigma(i) \sigma(s)\; ds
\end{aligned}
\tag{5.161}
$$

Similarly,

$$
\begin{aligned}
\Im_\sigma[V] \;\; = \;\; & \tau\, \sigma^t(t) Q_\sigma(i) \sigma(t) - \int_{-\tau}^{0} \sigma^t(t+\theta) Q_\sigma(i) \sigma(t+\theta)\; d\theta \\
+ & \sum_{m=1}^{s} \alpha_{im} \int_{-\tau}^{0} \int_{t+\theta}^{t} \sigma^t(s) Q_\sigma(i) \sigma(s)\; ds
\end{aligned}
\tag{5.162}
$$

It should be observed that

$$
\begin{aligned}
& \sum_{m=1}^{s} \alpha_{im} \int_{-\tau}^{0} \int_{t+\theta}^{t} \sigma^t(s) Q_\sigma(i) \sigma(s)\; ds \\
\leq & \; |\alpha_{ii}| \int_{-\tau}^{0} \int_{t+\theta}^{t} \sigma^t(s) Q_\sigma(i) \sigma(s)\; ds \\
= & \; \hat{\alpha}\, \tau \int_{t-\tau}^{t} \sigma^t(s) Q_\sigma(i) \sigma(s)\; ds
\end{aligned}
$$

$$+ \quad \hat{\alpha} \int_{t-\tau}^{t} \sigma^t(s) Q_\sigma(i) \sigma(s)(s-t+\tau)\, ds \tag{5.163}$$

In addition, algebraic manipulations show that:

$$\begin{aligned}
\Im_a[V] &= \hat{\alpha}\, \tau^2\, \sigma^t(t) Q_\sigma(i) \sigma(t) \\
&- \hat{\alpha} \int_{t-\tau}^{t} \sigma^t(s) Q_\sigma(i) \sigma(s)(s-t+\tau)\, ds \\
\Im_c[V] &= \hat{\alpha}\, \tau^2\, \sigma^t(t) Q_\sigma(i) \sigma(t) \\
&- \hat{\alpha} \int_{t-\tau}^{t} \sigma^t(t+\theta) Q_\sigma(i) \sigma(t+\theta)\, d\theta
\end{aligned} \tag{5.164}$$

Let

$$\chi(t) = [x^t(t) \;\; \sigma^t(t)]^t$$

By combining (5.161)-(5.164) and using (5.151)-(5.152), it follows that

$$\begin{aligned}
\Im[V] &= \Im_x[V] + \Im_\sigma[V] + \Im_a[V] + \Im_c[V] \\
&= \begin{bmatrix} x(t) \\ \sigma(t) \end{bmatrix}^t \Big\{ \mathbb{P}^t(i) \mathcal{A}_{oKd}(i) + \mathcal{A}_{oKd}^t(i) \mathbb{P}(i) \\
&+ \tau\, \mathbb{P}^t(i) \bar{A}_d(i) Q_\sigma^{-1}(i) \bar{A}_d^t(i) \mathbb{P}(i) \\
&+ \sum_{m=1}^{s} \alpha_{im} \bar{P}(m) + \tau_s\, E_2 Q_\sigma(i) E_2^t \Big\} \begin{bmatrix} x(t) \\ \sigma(t) \end{bmatrix} \\
&\triangleq \begin{bmatrix} x(t) \\ \sigma(t) \end{bmatrix}^t \Pi_{do}(i) \begin{bmatrix} x(t) \\ \sigma(t) \end{bmatrix}
\end{aligned} \tag{5.165}$$

Considering $\Pi_{do}(i)$ and using (5.151)-(5.152), some standard matrix manipulations convert it to LMIs (5.153).

In view of the fact that

$$C_K^t(i) C_K(i) > 0$$

it follows that

$$\Pi_{do}(i) < 0$$

is guaranteed for all $i \in \mathcal{S}$ and we conclude that

$$\Im[V] < 0 \qquad \forall\, \chi(t) \neq 0 \qquad \& \qquad \Im[V] \leq 0 \;\forall\, \chi(t)$$

Since

$$||\chi(t+\beta)|| \leq \varphi||\chi(t)|, \quad \forall \beta \in [-\tau, 0]$$

and some $\varphi > 0$ [85], it follows from (5.156) that

$$V(t,i) \leq \chi^t(t)\bar{P}(i)\chi(t) + \mu||\zeta||^2$$

where

$$\mu = \varphi\,\tau\big(\max_i \lambda_M[\bar{P}(i)] + \tau_s\,\lambda_M[E_2 Q_\sigma(i) E_2^t]\big)$$

Therefore, for all $\chi \neq 0$, we have

$$\begin{aligned} \frac{\Im[V]}{V(t,i)} &\leq \frac{\chi^t \Pi_{do}(i)\chi}{\chi^t \bar{P}(i)\chi + \mu||\chi||^2} \\ &\leq -\xi \\ &\triangleq -\min_{i\in\mathcal{S}} \left\{ \frac{\lambda_m[-\Pi_{do}(i)]}{\lambda_M[\bar{P}(i)] + \mu} \right\} \end{aligned} \qquad (5.166)$$

It is readily seen from (5.166) that $\xi > 0$ and hence we get

$$\Im[V] \leq -\xi\, V(t,i)$$

It follows from [78] by using the Gronwall-Bellman lemma [98] and letting $\chi(t = 0, \phi, \eta_o) = \chi_o$, one has

$$\mathbb{E}[V(t,i)\big|\phi, i] \leq e^{-\xi\, t}\, V(\chi_o, i) \qquad (5.167)$$

Since

$$\mathbb{E}\left\{ \int_{t-\tau}^{t} \int_{t+\theta}^{t} \chi^t(s) E_2 Q_x(i) E_2^t \chi(s)\, ds d\theta \middle| \phi, \eta_o \right\} \geq 0$$

it is easy to see from (5.167) that

$$
\begin{aligned}
& \mathbb{E}\left\{ \chi^t(t)\bar{P}(i)\chi(t) \Big| \phi, i \right\} \\
\leq \;& e^{-\xi\, t} V(\chi_o, i) \Longrightarrow \\
& \mathbb{E}\left\{ \int_0^{\mathcal{T}} \chi^t(t)\bar{P}(i)\chi(t)dt \Big| \phi, \eta_o = i \right\} \\
\leq \;& \left[\int_0^{\mathcal{T}} e^{-\xi\, t} dt \right] V(\chi_o, i) \\
= \;& \frac{1}{\zeta}[e^{-\xi\, \mathcal{T}} \; - \; 1] \; V(\chi_o, i) \Longrightarrow \\
& \lim_{\mathcal{T}\to\infty} \; \mathbb{E}\left\{ \int_0^{\mathcal{T}} \chi^t(t)\bar{P}(i)\chi(t)dt \Big| \phi, \eta_o = i \right\} \\
\leq \;& \frac{1}{\xi}\chi_o^t \bar{P}(\eta_o)\chi_o \\
+ \;& \frac{\mathcal{T}_s}{\xi}[E_2 Q_x((\eta_o))E_2^t] \\
+ \;& \chi(t+\theta)||_*^2, \quad \forall \theta \in [-\tau, 0]
\end{aligned}
\tag{5.168}
$$

where

$$
||\chi(t+\theta)||_*^2 \triangleq \sup_{\theta\in[-\tau,0]} ||\chi(t+\theta)||_2^2
$$

Let

$$
\Xi_P(i) \;=\; \max_{i\in\mathcal{S}} \left\{ \frac{\bar{P}(\eta_o)||\chi_o||^2 + \tau_s[E_2 Q_x(\eta_o)E_2^t]||\chi(t+\theta)||_*^2}{\xi[\bar{P}(\eta_o)]||\chi_o||^2} \right\}
$$

it follows from (5.168) for $i \in \mathcal{S}$ that

$$
\mathbb{E}\left\{ \int_0^{\infty} \chi^t(t)\chi(t)dt | \phi, \eta_o = i \right\} \leq \chi_o^t \lambda_M[\Xi_P(i)]\chi_o < +\infty,
$$

which, in the light of Chapter 3 shows that system (Σ_{Do}) is **SSSDD** under the control law (5.7).

Now by the Dynkin's formula and (5.155)

$$
\mathbb{E}[V(\chi(t), \eta_t)] - V(\chi(0), \eta_o)
$$

$$
\begin{aligned}
&= \mathbb{E}\left[\int_0^t \Im[V(\chi(v),\eta_v)]dv\right] \\
&\Longrightarrow \mathbb{E}[V(\chi(t_f),\eta_{t_f})] - V(\chi(0),\eta_o) \\
&\leq \mathbb{E}\left[\int_0^{t_f} \chi^t\, \Pi_t(\eta_v)\, \chi\, dv\right] \qquad (5.169)
\end{aligned}
$$

Letting $t_f \to \infty$ and in view the system stability, it follows that

$$
\begin{aligned}
& \mathbb{E}\left[\int_0^{\infty} y^t(r)y(r)dr\right] \leq V(\chi(0),\eta_o)] \\
&= \chi^t(0)\bar{P}(\eta_o)\chi(0) + \int_{-\tau(0)}^{0}\int_{t+\theta}^{t} \chi^t(s)E_2Q_x(\eta_o)E_2^t\chi(s)\, dsd\theta \Longrightarrow \\
||y|| &\leq \Big[\chi^t(0)\bar{P}(\eta_o)\chi(0) \\
&+ \tau(0)\int_{-\tau(0)}^{0} \chi^t(s)E_2Q_x(\eta_o)E_2^t\chi(s)\, ds\Big]^{1/2} \\
&= \mathsf{J}^+ \qquad (5.170)
\end{aligned}
$$

which completes the proof. ∇∇∇

The following theorem provides a method for computing the feedback gains

Theorem 5.24 *The feedback gain associated with the $\mathcal{H}_2$-optimal controller for system (Σ_{DKo}) is given by $K(i) = Z(i)Y_x^{-1}(i),\ i \in \mathcal{S}$ where the matrices $Y_x(i) = Y_x^t(i) > 0,\ Y_\sigma(i) = Y_\sigma^t(i) > 0,\ Y_d(i) = Y_d^t(i) > 0,\ i \in \mathcal{S}$, satisfy the system of LMIs for all $i \in \mathcal{S}$*

$$
\begin{bmatrix}
\cup_q(i) & \begin{matrix} Y_x^t(i)C^t(i)+ \\ L^t(i)D^t(i) \end{matrix} & \mathcal{R}(i) \\
\begin{matrix} C(i)Y_x(i)+ \\ D(i)L(i) \end{matrix} & -I & 0 \\
\mathcal{R}^t(i) & 0 & -\mathcal{Y}(i)
\end{bmatrix} < 0 \qquad (5.171)
$$

where

$$
\cup_q(i) = \begin{bmatrix}
\Lambda_1(i) & \Lambda_2(i) & \tau_s Y_d^t(i) \\
\Lambda_2^t(i) & -\Lambda_3(i) & 0 \\
\tau_s Y_d(i) & \tau_s Y_\sigma^t(i) & -\tau_s Q_\sigma(i)
\end{bmatrix}
$$

$$
\begin{aligned}
\Lambda_1(i) &= Y_d^t(i) + Y_d(i) + \alpha_{ii} Y_x^t(i) \\
\Lambda_2(i) &= Y_\sigma - Y_d^t(i) + Y_x^t(i) A_{od}^t(i) + L^t(i) B_o^t(i) \\
\Lambda_3(i) &= Y_\sigma^t(i) + Y_\sigma(i) - \tau A_d Q_\sigma^{-1}(i) A_d(i) \\
\mathcal{R}(i) &= \left[\sqrt{\alpha_{i1}} Y_x(i) ..... \sqrt{\alpha_{is}} Y_x(i) \right] \\
\mathcal{Y}(i) &= diag \left[Y_x^t(1)...Y_x^t(i-1) \;\; Y_x^t(i+1)...Y_x^t(s) \right]
\end{aligned} \tag{5.172}
$$

Proof: By **Fact 3**, condition (5.165) is equivalent to

$$
\begin{aligned}
& \mathcal{A}_{oKd}^t(i) \mathbb{P}(i) + \mathbb{P}^t(i) \mathcal{A}_{oKd}(i) + \sum_{m=1}^{s} \alpha_{im} \bar{P}(m) + \bar{C}_K^t(i) \bar{C}_K(i) \\
+ & \; \tau \mathbb{P}^t(i) \bar{A}_d(i) Q_\sigma^{-1}(i) \bar{A}_d^t(i) \mathbb{P}(i) + \tau_s \, E_2 Q_\sigma(i) E_2^t < 0
\end{aligned} \tag{5.173}
$$

Using (5.151)-(5.152), premultiplying (5.173) by $Y^t(i) = \mathbb{P}^{-t}(i)$ and postmultiplying the result by $Y(i)$, we obtain:

$$
\begin{aligned}
& Y^t(i) \mathcal{A}_{oKd}^t(i) + \mathcal{A}_{oKd}(i) Y(i) + Y^t(i) \left[\sum_{m=1}^{s} \alpha_{im} \bar{P}(m) \right] Y(i) \\
+ & \; Y^t(i) \bar{C}_K^t(i) \bar{C}_K(i) Y(i) + \tau \bar{A}_d(i) Q_\sigma^{-1}(i) \bar{A}_d^t(i) \\
+ & \; \tau_s \, Y^t(i) E_2 Q_\sigma(i) E_2^t Y(i) < 0
\end{aligned} \tag{5.174}
$$

Expanding inequality (5.174) and using $K(i) Y_x(i) = Z(i)$, LMIs (5.171) subject to (5.172) follow immediately. $\nabla\nabla\nabla$

The following theorems summarize the main results pertaining to the $\mathcal{H}_\infty$ performance:

Theorem 5.25 *Given a prescribed constant $\gamma > 0$. controller (5.7) renders system (Σ_{DKo})* **RSSSDD** *with a disturbance attenuation level γ for all $w(t) \in \mathcal{L}_2[0, \infty)$ if, given matrix sequence $Q_x(i) = Q_x^t(i) > 0$, $i \in \mathcal{S}$ there exist matrices $P_x(i) = P_x^t(i) > 0$, $P_\sigma(i) = P_\sigma^t(i) > 0$, $P_d(i) = P_d^t(i) > 0$, $i \in \mathcal{S}$, satisfying the* *system of LMIs for all* $i \in \mathcal{S}$

$$\Pi_t(i) \triangleq$$

$$\begin{bmatrix} \begin{array}{c}\Omega_1(i)+\\ G_K^t(i)G_K(i)\end{array} & \Omega_2(i) & \tau P_d^t(i)A_d(i) & \begin{array}{c}P^t(i)\Gamma(i)+\\ \bar{G}_K^t(i)\Phi(i)\end{array} \\ \Omega_2^t(i) & -\Omega_3(i) & \tau P_\sigma^t(i)A_d(i) & 0 \\ \tau A_d^t(i)P_d(i) & \tau A_d^t(i)P_d(i) & -Q_\sigma(i) & 0 \\ \begin{array}{c}\Gamma^t(i)P(i)+\\ \Phi^t(i)G_K(i)\end{array} & 0 & 0 & \begin{array}{c}-\gamma^2 I+\\ \Phi^t(i)\Phi(i)\end{array} \end{bmatrix} < 0 \quad (5.175)$$

Moreover

$$\begin{aligned} ||z(t)|| &= \left[\gamma^2 ||w(t)||^2 + x^t(0)P_x(\eta_o)x(0)\right. \\ &+ \left.\tau(0)\int_{-\tau(0)}^{0} \sigma^t(s)Q_\sigma(\eta_o)\sigma(s)ds\right]^{1/2} \end{aligned} \quad (5.176)$$

Proof: The stochastic stability follows as a result of **Theorem** 5.23. To show that system (Σ_{DKo}) has a disturbance attenuation γ, we let the Lyapunov functional $V(i)$ be given by (5.156). By evaluating the weak infinitesimal operator $\Im_m^x[\cdot]$ of the process $\{x(t), \eta_t = i, t \geq 0\}$ for system (5.147)-(5.149) at the point $\{t, x, \eta_t = i\}$ using (5.151)-(5.152) and manipulating we get

$$\begin{aligned} \Im_m^x[V] &\leq \Im_1^x[V] + x^t(t)\mathbb{P}(i)\Gamma(i)w(t) + w^t(t)\Gamma^t(i)\mathbb{P}(i)x(t) \\ &\triangleq \Im_+^x[V] \end{aligned} \quad (5.177)$$

By Dynkin's formula, one has

$$\mathbb{E}\left\{\int_0^T \Im_+^x[V(.)]dt\right\} = \mathbb{E}\left\{V(x(t),\eta_t)|_{t=T}\right\} - V(x(t),\eta_t)|_{t=0}$$

With standard matrix manipulations using (5.157), it follows from (5.177) that:

$$\begin{aligned} & z^t(t)z(t) - \gamma^2\, w^t(t)w(t) + \Im_+^x[V] \\ \leq & \begin{bmatrix} x(t) \\ x(t-\tau) \\ w(t) \end{bmatrix}^t \Pi_r(i) \begin{bmatrix} x(t) \\ x(t-\tau) \\ w(t) \end{bmatrix} \end{aligned} \quad (5.178)$$

Therefore from (8.7), we have:

$$
\begin{aligned}
\mathcal{J}_T &\leq \mathbb{E}\left\{ \int_0^T \begin{bmatrix} x(t) \\ x(t-\tau) \\ w(t) \end{bmatrix}^t \Pi_r(i) \begin{bmatrix} x(t) \\ x(t-\tau) \\ w(t) \end{bmatrix} dt \right\} \\
&- \mathbb{E}\left\{ V(x(t),\eta_t)\Big|_{t=T} + V(x(t),\eta_t)\Big|_{t=0} \right\}
\end{aligned} \tag{5.179}
$$

In view of (5.175) and the fact that

$$
\mathbb{E}\left\{ V(x(t),\eta_t)|_{t=T} \geq 0 \right\}
$$

inequality (5.179) leads to

$$
\begin{aligned}
\mathcal{J}_T &\leq V(x(0),\eta_0) \Longrightarrow \\
\mathcal{J}_\infty &\leq V(x(0),\eta_0) \Longrightarrow \\
||z(t)||^2 - \gamma^2 \, ||w(t)||^2 &\leq x_0^t P_x(\eta_o) x_0 + \tau(0) \int_{-\tau(0)}^0 \phi^t(\beta) Q_x(\eta_o) \phi(\beta) d\beta \Longrightarrow \\
||z(t)||^2 &\leq \gamma^2 \, ||w(t)||^2 + x_0^t P_x(\eta_o) x_0 \\
&+ \tau(0) \int_{-\tau(0)}^0 \phi^t(\beta) Q_x(\eta_o) \phi(\beta) d\beta
\end{aligned} \tag{5.180}
$$

which completes the proof. $\nabla\nabla\nabla$

Theorem 5.26 *Given a prescribed constant $\gamma > 0$. The feedback gain associated with the $\mathcal{H}_\infty$-controller for system (Σ_{DKo}) is given by $K(i) = Z(i)Y_x^{-1}(i)$, $i \in \mathcal{S}$ where the matrices $Y_x(i) = Y_x^t(i) > 0,\ Y_\sigma(i) = Y_\sigma^t(i) > 0,\ Y_d(i) = Y_d^t(i) > 0,\ i \in \mathcal{S}$, satisfy the system of LMIs for all $i \in \mathcal{S}$*

$$
\begin{bmatrix}
\Lambda_1(i) & \Lambda_2(i) & \tau_s Y_d^t(i) & \Lambda_g(i) \\
\Lambda_2^t(i) & -\Lambda_3(i) & \tau_s Y_\sigma^t(i) & 0 \\
\tau_s Y_d(i) & \tau_s Y_\sigma^t(i) & -\tau_s Q_\sigma(i) & 0 \\
\Lambda_g^t(i) & 0 & 0 & -\Lambda_k(i)
\end{bmatrix} < 0 \tag{5.181}
$$

$$
\Lambda_g(i) = [\Lambda_{g1}(i) \ \mathcal{R}(i) \ \Lambda_{g2}(i)]
$$

$$
\begin{aligned}
\Lambda_{g1}(i) &= Y_x^t(i)G_o^t(i) + L^t(i)F_o^t(i) \\
\Lambda_{g2}(i) &= [\Gamma(i) + Y^t(i)G_o^t(i)\Phi(i)] \\
\Lambda_k(i) &= diag[I \ \ \mathcal{Y}(i) \ \ -\gamma^2 I + \Phi^t(i)\Phi(i)]
\end{aligned}
$$

Proof: Follows from parallel development to **Theorem** 5.24. ∇∇∇

The solution to the simultaneous $\mathcal{H}_2/\mathcal{H}_\infty$ control problem is established by the following theorem:

Theorem 5.27 *Given a prescribed constant* $\gamma > 0$. *The feedback gain* $K(i) = Z(i)Y^{-1}(i)$, $i \in \mathcal{S}$ *is a* **simultaneous** $\mathcal{H}_2/\mathcal{H}_\infty$ **controller** *satisfying the performance measure (8.7) for system* (Σ_{nc}) *if there exist matrices* $Y(i) = Y^t(i) > 0$, $Z(i)$, $L(i) = L^t(i) > 0$, $W = W^t > 0$, $i \in \mathcal{S}$, *such the system of generalized eigenvalue problems*

$$
\begin{aligned}
&\min \left[\lambda \ + \ Tr(W)\right] \\
&subject \ \ to \ \ (5.171), (5.181) \ \ \ and \\
&\begin{bmatrix} -L(i) & I \\ I & -Q_x(i) \end{bmatrix} < 0 \, , \begin{bmatrix} -\lambda & \phi^t(0) \\ \phi(0) & -Y_x(i) \end{bmatrix} < 0 \\
&\begin{bmatrix} -W & X^t \\ X & -L(i) \end{bmatrix} < 0
\end{aligned}
\tag{5.182}
$$

has a feasible solution

Proof: On observing that

$$
\begin{aligned}
x^t(0)P_x(i)x(0) &\leq \max \left[x^t(0)P_x(1)x(0), ..., x^t(0)P_x(s)x(0)\right] \\
&\triangleq \lambda \Longrightarrow \\
&- \ \ \lambda + \phi^t(0)Y_x^{-1}(i)\phi(0) \ < \ 0
\end{aligned}
\tag{5.183}
$$

and on using the cyclic properties of matrix trace [25], we have

$$
\int_{-\tau(0)}^{0} x^t(s)Q(\eta_o)x(s)ds \ = \ \int_{-\tau(0)}^{0} Tr\left[x^t(s)Q(\eta_o)x(s)\right]ds
$$

$$
\begin{aligned}
&= Tr\Big[XX^t L^{-1}(i)\Big] \\
&= Tr\Big[X^t L^{-1}(i)X\Big] \; < Tr(W) \Longrightarrow \\
&- \; W + X^t L^{-1}(i)X < 0
\end{aligned}
\tag{5.184}
$$

where

$$
XX^t = \int_{-\tau(0)}^{0} x^t(s)x(s)ds
$$

The objective of simultaneous $\mathcal{H}_2/\mathcal{H}_\infty$ control leads to the convex minimization in (5.182) subject to (5.183)-(5.184) as expressed by LMIs (5.182). ∇∇∇

5.7 Simultaneous Uncertain Design

In this section, we consider the robust counterparts of **Theorems** 5.23-5.27 for the uncertain system (Σ_J) described by (5.46)-(5.48).

Under the control law (5.7) and invoking the descriptor transformation, system (Σ_J) becomes:

$$
\begin{aligned}
(\Sigma_{DK}): \quad \dot{x}(t) &= \sigma(t) \\
0 &= -\sigma(t) + A_{\Delta Kd}(i)x(t) \\
&- A_{\Delta d}(i) \int_{t-\tau(t)}^{t} \sigma(s)ds + \Gamma(i)w(t) && (5.185) \\
y(t) &= C_K(i)x(t) + \Psi(i)w(t) && (5.186) \\
z(t) &= G_K(i)x(t) + \Phi(i)w(t) && (5.187)
\end{aligned}
$$

where

$$
\begin{aligned}
A_{\Delta Kd}(i) &= A_{\Delta o}(i) + B_{\Delta d}(i)K(i) + A_{\Delta d}(i) \\
&= [A_{od}(i) + B_o(i)K(i)] \\
&+ M_a(i)\Delta(t,i)[N_a(i) + N_d(i) + N_b(i)K(i)] \\
&= A_{Kd}(i) + M_a(i)\Delta(t,i)N_{Kd}(i)
\end{aligned}
\tag{5.188}
$$

For convenience, we introduce the matrices:

$$
\begin{aligned}
\bar{M}_a(i) &= \begin{bmatrix} 0 \\ M_a(i) \end{bmatrix}, \ \bar{N}_{Kd}(i) = [N_{Kd} \quad 0], \ \bar{A}_d(i) = \begin{bmatrix} 0 \\ A_d(i) \end{bmatrix} \\
\mathcal{A}_{\Delta Kd}(t,i) &= \begin{bmatrix} 0 & I \\ A_{\Delta Kd}(i) & -I \end{bmatrix} \\
&= \mathcal{A}_{oKd}(i) + \bar{M}_a(i)\Delta(t,i)\bar{N}_{Kd}(i) \\
\bar{A}_{\Delta d}(t,i) &= \begin{bmatrix} 0 \\ A_{\Delta d}(i) \end{bmatrix} = \bar{M}_a(i)\Delta(t,i)N_d(i) \qquad (5.189)
\end{aligned}
$$

In line of the simultaneous nominal design, we have the following results:

Theorem 5.28 *In the absence of input disturbance $w(t) \equiv 0$, controller (5.7) is an $\mathcal{H}_2$-optimal controller for system (Σ_{DK}) minimizing the $\mathcal{H}_2$-performance measure (5.5) if, given matrix sequence $Q_\sigma(i) = Q_\sigma^t(i) > 0,\ i \in \mathcal{S}$, there exist matrices $P_x(i) = P_x^t(i) > 0,\ P_\sigma(i) = P_\sigma^t(i) > 0,\ P_d(i) = P_d^t(i) > 0,\ i \in \mathcal{S}$ and scalars $\varepsilon(i) > 0,\ \nu(i) > 0,\ \mu(i) > 0,\ i \in \mathcal{S}$ satisfying the system of LMIs for all $i \in \mathcal{S}$*

$$
\Pi_t(i) = \begin{bmatrix} \cup_m(i) & \Xi_1(\tau,i) \\ \Xi_1^t(\tau,i) & -\Xi_2(\tau,i) \end{bmatrix} < 0 \qquad (5.190)
$$

where

$$
\begin{aligned}
\cup_m(i) &= \begin{bmatrix} \hat{\Omega}_1(i) + C_K^t(i)C_K(i) & \Omega_2(i) & \tau P_d^t(i)A_d(i) \\ \Omega_2^t(i) & -\Omega_3(i) & \tau P_\sigma^t(i)A_d(i) \\ \tau A_d^t(i)P_d(i) & \tau A_d^t(i)P_d(i) & -Q_\sigma(i) \end{bmatrix} \\
\Xi_1(\tau,i) &= \begin{bmatrix} P_d^t(i)M_a(i) & P_d^t(i)M_a(i) & \tau P_d^t(i)A_d(i) \\ P_\sigma^t(i)M_a(i) & P_\sigma^t(i)M_a(i) & \tau P_\sigma^t(i)A_d(i) \end{bmatrix} \\
\Xi_2(\tau,i) &= diag[\varepsilon(i)I \quad \nu(i)I \quad Q_x(i) - \mu(i)N_d^t(i)N_d(i)] \\
\hat{\Omega}_1(i) &= P_d^t(i)A_{Kd}(i) + A_{Kd}^t(i)P_d(i) + \sum_{m=1}^{s} \alpha_{im}P_x(m) \\
&+ \varepsilon(i)\bar{N}_{Kd}^t(i)\bar{N}_{Kd}(i) + \nu(i)N_d^t(i)N_d(i) \qquad (5.191)
\end{aligned}
$$

for all admissible uncertainties satisfying (5.49)-(5.50). An upper bound on the

$\mathcal{H}_2$ performance measure is given by

$$
\begin{aligned}
\mathcal{J}_2 \quad &\leq \quad \mathsf{J}^+ \triangleq \Big[x^t(0) P_x(\eta_o) x(0) \\
&+ \quad \tau(0) \int_{-\tau(0)}^{0} \sigma^t(s) Q_\sigma(\eta_o) \sigma(s)\, ds \Big]^{1/2}
\end{aligned} \tag{5.192}
$$

Proof: By similarity to **Theorem** 5.23, it follows that system (5.185)-(5.186) is **RSSSDD** if the inequality

$$
\begin{aligned}
&\mathcal{A}^t_{\Delta K d}(i)\mathbb{P}(i) + \mathbb{P}^t(i)\mathcal{A}_{\Delta K d}(i) + \sum_{m=1}^{s} \alpha_{im} \bar{P}(m) + \bar{C}^t_K(i)\bar{C}_K(i) + \\
&\tau \mathbb{P}^t(i) \bar{A}_{\Delta d}(i) Q_\sigma^{-1}(i) \bar{A}^t_{\Delta d}(i) \mathbb{P}(i) + \tau_s E_2 Q_\sigma(i) E_2^t < 0
\end{aligned} \tag{5.193}
$$

holds for all $i \in \mathcal{S}$. By **Facts 2** and **3**, we have:

$$
\begin{aligned}
& \mathcal{A}^t_{\Delta K d}(i)\mathbb{P}(i) + \mathbb{P}^t(i)\mathcal{A}_{\Delta K d}(i) + \tau \mathbb{P}^t(i)\bar{A}_{\Delta d}(i) Q_\sigma^{-1}(i) \bar{A}^t_{\Delta d}(i)\mathbb{P}(i) \\
\leq \; & \mathcal{A}^t_{oKd}(i)\mathbb{P}(i) + \mathbb{P}^t(i)\mathcal{A}_{oKd}(i) \\
+ \; & [\varepsilon^{-1}(i) + \tau \mu^{-1}(i)] \mathbb{P}^t(i) \bar{M}_a(i) \bar{M}^t_a(i) \mathbb{P}(i) + \varepsilon(i) \bar{N}^t_{ad}(i) \bar{N}_{ad}(i) \\
+ \; & \tau \mathbb{P}^t(i) \bar{A}_d(i) [Q_\sigma(i) - \mu(i) N^t_d(i) N_d(i)]^{-1} \bar{A}^t_d(i) \mathbb{P}(i)
\end{aligned} \tag{5.194}
$$

LMI (5.190) follows immediately from substituting (5.194) into (5.193) and using (5.151)-(5.152). ∇∇∇

Theorem 5.29 *The feedback gain associated with the $\mathcal{H}_2$-optimal controller for system (Σ_{DK}) is given by $K(i) = Z(i) Y_x^{-1}(i)$, $i \in \mathcal{S}$ if there exist matrices $Y_x(i) = Y^t_x(i) > 0$, $Y_\sigma(i) = Y^t_\sigma(i) > 0$, $Y_d(i) = Y^t_d(i) > 0$, $i \in \mathcal{S}$ and scalars scalars $\varepsilon(i) > 0$, $\mu(i) > 0$, $\delta(i) > 0$, $i \in \mathcal{S}$ satisfying the system of LMIs for all $i \in \mathcal{S}$*

$$
\begin{bmatrix}
\Lambda_1(i) & \Lambda_2(i) & \tau_s Y^t_d(i) & \Xi_3(i) & \mathcal{R}(i) \\
\Lambda^t_2(i) & -\bar{\Lambda}_3(i) & \tau_s Y^t_\sigma(i) & 0 & 0 \\
\tau_s Y_d(i) & \tau_s Y^t_\sigma(i) & -\tau_s Q_\sigma(i) & 0 & 0 \\
\Xi^t_3 & 0 & 0 & -\Xi_4(i) & 0 \\
\mathcal{R}^t(i) & 0 & 0 & 0 & -\mathcal{Y}(i)
\end{bmatrix} < 0
$$

$$\begin{bmatrix} \mu(i) & 1 \\ 1 & \delta(i) \end{bmatrix} \geq 0 \tag{5.195}$$

where

$$\begin{aligned}
\bar{\Lambda}_3(i) &= Y_\sigma^t(i) + Y_\sigma(i) - [\varepsilon(i) + \mu(i)] M_a(i) M_a^t(i) \\
&- \tau A_d [Q_\sigma(i) - \delta(i) N_d^t(i) N_d(i)]^{-1} A_d(i) \\
\Xi_3(i) &= [\Xi_{31} \quad \Xi_{32}(i) \quad \Xi_{33}(i)] \\
\Xi_{31}(i) &= Y_x^t(i) C_o^t(i) + Z^t(i) D_o^t(i) \\
\Xi_{32}(i) &= Y_x^t(i) N_a^t(i) + Z^t(i) N_d^t(i) \\
\Xi_{33}(i) &= \tau Y_x^t(i) N_d^t(i) \\
\Xi_4(i) &= diag[I \quad \varepsilon(i) I]
\end{aligned} \tag{5.196}$$

for all admissible uncertainties satisfying (5.185). Moreover

$$\begin{aligned}
||z(t)|| &= \Big[\gamma^2 ||w(t)||^2 + x^t(0) P_x(\eta_o) x(0) \\
&+ \tau(0) \int_{-\tau(0)}^{0} \sigma^t(s) Q_\sigma(\eta_o) \sigma(s) ds \Big]^{1/2}
\end{aligned} \tag{5.197}$$

We now direct attention to the case of $\mathcal{H}_\infty$ performance. The results are summarized by the following theorems.

Theorem 5.30 *Given a prescribed constant $\gamma > 0$. Controller (5.7) renders system (Σ_{DK})* **RSSSDD** *with a disturbance attenuation level γ for all $w(t) \in \mathcal{L}_2[0, \infty)$ if, given matrix sequence $Q_x(i) = Q_x^t(i) > 0,\ i \in \mathcal{S}$ there exist matrices $P_x(i) = P_x^t(i) > 0,\ P_\sigma(i) = P_\sigma^t(i) > 0,\ P_d(i) = P_d^t(i) > 0,\ i \in \mathcal{S}$ and scalars $\varepsilon(i) > 0,\ \nu(i) > 0,\ \mu(i) > 0,\ i \in \mathcal{S}$ satisfying the system of LMIs for all $i \in \mathcal{S}$*

$$\begin{bmatrix} \cup_{aa}(i) & \cup_{ab}(i) \\ \cup_{ba}^t(i) & \cup_{bb}(i) \end{bmatrix} < 0 \tag{5.198}$$

where

$$\mho_{aa}(i) = \begin{bmatrix} \Omega_1(i)+ & & \\ G_K^t(i)G_K(i) & \Omega_2(i) & \tau P_d^t(i)A_d(i) \\ \Omega_2^t(i) & -\Omega_3(i) & \tau P_\sigma^t(i)A_d(i) \\ \tau A_d^t(i)P_d(i) & \tau A_d^t(i)P_d(i) & -Q_\sigma(i) \end{bmatrix}$$

$$\mho_{ab}(i) = \begin{bmatrix} \Xi_1(\tau,i) & \begin{array}{c} P^t(i)\Gamma(i)+ \\ \bar{G}_K^t(i)\Phi(i) \end{array} \\ 0 & 0 \\ 0 & 0 \end{bmatrix}$$

$$\mho_{bb}(i) = \begin{bmatrix} -\Xi_2(\tau,i) & 0 \\ 0 & \begin{array}{c} -\gamma^2 I+ \\ \Phi^t(i)\Phi(i) \end{array} \end{bmatrix} \tag{5.199}$$

Moreover

$$\begin{aligned} ||z(t)|| &= \Big[\gamma^2||w(t)||^2 + x^t(0)P_x(\eta_o)x(0) \\ &+ \tau(0)\int_{-\tau(0)}^{0} \sigma^t(s)Q_\sigma(\eta_o)\sigma(s)ds\Big]^{1/2} \end{aligned} \tag{5.200}$$

Proof: Can be easily derived by extending on **Theorem** 5.25 and following parallel development to **Theorem** 5.28.

A method for computing the desired state-feedback gain is given by the following theorem:

Theorem 5.31 *Given a prescribed constant $\gamma > 0$. The feedback gain associated with the $\mathcal{H}_\infty$-controller for system (Σ_{DK}) is given by $K(i) = Z(i)Y_x^{-1}(i)$, $i \in \mathcal{S}$ if there exist matrices $Y_x(i) = Y_x^t(i) > 0$, $Y_\sigma(i) = Y_\sigma^t(i) > 0$, $Y_d(i) = Y_d^t(i) > 0$, $i \in \mathcal{S}$ and scalars scalars $\varepsilon(i) > 0$, $\nu(i) > 0$, $\mu(i) > 0$, $\delta(i)$, $i \in \mathcal{S}$ satisfying the system of LMIs for all $i \in \mathcal{S}$*

$$\begin{bmatrix} \mho_{11}(i) & \mho_{12}(i) \\ \mho_{12}^t(i) & \mho_{22}(i) \end{bmatrix} < 0 \quad , \quad \begin{bmatrix} \mu(i) & 1 \\ 1 & \delta(i) \end{bmatrix} \geq 0 \tag{5.201}$$

where

$$\mho_{11}(i) = \begin{bmatrix} \Lambda_1(i) & \Lambda_2(i) & \tau_s Y_d^t(i) \\ \Lambda_2^t(i) & -\bar{\Lambda}_3(i) & \tau_s Y_\sigma^t(i) \\ \tau_s Y_d(i) & \tau_s Y_\sigma^t(i) & -\tau_s Q_\sigma(i) \end{bmatrix}$$

$$\mho_{12}(i) = \begin{bmatrix} \Xi_5(i) & \mathcal{R}(i) & \begin{matrix} P^t(i)\Gamma(i)+ \\ \bar{G}^t_K(i)\Phi(i) \end{matrix} \\ 0 & 0 & 0 \\ 0 & 0 & 0 \end{bmatrix}$$

$$\mho_{22}(i) = \begin{bmatrix} -\Xi_4(i) & 0 & 0 \\ 0 & -\mathcal{Y}(i) & 0 \\ 0 & 0 & \begin{matrix} -\gamma^2 I+ \\ \Phi^t(i)\Phi(i) \end{matrix} \end{bmatrix} \quad (5.202)$$

and

$$\begin{aligned} \bar{\Lambda}_3(i) &= Y^t_\sigma(i) + Y_\sigma(i) - [\varepsilon(i) + \nu(i) + \mu(i)]M_a(i)M^t_a(i) \\ &- \tau A_d[Q_\sigma(i) - \delta(i)N^t_d(i)N_d(i)]^{-1}A_d(i) \\ \Xi_5(i) &= [\Xi_{51} \quad \Xi_{32}(i) \quad \Xi_{33}(i)] \\ \Xi_{31}(i) &= Y^t_x(i)G^t_o(i) + Z^t(i)F^t_o(i) \end{aligned} \quad (5.203)$$

for all admissible uncertainties satisfying (5.49)-(5.50).

Proof: Follows from parallel development to **Theorem** 5.24. ∇∇∇

Finally, the solution to the simultaneous $\mathcal{H}_2/\mathcal{H}_\infty$ control problem is established by the following theorem:

Theorem 5.32 *Given a prescribed constant $\gamma > 0$. The feedback gain $K(i) = Z(i)Y_x^{-1}(i)$, $i \in \mathcal{S}$ is a* **simultaneous $\mathcal{H}_2/\mathcal{H}_\infty$ controller** *satisfying the performance measure (8.7) for system (Σ_{DK}) if there exist matrices $Y_x(i) = Y^t_x(i) > 0$, $Y_\sigma(i) = Y^t_\sigma(i) > 0$, $Y_d(i) = Y^t_d(i) > 0$, $i \in \mathcal{S}$ such the system of generalized eigenvalue problems*

$$\min \Big[\lambda + Tr(W)\Big]$$

$$\textit{subject to} \quad (5.195),\ (5.203) \quad \textit{and}$$

$$\begin{bmatrix} -L(i) & I \\ I & -Q_x(i) \end{bmatrix} < 0\,, \begin{bmatrix} -\lambda & \phi^t(0) \\ \phi(0) & -Y_x(i) \end{bmatrix} < 0$$

$$\begin{bmatrix} -W & X^t \\ X & -L(i) \end{bmatrix} < 0 \quad (5.204)$$

has a feasible solution

Proof: On observing that

$$
\begin{aligned}
& x^t(0)P_x(i)x(0) \\
\leq\ & \max\left[x^t(0)P_x(1)x(0), \ldots\ldots, x^t(s)P_x(s)x(0))x(0)\right] \\
\triangleq\ & \lambda \Longrightarrow \\
-\ & \lambda + \phi^t(0)Y_x^{-1}(i)\phi(0) \ < \ 0
\end{aligned} \tag{5.205}
$$

and in similar way

$$
\begin{aligned}
\int_{-\tau(0)}^{0} x^t(s)Q(\eta_o)x(s)ds &= \int_{-\tau(0)}^{0} Tr\left[x^t(s)Q(\eta_o)x(s)\right]ds \\
&= Tr\left[XX^tL^{-1}(i)\right] \\
&= Tr\left[X^tL^{-1}(i)X\right] \\
&< Tr(W) \Longrightarrow \\
&- W + X^tL^{-1}(i)X < 0
\end{aligned} \tag{5.206}
$$

where

$$
XX^t = \int_{-\tau(0)}^{0} x^t(s)x(s)ds
$$

Achieving the objective of simultaneous $\mathcal{H}_2/\mathcal{H}_\infty$ control leads to the convex minimization in (5.204) subject to (5.205)-(5.206) as expressed by LMIs in (5.204). ∇∇∇

5.8 Examples

Here we provide two examples regarding the simultaneous $\mathcal{H}_2/\mathcal{H}_\infty$ controller design for the nominal and the uncertain models, respectively.

5.8.1 Example 5.5

In order to illustrate **Theorem** (3.3), we consider a pilot-scale multi-reach water quality system which can fall into the type (8.2)-(8.5). Let the Markov process governing the mode switching has generator

$$\Im = \begin{bmatrix} -4 & 3 & 1 \\ 2 & -6 & 4 \\ 4 & 4 & -8 \end{bmatrix}$$

For the three operating conditions (modes), the associated date are:

Mode 1:

$$\begin{aligned} A_o(1) &= \begin{bmatrix} -0.3 & 0.05 \\ 0 & -0.1 \end{bmatrix}, \; A_d(1) = \begin{bmatrix} -0.1 & -0.1 \\ 0 & -0.2 \end{bmatrix} \\ \Gamma(1) &= \begin{bmatrix} 1 \\ 2 \end{bmatrix}, \; \Psi(1) = \begin{bmatrix} 0.5 \\ 0.4 \end{bmatrix} \\ B_o(1) &= \begin{bmatrix} 1 & 0 \\ 0 & 1 \end{bmatrix}, \; C(1) = \begin{bmatrix} 1 & 0 \\ 0 & 1 \end{bmatrix}, \; D(1) = \begin{bmatrix} 2 & 0 \\ 0 & 2 \end{bmatrix} \\ G(1) &= [0.1 \;\; 0.2] \, , \; \Phi(1) = 0.4 \, , \; F(1) = [2 \;\; 1] \end{aligned}$$

Mode 2:

$$\begin{aligned} A_o(2) &= \begin{bmatrix} -3 & 0 \\ 0 & -3 \end{bmatrix}, \; A_d(2) = \begin{bmatrix} 1 & 0 \\ 0 & 1 \end{bmatrix} \\ \Gamma(2) &= \begin{bmatrix} 2 \\ 1 \end{bmatrix}, \; \Psi(2) = \begin{bmatrix} 0.3 \\ 0.4 \end{bmatrix} \\ B_o(2) &= \begin{bmatrix} 1 & 0 \\ 0 & 1 \end{bmatrix}, \; C(2) = \begin{bmatrix} 2 & 0 \\ 0 & 1 \end{bmatrix}, \; D(2) = \begin{bmatrix} 1 & 0 \\ 0 & 1 \end{bmatrix} \\ G(2) &= [0.1 \;\; 0.1] \, , \; \Phi(2) = 0.3 \, , \; F(2) = [1 \;\; 2] \end{aligned}$$

Mode 3:

$$\begin{aligned} A_o(3) &= \begin{bmatrix} -2.2 & -0.1 \\ 0 & -1.2 \end{bmatrix}, \; A_d(3) = \begin{bmatrix} -0.9 & -1 \\ 0 & -1.1 \end{bmatrix} \\ \Gamma(3) &= \begin{bmatrix} 1 \\ 1 \end{bmatrix}, \; \Psi(3) = \begin{bmatrix} 0.2 \\ 0.3 \end{bmatrix} \end{aligned}$$

$$
\begin{aligned}
B_o(3) &= \begin{bmatrix} 2 & 0 \\ 0 & 1 \end{bmatrix}, \; C(3) = \begin{bmatrix} 2 & 0 \\ 0 & 2 \end{bmatrix}, \; D(3) = \begin{bmatrix} 1 & 0 \\ 0 & 1 \end{bmatrix} \\
G(3) &= [0.2 \;\; 0.2] \; , \; \Phi(3) = 0.5 \; , \; F(2) = [1 \;\; 1]
\end{aligned}
$$

Invoking the software environment [57], we solve the system of LMIs (5.182) using

$$
\begin{aligned}
Q_\sigma(1) &= \begin{bmatrix} 2 & 0 \\ 0 & 1 \end{bmatrix} \; , \; Q_\sigma(2) = \begin{bmatrix} 2 & 0 \\ 0 & 2 \end{bmatrix} \\
Q_\sigma(3) &= \begin{bmatrix} 1 & 0 \\ 0 & 2 \end{bmatrix}, \; \phi = \begin{bmatrix} e^{0.4\,t} \\ e^{0.4\,t} \end{bmatrix} \; , \; \tau \; = \; 0.7
\end{aligned}
$$

The feasible solution is given by:

$$
\begin{aligned}
Y_x(1) &= \begin{bmatrix} 2.1954 & 1.1382 \\ 1.1382 & 3.9872 \end{bmatrix}, \; Y_\sigma(1) \; = \; \begin{bmatrix} 1.1543 & 0.2637 \\ 0.2637 & 3.0775 \end{bmatrix} \\
Y_d(1) &= \begin{bmatrix} 2.1044 & 1.2445 \\ 1.2445 & 3.0294 \end{bmatrix} \\
Y_x(2) &= \begin{bmatrix} 4.8754 & 1.3753 \\ 1.3753 & 3.6789 \end{bmatrix}, \; Y_\sigma(2) \; = \; \begin{bmatrix} 1.2864 & 0.4483 \\ 0.4483 & 2.7765 \end{bmatrix} \\
Y_d(2) &= \begin{bmatrix} 1.5594 & 1.5120 \\ 1.5120 & 6.3376 \end{bmatrix} \\
Y_x(3) &= \begin{bmatrix} 8.0886 & 2.3972 \\ 2.3972 & 6.0277 \end{bmatrix}, \; Y_\sigma(3) \; = \; \begin{bmatrix} 2.1155 & 1.7944 \\ 1.7944 & 4.9834 \end{bmatrix} \\
Y_d(3) &= \begin{bmatrix} 1.8210 & 0.7788 \\ 0.7788 & 2.2275 \end{bmatrix} \\
Z(1) &= \begin{bmatrix} 1.0154 & -0.2441 \\ -0.6125 & 1.1332 \end{bmatrix}, \; Z(2) = \begin{bmatrix} 1.5644 & -0.6112 \\ -0.7153 & 1.4312 \end{bmatrix} \\
Z(3) &= \begin{bmatrix} 2.1204 & -0.8433 \\ -0.2695 & 1.3342 \end{bmatrix} \\
W &= \begin{bmatrix} 4.9811 & 2.3002 \\ 2.3002 & 4.8005 \end{bmatrix}, \; \lambda = 1.2213 \; , \; \gamma = 2.1189
\end{aligned}
$$

The feedback gains and the associated performance bound are

$$
\begin{aligned}
K(1) &= \begin{bmatrix} 0.5231 & -0.2224 \\ -0.3365 & 0.3814 \end{bmatrix}, \; K(2) = \begin{bmatrix} 0.3816 & -0.2771 \\ -0.2334 & 0.4038 \end{bmatrix} \\
K(3) &= \begin{bmatrix} 0.3469 & -0.2775 \\ -0.1207 & 0.2397 \end{bmatrix} \; , \; \mathcal{J} = 10.9876
\end{aligned}
$$

This show that the water quality model is **SSSDD** with a disturbance attenuation level of $\gamma = 2.1189$.

5.8.2 Example 5.6

In order to illustrate **Theorem** (5.32), we use the data of Example 1 in addition to

$$
\begin{aligned}
M_a(1) &= \begin{bmatrix} 0.1 \\ 0.2 \end{bmatrix}, \; N_a(1) = [0.3 \quad 0.3] \\
N_d(1) &= [0.1 \quad 0.3], \; N_b(1) = [0.2 \quad 0.3] \\
M_a(2) &= \begin{bmatrix} 0.1 \\ 0.1 \end{bmatrix}, \; N_a(2) = [0.2 \quad 0.1] \\
N_d(2) &= [0.1 \quad 0.2], \; N_b(2) = [0.1 \quad 0.2] \\
M_a(3) &= \begin{bmatrix} 0.2 \\ 0.1 \end{bmatrix}, \; N_a(3) = [0.3 \quad 0.3] \\
N_d(3) &= [0.1 \quad 0.1], \; N_b(3) = [0.3 \quad 0.1]
\end{aligned}
$$

Using the software LMILab [57], the feasible solution is summarized by

$$
\begin{aligned}
K(1) &= \begin{bmatrix} 0.7942 & 0.8876 \\ -1.3357 & -1.4298 \end{bmatrix}, \; K(2) = \begin{bmatrix} 0.8845 & -0.5528 \\ 0.6744 & 2.1245 \end{bmatrix} \\
K(3) &= \begin{bmatrix} 1.1414 & 0.6563 \\ -0.3345 & 0.4982 \end{bmatrix}
\end{aligned}
$$

for

$$\varepsilon(1) = 1.4992 \,, \; \nu(1) = 0.6634 \,, \; \mu(1) = 2.4567$$

$$\varepsilon(2) = 2.3895 \,, \; \nu(2) = 2.5553 \,, \; \mu(2) = 1.5321$$

$$\varepsilon(3) = 4.5578 \,, \; \nu(3) = 0.9774 \,, \; \mu(3) = 2.2296$$

and

$$W = \begin{bmatrix} 4.9554 & 1.3373 \\ 1.3373 & 4.0589 \end{bmatrix}, \; \lambda = 2.6619 \,, \; \gamma = 1.675 \,, \; \mathcal{J} = 10.9847$$

5.9 Notes and References

We have presented three different approaches to the simultaneous $\mathcal{H}_2/\mathcal{H}_\infty$ control design of a class of uncertain JTDS. The main target has been to focus on delay-dependent behavior while performing the analysis and/or constructing the controllers. The first approach is a direct one and it is based on mode-dependent treatment and the remaining two approaches rely on model transformation. The problems of interest to be investigated would be towards exploring other methodologies to simultaneous $\mathcal{H}_2/\mathcal{H}_\infty$ control and/or refining the convex optimization analysis using ideas like those presented in [74, 143]. In the present book, we have not attempted to talk about optimal control of jump linear systems since there have been no research investigations into optimal control of JTD system following the work of [2, 3, 41, 158] for JLS.

Chapter 6

Robust Filtering

6.1 Introduction

Filtering (equivalently estimation) is perhaps among the oldest problems studied in systems theory and engineering. Intuitively, estimating the state-variables of a dynamic models is of paramount importance as it helps in building better models and improving our knowledge about system forms and behavior. In this regard, the celebrated Kalman filtering [6, 68] is the optimal estimator over all possible linear ones and gives unbiased estimates of the unknown state vector under the conditions that the system and measurement noise processes are mutually-independent Gaussian distributions. Robust state-estimation arises out of the desire to estimate unmeasurable state variables when the plant model has uncertain parameters. In [15] , a Kalman filtering approach has been studied with an H_∞-norm constraint. For linear systems with norm-bounded parameter uncertainty, the robust estimation problem has been addressed in [55, 160, 142, 148], where H_∞-estimators have been constructed in [55]. A robust Kalman filter design is developed in [160] and an alternative approach, based on guaranteed cost, is presented in [142]. In [148], the design of robust filters is considered to yield an estimation error variance with a guaranteed upper bound for all admissible uncertainties.

On another front of research, the problem of estimating the state of uncertain system with state-delay has received increasing interests in recent years [97, 98]. Indeed, the situation becomes compounded which the system parameters undergo jump behavior which is a typical feature of JTDS.

The purpose of this chapter is to consider the filtering problem for a class of linear continuous JTDS with norm-bounded uncertainties. We present three different approaches to this problem:

1) Robust Kalman Filtering

2) Robust $\mathcal{H}_\infty$ Filtering

3) Robust Mode-Dependent $\mathcal{H}_\infty$ Filtering

Each approach will be fully and separately analyzed and later on we will provide some relevant comparisons.

6.2 System Description

Consider the following class of uncertain time-lag systems in a fixed complete probability space $(\Omega,\ \mathcal{F},\ P)$ for all $\eta_t = i \in \mathcal{S}$:

$$\begin{aligned} \dot{x}(t) &= [A_o(i) + \Delta A_o(t,i)]x(t) + A_d(i)x(t-\tau) + w(t) \\ &= A_{\Delta o}(t,i)x(t) + A_d(i)x(t-\tau) + w(t) \end{aligned} \tag{6.1}$$

$$\begin{aligned} x(s) &= \phi(s), \quad s \in [-\tau, 0] \\ y(t) &= [C_o(i) + \Delta C_o(t,i)]x(t) + v(t) \\ &= C_{\Delta o}(t,i)x(t) + v(t) \end{aligned} \tag{6.2}$$

where $x(t) \in \mathbb{R}^n$ is the system state, $y(t) \in \mathbb{R}^m$ is the measurement, $w(t) \in \mathbb{R}^n$ and $v(t) \in \mathbb{R}^m$ are the process and measurement noises, respectively. Here,

$A(i) \in \mathbb{R}^{n\times n}$, $A_d(i) \in \mathbb{R}^{n\times n}$ and $C(i) \in \mathbb{R}^{m\times n}$ are constant matrices that describe the nominal system for every $\eta_t = i,\ i \in \mathcal{S}$. The factor τ is a constant scalar representing the amount of time-lag in the state with

$$0 \leq \tau < \tau^+$$

The form process $\{\eta_t,\ t \geq 0\}$ is a time homogeneous Markov process with right continuous trajectories and taking values in a finite set $\mathcal{S} = \{1, 2, \cdots, s\}$, with stationary transition probabilities

$$\begin{aligned} p_{ij} &= Pr(\eta_{t+\delta} = j \mid \eta_t = i) \\ &= \begin{cases} \alpha_{ij}\delta + o(\delta), & if\ \ i \neq j, \\ 1 + \alpha_{ij}\delta + o(\delta), & if\ \ i = j \end{cases} \end{aligned} \tag{6.3}$$

with transition probability rates $\alpha_{ij} \geq 0$ for $i, j \in \mathcal{S}, i \neq j$ and $\alpha_{ii} = -\sum_{m=1, m\neq i}^{s} \alpha_{im}$ where $\delta > 0$ and $\lim_{\delta\downarrow 0} o(\delta)/\delta = 0$. The set $\mathcal{S}$ comprises the various

The uncertain matrices $\Delta A(t,i)$ and $\Delta C(t,i)$, for any $i \in \mathcal{S}$, are unknown matrix functions which represent time-varying parametric uncertainties and are assumed to be of the form

$$\begin{aligned} \Delta A_o(t,i) &= M_a(i)\, \Delta(t,i)\, N_a(i) \\ \Delta C_o(t,i) &= M_c(i)\, \Delta(t,i)\, N_a(i) \end{aligned} \tag{6.4}$$

where $M_a(i) \in \mathbb{R}^{n\times\alpha}$, $M_c(i) \in \mathbb{R}^{m\times\alpha}$ and $N_a(i) \in \mathbb{R}^{\beta\times n}$, for any $i \in \mathcal{S}$, are known constant matrices and $\Delta(t,i)$, for any $i \in \mathcal{S}$, is an unknown matrix with Lebesgue measurable elements satisfying

$$\Delta^t(t,i)\, \Delta(t,i) \leq I, \quad \forall t \geq 0, \quad i \in \mathcal{S}. \tag{6.5}$$

6.3 Robust Kalman Filtering

In this section, we address the state-estimator design problem such that the estimation error covariance has a guaranteed bound for all admissible uncertainties. The main tool for solving the foregoing problem is the linear matrix inequality (equivalently the Riccati equation) approach. It will be established that the solution of robust steady-state Kalman filtering is expressed in terms of two Riccati equations involving scaling parameters.

The initial condition is specified as $\zeta_0 \stackrel{\Delta}{=} \langle x(0), x(s)\rangle = \langle x_o, \phi(s)\rangle$, where $\phi(\cdot) \in \mathcal{L}_2[-\tau, 0]$ which is assumed to be a zero-mean Gaussian random vector. The following standard assumptions on noise statistics are recalled:

Assumption 1: $\forall\, t\,,\ s \geq 0$

$$(\mathbf{a})\mathbb{E}[w(t)] = 0\ ;\ \mathbb{E}[w(t)w^t(s)] \;=\; W\delta(t-s)\,; W > 0 \qquad (6.6)$$

$$(\mathbf{b})\mathbb{E}[v(t)] = 0\ ;\ \mathbb{E}[v(t)v^t(s)] \;=\; V\delta(t-s)\,; V > 0 \qquad (6.7)$$

$$(\mathbf{c})\mathbb{E}[x(0)w^t(t)] \;=\; 0\,; \mathbb{E}[x(0)v^t(t)] = 0 \qquad (6.8)$$

$$(\mathbf{d})\mathbb{E}[w(t)v^t(s)] = 0\,;\ \mathbb{E}[x(0)x^t(0)] = R_o \qquad (6.9)$$

6.3.1 Preliminary Results

Motivated by the concept of quadratic stability in deterministic system [141] and following the stability results of Chapter 3, we introduce the following definition of stochastic quadratic stability for system (6.1).

Definition 6.1 *System (6.1) without disturbance ($w(t) \equiv 0$) is said to be* **stochastically quadratically stable (SQS)** *if given sequence of matrices* $\{Q(i) = Q^t(i) > 0,\ i \in \mathcal{S}\}$ *, there exists a set of matrices* $\{P(i) = P^t(i) > 0,\ \ i \in \mathcal{S}\}$ *and satisfying the ARIs for all* $i \in \mathcal{S}$

$$\Pi(i) \;\stackrel{\Delta}{=}\; A^t_{\Delta o}(i)P(i) + P(i)A_{\Delta o}(i) + \sum_{j=1}^{s} \alpha_{ij}P(j)$$

$$+ \quad \hat{Q}(i) + P(i)A_d(i)\bar{Q}^{-1}(i)A_d^t(i)P(i) < 0 \tag{6.10}$$

for all admissible parameter uncertainties $\Delta A_o(t,i),\ i \in \mathcal{S}$

where the matrices $\bar{Q}(i), \hat{Q}(i),\ i \in \mathcal{S}$ are given by in Chapter 3.

Remark 6.1 *In [45], it has been proved that for system (6.1), the conditions of* stochastically stable, mean square stable *and* exponentially mean square stable *are equivalent, and any of them imply the* almost surely stable *condition. Therefore,* **Theorem** *3.1 of Chapter 3 also provides the necessary and sufficient conditions for mean square stability and exponential mean square stability, and sufficient conditions for almost sure (asymptotic) stability of system (6.1).*

Now we show that for system (6.1), **SQS** implies **RSSWDD**.

Theorem 6.1 *System (6.1) without disturbance (setting $w(t) \equiv 0$) is* **RSSWDD** *for all admissible uncertainties $\Delta A_o(t,i),\ i \in \mathcal{S}$ if it is* **SQS**.

Proof: Since system (6.1) is **SQS**, there exists a set of matrices $\{P(i) = P^t(i) > 0,\ i \in \mathcal{S}\}$ satisfying (6.10) for all admissible parameter uncertainties $\Delta A_o(t,i),\ i \in \mathcal{S}$.

Select $\eta_t = i, i \in \mathcal{S}$, and given $Q(i) = Q^t(i) > 0$, let the stochastic Lyapunov functional $V(\cdot) : \Re_+ \times \Re^n \times \mathcal{S} \rightarrow \Re_+$ be selected as

$$\begin{aligned} V(t, x, \eta_t = i) &= V(t, x, i) \\ &= x^t(t)P(i)x(t) \\ &+ \int_{-\tau}^{0} x^t(t+\theta)Q(i)x(t+\theta)d\theta \end{aligned} \tag{6.11}$$

along all trajectories of system (6.1). The weak infinitesimal operator $\Im^x[\cdot]$ of the process $\{x(t), \eta_t, t \geq 0\}$ for system (6.1) at the point $\{t, x, \eta_t\}$ is given by

[78]:

$$\Im_1^x[V] = \partial V/\partial t + \dot{x}^t(t)\, \partial V/\partial x \mid_{\eta_t = i} + \sum_{m=1}^{s} \alpha_{im} V(t, x, i, m) \tag{6.12}$$

From (6.11)-(6.12) and considering (6.1) with some standard manipulations in the manner of [113], we have

$$\begin{aligned}
\Im^x[V] &= x^t(t)\Big\{ A_\Delta^t(i)P(i) + P(i)A_\Delta(i) + \sum_{m=1}^{s} \alpha_{im} P(m) + Q(i) \Big\} x(t) \\
&+ \Big\{ x^t(t-\tau)A_d^t(i)P(i)x(t) \;+\; x^t(t)P(i)A_d(i)x(t-\tau) \\
&- x^t(t-\tau)Q(i)x(t-\tau) \Big\} + \sum_{m=1}^{s} \alpha_{im} \int_{-\tau}^{0} x^t(t+\theta)Q(m)x(t+\theta)d\theta \\
&\leq x^t(t)\Big\{ A_\Delta^t(i)P(i) + P(i)A_\Delta(i) + \sum_{m=1}^{s} \alpha_{im} P(m) + \hat{Q}(i) \Big\} x(t) \\
&+ x^t(t)P(i)A_d(i)\bar{Q}^{-1}(i)A_d^t(i)P(i)x(t) \\
&- \Big\{ x^t(t-\tau) - x^t(t)P(i)A_d(i)\bar{Q}^{-1}(i) \Big\} \bar{Q}(i) \\
&\cdot \Big\{ x(t-\tau) - \bar{Q}^{-1}(i)A_d^t(i)P(i)x(t) \Big\} \\
&\leq x^t(t)\Big\{ A_\Delta^t(i)P(i) + P(i)A_\Delta(i) + \sum_{m=1}^{s} \alpha_{im} P(m) + \hat{Q}(i) \\
&+ P(i)A_d(i)\bar{Q}^{-1}(i)A_d^t(i)P(i) \Big\} x(t) \\
&\triangleq x^t(t)\Pi(i)x(t)
\end{aligned} \tag{6.13}$$

for some scalars $\xi(i) > 0,\ i \in \mathcal{S}$.

In view of (6.10), it follows that $\Pi(i) < 0,\ \forall\, i \in \mathcal{S}$. Following the stability analysis of Chapter 3, it can be easily shown that there exists a matrix $L = L^t > 0$ such that $\Pi(i) \leq L$ for any $i \in \mathcal{S}$. Hence, for any $x \neq 0$ and $\eta_t = i$, we have:

$$\begin{aligned} \Im^x[V(x,i)] &\le -\frac{x^t(t)Lx(t)}{V(x(t),i)} V(x(t),i) \\ &\le -\zeta \, V(t,x,i) \end{aligned} \tag{6.14}$$

where

$$\begin{aligned} \zeta &= \min_{x(t),i\in\mathcal{S}} \frac{x^t(t)Lx(t)}{V(x(t),i)} = \min_{x(t),i\in\mathcal{S}} \frac{x^t(t)Lx(t)}{x^t(t)P(i)x(t)} \\ &= \frac{\lambda_m[L]}{\max_{i\in\mathcal{S}} \lambda_M[P(i)]} \end{aligned} \tag{6.15}$$

It follows from [78], by using the Grownwell-Bellman lemma [98] and letting $x(t=0,\phi,i) = x_o$ that

$$\mathbb{E}[V(x,i)|\phi, \eta_o = i] \;\le\; e^{-\zeta\, t}\, V(x_o,i) \tag{6.16}$$

and therefore system (6.1) is exponentially mean square stable and hence stochastic stability via **Remark** 6.1. $\nabla\nabla\nabla$

Finally, by making use of **Fact 3** and **Theorem** (3.3), we have the following result.

Theorem 6.2 *System (6.1) without disturbance (setting $w(t) \equiv 0$) is* **SQS** *if for some matrices $Q(i) = Q^t(i) > 0,\ i \in \mathcal{S}$, there exist a sequence $\{\varepsilon(i) > 0,\ i \in \mathcal{S}\}$ and a set of matrices $\{P(i) = P^t(i) > 0,\ i \in \mathcal{S}\}$ satisfying the system of LMIs for all $i \in \mathcal{S}$*

$$\begin{bmatrix} \begin{array}{c} P(i)A(i) + A^t(i)P(i) \\ +\sum_{m=1}^{s} \alpha_{im} P(m) + \hat{Q}(i) \\ +\varepsilon(i)P(i)M(i)M^t(i)P(i) \end{array} & P(i)A_d(i) & E^t(i) \\ A_d^t(i)P(i) & -\bar{Q}(i) & 0 \\ E(i) & 0 & -\varepsilon(i)I \end{bmatrix} < 0 \tag{6.17}$$

6.3.2 Robust Filter

Our objective in this section is to design a stochastically stable estimator such that the error covariance of state $x(t)$ and its estimate $\hat{x}(t)$ is bounded for all admissible uncertainties.

Definition 6.2 *Given system (6.1)-(6.2), the state equations, for all* $\eta_t = i \in \mathcal{S}$,

$$\dot{\hat{x}}(t) = G(i)\,\hat{x}(t) + K(i)\,y(t),\ \hat{x}(0) = x_0 \tag{6.18}$$

are said to define a **guaranteed cost state estimator** *for this system if there exists a constant matrix* $P = P^t \geq 0$ *such that*

$$\begin{aligned} \mathbb{E}\{(x-\hat{x})(x-\hat{x})^t\} &\leq P \Longrightarrow \\ \mathbb{E}\{(x-\hat{x})^t(x-\hat{x})\} &\leq Tr(P) \end{aligned} \tag{6.19}$$

for all admissible uncertainties $\Delta A_o(t,i)$ *and* $\Delta C_o(t,i)$, $i \in \mathcal{S}$.

In this situation, the estimator (6.18) is said to provide **a guaranteed cost matrix** P.

Examination of the proposed estimator proceeds by analyzing the estimation error for all $\eta_t = i \in \mathcal{S}$

$$e(t) \;\; = x(t) \; - \; \hat{x}(t) \tag{6.20}$$

Substituting (6.1)-(6.2) and (6.18) into (6.20), we express the dynamics of the error in the form:

$$\begin{aligned} \dot{e}(t) &= G(i)\,e(t) + [A_o(i) - G(i) - K(i)C_o(i)]x(t) \\ &+ [\Delta A_o(t,i) - K(i)\Delta C_o(t,i)]x(t) \\ &+ A_d(i)x(t-\tau) + [w(t) - K(i)v(t)] \end{aligned} \tag{6.21}$$

By introducing the extended state-vector

$$\xi(t) = \begin{bmatrix} x(t) \\ e(t) \end{bmatrix} \in \Re^{2n} \tag{6.22}$$

it follows from (6.1)-(6.2) and (6.21) for $i \in \mathcal{S}$ that

$$\begin{aligned} \dot{\xi}(t) &= [\widehat{A}(i) + \widehat{M}(i)\Delta(t,i)\widehat{E}(i)]\xi(t) + \widehat{D}(i)\xi(t-\tau) + \widehat{B}(i)\beta(t) \\ &= \widehat{A}_\Delta(i)\xi(t) + \widehat{D}(i)\xi(t-\tau) + \widehat{B}(i)\beta(t) \end{aligned} \tag{6.23}$$

where $\beta(t)$ is a stationary zero-mean noise signal with identity covariance matrix and

$$\widehat{A}(i) = \begin{bmatrix} A(i) & 0 \\ A(i) - G(i) - K(i)C(i) & G(i) \end{bmatrix} \tag{6.24}$$

$$\widehat{M}(i) = \begin{bmatrix} M_a(i) \\ M_a(i) - K(i)M_c(i) \end{bmatrix}, \quad \widehat{E}(t) = [N_a(i) \quad 0] \tag{6.25}$$

$$\widehat{B}\widehat{B}^t(i) = \begin{bmatrix} W & W \\ W & W + K(i)VK^t(i) \end{bmatrix}, \tag{6.26}$$

$$\widehat{D}(i) = \begin{bmatrix} A_d(i) & 0 \\ A_d(i) & 0 \end{bmatrix}, \quad \beta(t) = \begin{bmatrix} w(t) \\ v(t) \end{bmatrix} \tag{6.27}$$

Definition 6.3 *Estimator (6.18) is said to be a* **stochastically stable quadratic state estimator (SSQSE)** *associated with a sequence of matrices $0 < \Omega(i) \in \Re^{2n}$, $i \in \mathcal{S}$ for system (6.1)-(6.2) if there exist a sequence of scalars $\varrho(i) > 0$, $i \in \mathcal{S}$ and a matrix block*

$$0 < \Omega(i) = \begin{bmatrix} \Omega_1(i) & \Omega_3(i) \\ \Omega_3^t(i) & \Omega_2(i) \end{bmatrix}, \quad i \in \mathcal{S} \tag{6.28}$$

satisfying the algebraic inequality for all $i \in \mathcal{S}$:

$$\begin{aligned} &\widehat{A}_\Delta(t,i)\,\Omega(i) + \Omega(i)\,\widehat{A}_\Delta^t(t,i) + \varrho(i)\,\Omega(i) + \sum_{m=1}^{s} \alpha_{im}\Omega(m) \\ &+\varrho^{-1}(i)\,\widehat{D}(i)\,\Omega(i)\,\widehat{D}^t(i) + \widehat{B}(i)\,\widehat{B}^t(i) \leq 0 \end{aligned} \tag{6.29}$$

The next result shows that if (6.18) is **SSQES** for system (6.1)-(6.2) with cost matrix $\Omega(i),\ i \in \mathcal{S}$, then $\Omega(i),\ i \in \mathcal{S}$ defines an upper bound for the filtering error covariance, that is,

$$\mathbb{E}[e(t)e^t(t)] \ \leq \ \Omega_2(i) \qquad \forall t$$

for all admissible uncertainties satisfying (6.4)-(6.5).

Theorem 6.3 *Consider the time-lag system (6.1)-(6.2) satisfying (6.4)-(6.5) and with known initial state* ζ_0. *Suppose there exists a solution* $\Omega(i) \ \geq \ 0,\ i \in \mathcal{S}$ *to inequality (6.29) for some* $\alpha(i) \ > \ 0,\ i \in \mathcal{S}$ *and for all admissible uncertainties. Then the estimator (6.18) provides an upper bound for the filtering error covariance, that is,*

$$\mathbb{E}[e(t)e^t(t)] \ \leq \ \bar{\Omega}_2$$

where $\bar{\Omega}_2 = \max_{i \in \mathcal{S}} \ \Omega_2(i)$.

Proof: Suppose that the estimator (6.18) is **SSQES** with cost matrix $\Omega(i),\ i \in \mathcal{S}$. It follows that

$$\begin{aligned} \mathbb{E}[\xi(t)\ \xi^t(t)] \ &\triangleq \ \Upsilon(t,i) \\ &= \ \sum_{j=1}^{s} \alpha_{ij}\ \Sigma(t,j). \end{aligned}$$

It is a straightforward task to show that

$$\Upsilon(t,i) = \Upsilon^t(t,i) \ \geq \ 0,\ \forall\, t \ \geq \ 0$$

By evaluating the derivative of the covariance matrix $\Upsilon(t,i)$ we get:

$$\begin{aligned} \dot{\Upsilon}(t,i) \ &= \ \widehat{A}_\Delta\,(t,i)\Upsilon(t,i) + \Upsilon(t,i)\ \widehat{A}_\Delta^t\,(t,i) + \widehat{D}\,(i)\mathbb{E}[\xi(t-\tau)\xi^t(t)] \\ &+ \ \sum_{m=1}^{s} \alpha_{im}\Upsilon(t,m) + \mathbb{E}[\xi(t)\ \xi^t(t-\tau)]\ \widehat{D}^t\,(t) \\ &+ \ \mathbb{E}[\beta(t)\ \xi^t(t)] + \mathbb{E}[\xi(t)\ \beta^t(t)] \end{aligned} \qquad (6.30)$$

In the light of [129] and using **Fact 1** of the Appendix, we get the inequality:

$$\begin{aligned} & \widehat{D}\,(i)\mathbb{E}[\xi(t-\tau)\;\xi^t(t)] + \mathbb{E}[\xi(t)\;\xi^t(t-\tau)]\;\widehat{D}^t\,(i) \\ \leq\;& \varrho(i)\;\widehat{D}\,(i)\mathbb{E}[\xi(t-\tau)\;\xi^t(t-\tau)]\;\widehat{D}^t\,(i) + \varrho^{-1}(i)\;\mathbb{E}[\xi(t)\;\xi^t(t)] \\ =\;& \varrho(i)\;\widehat{D}\,(i)\;\Upsilon(t-\tau,i)\;\widehat{D}^t\,(i)\;+\;\varrho^{-1}(i)\;\Upsilon(t,i) \end{aligned} \tag{6.31}$$

Substituting (6.31) into (6.30) and arranging terms, we obtain:

$$\begin{aligned} \dot{\Upsilon}(t,i) &= \widehat{A}_\Delta\,(t,i)\Upsilon(t,i) + \Upsilon(t,i)\;\widehat{A}_\Delta^t\,(t,i)\;+\;\sum_{m=1}^{s}\alpha_{im}\Upsilon(t,m) \\ &+\;\varrho(i)\;\widehat{D}\,(i)\;\Upsilon(t-\tau,i)\;\widehat{D}^t\,(i) + \varrho^{-1}(i)\;\Upsilon(t,i) \\ &+\;\widehat{B}\widehat{B}^t\,(i) \end{aligned} \tag{6.32}$$

Combining (6.29) and (6.32) and letting

$$\Xi(t,i) = \Upsilon(t,i)\;-\;\Omega(i)$$

we obtain:

$$\begin{aligned} \dot{\Xi}(t,i) &\leq \widehat{A}_\Delta\,(t,i)\;\Xi(t,i)\;+\;\Xi(t,i)\;\widehat{A}_\Delta^t\,(t,i)\;+\;\varrho^{-1}(i)\;\Xi(t,i) \\ &+\;\varrho(i)\;\widehat{D}\,(i)\;\Xi(t-\tau,i)\;\widehat{D}^t\,(i)\;+\;\sum_{m=1}^{s}\alpha_{im}\Xi(t,m) \end{aligned} \tag{6.33}$$

On considering that the state is known over the period $[-\tau,0]$ it justifies letting

$$\Upsilon(0,i) = 0\;,\;i\in\mathcal{S}$$

Hence, inequality (6.33) implies that

$$\begin{aligned} \Xi(t,i) &\leq 0\;,\;i\in\mathcal{S}\Longrightarrow \\ \Upsilon(t,i) &\leq \Omega(i)\;\forall\,t\,>\,0 \end{aligned}$$

Finally, it is obvious that

$$\mathbb{E}[e(t)e^t(t)]\;=\;[0\;\;I]\Xi(t)\left[\begin{array}{c}0\\I\end{array}\right]\;=\;\Omega_2(i)\;\leq\;\bar{\Omega}_2(i)$$

$\nabla\nabla\nabla$

Remark 6.2 *It should be noted that* **Theorem** *6.3 is an analytical result. The design principles will be laid down by the subsequent theorems.*

6.3.3 Robust Steady-State Filter

Now, we investigate the asymptotic properties of the Kalman filter developed previously. It is assumed that the sequence of matrices $(A(i),\ i \in \mathcal{S})$ is Hurwitz. The objective is to design a time-invariant *a priori* estimator for every $i \in \mathcal{S}$ of the form :

$$\begin{aligned} \dot{\hat{x}}(t) &= \hat{A}(i)\,\hat{x}(t) + K(i)\,[y(t) - C(i)\hat{x}(t)] \\ \hat{x}(t_o) &= 0 \end{aligned} \tag{6.34}$$

that achieves the following asymptotic performance bound

$$\lim_{t\to\infty} \mathbb{E}\left\{[\hat{x}(t) \ - \ x(t)][\hat{x}(t) \ - \ x(t)]^t\right\} \ \leq \ L(i) \tag{6.35}$$

for a given set of matrices $\{L(i) > 0,\ i \in \mathcal{S}\}$.

The following theorem summarizes the main result.

Theorem 6.4 *Consider the composite time-lag system (6.23). If for some sequences of scalars $\mu(i) > 0,\ \rho(i) > 0,\ i \in \mathcal{S}$ there exist stabilizing solutions*[1]

[1] A solution $P(i),\ \eta_t = i \in \mathcal{S}$ of the following algebraic Riccati-like coupled equations

$$A^t(i)P(i) + P(i)A(i) + \sum_{j=1}^{s} \alpha_{ij}P(j) + \gamma^{-2}P(i)B(i)^T B(i)P(i) + C^T(i)C(i) = 0,\ t \geq 0,\ i \in \mathcal{S}$$

is said to be stabilizing if the system

$$\dot{x}(t) = \left[A(i) + \gamma^{-2}B(i)^T B(i)P(i)\right]x(t)$$

is stochastically stable.

$P(i) > 0$ and $L(i) > 0$ for the following AREs for all $i \in \mathcal{S}$

$$A(i)P(i) + P(i)A^t(i) + \rho(i)P(i) + \hat{W}(i) + \rho^{-1}(i)A_d(i)P(i)A_d^t(i) + \sum_{m=1}^{s} \alpha_{im}P(m) + \mu(i)P(i)E^t(i)E(i)P = 0 \tag{6.36}$$

$$A(i)L(i) + L(i)A^t(i) + \rho(i)L(i) + \hat{W}(i) + \rho^{-1}(i)A_d(i)P(i)A_d^t(i) + \sum_{m=1}^{s} \alpha_{im}L(m) + \mu(i)L(i)E^t(i)E(i)L(i) - \left\{L(i)C^t(i) + \mu^{-1}(i)M(i)N^t(i)\right\}\hat{V}^{-1}(i) \cdot \left\{C(i)L(i) + \mu^{-1}(i)N(i)M^t(i)\right\} = 0 \ , \quad i \in \mathcal{S} \tag{6.37}$$

then the estimator (6.34) is a **SSQSE** *and achieves the bound (6.35) where for any $i \in \mathcal{S}$*

$$\begin{aligned} \hat{W}(i) &= W + \mu^{-1}(i)M(i)M^t(i) \\ \hat{V}(i) &= V + \mu^{-1}N(i)N(i)^t \\ K(i) &= \left\{L(i)C^t(i) \ + \ \mu^{-1}(i)M(i)N^t(i)\right\} \hat{V}^{-1}(i) \end{aligned} \tag{6.38}$$

$$\begin{aligned} \hat{A}(i) &\triangleq A(i) \ + \ \delta A(i) \\ &= A(i) \ + \ \mu^{-1}(i)L^t(i)E^t(i)E(i) \end{aligned} \tag{6.39}$$

Proof: To examine the stability of the composite-loop system (6.23) ($w(t) = 0, v(t) = 0$), to obtain

$$\dot{\xi}(t) = \widehat{A}_\Delta(i)\xi(t) + \widehat{D}(i)\xi(t-\tau) \tag{6.40}$$

By a similar argument as in the proof of **Theorem** 6.3, it is easy to see ([113]) that for some given $\varrho(i) > 0, \ i \in \mathcal{S}$,

$$\widehat{A}_\Delta(t)\Theta(i) + \Theta(i)\widehat{A}_\Delta^t(t) + \varrho^{-1}(i)\,\Theta(i) + \varrho(i)\,\widehat{D}(t)\,\Theta(i)\,\widehat{D}^t(t) + \widehat{B}\widehat{B}^t(i) + \sum_{m=1}^{s} \alpha_{im}\Theta(m) < 0 \tag{6.41}$$

where

$$\Theta(i) = \begin{bmatrix} P(i) & L(i) \\ L(i) & L(i) \end{bmatrix} \tag{6.42}$$

Introducing a stochastic Lyapunov functional for any $i \in \mathcal{S}$

$$\begin{aligned} V(x,i) &= \xi^t(t)\Theta(i)\xi(t) \\ &+ \int_{t-\tau}^{t} \xi^t(q)\varphi^{-1} \widehat{D}\,\Theta(i)\,\widehat{D}^t\,\xi(q)\;dq \end{aligned} \tag{6.43}$$

and observe that $V(x,i) > 0,\ i \in \mathcal{S}$ for $\xi(t) \neq 0$, for some $\varphi > 0$ and $V(x,i) = 0,\ i \in \mathcal{S}$ when $\xi = 0$. By evaluating the weak infinitesimal of the Lyapunov functional (6.43) along the trajectories of system (6.40), we get:

$$\begin{aligned} \Im^x[V] &= \xi^t(t)[\Theta(i)\,\widehat{A}_\Delta + \widehat{A}_\Delta^t\,\Theta(i) + \varrho^{-1}(i)\,\widehat{D}\,\Theta(i)\,\widehat{D}^t + \sum_{m=1}^{s} \alpha_{im}\Theta(m)]\xi(t) \\ &+ \xi^t(t)\Theta(i)\,\widehat{D}\,\xi(t-\tau) + \xi^t(t-\tau)\,\widehat{D}^t\,\Theta(i)\xi(t) \\ &- \varrho^{-1}(i)\xi^t(t-\tau)\,\widehat{D}\,\Theta(i)\,\widehat{D}^t\,\xi(t-\tau) \\ &\leq \xi^t(t)[\Theta(i)\,\widehat{A}_\Delta + \widehat{A}_\Delta^t\,\Theta(i) + \varrho^{-1}(i)\,\widehat{D}\,\Theta(i)\,\widehat{D}^t + \sum_{m=1}^{s} \alpha_{im}\Theta(m)]\xi(t) \\ &+ \varrho(i)\xi^t(t)\Theta(i)\xi(t) \\ &= \xi^t(t)[\Theta(i)\,\widehat{A}_\Delta + \widehat{A}_\Delta^t\,\Theta(i) + \varrho^{-1}(i)\,\widehat{D}\,\Theta(i)\,\widehat{D}^t \\ &+ \sum_{m=1}^{s} \alpha_{im}\Theta(m) + \varrho(i)\Theta(i)]\xi(t) \\ &< 0, \qquad \xi(t) \neq 0 \end{aligned} \tag{6.44}$$

which means that the augmented system (6.23) is stochastically stable. In turn, this implies that (6.34) is **SSQES**. The guaranteed performance

$$\mathbb{E}[e(t)e^t(t)] \leq L(i)\ ,\ i \in \mathcal{S}$$

follows from similar lines of argument as in the proof of **Theorem** 6.3. $\nabla\nabla\nabla$

The next theorem provides an LMI-based solution to the steady-state robust Kalman filter.

Theorem 6.5 *Consider the uncertain time-lag system (6.1)-(6.2). For every $i \in \mathcal{S}$, the estimator*

$$\begin{aligned} \dot{\hat{x}}(t) &= [A(i) + \mu^{-1}(i)L^t(i)E^t(i)E(i)]\hat{x}(t) \\ &+ [L(i)C^t(i) + \mu^{-1}(i)M(i)N(i)^t]\hat{V}^{-1} \\ &\cdot [y(t) - C(i)\hat{x}(t)] \end{aligned} \tag{6.45}$$

where

$$\hat{V} = V + \mu^{-1}(i)N(i)N^t(i) \tag{6.46}$$

is a **SSQES** *and achieves the bound (6.35) for some $L(i) \geq 0$, $i \in \mathcal{S}$ if for some scalars $\mu(i) > 0$, $\rho(i) > 0$, $i \in \mathcal{S}$ there exist matrices $0 < Y(i) = Y^t(i)$ and $0 < X(i) = X^t(i)$, $i \in \mathcal{S}$ satisfying the AMIs for all $i \in \mathcal{S}$*

$$\begin{bmatrix} \Pi_y(i) & A_d(i)Y(i) & M(i) & Y(i)E^t(i) \\ Y(i)A_d^t(i) & -\rho(i)I & 0 & 0 \\ M^t(i) & 0 & -\mu(i)I & 0 \\ E(i)Y(i) & 0 & 0 & -\mu^{-1}(i)I \end{bmatrix} < 0 \tag{6.47}$$

$$\begin{bmatrix} \Pi_x(i) & A_d(i)Y(i) & M(i) & X(i)E^t(i) \\ Y(i)A_d^t(i) & -\rho(i)I & 0 & 0 \\ M^t(i) & 0 & -\mu(i)I & 0 \\ E(i)X(i) & 0 & 0 & -\mu^{-1}(i)I \end{bmatrix} < 0 \tag{6.48}$$

where

$$\begin{aligned} \Pi_y(i) &= A(i)Y(i) + Y(i)A^t(i) + \rho(i)Y(i) + W \\ &+ \sum_{m=1}^{s} \alpha_{im}Y(m) \\ \Pi_x(i) &= A(i)X(i) + X(i)A^t(i) + \rho(i)X(i) + W \\ &+ \sum_{m=1}^{s} \alpha_{im}X(m) + Q_x(X,\rho,i) \\ Q_x(X,\rho,i) &= -\Big[X(i)C^t(i) + \mu^{-1}(i)M(i)N^t(i)]\hat{V}^{-1} \\ &\cdot [C(i)X(i) + \mu^{-1}N(i)M^t(i)\Big] \end{aligned} \tag{6.49}$$

Proof: By **Fact 3** and (6.36)-(6.37), it follows that there exist matrices $0 < Y(i) = Y^t(i),\ i \in \mathcal{S}$ and $0 < X(i) = X^t(i),\quad i \in \mathcal{S}$ satisfying the ARIs for all $i \in \mathcal{S}$:

$$A(i)Y(i) + Y(i)A^t(i) + \rho(i)Y(i) + \hat{W} + \sum_{m=1}^{s} \alpha_{im} Y(m) +$$
$$\rho^{-1}(i)A_d(i)Y(i)A_d^t(i) + \mu Y(i)E^t(i)E(i)Y(i) \ < \ 0 \tag{6.50}$$
$$A(i)X(i) + X(i)A^t(i) + \rho(i)X(i) + \hat{W} + \sum_{m=1}^{s} \alpha_{im} X(m) +$$
$$\rho^{-1}(i)A_d(i)Y(i)A_d^t(i) + \mu X(i)E^t(i)E(i)X(i) -$$
$$[X(i)C^t(i) + \mu^{-1}(i)M(i)N^t(i)]\hat{V}^{-1}$$
$$\cdot\ [C(i)X(i) + \mu^{-1}(i)N(i)M^t(i)] \ < \ 0 \tag{6.51}$$

such that $Y(i) > P(i), X(i) > L(i),\ i \in \mathcal{S}$. Application of **Fact 2** to the ARIs (6.50)-(6.51) yields the AMIs (6.47)-(6.48). $\nabla\nabla\nabla$

Remark 6.3 *It should be emphasized the AREs (6.36)-(6.37) do not have clear-cut monotonicity properties enjoyed by standard AREs. The main reasons for this are the presence of the delay term*

$$A_d(i)P(i)A_d^t(i)$$

and the coupling term

$$\sum_{m=1}^{s} \alpha_{im} P(m)$$

In view of the stability properties, it can be argued that there exist some sequences $\mu(i) > 0,\ \rho(i) > 0,\ i \in \mathcal{S}$ such that the ARE (6.36) admits a positive-definite solution $P(i) > 0,\ i \in \mathcal{S}$ for some $0 < R(i) = R^t(i),\ i \in \mathcal{S}$. Furthermore, for a given $\mu(i) > 0, \rho(i) > 0,\ i \in \mathcal{S}$ and $R(i) > 0,\ i \in \mathcal{S}$, if there exist $\bar{\mu}(i) > 0,\ \bar{\rho}(i) > 0,\ i \in \mathcal{S}$ such that (6.36) admits a positive-definite solution $0 < \bar{P}(i) = \bar{P}^t(i),\ i \in \mathcal{S}$ for some $0 < \bar{R}(i) = \bar{R}^t(i)$, then for any $\mu(i) \in$

$(0, \bar{\mu}(i)]$, $\rho(i) \in (0, \bar{\rho}(i)]$, $i \in \mathcal{S}$, *the solution of (6.37)* $P(i) > 0$, $i \in \mathcal{S}$ *satisfies* $0 < P(i) \leq \left\{\mu(i)/\bar{\mu}(i)\right\}\bar{P}(i)$, $i \in \mathcal{S}$ *for some* $0 < R(i) \leq \left\{\rho(i)/\bar{\rho}(i)\right\}\bar{R}(i)$, $i \in \mathcal{S}$.

6.3.4 Example 6.1

For the purpose of illustrating the developed theory, we proceed to determine the steady-state estimator gains. Essentially, we seek to solve (6.36)-(6.39) when $\rho(i) \in [\rho_1(i) \to \rho_2(i)]$, $i \in \mathcal{S}$, $\mu(i) \in [\mu_1(i) \to \mu_2(i)]$, $i \in \mathcal{S}$, where $\rho_1(i), \rho_2(i), \mu_1(i), \mu_2(i)$, $i \in \mathcal{S}$ are some constants to be chosen (for instance, by trial and error). Note that (6.36) depends on $P(i)$, $i \in \mathcal{S}$ only and it is not of the standard-forms of AREs. On the other hand, (6.37) depends on both $L(i)$, $i \in \mathcal{S}$ and $P(i)$, $i \in \mathcal{S}$ and it can be put into the standard ARE form. For efficient numerical computations, we employ a Kronecker Product-like technique to reduce (6.36) into a system of nonlinear algebraic equations of the form

$$f(\omega) = G\,\omega + h(\omega) + q \tag{6.52}$$

where $\omega \in \Re^{n(n+1)/2}$ is a vector of the unknown elements of the $P(i)$-matrix , $i \in \mathcal{S}$. The algebraic equation (6.52) can then be solved using an iterative Newton Raphson technique according to the rule:

$$\omega_{(m+1)} = \omega_{(m)} - \gamma_{(m)}[G + \nabla_\varepsilon h(\varepsilon_{(m)}]^{-1} f(\varepsilon_{(m)}) \tag{6.53}$$

where m is the iteration index, $\omega_{(o)} = 0$, $\nabla_\varepsilon h(\varepsilon)$ is the Jacobian of $h(\varepsilon)$ and the step-size $\gamma_{(m)}$ is given by $\gamma_{(m)} = 1/[||f(\varepsilon(m))|| + 1]$. It is well-known that the iterative scheme (6.53) has super-linear convergence properties [17].

Given the solution of (6.36), we proceed to solve ((6.37)) using a standard Hamiltonian/Eigenvector method which is quite fast and efficient. All the computations are conveniently carried out using the Linear Algebra and System (L-A-S) software [17]. As a typical case, consider a two-mode jump time-lag system of the type (6.1)-(6.2) with

Mode 1

$$\begin{aligned} A &= \begin{bmatrix} -2 & 0.5 \\ 1 & -3 \end{bmatrix}, \; A_d = \begin{bmatrix} -0.2 & -0.1 \\ 0.1 & 0.4 \end{bmatrix} \\ W &= \begin{bmatrix} 1 & 0 \\ 0 & 1 \end{bmatrix}, \; H = \begin{bmatrix} 0.5 \\ 0.5 \end{bmatrix} \\ C &= [1 \;\; -3] \;, \; E = [0.5 \;\; 1] \\ H_c &= 2 \;, \; V = 1 \end{aligned}$$

Mode 2

$$\begin{aligned} A &= \begin{bmatrix} -4 & 0.6 \\ 1.3 & -2 \end{bmatrix}, \; A_d = \begin{bmatrix} -0.3 & -0.1 \\ 0.1 & 0.3 \end{bmatrix} \\ W &= \begin{bmatrix} 1 & 0 \\ 0 & 1 \end{bmatrix}, \; H = \begin{bmatrix} 0.4 \\ 0.6 \end{bmatrix} \\ C &= [1 \;\; -2] \;, \; E = [0.5 \;\; 1] \\ H_c &= 2 \;, \; V = 1 \end{aligned}$$

The associated transition probability matrix is given by

$$\{\alpha_{ij}\} = \begin{bmatrix} -1 & 1 \\ 1 & -1 \end{bmatrix}$$

Past computational experience of parameter-dependent algorithms [18] has indicated that the region covered by $\{\rho \in [0.1 \to 0.9] \cap \mu \in [0.55 \to 0.9]$ yields a stable estimator. Based thereon, the computational results for the two modes are presented in Table 1. By changing the values of $\{\rho, \mu\}$ over the prescribed has resulted in small change in the values of K and δA. Therefore we conclude that the stable-estimator gains are practically insensitive to the (λ, μ)-parameters. Indeed, there is a finite range for (λ, μ) that guarantees stable performance of the developed Kalman filter.

$Mode$	$\{\rho, \mu\}$	P		L		δA	
1	{0.1,0.9}	0.5149	0.1524	0.479	0.124	0.2018	0.4039
		0.1524	0.4248	0.124	0.409		
2	{0.8,0.6}	0.735	0.321	0.6498	0.281	0.507	1.015
		0.321	0.392	0.283	0.384		
$Mode$	$\{\rho, \mu\}$	K^t		$\lambda(\hat{A})$			
1	{0.1,0.9}	0.3149	0.3018	-3.2579	-1.0173		
		0.261	0.523				
2	{0.8,0.6}	0.340	0.304	-4.3603	-0.2587		
		0.437	0.874				

Table 6.1: Summary of the Computational Results

6.4 Robust $\mathcal{H}_\infty$ Filtering

In this section, we examine another robust filtering technique for linear, uncertain multi-state-delay systems with Markovian jump parameters. Of particular interest to this section, is the $\mathcal{H}_\infty$ filtering in which the design is based on minimizing the $\mathcal{H}_\infty$-norm of the system. This design reflects a worst-case gain of the transfer function from the disturbance inputs to the estimation error output. In addition, it has been generally quoted [81] that $\mathcal{H}_\infty$ filtering is superior to standard $\mathcal{H}_2$ filtering since no statistical assumption is made on the input signals.

Thus we proceed to the further development of robust state estimation techniques of linear uncertain systems with multi-state-delay and Markovian jump parameters. In this regard, we build upon the results of [165, 110] for continuous time-delay systems. It provides a linear matrix inequality procedure for the design of a Markovian jump filter which guarantees the robust stochastic stability with a prescribed performance measure and it generalizes previous existing results.

6.4.1 Problem Formulation

We consider a class of multi-state-delayed dynamical systems with Markovian jump parameters described over the space $(\Omega, \mathcal{F}, \mathbf{P})$ by:

$$\begin{aligned}
\dot{x}(t) &= \big[A_o(\eta_t) + \Delta A_o(t, \eta_t)\big] x(t) \\
&+ \sum_{k=1}^{q} \big[A_{d_k}(\eta_t) + \Delta A_{d_k}(t, \eta_t)\big] x(t - \tau_k) \\
&+ \big[B_o(\eta_t) + \Delta B_o(t, \eta_t)\big] w(t) \\
&= A_{\Delta o}(t, \eta_t) x(t) + [\mathcal{E}_t + \Delta \mathcal{E}_t](\eta_t)\ \chi(t, T, q) \\
&+ B_{\Delta o}(t, \eta_t) w(t) \ , \\
x(t) &= \phi(t) \ \in \mathcal{L}_2(-\max\{\tau_k, k \in J_q\}, 0; \Re^n) \\
&\quad \eta_o = j \ \in \mathcal{S}, \ t \geq 0, \qquad (6.54) \\
y(t) &= C_o(\eta_t) x(t) + D_o(\eta_t) w(t) \qquad (6.55) \\
z(t) &= L(\eta_t) x(t) + \Gamma(\eta_t) w(t) \qquad (6.56)
\end{aligned}$$

where

$$J_r = \{1, ..., r\}$$

is the set of the r first positive integers. Given a set of constant matrices

$$A_{d_1}, ..., A_{d_q} \ \in \mathbb{R}^{n \times n}$$

we denote

$$\mathcal{E}_t = \big[A_{d_1} \ .... \ A_{d_q}\big] \in \mathbb{R}^{n \times nq} \quad , \quad \mathcal{E}_s = \big[A_{d_1} + ... + A_{d_q}\big] \in \mathbb{R}^{n \times n}$$

Also, for some positive constants

$$(a_1, ..., a_q)$$

with the vector

$$x(t - a_j) \in \mathbb{R}^n, \ \forall j \in J_q$$

we let,

$$\chi(t, a, q) = [x^t(t - a_1),\ x^t(t - a_2),\ ....x^t(t - a_q)]^t \in \mathbb{R}^{nq \times 1}$$

Also, we use in the sequel

$$\mathcal{T} = (\tau_1, ..., \tau_q)^t$$

As usual, $x(t) \in \Re^n$ is the state vector and $w(t) \in \Re^m$ is the input noise which is assumed to be an arbitrary signal in $\mathcal{L}_2$. Here, $\tau_k \geq 0,\ k \in J_q$ are unknown time-varying factors representing the amount of delays in the state of the system where $\tau_k(t) \in (0, \tau_k^*]$ is such that $\dot{\tau}_k(t) \leq \tau_k^+ < 1$ with $\tau_k^*,\ \tau_k^+;\ j \in \mathcal{J}_q$ being finite known constants. In (6.54)-(6.55), for $\eta_t = j\ \in \mathcal{S}$, $A_o(\eta_t)$, $B_o(\eta_t)$ $C_o(\eta_t)$, $D_o(\eta_t)$, $L(\eta_t)$, $\Gamma(\eta_t)$ and $\{A_{d_k}(\eta_t);\ k \in \mathcal{J}_q\}$ are known constant matrices of appropriate dimensions and $\Delta A_o(\eta_t)$, $\Delta B_o(\eta_t)$ and $\{\Delta A_{d_k}(\eta_t);\ k \in \mathcal{J}_q\}$ are unknown matrices which represent time-varying parametric uncertainties and assumed to belong to certain bounded compact sets. The initial vector function is specified as $\xi_0 \equiv \langle x(0), x(s)\rangle = \langle x_o, \phi(s)\rangle$, and will be assumed, throughout this section, that it is independent of the process $\{\eta_t,\ t \in [0, \mathcal{T}]\}$. The admissible parameter uncertainties are assumed to be of the following forms

$$\begin{aligned} &\left[\ \Delta A_o(t, \eta_t) \quad \Delta A_{d_1}(t, \eta_t) \cdots \Delta A_{d_q}(t, \eta_t) \quad \Delta B_o(t, \eta_t)\ \right] = \\ &M_a(\eta_t)\Delta(t, i)[N_a(\eta_t) \quad N_{d_1}(\eta_t) \cdots N_{d_q}(\eta_t) \quad N_b(\eta_t)] \end{aligned} \tag{6.57}$$

where for any $\eta_t = i,\ i \in \mathcal{S}$, $M_a(i), N_a(i), N_b(i)$ and $N_{d_1}(i) \cdots N_{d_q}(i)$ are known constant matrices of appropriate dimensions and $\Delta(t, \eta_t)$, is an unknown time-varying matrix satisfying

$$\Delta^t(t, i)\Delta(t, i)\ \leq I\ ,\ = i \in \mathcal{S}\ ,\ t \geq 0 \tag{6.58}$$

For each value $\eta_t = j \in \mathcal{S}$, we denote the matrices of system (6.54)-(6.56) associated with mode j by

$$A_o(\eta_t) \triangleq A_o(i)\ ,\ B_o(\eta_t) \triangleq B_o(i)\ ,\ C_o(\eta_t) \triangleq C_o(i)$$

$$
\begin{aligned}
N_a(\eta_t) &\triangleq N_a(i)\ ,\ N_b(\eta_t) \triangleq N_b(i)\ ,\ M_a(\eta_t) \triangleq M_a(i) \\
D_o(\eta_t) &\triangleq D_o(i)\ ,\ \Gamma(\eta_t) \triangleq \Gamma(j)\ ,\ L(\eta_t) \triangleq L(i) \\
A_{d_1}(\eta_t) &\triangleq A_{d_1}(i) \cdots A_{d_q}(\eta_t) \triangleq A_{d_q}(i) \\
N_{d_1}(\eta_t) &\triangleq N_{d_1}(i) \cdots N_{d_q}(\eta_t) \triangleq N_{d_q}(i) \qquad (6.59)
\end{aligned}
$$

where $A_o(i) \cdots\cdots N_{d_q}(i)$ are known, real, piecewise-constant between each jump, matrices of appropriate dimensions describing the nominal system. Let $\mathsf{X}(\xi_0, \eta_0)$ denote the state trajectory in system (6.54) from the initial state (ξ_0, η_0). In the sequel, it is assumed that no *a priori* estimate of the initial state, x_0, is available and the jumping process $\{\eta_t\}$, is accessible, that is the operation mode of system (Σ_J) is known for every $t \geq 0$.

6.4.2 Preliminary Results

Building upon the stability results and definitions of Chapter 3, we introduce the following definition and theorem which are appropriate to the subsequent analysis.

Definition 6.4 *System (6.54)-(6.56) is said to be* **robustly stochastically stable and weakly delay-dependent (RSSWDD) with a disturbance attenuation** γ *if for all finite initial vector function* $\zeta_0 \in \Re^n$ *defined on the interval* $(\max[-\tau_k^*, 0],\ k \in \mathcal{J}_q)$ *and initial mode* $\eta_o \in \mathcal{S}$*, the following inequality*

$$
||z(t)||_{E_2} \triangleq \mathsf{E}\Big[\int_0^T z^t(t) z(t) dt\Big]^{1/2} \leq
$$
$$
\gamma \left(||w(t)||_2^2 + \mathcal{Y}(\zeta_0, \eta_0) \right)^{1/2} \qquad (6.60)
$$

holds for all $0 \neq w(t) \in \mathcal{L}_2[0, \infty)$ *and for all admissible parameter uncertainties satisfying (6.57)-(6.58) where* $\mathcal{Y}(\zeta_0, \eta_0)$ *is a nonnegative function of the initial conditions satisfying* $\mathcal{Y}(0, 0) = 0$.

In the sequel, for some $\xi_k(i) > 0,\ k \in \mathcal{J}_k$ and for all $j \in \mathcal{S}$ we associate with the sequence of matrices $Q_1(i) = Q_1^t(i) \cdots Q_q(i) = Q_q^t(i)$ the following matrices:

$$
\begin{aligned}
\bar{\Omega}_q(i) &= diag[(1-\tau_k^+)Q_k(i) - \xi_k^{-1}(i)\sum_{m=1}^{s}\alpha_{im}Q_k(m)] \\
\mathcal{N}(i) &= [E_{d_1}(i)\ \ E_{d_q}(i)] \\
\hat{\Omega}_q(i) &= \sum_{k=1}^{q}[Q_k(i) + \xi_k(i)\sum_{m=1}^{s}\alpha_{im}Q_k(m)]
\end{aligned}
\tag{6.61}
$$

Theorem 6.6 *System (6.54)-(6.56) is* **RSSWDD with a disturbance attenuation** γ, *if there exist matrices* $Q_k(i) = Q_k^t(i) > 0,\ i \in \mathcal{S},\ k \in \mathcal{J}_k$ *and scalars* $\gamma > 0,\ \tau_k^+ < 1,\ k \in \mathcal{J}_k,$ *such that there exist scalars* $\xi_k(i) > 0,\ k \in \mathcal{J}_k,\ \mu(i) > 0,\ \epsilon(i) > 0,\ i \in \mathcal{S}$ *and matrices* $P(i) = P^t(i) > 0,\ j \in \mathcal{S}$, *satisfying the system of LMIs for all* $i \in \mathcal{S}$

$$
\begin{bmatrix} \Pi_w(i) & \Upsilon(i) & \Psi(i) \\ \Upsilon^t(i) & -\Theta_a(i) & 0 \\ \Psi^t(i) & 0 & -\Theta_b(i) \end{bmatrix} < 0 \tag{6.62}
$$

with

$$
\begin{aligned}
\Pi_w(i) &= A_o^t(i)P(i) + P(i)A_o(i) + \sum_{m=1}^{s}\alpha_{im}P(m) + \\
&\quad \hat{\Omega}_q(i) + \epsilon(i)N_a^t(i)N_a(i) + L^t(i)L(i) \\
\Lambda_o(i) &= [\bar{\Omega}_q(i) - \mu(j)\mathcal{N}^t(i)\mathcal{N}(i)] > 0 \qquad\qquad (6.63) \\
\Upsilon(i) &= [P(i)M_a(i)\ \ P(i)M_a(i)\ \ P(i)\mathbb{E}_t(i)] \\
\Theta_a(i) &= diag[\epsilon(i)I\ \ \mu(i)I\ \ \Lambda_o(i)] \\
\Psi(i) &= [P(i)B_o(i)\quad L^t(i)] \\
\Theta_b(i) &= \begin{bmatrix} -\gamma^2 I & \Gamma(i) \\ \Gamma^t(i) & -I \end{bmatrix} > 0 \qquad\qquad (6.64)
\end{aligned}
$$

Proof: Can be readily adapted from [110]. ∇∇∇

Remark 6.4 *In [66], it has been established that for the free delayless portion of system (6.54), the terms "stochastically stable", "exponentially mean-square stable", and "asymptotically mean-square stable" are equivalent, and any of them can imply "almost surely asymptotically stable". In the sequel, we will use for system (6.54), the equivalent terms "stochastically stable", "exponentially mean-square stable" and "asymptotically mean-square stable" interchangeably.*

6.4.3 Linear Filtering

We consider the problem of obtaining an estimate, $\hat{z}(t)$, of $z(t)$ via a causal Markovian jump linear filter which provides a uniformly small estimation error, $\tilde{z}(t) \stackrel{\Delta}{=} z(t) - \hat{z}(t)$, for all $w(t) \in \mathcal{L}_2$ and for all admissible uncertainties. In order to cast our problem into a stochastic setting, we introduce the space $\mathcal{L}_2[\Omega, \mathcal{G}, \mathbf{P}]$ of $\mathcal{G}$-measurable processes, $\tilde{z}(t) = z(t) - \hat{z}(t)$ for which

$$||\tilde{z}(t)||_2 \stackrel{\Delta}{=} \left\{\mathbb{E}[\int_0^\infty (\tilde{z}(t))^t (\tilde{z}(t))dt]\right\}^{1/2} \leq \infty \tag{6.65}$$

We focus attention on the design of a linear Markovian jump n^{th}-order filter for which the jumping process $\{\eta_t\}$ is available for $t \geq 0$ and has the following state-space realization:

$$\begin{aligned} \dot{\hat{x}}(t) &= A_f(i)\,\hat{x}(t) + K_f(i)y(t),\ \hat{x}(0) = 0 \\ \hat{z}(t) &= L(i)\hat{x}(t) \end{aligned} \tag{6.66}$$

where the matrices $A_f(i),\ K_f(i), i \in \mathcal{S}$ are to be determined in the course of the design, such that the estimation error $\tilde{z}(t)$ is robustly stochastically stable for all admissible uncertainties in the sense of **Definition** 6.4.

Remark 6.5 *One standard characterization of $\mathcal{Y}(\zeta_0, \eta_0)$ is the quadratic form $\zeta_0^t R \zeta_0$ where $R > 0$ is a given weighting matrix of the initial function and reflects the uncertainty in ζ_0 relative to the uncertainty in w. A "large" value of R indicates that ζ_0 is very close to zero.*

6.4.4 Augmented System

In terms of the state error $\tilde{x}(t) = x(t) - \hat{x}(t)$, it follows from system (6.54)-(6.56) and the filter (6.66) that :

$$\begin{aligned} \dot{\tilde{x}}(t) &= A_f(i)\tilde{x}(t) + [\mathcal{E}_t + \Delta\mathcal{E}_t](i)\ \chi(t,T,q) \\ &+ \Big[A_{\Delta o}(i) - K_f(i)C_o(i) - A_f(i)\Big]x(t) \\ &+ \Big[B_{\Delta o}(i) - K_f(i)C_o(i)\Big]w(t)\ ,\ \tilde{x}(0) = \xi_0 \end{aligned} \tag{6.67}$$

Therefore, a state-space augmented model of the estimation error, $\tilde{z}(t) = z(t) - \hat{z}(t)$, can be constructed as follows:

$$\begin{aligned} \dot{\zeta}(t) &\triangleq \begin{bmatrix} \dot{\tilde{x}}(t) \\ \dot{x}(t) \end{bmatrix} = \mathcal{A}_{\Delta o}(\eta_t)\zeta(t) + \mathcal{D}_{\Delta o}(\eta_t)\zeta(t-\tau_\zeta) \\ &+ \mathcal{B}_{\Delta o}(i)w(t) \end{aligned} \tag{6.68}$$

$$\tilde{z}(t) = \hat{L}(i)\zeta(t) + \Gamma(i)w(t) \tag{6.69}$$

where for $\eta_t = i,\ i \in \mathcal{S}$:

$$\begin{aligned} \mathcal{A}_{\Delta o}(i) &= \Big[\mathsf{A}(i) + M_A(i)\Delta(t,i)N_A(i)\Big] \\ \mathcal{D}_{\Delta o}(i) &= \Big[\mathsf{D}(i) + M_A(i)\Delta(t,i)N_D(i)\Big] \\ \mathcal{B}_{\Delta o}(i) &= \Big[\mathsf{B}(i) + M_A(i)\Delta(t,i)N_b(i)\Big] \\ \mathsf{A}(i) &= \begin{bmatrix} A_f(i) & A_o(i) - K_f(i)C_o(j) - A_f(i) \\ 0 & A_o(i) \end{bmatrix} \\ \zeta(t-\tau_\zeta) &= \begin{bmatrix} \chi(t,T,q) \\ 0 \end{bmatrix}, \hat{L}(i) = [L(i) \quad 0] \\ \mathsf{D}(i) &= \begin{bmatrix} \mathcal{E}_t(i) & 0 \\ \mathcal{E}_t(j) & 0 \end{bmatrix}, N_D(i) = [\mathcal{N} \quad 0] \\ N_A(i) &= [0 \quad E_a(i)],\ M_A(i) = \begin{bmatrix} M_a(i) \\ M_a(i) \end{bmatrix} \\ \mathsf{B}(i) &= \begin{bmatrix} B_o(i) - K_f(i)D_o(i) \\ B_o(i) \end{bmatrix} \end{aligned}$$

The following theorem provides a weakly delay-dependent sufficient condition for the robust stability of the filtering error system (6.68)-(6.69).

Theorem 6.7 *Consider system (6.68)-(6.69) and let* $\gamma > 0$ *be a given scalar. If there exist scalars* $\gamma > 0,\ \tau_k^+ < 1,\ k \in \mathcal{J}_k,\ \xi_k(i) > 0,\ k \in \mathcal{J}_k,\ \mu(i) > 0,\ \epsilon(j) > 0,\ i \in \mathcal{S}$ *and matrices* $0 < \mathcal{Q}_k(i) = \mathcal{Q}_k^t(i) \in \Re^{2n \times 2n},\ i \in \mathcal{S},\ k \in \mathcal{J}_k,\ 0 < \mathcal{P}(i) = \mathcal{P}^t(i) \in \Re^{2n \times 2n},\ i \in \mathcal{S}$ *, satisfying the system of LMIs for all* $i \in \mathcal{S}$

$$\begin{bmatrix} \Pi_f(i) & \bar{\Upsilon}(j) & \bar{\Psi}(i) \\ \bar{\Upsilon}^t(i) & -\bar{\Theta}_A(i) & 0 \\ \bar{\Psi}^t(i) & 0 & -\Theta_b(i) \end{bmatrix} < 0 \tag{6.70}$$

$$\begin{aligned}
\Pi_f(i) &= \mathsf{A}^t(j)\mathcal{P}(i) + \mathcal{P}(i)\mathsf{A}(i) + \sum_{m=1}^{s} \alpha_{im}\mathcal{P}(m) + \hat{\Xi}_q(i) \\
&+ \epsilon(j)N_A^t(i)N_A(i) + \hat{L}^t(i)\hat{L}(i) \\
\bar{\Upsilon}(i) &= [\mathcal{P}(i)M_A(i)\ \ \mathcal{P}(i)M_A(i)\ \ \mathcal{P}(i)\mathsf{D}(i)] \qquad (6.71) \\
\hat{\Xi}_q(i) &= \sum_{k=1}^{q}[\mathcal{Q}_k(i) + \xi_k(i)\sum_{m=1}^{s}\alpha_{im}\mathcal{Q}_k(m)] \\
\bar{\Xi}_q(i) &= diag[(1-\tau_k^+)\mathcal{Q}_k(i) - \xi_k^{-1}(i)\sum_{m=1}^{s}\alpha_{im}\mathcal{Q}(m)] \\
\bar{\Theta}_A(i) &= diag[\epsilon(i)I\ \ \mu(i)I\ \ \bar{\Lambda}_o(i)] \qquad (6.72) \\
\bar{\Psi}(i) &= [\mathcal{P}(i)\mathsf{B}(i) \quad \hat{L}^t(i)] \\
\bar{\Lambda}_o(i) &= [\bar{\Xi}_q(i) - \mu(i)N_D^t(i)N_D(i)] > 0 \qquad (6.73)
\end{aligned}$$

then system (6.68)-(6.69) is **RSSWDD with a disturbance attenuation** γ.

Proof: Follows easily from **Theorem** 6.6.

6.4.5 Design Procedure

Now, we provide expressions for the gains of the Markovian jump filter (6.66). The following theorem summarizes a design method for the robust Markovian filter by means of an LMI-based feasibility test.

Theorem 6.8 *Consider system (6.68)-(6.69) and let $\gamma > 0$ be a given scalar and $R > 0$ be a given initial state weighting matrix. Then there exists a Markovian jump filter of the type (6.66) such that the estimation error system is RSS-WDD and*

$$||z(t) - \hat{z}(t)||_{E_2} \leq \gamma \left(||w(t)||_2^2 + x_0^t R x_0) \right)^{1/2}$$

for all $0 \neq w(t) \in \mathcal{L}_2[0, \infty)$ and for all admissible parameter uncertainties satisfying (6.57)-(6.58) if for all ϕ there exist scalars $\tau_k^+ < 1,\ k \in \mathcal{J}_k,\ \xi_k(i) > 0,\ k \in \mathcal{J}_k,\ \mu(j) > 0,\ \epsilon(i) > 0,\ i \in \mathcal{S}$ and matrices $\mathcal{Z}(i) > 0,\ \mathcal{W}(i) > 0,\ 0 < \mathcal{Q}_{ke}(i) = \mathcal{Q}_{ke}^t(i) \in \Re^{n\times n},\ 0 < \mathcal{Q}_{ks}(i) = \mathcal{Q}_{ks}^t(i) \in I\!R^{n\times n},\ k \in \mathcal{J}_k,\ 0 < \mathcal{P}_e(j) = \mathcal{P}_e^t(i) \in I\!R^{n\times n},\ 0 < \mathcal{P}_e(i) = \mathcal{P}_e^t(i) \in I\!R^{n\times n},\ i \in \mathcal{S}$, satisfying the system of LMIs for all $i \in \mathcal{S}$

$$\begin{bmatrix} \Pi_e(i) & \bar{\Upsilon}_e(i) & \Sigma_e(i) & \Psi_e(i) \\ \bar{\Upsilon}_e^t(j) & -\Pi_s(i) & \Sigma_s(i) & \Psi_s(i) \\ \Sigma_e^t(i) & \Sigma_(^t i) & -\Theta_t(i) & 0 \\ \Psi_e^t(i) & \Psi_s^t(i) & 0 & -\Theta_b(i) \end{bmatrix} < 0 \tag{6.74}$$

$$\mathcal{P}_e(0) + \mathcal{P}_s(0) - \gamma^2 R \leq 0 \tag{6.75}$$

where

$$\begin{aligned} \Pi_e(i) &= \mathcal{W}(u) + \mathcal{W}^t(i) + \sum_{m=1}^{s} \alpha_{im}\mathcal{P}_e(m) \\ &+ L^t(i)L(i) + \hat{\Xi}_{qe}(i) \end{aligned} \tag{6.76}$$

$$\bar{\Upsilon}_e(i) = \mathcal{P}_e(i)A_o(i) + \mathcal{Z}(i)C_o(i) - \mathcal{W}(i) \tag{6.77}$$

$$\begin{aligned} \Pi_s(i) &= \mathcal{P}_s(i)A_o(i) + A_o^t(i)\mathcal{P}_s(i) + \sum_{m=1}^{s} \alpha_{im}\mathcal{P}_s(m) \\ &+ \epsilon(i)N_a^t(j)N_a(i) + \hat{\Xi}_{qs}(i) \end{aligned} \tag{6.78}$$

$$\begin{aligned} \Sigma_e(i) &= [\mathcal{P}_e(i)M_a(i)\ \ \mathcal{P}_e(i)M_a(i)\ \ \mathcal{P}_e(i)\mathcal{E}_t(i)] \\ \Sigma_s(i) &= [\mathcal{P}_s(i)M_a(i)\ \ \mathcal{P}_s(i)M_a(i)\ \ \mathcal{P}_s(i)\mathcal{E}_t(i)] \\ \Theta_t(i) &= \left[\epsilon(i)I\ \ \mu(i)I\ \ \hat{\Lambda}(i)\right] \end{aligned}$$

$$\hat{\Lambda}(i) = \begin{bmatrix} \bar{\Xi}_{qe}(i) - \mu(i)\mathcal{N}^t(i)\mathcal{N}(i) & 0 \\ 0 & \bar{\Xi}_{qs}(i) \end{bmatrix} \quad (6.79)$$

$$\Psi_e(i) = [\mathcal{P}_e(i)B_o(i) - \mathcal{Z}(i)D_o(i) \;\; L(i)]$$

$$\Psi_s(i) = [\mathcal{P}_e(i)B_o(i) \;\; 0] \quad (6.80)$$

Moreover, a suitable filter has the gains

$$A_f(i) = \mathcal{P}_e^{-1}(i)\mathcal{W}(i) \; , \;\; K_f(i) = \mathcal{P}_e^{-1}\mathcal{Z}(i) \quad (6.81)$$

Proof: It follows from Theorem (6.7) that system (6.68)-(6.69) is **RSSWDD with disturbance attenuation** $\gamma > 0$ when inequality (6.70) using (6.71)-(6.73) is satisfied. Now introduce

$$\mathcal{P}(i) = \begin{bmatrix} \mathcal{P}_e(i) & 0 \\ 0 & \mathcal{P}_s(i) \end{bmatrix} , \; \mathcal{Q}_k(i) = \begin{bmatrix} \mathcal{Q}_{ke}(i) & 0 \\ 0 & \mathcal{Q}_{ks}(i)] \end{bmatrix} \quad (6.82)$$

Hence, we can write

$$\bar{\Xi}_{qe}(i) = \begin{bmatrix} \hat{\Xi}_{qe}(i) & 0 \\ 0 & \hat{\Xi}_{qe}(i) \end{bmatrix} , \; \bar{\Xi}_{qe}(i) = \begin{bmatrix} \bar{\Xi}_{qe}(i) & 0 \\ 0 & \bar{\Xi}_{qs}(i)] \end{bmatrix} \quad (6.83)$$

Denoting

$$\mathcal{W}(i) = \mathcal{P}_e(i)A_f(i) \; , \;\; \mathcal{Z}(i) = \mathcal{P}_e(i)K_f(i) \quad (6.84)$$

it is a straightforward task to show that (6.74)-(6.75) using (6.76)-(6.80) are equivalent to (6.70) with (6.71)-(6.73). By (6.84), we obtain the gains (6.81) and therefore conclude that the filter synthesis is solved.

Remark 6.6 *It is obvious that the problem of finding the robust* $\mathcal{H}_\infty$ *filter of* **Theorem** *6.8 for the smallest possible* $\gamma \geq 0$ *can be easily solved in terms of the following linear programming problem:*

$$\begin{array}{ll} \textit{minimize} & \gamma \\ \textit{subject to} & \begin{array}{lll} \gamma > 0, & \mathcal{P}_e(i) > 0, & \mathcal{P}_s(i) > 0, \\ \mathcal{Z}(i) > 0, & \mathcal{W}(i) > 0, & \mu(i) > 0, \\ \epsilon(i) > 0, & \xi_k(i) > 0, & \\ (6.74), & (6.75) & \end{array} \end{array}$$

The robust filter design of linear uncertain multi-state-delay system without jumping parameters can be established as a special case of **Theorem** 6.8 by setting $s = 1,\ \alpha_{ij} \equiv 0, \xi_k \equiv 0 \forall k \in \mathcal{J}_k$. The following corollary provides the main result.

Corollary 6.1 *Consider system (6.54)-(6.56) and filter (6.66) without jumping parameters and let $\gamma > 0$ be a given scalar and $R > 0$ be a given initial state weighting matrix. Then there exists a Markovian jump filter of the type (6.66) such that the estimation error system is* **RSSWDD** *and*

$$||z(t) - \hat{z}(t)||_{E_2} \leq \gamma \left(||w(t)||_2^2 + x_0^t R x_0) \right)^{1/2}$$

for all $0 \neq w(t) \in \mathcal{L}_2[0, \infty)$ and for all admissible parameter uncertainties satisfying (6.4)-(6.5) if for all ϕ there exist scalars $\tau_k^+ < 1,\ k \in \mathcal{J}_k,\ \mu > 0,\ \epsilon > 0,$ and matrices $\mathcal{Z} > 0,\ \mathcal{W} > 0,\ 0 < \mathcal{Q}_{ke} = \mathcal{Q}_{ke}^t \in \Re^{n \times n},\ 0 < \mathcal{Q}_{ks} = \mathcal{Q}_{ks}^t \in \Re^{n \times n},\ k \in \mathcal{J}_k,\ 0 < \mathcal{P}_e = \mathcal{P}_e^t \in \Re^{n \times n},\ 0 < \mathcal{P}_e = \mathcal{P}_e^t \in \Re^{n \times n}$, satisfying the system of LMIs

$$\begin{bmatrix} \Pi_e & \bar{\Upsilon}_e & \Sigma_e & \Psi_e \\ \bar{\Upsilon}_e^t & -\Pi_s & \Sigma_s & \Psi_s \\ \Sigma_e^t & \Sigma_s^t & -\Theta_t & 0 \\ \Psi_e^t & \Psi_s^t & 0 & -\Theta_b \end{bmatrix} < 0 \tag{6.85}$$

$$\mathcal{P}_e + \mathcal{P}_s - \gamma^2 R \leq 0 \tag{6.86}$$

where

$$\hat{\Xi}_q = \sum_{k=1}^{q} \mathcal{Q}_k = diag[\hat{\Xi}_{qe} \quad \hat{\Xi}_{qs}]$$

$$\bar{\Xi}_q = diag[(1 - \tau_k^+)\mathcal{Q}_k] = diag[\bar{\Xi}_{qe} \quad \bar{\Xi}_{qs}]$$

$$\Pi_e = \mathcal{W} + \mathcal{W}^t + L^t(j)L(j) + \hat{\Xi}_{qe}(j) \tag{6.87}$$

$$\bar{\Upsilon}_e = \mathcal{P}_e A_o + \mathcal{Z} C_o - \mathcal{W} \tag{6.88}$$

$$\Pi_s = \mathcal{P}_s A_o + A_o^t \mathcal{P}_s + \epsilon E_a^t E_a + \hat{\Xi}_{qs} \tag{6.89}$$

$$
\begin{aligned}
\Sigma_e &= [\mathcal{P}_e H_a \;\; \mathcal{P}_e H_a \;\; \mathcal{P}_e(j)\mathcal{E}_t] \\
\Sigma_s &= [\mathcal{P}_s H_a \;\; \mathcal{P}_s H_a \;\; \mathcal{P}_s \mathcal{E}_t] \\
\Theta_t &= \left[\epsilon I \;\; \mu I \;\; \hat{\Lambda}\right] \\
\hat{\Lambda} &= \begin{bmatrix} \bar{\Xi}_{qe} - \mu \mathcal{N}^t \mathcal{N} & 0 \\ 0 & \bar{\bar{\Xi}}_{qs} \end{bmatrix} \qquad (6.90) \\
\Psi_e &= [\mathcal{P}_e B_o - \mathcal{Z} D_o \;\; L] \\
\Psi_s &= [\mathcal{P}_e B_o \;\; 0] \qquad (6.91)
\end{aligned}
$$

Moreover, a suitable filter has the gains

$$
A_f = \mathcal{P}_e^{-1} \mathcal{W} \;\; , \;\; K_f = \mathcal{P}_e^{-1} \mathcal{Z} \qquad (6.92)
$$

This corollary provides a dual result to [165].

6.4.6 Example 6.2

We consider a two-mode JTD system with two time-delays

$$
\tau_1 = 0.4 \;\; , \;\; \tau_2 = 0.7
$$

and having mode-switching generator

$$
\Im = \begin{bmatrix} -2 & 2 \\ 3 & -3 \end{bmatrix}
$$

The associate date are given by:

Mode 1

$$
\begin{aligned}
A_o(1) &= \begin{bmatrix} 0 & 2 \\ -3 & -4 \end{bmatrix} \; , \; A_{d_1}(1) = \begin{bmatrix} -0.1 & 0 \\ 0.2 & -0.3 \end{bmatrix} \\
A_{d_2}(1) &= \begin{bmatrix} -0.1 & 0 \\ -0.2 & -0.1 \end{bmatrix} \; , \; W(1) = \begin{bmatrix} 1 & 0 \\ 0 & 2 \end{bmatrix} \; , \; M_a(1) = \begin{bmatrix} 0 \\ 0.5 \end{bmatrix} \\
C_o(1) &= [1 \;\; 0] \; , \; N_a(1) = [0 \;\; 2] \; , \; V = 1 \\
M_c(1) &= 2 \; , \; N_{d_1}(1) = [0 \;\; 2] \; , \; N_{d_2}(1) = [0 \;\; 1] \\
D_o(1) &= 2 \; , \; \Gamma(1) = 1
\end{aligned}
$$

Mode 2

$$\begin{aligned}
A_o(1) &= \begin{bmatrix} 0 & 2 \\ -3 & -2 \end{bmatrix}, \; A_{d_1}(1) = \begin{bmatrix} -0.1 & 0 \\ 0.2 & -0.1 \end{bmatrix} \\
A_{d_2}(1) &= \begin{bmatrix} -0.1 & 0 \\ -0.2 & -0.3 \end{bmatrix}, \; W(1) = \begin{bmatrix} 2 & 0 \\ 0 & 1 \end{bmatrix}, \; M_a(1) = \begin{bmatrix} 0 \\ 1 \end{bmatrix} \\
C_o(1) &= [1 \;\; 0] \;, \; N_a(1) = [0 \;\; 0.5] \;, \; V = 1 \\
M_c(1) &= 3 \;, \; N_{d_1}(1) = [0 \;\; 1] \;, \; N_{d_2}(1) = [0 \;\; 2] \\
D_o(1) &= 2 \;, \; \Gamma(1) = 1
\end{aligned}$$

Application of **theorem** 6.8 and solving LMIs (6.74)-(6.75) with

$$\gamma = 1.25 \quad , \quad R = \begin{bmatrix} 1.4 & 0 \\ 0 & 2.2 \end{bmatrix}$$

the state-space matrices of the Markovian filter (6.66)are given by:

$$\begin{aligned}
A_f(1) &= \begin{bmatrix} -0.1922 & 1.5456 \\ -4.4155 & -2.4908 \end{bmatrix}, \; K_f(1) = \begin{bmatrix} 0.1798 \\ 1.6187 \end{bmatrix} \\
A_f(2) &= \begin{bmatrix} 0.1242 & 1.7688 \\ -5.1156 & -4.0284 \end{bmatrix}, \; K_f(2) = \begin{bmatrix} 0.1236 \\ 1.8007 \end{bmatrix}
\end{aligned}$$

6.5 Filtering with Mode-dependent Delays

In this section, we move to study the filtering problem for a class of linear uncertain systems with Markovian jump parameters when the time-delays are mode-dependent. Here also the uncertainties are time-varying and norm-bounded parametric uncertainties and the delay factor depends functionally on the mode of operation. Recall that the stability analysis of this class of systems were discussed in Chapter 3 and the corresponding control design was examined in Chapter 4.

Design of robust state estimators and observers to different classes of continuous-time systems with parametric uncertainties and state-delay have been pursued in [5, 97, 101, 102, 103, 104, 109, 110, 112]. Looked at in this light and since

Markovian jump systems emerge when the physical models under consideration are subject to random changes [32, 66, 78, 155, 157], this section contributes to the further development of robust filters of a class of uncertain jump time-delay systems and establish new results for the case in which the delay factor depends on the mode of operation. We design a linear Markovian filter which ensures that the augmented filtering system is mean-square quadratically stable for all admissible uncertainties. The results are then extended to $\mathcal{H}_\infty$-filtering.

6.5.1 Problem Formulation

We consider a class of Markovian jump dynamical systems with mode-dependent state-delay described over the space $(\Omega, \mathcal{G}, \mathbf{P})$ by the following model:

$$\begin{aligned}
(\Sigma_J): \quad \dot{x}(t) &= \Big[A_o(\eta_t) + \Delta A_o(t,\eta_t)\Big] x(t) \\
&+ \Big[A_d(\eta_t) + \Delta A_d(t,\eta_t)\Big] x(t-\tau_{\eta_t}) \\
&+ \Big[B_o(\eta_t) + \Delta B_o(t,\eta_t)\Big] w(t) \\
&= A_{\Delta o}(t,\eta_t)x(t) + A_{\Delta d}(t,\eta_t)x(t-\tau_{\eta_t}) \\
&+ B_{\Delta o}(t,\eta_t)w(t) \qquad (6.93)\\
x(t) &= \phi(t) \ \in \mathcal{L}_2(-\max\{\tau_{\eta_t}\}, 0; \mathbb{R}^n), \ \eta_o = i \ \in \mathcal{S} \qquad (6.94)\\
y(t) &= \Big[C_o(\eta_t) + \Delta C_o(t,\eta_t)\Big] x(t) \\
&+ \Big[D_d(\eta_t) + \Delta D_d(t,\eta_t)\Big] x(t-\tau_{\eta_t}) \\
&+ \Big[B_d(\eta_t) + \Delta B_d(t,\eta_t)\Big] w(t) \\
&= C_{\Delta o}(\eta_t)x(t) + D_{\Delta o}(\eta_t)x(t-\tau_{\eta_t}) \\
&+ B_{\Delta d}(\eta_t)w(t) \qquad (6.95)\\
z(t) &= L(\eta_t)x(t) + \Gamma(\eta_t)w(t) \qquad (6.96)
\end{aligned}$$

where $x(t) \in \mathbb{R}^n$ is the state vector; $w(t) \in \mathbb{R}^q$ is the disturbance input which belongs to $\mathcal{L}_2[0, \infty)$; $y(t) \in \mathbb{R}^p$ is the measured output; $z(t) \in \mathbb{R}^r$ is the controlled output which belongs to $\mathcal{L}_2\big[(\Omega, \mathcal{G}, \mathbf{P}), [0, \mathcal{T}]\big]$ and τ_{η_t} denotes the time-delay in the jumping system when the mode is in η_t with

$$\hat{\tau} \stackrel{\Delta}{=} \max_j\{\tau_j, j \in \mathcal{S}\} \ , \ \check{\tau} \stackrel{\Delta}{=} \min_j\{\tau_j, j \in \mathcal{S}\}$$

Note in general that the mode-time delay functional relationship could be expressed analytically or presented in table form.

In (6.93)-(6.96), for $\eta_t = i \ \in \mathcal{S}$, $A_o(\eta_t)$, $B_o(\eta_t)$ $C_o(\eta_t)$, $D_o(\eta_t)$ and $\{A_{d_k}(\eta_t);$ $k \in \mathcal{J}_q\}$ are known constant matrices of appropriate dimensions and the matrices $\Delta A_o(t, \eta_t)$, $\Delta A_d(t, \eta_t)$, $\Delta B_o(t, \eta_t)$, $\Delta C_o(t, \eta_t)$,$\Delta B_d(t, \eta_t)$ and $\Delta D_o(t, \eta_t)$ are real, time-varying matrix functions representing the norm-bounded parameter uncertainties and assumed to belong to certain bounded compact sets.

The initial vector function is specified as $\xi_0 \equiv \langle x(0), x(s)\rangle = \langle x_o, \phi(s)\rangle$, and will be assumed, throughout this paper, that it is independent of the process $\{\eta_t, \ t \in [0, \mathcal{T}]\}$. For $\eta_t = i$, the admissible uncertainties are assumed to be modeled in the form:

$$\begin{bmatrix} \Delta A_o(t,i) & \Delta A_d(t,i) & \Delta B_o(t,i) \\ \Delta C_o(t,i) & \Delta D_d(t,i) & \Delta B_d(t,i) \end{bmatrix} = \begin{bmatrix} M_a(i) \\ M_c(i) \end{bmatrix} \Delta(t,i) \begin{bmatrix} N_a(i) & N_d(i) & N_b(i) \end{bmatrix} \tag{6.97}$$

where $M_a(i) \in \mathbb{R}^{n\times\alpha}$, $M_c(i) \in \mathbb{R}^{p\times\alpha}$, $N_a(i) \in \mathbb{R}^{\beta\times n}$, $N_b(i) \in \mathbb{R}^{\beta\times m}$ and $N_d(i) \in \mathbb{R}^{\beta\times n}$ are known real constant matrices, with $\Delta(t,i)$ being unknown, time-varying matrix function satisfying

$$||\Delta(t,i)||_2 \ \le \ 1 \tag{6.98}$$

where the elements of $\Delta(t,i)$ are Lebesgue measurable for any $i \in \mathcal{S}$. For each possible value $\eta_t = i \in \mathcal{S}$, we will denote the matrices of system (Σ_J) associated with mode i by

$$
\begin{aligned}
A_o(\eta_t) &\triangleq A_o(i)\ ,\ B_o(\eta_t) \triangleq B_o(i)\ ,\ C_o(\eta_t) \triangleq C_o(i) \\
A_d(\eta_t) &\triangleq A_d(i)\ ,\ M_a(\eta_t) \triangleq M_a(i)\ ,\ M_c(\eta_t) \triangleq M_c(i) \\
N_a(\eta_t) &\triangleq N_a(i)\ ,\ N_d(\eta_t) \triangleq N_d(i)\ ,\ N_b(\eta_t) \triangleq N_b(i) \\
B_d(\eta_t) &\triangleq B_d(i)\ ,\ D_d(\eta_t) \triangleq D_d(i)\ ,\ L(\eta_t) \triangleq L(i) \\
\Gamma(\eta_t) &\triangleq \Gamma(i) \qquad (6.99)
\end{aligned}
$$

where $A_o(i), \cdots\cdots, \Gamma(j)$ are known, real, piecewise-constant between each jump, matrices of appropriate dimensions describing the nominal system. Let $\mathsf{X}(\xi_0, \eta_o)$ denote the state trajectory in system (6.93) from the initial state (ξ_0, η_0). In the sequel, it is assumed that no *a priori* estimate of the initial state, x_0, is available and the jumping process $\{\eta_t\}$, is accessible, that is the operation mode of system (Σ_J) is known for every $t \geq 0$.

We extract from system Σ_J the free system:

$$
\begin{aligned}
(\Sigma_{Jo}): \quad \dot{x}(t) &= A_{\Delta o}(t, \eta_t)x(t) + A_{\Delta d}(t, \eta_t)x(t - \tau_{\eta_t}) \qquad (6.100) \\
x(t) &= \phi(t) \ \in \mathcal{L}_2(-\max\{\tau_{\eta_t}\}, 0; \mathbb{R}^n),\ \eta_o = i\ \in \mathcal{S} \qquad (6.101)
\end{aligned}
$$

for which we introduce the following definition:

Definition 6.5 *System Σ_{Jo} is said to be* **robustly mean square quadratically stable (RMSQS)** *if there exist matrices $W = W^t > 0, P(i) = P^t(i) > 0,\ i \in \mathcal{S}$ such that the LMIs for all $i \in \mathcal{S}$*

$$
\begin{aligned}
\Upsilon(\eta_t, t) &\triangleq \begin{bmatrix} \Pi_t(\eta_t, t) & P(\eta_t)A_{\Delta d}(t, \eta_t) \\ A^t_{\Delta d}(t, \eta_t)P(\eta_t) & -W \end{bmatrix} < 0 \qquad (6.102) \\
\Pi_t(\eta_t, t) &= P(\eta_t)A_{\Delta o}(t, \eta_t) + A^t_{\Delta o}(t, \eta_t)P(\eta_t) \\
&+ \sum_{k=1}^{s} \alpha_{ik}P(k) + \Big(1 + [\hat{\tau} - \check{\tau}]\Big)\hat{\alpha}W \qquad (6.103)
\end{aligned}
$$

hold for all admissible uncertainties satisfying (6.97)-(6.98)

6.5.2 Linear Markovian Filter

In this work, we consider the problem of obtaining an estimate, $\hat{z}(t)$, of $z(t)$ via a causal Markovian jump linear filter which provides a uniformly small estimation error, $\tilde{z}(t) \stackrel{\Delta}{=} z(t) - \hat{z}(t)$, for all $w(t) \in \mathcal{L}_2[0, \infty)$ and for all admissible uncertainties. In order to cast our problem into a stochastic setting, we introduce the space $\mathcal{L}_2[\Omega, \mathcal{G}, \mathbf{P}]$ of $\mathcal{G}$-measurable processes, $\tilde{z}(t) \stackrel{\Delta}{=} z(t) - \hat{z}(t)$ for which

$$||\tilde{z}(t)||_2 \stackrel{\Delta}{=} \left\{ \mathbb{E}\left[\int_0^\infty \tilde{z}^t(t)\tilde{z}(t)dt \right] \right\}^{1/2} \leq \infty \tag{6.104}$$

We focus attention on the design of a linear Markovian jump n^{th}-order filter for which the jumping process $\{\eta_t\}$ is available for $t \geq 0$ and has the following state-space model for all $i \in \mathcal{S}$:

$$\begin{aligned} \dot{\hat{x}}(t) &= A_f(i)\,\hat{x}(t) + K_f(i)y(t)\ , \quad \hat{x}(0) = 0 \\ \hat{z}(t) &= L(i)\hat{x}(t) \end{aligned} \tag{6.105}$$

where the matrices $A_f(i)$, $K_f(i)$, $forall\ i \in \mathcal{S}$ are to be determined in the course of the design, such that the estimation error $\tilde{z}(t)$ is robustly mean square quadratically stable for all admissible uncertainties in the sense of **Definition** (6.5).

6.5.3 State Error Dynamics

In terms of the state error $\tilde{x}(t) \stackrel{\Delta}{=} x(t) - \hat{x}(t)$, it follows from system (6.93)-(6.96) and filter (6.105) that the state error dynamics has the form:

$$\begin{aligned} \dot{\tilde{x}}(t) &= A_f(\eta_t)\,\tilde{x}(t) + \Big[A_{\Delta o}(t,\eta_t) - A_f(\eta_t) - K_f(\eta_t)C_{\Delta o}(\eta_t)\Big]x(t) \\ &+ \Big[A_{\Delta d}(t,\eta_t) - K_f(\eta_t)C_{\Delta o}(t,\eta_t)\Big]x(t-\tau_{\eta_t}) \\ &+ \Big[B_{\Delta o}(t,\eta_t) - K_f(\eta_t)D_{\Delta d}(\eta_t)\Big]w(t) \end{aligned} \tag{6.106}$$

A state-space augmented model of the estimation error, $\tilde{z}(t) = z(t) - \hat{z}(t)$, can then be constructed using (6.97)-(6.98) and (6.106) as follows:

$$
\begin{aligned}
(\Sigma_A): \quad \dot{\zeta}(t) &\triangleq \begin{bmatrix} \dot{\tilde{x}}(t) \\ \dot{x}(t) \end{bmatrix} \\
&= \mathcal{A}_{\Delta o}(\eta_t)\zeta(t) + \mathcal{D}_{\Delta o}(\eta_t)\zeta(t - \tau_{\eta_t}) \\
&+ \mathcal{B}_{\Delta o}(\eta_t)w(t) \qquad (6.107) \\
\tilde{z}(t) &= L_f(\eta_t)\zeta(t) + \Gamma(\eta_t)w(t) \qquad (6.108)
\end{aligned}
$$

where for $\eta_t = i, \ i \in \mathcal{S}$:

$$
\begin{aligned}
\mathcal{A}_{\Delta o}(i) &= \left[\mathsf{A}(i) + \bar{M}_A(i)\Delta(t,i)\bar{N}_A(i)\right] \\
\mathcal{D}_{\Delta o}(i) &= \left[\mathsf{D}(i) + \bar{M}_A(i)\Delta(t,i)\bar{N}_D(i)\right] \\
\mathcal{B}_{\Delta o}(i) &= \left[\mathsf{B}(i) + \bar{M}_A(i)\Delta(t,i)N_b(i)\right] \\
L_f(i) &= \left[L(i) \quad 0\right] \qquad (6.109)
\end{aligned}
$$

and

$$
\begin{aligned}
\mathsf{A}(i) &= \begin{bmatrix} A_f(i) & A_o(i) - K_f(i)C_o(i) - A_f(i) \\ 0 & A_o(i) \end{bmatrix} \\
\zeta(t - \tau_{\eta_t}) &= \begin{bmatrix} x(t - \tau_{\eta_t}) \\ 0 \end{bmatrix} \qquad (6.110) \\
\mathsf{D}(i) &= \begin{bmatrix} A_d(i) - K_f(i)C_o(i) & 0 \\ A_d(i) & 0 \end{bmatrix} \\
\mathsf{B}(i) &= \begin{bmatrix} B_o(i) - K_f(i)D_o(i) \\ B_o(i) \end{bmatrix} \qquad (6.111) \\
\bar{M}_A(i) &= \begin{bmatrix} M_a(i) - K_f(i)M_c(i) \\ M_a(i) \end{bmatrix}, \ \bar{N}_A(i) = [0 \quad N_a(i)] \\
\bar{N}_D(i) &= [N_d(i) \quad 0] \qquad (6.112)
\end{aligned}
$$

The following theorem establishes the behavior of the augmented system

Theorem 6.9 *If the augmented system* (Σ_A) *is* **RMSQS**, *then it is* **robustly stochastically stable with weak-delay dependence (RSSWDD)**.

Proof: Let $\zeta_s(t) \stackrel{\Delta}{=} \zeta(s+t),\ t-\tau_{\eta_t} \ \leq\ s\ \leq\ t$ and define the process $\{(\zeta(t),\eta_t),\ t\ \geq\ 0\}$ over the state space $\bar{\mathbb{C}}$. It should be observed that $\{(\zeta(t),\eta_t),\ t\ \geq\ 0\}$ is strong Markovian [78]. Now introduce the following Lyapunov functional :

$$\begin{aligned}
V_t(\zeta(t),\eta_t) &= V_n(\zeta(t),\eta_t) + V_d(\zeta(t),\eta_t) + V_f(\zeta(t),\eta_t) \\
V_d(\zeta(t),\eta_t) &= \int_{t-\tau_{\eta_t}}^{t} \zeta^t(r)\mathbb{L}\zeta(r)dr \\
V_f(\zeta(t),\eta_t) &= \int_{-\hat{\tau}}^{-\check{\tau}} \int_{t+\beta}^{t} \zeta(\beta)\hat{\alpha}\mathbb{L}\zeta(\beta)\, d\beta\, dr \\
V_n(\zeta(t),\eta_t) &= \zeta^t(t)\mathbb{P}(\eta_t)\zeta(t)
\end{aligned} \tag{6.113}$$

where $0 < \mathbb{L} \in \mathbb{R}^{2n\times 2n}, 0 < \mathbb{P}(i) = \mathbb{P}^t(i) \in \mathbb{R}^{2n\times 2n},\ i \in \mathcal{S}$. The weak infinitesimal operator $\Im_t^\zeta[\cdot]$ of the process $\{(\zeta(t),\eta_t),\ t\ \geq\ 0\}$ for system (Σ_A) at the point $\{t,\zeta,\eta_t\}$ is given by [78]:

$$\begin{aligned}
\Im_t^\zeta[V_t] &= \sum_{m=n,d,f} \Im_m^\zeta[V_m] \\
\Im_m^\zeta[V_m] &= \lim_{\Delta\to 0}\frac{1}{\Delta}\Big\{\mathbb{E}[V_m(t+\Delta,\zeta(t+\Delta),\eta_{t+\Delta})|\zeta(t),\eta_t=i] \\
&- V_m(t,\zeta(t),\eta_t=i)\Big\}
\end{aligned} \tag{6.114}$$

It has been shown in [78] that

$$\Im_n^\zeta[V_n] = \partial V_n/\partial t + \dot{\zeta}^t(t)\ \partial V_n/\partial\zeta\ |_{\eta_t=i} + \sum_{m=1}^{s}\alpha_{im}V_n(t,\zeta,i,m) \tag{6.115}$$

Upon applying (6.113) and (6.115) to system (6.107) with $w(.)\equiv 0$, it yields:

$$\begin{aligned}
\Im_n^\zeta[V_n] &= \zeta^t(t)\Big\{\mathbb{P}(\eta_t)\mathcal{A}_{\Delta o}(t,\eta_t) + \mathcal{A}_{\Delta o}^t(t,\eta_t)\mathbb{P}(\eta_t) + \sum_{k=1}^{s}\alpha_{\eta_t k}\mathbb{P}(k)\Big\}\zeta(t) \\
&+ 2\zeta^t(t)\mathbb{P}(\eta_t)\mathcal{A}_{\Delta d}(t,\eta_t)\zeta(t-\tau_{\eta_t})
\end{aligned} \tag{6.116}$$

Select $\eta_t = i \in \mathcal{S},\ \zeta \in \mathbb{C}[-\tau_i, 0]$, thus:

$$\begin{aligned}
\Im_d^\zeta[V_d] &= \lim_{\Delta\to 0}\frac{1}{\Delta}\Big\{\mathbb{E}\Big[V_d(\zeta(t+\Delta),\eta_{t+\Delta}) \mid \zeta(t),\eta_t=i\Big] \\
&- V_m(\zeta(t),\eta_t=i)\Big\} \\
&= \sum_{k\neq j}\mathbb{E}\Big[\mathcal{I}_{\eta_{t+\Delta}=k}\int_t^{t+\Delta}\zeta^t(r)\mathbb{L}\zeta(r)dr|\zeta(t),\eta_t=i\Big] \\
&+ \sum_{k\neq j}\mathbb{E}\Big[\mathcal{I}_{\eta_{t+\Delta}=k}\int_{t+\Delta-\tau_i}^{t}\zeta^t(r)\mathbb{L}\zeta(r)dr|\zeta(t),\eta_t=i\Big] \\
&+ \sum_{k\neq i}\mathbb{E}\Big[\mathcal{I}_{\eta_{t+\Delta}=k}\int_{t+\Delta-\tau_i}^{t+\Delta}\zeta^t(r)\mathbb{L}\zeta(r)dr|\zeta(t),\eta_t=i\Big] \\
&- \int_{t-\tau_{\eta_t}}^{t}\zeta^t(r)\mathbb{L}\zeta(r)dr \qquad (6.117)
\end{aligned}$$

Let $\mathcal{I}_{\{.\}}$ be an indicator function, then it is easy to show that

$$\begin{aligned}
&\sum_{k\neq j}\mathbb{E}\Big[\mathcal{I}_{\eta_{t+\Delta}=k}\int_t^{t+\Delta}\zeta^t(r)\mathbb{L}\zeta(r)dr|\zeta(t),\eta_t=i\Big] \\
&\cong O(\Delta^2) \\
&\sum_{k\neq i}\mathbb{E}\Big[\mathcal{I}_{\eta_{t+\Delta}=k}\int_{t+\Delta-\tau_j}^{t}\zeta^t(r)\mathbb{L}\zeta(r)dr|\zeta(t),\eta_t=i\Big] \\
&\cong \sum_{k\neq i}\Big(\alpha_{jk}\Delta + O(\Delta)\Big)\int_{t+\Delta-\tau_i}^{t}\zeta^t(r)\mathbb{L}\zeta(r)dr \\
&\sum_{k\neq i}\mathbb{E}\Big[\mathcal{I}_{\eta_{t+\Delta}=k}\int_{t+\Delta-\tau_i}^{t+\Delta}\zeta^t(r)\mathbb{L}\zeta(r)dr|\zeta(t),\eta_t=i\Big] \\
&\cong \sum_{k\neq j}\Big(1+\alpha_{ik}\Delta + O(\Delta)\Big)\int_{t+\Delta-\tau_i}^{t+\Delta}\zeta^t(r)\mathbb{L}\zeta(r)dr \qquad (6.118)
\end{aligned}$$

from which it follows that

$$\Im_d^\zeta[V_d] = \zeta^t(t)\mathbb{L}\zeta(t) - \zeta^t(t-\tau_{\eta_t})\mathbb{L}\zeta(t-\tau_{\eta_t})$$

$$+ \sum_{k \in \mathcal{S}} \alpha_{\eta_t k} \int_{t-\tau_k}^{t} \zeta^t(r) \mathbb{L} \zeta(r) dr \tag{6.119}$$

In a similar way, it can be easily shown that

$$\begin{aligned} \Im_f^{\zeta}[V_f] &= \hat{\alpha} \Big(\hat{\tau} - \check{\tau} \Big) \zeta^t(t) \mathbb{L} \zeta(t) \\ &- \hat{\alpha} \int_{t-\hat{\tau}}^{t-\check{\tau}} \zeta^t(r) \mathbb{L} \zeta(r) dr \end{aligned} \tag{6.120}$$

Observe that

$$\sum_{k \in \mathcal{S}} \alpha_{\eta_t k} \int_{t-\tau_k}^{t} \zeta^t(r) \mathbb{L} \zeta(r) dr \leq \hat{\alpha} \int_{t-\hat{\tau}}^{t-\check{\tau}} \zeta^t(r) \mathbb{L} \zeta(r) dr \tag{6.121}$$

Now by combining (6.113) through (6.121), we obtain

$$\begin{aligned} \Im_t^{\zeta}[V_t] &= \zeta^t(t) \Big\{ \mathbb{P}(\eta_t) \mathcal{A}_{\Delta o}(t, \eta_t) + \mathcal{A}_{\Delta o}^t(t, \eta_t) \mathbb{P}(\eta_t) \\ &+ \sum_{k=1}^{s} \alpha_{jk} \mathbb{P}(k) + \Big(1 + [\hat{\tau} - \check{\tau}] \Big) \hat{\alpha} \mathbb{L} \Big\} \zeta(t) \\ &+ 2 \zeta^t(t) \mathbb{P}(\eta_t) \mathcal{A}_{\Delta d}(t, \eta_t) \zeta(t - \tau_{\eta_t}) - \zeta^t(t - \tau_{\eta_t}) \mathbb{L} \zeta(t - \tau_{\eta_t}) \\ &\triangleq [\zeta^t(t) \quad \zeta^t(t - \tau_{\eta_t})] \Upsilon(i, t) \left[\begin{array}{c} \zeta(t) \\ \zeta(t - \tau_{\eta_t}) \end{array} \right] \end{aligned} \tag{6.122}$$

which is negative from (6.111)-(6.112). We conclude that

$$\Im_t^{\zeta}[V_t] < 0 \quad \forall \zeta \neq 0 \ , \ \Im_t^{\zeta}[V_t] \leq 0 \ \forall \ \zeta$$

Therefore, from

$$V_t(t, \zeta, i) \geq \zeta^t(t) \mathbb{P}(i) \zeta(t)$$

and in particular, for all $\zeta \neq 0$, we have

$$\begin{aligned} \frac{\Im_t^{\zeta}[V_t]}{V_t(\zeta, i)} \leq \frac{\zeta^t \Upsilon(i) \zeta}{\zeta^t \mathbb{P}(i) \zeta} &\leq - \varrho \\ &\triangleq - \min_{i \in \mathcal{S}} \left\{ \frac{\lambda_m[-\Upsilon(i)]}{\lambda_M[\mathbb{P}(i)]} \right\} \end{aligned} \tag{6.123}$$

It is readily seen from (6.123) that $\varrho > 0$ and hence we get

$$\Im_t^{\zeta}[V_t] \;\; \leq -\varrho\, V_t(t, \zeta, i)$$

It follows from [78], by using the Gronwall-Bellman lemma [98] and letting $x(t = 0, \phi, i) = \zeta_o$, one has

$$\mathbb{E}[V_t(\zeta, i)|\phi, \eta_o = i] \;\; \leq \;\; e^{-\varrho\, t}\, V_t(\zeta_o, i) \tag{6.124}$$

Therefore

$$\begin{aligned} \mathbb{E}[V_t(x, i)|\phi, \eta_o = i] &= \mathbb{E}\Big\{\zeta^t(t)\mathbb{P}(i)\zeta(t) \\ &+ \int_{t-\tau_{\eta_t}}^{t} \zeta^t(t+\theta)\mathbb{L}(i)x(t+\theta)d\theta|\phi, \eta_o \\ &+ \int_{-\hat{\tau}}^{-\check{\tau}} \int_{t+\beta}^{t} \zeta(\beta)\hat{\alpha}\mathbb{L}\zeta(\beta)\, d\beta\, dr\Big\} \\ &= \mathbb{E}\Big\{\zeta^t(t)\mathbb{P}(i)\zeta(t)|\phi, \eta_o\Big\} \\ &+ \mathbb{E}\Big\{\int_{t-\tau_{\eta_t}}^{t} \zeta^t(t+\theta)\mathbb{L}(i)\zeta(t+\theta)d\theta|\phi, \eta_o\Big\} \\ &+ \mathbb{E}\Big\{\int_{-\hat{\tau}}^{-\check{\tau}} \int_{t+\beta}^{t} \zeta(\beta)\hat{\alpha}\mathbb{L}\zeta(\beta)\, d\beta\, dr\Big\} \\ &\leq e^{-\varrho\, t}\, V_t(\zeta_o, i) \end{aligned} \tag{6.125}$$

Since the foregoing analysis entails that

$$\mathbb{E}\Big\{\int_{-\tau}^{0} \zeta^t(t+\theta)\mathbb{L}(i)\zeta(t+\theta)d\theta|\phi, \eta_o\Big\} \geq 0$$
$$\mathbb{E}\Big\{\int_{-\hat{\tau}}^{-\check{\tau}} \int_{t+\beta}^{t} \zeta(\beta)\hat{\alpha}\mathbb{L}\zeta(\beta)\, d\beta\, dr\Big\} \geq 0$$

then by some algebraic manipulation of (6.125) it yields:

$$\mathbb{E}\Big\{\zeta^t(t)\mathbb{P}(i)\zeta(t)|\phi, \eta_o\Big\} \leq e^{-\varrho\, t}V(\zeta_o, i) \Longrightarrow$$

$$
\begin{aligned}
&\mathbb{E}\Big\{\int_0^{\mathcal{T}} \zeta^t(t)\mathbb{P}(i)\zeta(t)dt|\phi,\eta_o = i\Big\} \\
&\leq \Big[\int_0^{\mathcal{T}} e^{-\varrho\, t}dt\Big] V_t(\zeta_o,i) - \frac{1}{\varrho}[e^{-\varrho\,\mathcal{T}} \; - \; 1]\, V(\zeta_o,i) \Longrightarrow \\
&\lim_{\mathcal{T}\to\infty}\; \mathbb{E}\Big\{\int_0^{\mathcal{T}} \zeta^t(t)\mathbb{P}(i)\zeta(t)dt|\phi,\eta_o = i\Big\} \\
&\leq \; \frac{1}{\varrho}\zeta_o^t\mathbb{P}(i)\zeta_o + \frac{\kappa}{\varrho}\lambda_M[\mathbb{L}(i)]||\zeta(t+\theta)||_*^2, \;\; \forall\theta\in[-\tau_{\eta_t},0] \qquad (6.126)
\end{aligned}
$$

where κ can be taken a finitely large real value and

$$
||\zeta(t+\theta)||_*^2 \stackrel{\Delta}{=} \sup_{\theta\in[-\tau_{\eta_t},0]} ||\zeta(t+\theta)||_2^2
$$

Let

$$
\bar{\mathbb{P}}(i) \;=\; \max_{i\in\mathcal{S}}\left\{\frac{\lambda_M[\mathbb{P}(i)]||x_o||^2 + \hat{\tau}\lambda_M[\mathbb{L}(i)]||\zeta(t+\theta)||_*^2}{\varrho\lambda_m[\mathbb{P}(i)]||\zeta_o||^2}\right\}
$$

it follows from (6.126) for $i\in\mathcal{S}$ that

$$
\begin{aligned}
&\lim_{\mathcal{T}\to\infty}\; \mathbb{E}\Big\{\int_0^{\mathcal{T}} \zeta^t(t)\zeta(t)dt|\phi,\eta_o = i\Big\} \\
&\leq \zeta_o^t\lambda_M(\bar{\mathbb{P}}(i))\zeta_o < +\infty,
\end{aligned}
$$

which, in the light of **Definition** 6.5, shows that system (Σ_A) is **RSSWDD**. $\nabla\nabla\nabla$

The next theorem provides the stochastic stability condition as an LMI-feasibility criterion.

Theorem 6.10 *System Σ_A is* **RSSWDD**, *if there exist matrices $0 < I\!\!L \in I\!\!R^{2n\times 2n}, 0 < I\!\!P(i) = I\!\!P^t(i) \in I\!\!R^{2n\times 2n}$, $i\in\mathcal{S}$ and scalars $\epsilon(i) > 0,\ \mu(i) > 0,\ i\in\mathcal{S}$ satisfying the LMIs forall $i\in\mathcal{S}$*

$$
\begin{bmatrix}
\Pi_s(i) & I\!\!P(i)\bar{M}_A(i) & I\!\!P(i)\bar{M}_A(i) & I\!\!P(i)\mathsf{D}(i) \\
\bar{M}_A^t(i)I\!\!P(i) & -\epsilon(i)I & 0 & 0 \\
\bar{M}_A^t(i)I\!\!P(i) & 0 & -\mu(i)I & 0 \\
\mathsf{D}^t(i)I\!\!P(i) & 0 & 0 & \begin{matrix} -I\!\!L+ \\ \mu(i)\bar{N}_D^t(i)\bar{N}_D(i)\end{matrix}
\end{bmatrix} < 0 \, (6.127)
$$

$$\begin{aligned}\Pi_s(i) &= \mathsf{A}^t(i)I\!P(i) + I\!P(i)\mathsf{A}(i) + \sum_{k=1}^{s} \alpha_{ik} I\!P(k) \\ &+ \Big(1 + [\hat{\tau} - \check{\tau}]\Big)\hat{\alpha} I\!L + \epsilon(i)\bar{N}_A^t(i)\bar{N}_A(i) \end{aligned} \tag{6.128}$$

Proof: By **Definition** 6.5 and the Schur complements, it follows that for all $\eta_t = i \in \mathcal{S}$

$$\begin{aligned} & [\zeta^t(t) \quad \zeta^t(t-\tau_{\eta_t})]\Upsilon(i,t)\left[\begin{array}{c} \zeta(t) \\ \zeta(t-\tau_{\eta_t}) \end{array}\right] \\ = \;& \zeta^t(t)\Lambda(i,t)\zeta(t) \\ = \;& \zeta^t(t)\Big\{\mathbb{P}(i)\mathcal{A}_{\Delta o}(t,i) + \mathcal{A}_{\Delta o}^t(t,i)\mathbb{P}(i) + \sum_{k=1}^{s}\alpha_{ik}\mathbb{P}(k) \\ + \;& \mathbb{P}(i)\mathcal{D}_{\Delta d}(t,i)\mathbb{L}^{-1}\mathcal{D}_{\Delta d}^t(t,i)\mathbb{P}(i) \\ + \;& \Big(1 + [\hat{\tau} - \check{\tau}]\Big)\hat{\alpha}\mathbb{L}\Big\}\zeta(t) < 0 \end{aligned} \tag{6.129}$$

By considering (6.97)-(6.98) and applying **Facts 1,2** to (6.129) with some algebraic manipulations, it follows that:

$$\begin{aligned} \zeta^t(t)\Lambda(i,t)\zeta(t) &= \zeta^t(t)\Big\{\mathsf{A}^t(i)\mathbb{P}(i) + \mathbb{P}(i)\mathsf{A}(i) \\ &+ \sum_{k=1}^{s}\alpha_{im}\mathbb{P}(m) + \epsilon(i)\bar{N}_A^t(i)\bar{N}_A \\ &+ \epsilon^{-1}(i)\mathbb{P}(i)\bar{M}_A(i)\bar{M}_A^t(i)\mathbb{P}(i) \\ &+ \mu^{-1}(i)\mathbb{P}(i)\bar{M}_A(i)\bar{M}_A^t(i)\mathbb{P}(i) \\ &+ \mathbb{P}(i)\mathsf{D}(i)[\mathbb{L}(i) - \mu(i)\bar{N}_D^t(i)\bar{N}_D(i)]^{-1}\mathsf{D}^t(i)\mathbb{P}(i) \\ &+ \Big(1 + [\hat{\tau} - \check{\tau}]\Big)\hat{\alpha}\mathbb{L}\Big\}\zeta(t) < 0 \end{aligned} \tag{6.130}$$

hold for some scalars $\epsilon(i) > 0, \; \mu(i) > 0, \; i \in \mathcal{S}$. By **Facts 1**, we obtain LMIs (6.127). $\nabla\nabla\nabla$

To facilitate further development, we introduce the following matrix expressions for some scalars $\epsilon(i) > 0,\ \mu(i) > 0,\ i \in \mathcal{S}$:

$$\begin{aligned}
\mathbb{P}(i) &= \begin{bmatrix} \mathbb{P}_e(i) & 0 \\ 0 & \mathbb{P}_x(i) \end{bmatrix}, \quad \mathbb{L} = \begin{bmatrix} \mathbb{L}_e & 0 \\ 0 & \mathbb{L}_x \end{bmatrix} \\
\mathbb{L}_a(i) &= \mathbb{L}_e - \mu(i) N_d^t(i) N_d(i) \\
\hat{A}_o(i) &= A_o(i) + M_a(i) M_a^t(i) \mathbb{P}_x(i) + A_d(i) \mathbb{L}_a^{-1} A_d^t(i) \mathbb{P}_x(i) \qquad (6.131) \\
\hat{C}_o(i) &= C_o(i) + M_a(i) M_a^t(i) \mathbb{P}_x(i) + C_o(i) \mathbb{L}_a^{-1} A_d^t(i) \mathbb{P}_x(i) \\
\Omega(i) &= \Big[A_d(i) \mathbb{L}_a^{-1}(i) + [\mu^{-1}(i) + \epsilon^{-1}(i)] M_a(i) \Big] \\
\mathbb{H}(i) &= \mathbb{L}_a^{-1}(i) + \Big(\epsilon^{-1}(i) + \mu^{-1}(i) \Big) I \qquad (6.132)
\end{aligned}$$

The following theorem gives the expressions for the filter gains

Theorem 6.11 *System Σ_A is* **RSSWDD**, *if there exist matrices $0 < I\!\!L_e \in I\!\!R^{n \times n}$, $0 < I\!\!L_x \in I\!\!R^{n \times n}$, $0 < I\!\!P_e(i) = I\!\!P_e^t(i) \in I\!\!R^{n \times n}$, $0 < I\!\!P_x(i) = I\!\!P_x^t(i) \in I\!\!R^{n \times n}$, $i \in \mathcal{S}$ and scalars $\epsilon(i) > 0,\ \mu(i) > 0,\ i \in \mathcal{S}$ satisfying the LMIs for all $i \in \mathcal{S}$*

$$\begin{bmatrix} \Pi_e(i) & I\!\!P_e(i) M_a(i) & I\!\!P_e(i) M_a(i) & I\!\!P_e(i) A_d(i) \\ M_a^t(i) I\!\!P_e(i) & -\epsilon(i) I & 0 & 0 \\ M_a^t(i) I\!\!P_e(i) & 0 & -\mu(i) I & 0 \\ A_d^t(i) I\!\!P_e(i) & 0 & 0 & \begin{matrix} -I\!\!L_e(i) + \\ \mu(i) N_d^t(i) N_d(i) \end{matrix} \end{bmatrix} < 0 \qquad (6.133)$$

$$\begin{aligned}
\Pi_e(i) &= [\hat{A}_o(i) - \Omega(i) I\!\!H^{-1}(i)]^t I\!\!P_e(i) + I\!\!P_e(i) [\hat{A}_o(i) - \Omega(i) I\!\!H^{-1}(i)] \\
&+ \sum_{k=1}^{s} \alpha_{ik} I\!\!P_e(k) + \Big(1 + [\hat{\tau} - \check{\tau}] \Big) \hat{\alpha} I\!\!L_e \\
&- I\!\!P_e(i) \Omega(i) I\!\!H^{-1}(i) \Omega^t(i) I\!\!P_e(i) \qquad (6.134)
\end{aligned}$$

$$\begin{bmatrix} \Pi_x(i) & I\!\!P_x(i) M_a(i) & I\!\!P_x(i) M_a(i) & I\!\!P_x(i) A_d(i) \\ M_a^t(i) I\!\!P_x(i) & -\epsilon(i) I & 0 & 0 \\ M_a^t(i) I\!\!P_x(i) & 0 & -\mu(i) I & 0 \\ A_d^t(i) I\!\!P(i) & 0 & 0 & \begin{matrix} -I\!\!L_e + \\ \mu(i) N_d^t(i) N_d(i) \end{matrix} \end{bmatrix} < 0 \quad (6.135)$$

$$\Pi_x(i) = A_o^t(i)\mathbb{P}_x(i) + \mathbb{P}_x(i)A_o(i) + \sum_{k=1}^{s} \alpha_{ik}\mathbb{P}_x(k)$$
$$+ \epsilon(i)N_a^t(i)N_a(i) + \Big(1 + [\hat{\tau} - \check{\tau}]\Big)\hat{\alpha}\mathbb{L}_x \quad (6.136)$$

Moreover, the filter gains are given by:

$$K_f(i) = \Omega(i)C_o^t(i)\Big[C_o(i)\mathbb{H}C_o^t(i)\Big]^{-1}$$
$$A_f(i) = \hat{A}_o(i) - K_f(i)\hat{C}_o(i) \quad (6.137)$$

Proof: Extending on **Theorem** (6.10) by using (6.131)-(6.132) into (6.130) and expanding terms we express it into the block form

$$\begin{bmatrix} \Sigma_{11} & \Sigma_{12} \\ \Sigma_{12}^t & \Sigma_{22} \end{bmatrix}$$

Applying **Fact 1** to the matrix block Σ_{22}, we can readily obtain the LMI (6.135). The substitution of (6.137) yields $\Sigma_{12} \equiv 0$. Finally, using (6.131)-(6.132) and (6.137) into Σ_{11} with some matrix manipulations and applying the Schur complements we get the LMI (6.133). $\nabla\nabla\nabla$

6.5.4 $\mathcal{H}_\infty$ Filtering

A natural extension of the foregoing results to an $\mathcal{H}_\infty$ setting is now considered. We first recall the following definition:

Definition 6.6 *System Σ_A is said to be* **stochastically stable with weak-delay dependence (SSWDD) with a disturbance attenuation level** γ *if for all finite initial vector function $\phi \in \mathbb{R}^n$ defined on the interval $[-\tau_{\eta_t}, 0]$ and initial mode $\eta_o \in \mathcal{S}$ there exists a constant $\mathcal{G}$ such that the following inequality holds*

$$||\tilde{z}(t)||_{E_2} \triangleq \left[\int_0^\infty \mathbb{E}\Big(\tilde{z}^t(t)\tilde{z}(t)dt\Big)\right]^{1/2}$$
$$\leq \Big[\gamma^2\, ||w(t)||_2^2 \,+\, \mathcal{G}\,\Psi\big(\eta_o, \phi(.)\big)\Big]^{1/2} \quad (6.138)$$

for all $0 \neq w(t) \in \mathcal{L}_2[0,\infty)$*, where* $\gamma > 0$ *is a prescribed level of disturbance attenuation ,* $||.||_{E_2}$ *denotes the norm in* $\mathcal{L}_2((\Omega,\mathcal{G},\mathbf{P}),[0,\infty))$ *and* $\Psi\big(\eta_o,\phi(.)\big)$ *is a nonnegative function of the initial values with* $\Psi\big(0,0\big) = 0$*.*

Based thereon, the following theorems are established:

Theorem 6.12 *System* Σ_A *is* **RSSWDD with a disturbance attenuation** γ*, if there exist matrices* $0 < I\!\!L \in I\!\!R^{2n\times 2n}, 0 < I\!\!P(i) = I\!\!P^t(i) \in I\!\!R^{2n\times 2n},\ i \in \mathcal{S}$ *and scalars* $\gamma > 0,\ \epsilon(i) > 0,\ \mu(i) > 0,\ \sigma(i) > 0,\ i \in \mathcal{S}$ *satisfying the LMIs for all* $i \in \mathcal{S}$

$$\begin{bmatrix} \Pi_d(i) & \Xi_A(i) & I\!\!P_e(i)\mathsf{D}(i) & \Xi_B(i) \\ \Xi_A^t(i) & -\Psi_A(i) & 0 & 0 \\ \mathsf{D}^t(i)I\!\!P_e(i) & 0 & \begin{matrix} -I\!\!L(i)+ \\ \mu(i)\bar{N}_D^t(i)\bar{N}_D(i) \end{matrix} & 0 \\ \Xi_B^t(i) & 0 & 0 & -\Psi_B(i) \end{bmatrix} < 0 \qquad (6.139)$$

$$\Pi_d(i) = \Pi_s(i) + L_f^t(i)L_f(i)$$

$$\Xi_A(i) = [I\!\!P\bar{M}_A(i) \quad I\!\!P\bar{M}_A(i) \quad I\!\!P\bar{M}_A(i)]$$

$$\Psi_A(i) = diag[\epsilon(i)I \quad \mu(i)I \quad \sigma(i)I]$$

$$\Xi_B(i) = [I\!\!P(i)\mathsf{B}(i) \quad L_f^t(i)] \qquad (6.140)$$

$$\Psi_B(i) = \begin{bmatrix} \begin{matrix} -\gamma^2 I+ \\ \sigma(i)N_b^t(i)N_b(i) \end{matrix} & \Gamma^t(i) \\ \Gamma(i) & -I \end{bmatrix} \qquad (6.141)$$

Proof: The stochastic stability of system (Σ_A) follows as a result of **Theorem** (6.10). We need to show here is that system (Σ_A) has a disturbance attenuation γ. Without loss of generality, we assume that $\phi(t) = 0$. Let the stochastic Lyapunov function $V_t(\zeta(t),\eta_t)$, for $\eta_t = i$ be given by (6.113). By evaluating the weak infinitesimal operator $\Im_w^\zeta[V_t]$ of the process $\{\zeta(t),\eta_t,t \geq 0\}$ for system (6.107)-(6.108) at the point $\{t,\zeta,\eta_t\}$ using (6.122) we get

$$\begin{aligned} \Im_w^\zeta[V_t] &= \Im_w^\zeta[V_t] \\ &+ \zeta^t(t)\mathbb{P}(\eta_t)\mathcal{B}_{\Delta o}(t,\eta_t)w(t)w^t(t)\mathcal{B}_{\Delta o}^t(t,\eta_t)\mathbb{P}(\eta_t)\zeta(t) \end{aligned} \qquad (6.142)$$

Now, we introduce

$$\mathcal{J}(\zeta) \stackrel{\Delta}{=} \mathbb{E}\Big\{ \int_0^\infty [\tilde{z}^t(t)\tilde{z}(t) \; - \; \gamma^2 w^t(t) w(t) \;]dt \Big\}$$

By Dynkin's formula [78], one has

$$E\Big\{ \int_0^\infty \Im_w^\zeta[V_t] \, dt \Big\} = \mathbb{E}\Big\{ V_t(\zeta(t), \eta_t)|_{t=\infty} \Big\} - V_t(\zeta(t), \eta_t)|_{t=0} \; \geq 0$$

Standard matrix manipulations using (6.108), (6.130) and **Fact 2** show that:

$$\begin{aligned}
\mathcal{J}(\zeta) \;\; &= \;\; \mathbb{E}\Big\{ \int_0^\infty [\tilde{z}^t(t)\tilde{z}(t) \; - \; \gamma^2 w^t(t) w(t) \; + \Im_w^\zeta[V_t] - \Im_w^\zeta[V_t]]dt \Big\} \\
&\leq \;\; \mathbb{E}\Big\{ \int_0^\infty [\tilde{z}^t(t)\tilde{z}(t) \; - \; \gamma^2 w^t(t) w(t) \; + \Im_+[V]]dt \Big\} \\
&\leq \;\; \mathbb{E}\Big\{ \int_0^\mathcal{T} \zeta^t(t) \Big\{ \Lambda(i) + L_f^t(i) L_f(i) \\
&+ \;\; \Big[\mathbb{P}(i)\mathcal{B}_{\Delta o}(i) + \Gamma^t(i) L_f(i) \Big] \Big[\gamma^2 I - \Gamma(i)\Gamma^t(i) \Big]^{-1} \Big[\mathbb{P}(i)\mathcal{B}_{\Delta o}(i) \\
&+ \;\; \Gamma^t(i) L_f(i) \Big]^t \Big\} \zeta(t) \\
&\leq \;\; \mathbb{E}\Big\{ \int_0^\mathcal{T} \zeta^t(t) \Big\{ \Lambda(i) + \sigma^{-1}(i)\mathbb{P}(i)\bar{M}_A(i)\bar{M}_A^t(i)\mathbb{P} + L_f^t(i) L_f(i) \\
&+ \;\; \Big[\mathbb{P}(i)\mathsf{B}(i) + L_f^t(i)\Gamma(i) \Big] \Big[\gamma^2 I - \Gamma^t(i)\Gamma(i) - \sigma(i) N_b^t(i) N_b(i) \Big]^{-1} \\
&\cdot \;\; \Big[\mathsf{B}^t(i)\mathbb{P}(i) + \Gamma^t(i) L_f(i) \Big]^t \Big\} \zeta(t) < 0
\end{aligned} \tag{6.143}$$

By using (6.139)-(6.141) and the results of **Theorem** (6.10), it follows from inequality (3.838) that $\mathcal{J}(x) \; < \; 0$ and hence by **Definition** (6.6) the proof is completed. $\nabla\nabla\nabla$

Theorem 6.13 *If there exist matrices* $0 < \mathbb{L}_e \in \mathbb{R}^{n \times n}$, $0 < \mathbb{L}_x \in \mathbb{R}^{n \times n}$, $0 < \mathbb{P}_e(i) = \mathbb{P}_e^t(i) \in \mathbb{R}^{n \times n}$, $0 < \mathbb{P}_x(i) = \mathbb{P}_x^t(i) \in \mathbb{R}^{n \times n}$, $i \in \mathcal{S}$ *and scalars* $\epsilon(i) > 0$, $\mu(i) > 0$, $i \in \mathcal{S}$ *satisfying the LMIs for all* $i \in \mathcal{S}$

$$\begin{bmatrix} \Pi_w(i) & \Xi_C(i) & I\!\!P_e(i)A_d(i) & \Xi_D(i) \\ \Xi_C^t(i) & -\Psi_A(i) & 0 & 0 \\ A_d^t(i)I\!\!P_e(i) & 0 & \begin{array}{c} -I\!\!L_e(i)+ \\ \mu(i)N_d^t(i)N_d(i) \end{array} & 0 \\ \Xi_D^t(i) & 0 & 0 & -\Psi_B(i) \end{bmatrix} < 0 \tag{6.144}$$

$$\Pi_w(i) = \Pi_e(i) + L^t(i)L(i)$$

$$\Xi_C(i) = [I\!\!P_e M_a(i) \quad I\!\!P_e M_a(i) \quad I\!\!P_e M_a(i)]$$

$$\Xi_D(i) = [I\!\!P_e(i)B_o(i) \quad L^t(i)] \tag{6.145}$$

$$\begin{bmatrix} \Pi_x(i) & \Xi_C(i) & I\!\!P_x(i)A_d(i) & I\!\!P_x(i)B_o(i) \\ \Xi_C^t(i) & -\Psi_A(i) & 0 & 0 \\ A_d^t(i)I\!\!P_x(i) & 0 & \begin{array}{c} -I\!\!L_e+ \\ \mu(i)N_d^t(i)N_d(i) \end{array} & 0 \\ B_o^t(i)I\!\!P_x(i) & 0 & 0 & -\Omega_z(i) \end{bmatrix} < 0 \tag{6.146}$$

$$\Omega_z(i) = \left[\gamma^2 I - \Gamma^t\Gamma(i) - \sigma(i)N_b^t(i)N_b(i)\right] \tag{6.147}$$

then the augmented filter system (6.107)-(6.108) is **RSSWDD with a disturbance attenuation** γ *and the associated filter gains are given by:*

$$\begin{aligned} \bar{A}_o(i) &= B_o^t(i)\Omega_z(i)B_o(i)I\!\!P_x(i) + I\!\!P_e^{-1}(i)L^t(i)\Gamma(i)\Omega_z(i)B_o^t(i)I\!\!P_x(i) \\ \bar{C}_o(i) &= D_o(i)\Omega_z(i)B_o^t(i)I\!\!P_x(i) + M_c(i)M_a^t(i))I\!\!P_x(i) \qquad (6.148) \\ I\!\!M &= C_o(i)I\!\!H(i)C_o^t(i) + D_o(i)\Omega_z(i)D_o^t(i) \\ K_f(i) &= \left\{\Omega(i)C_o^t(i) + D_o(i)\Omega_z(i)[B_o^t(i)I\!\!P_x(i) + \Gamma^t(i)L(i)]\right\}I\!\!M^{-1} \\ A_f(i) &= [\hat{A}_o(i) + \bar{A}_o(i)] - K_f(i)[\hat{C}_o(i) + \bar{C}_o(i)] \qquad (6.149) \end{aligned}$$

Proof: Follows parallel development to **Theorem** 6.11. ∇∇∇

6.5.5 Example 6.3

Consider the uncertain JTD system with bounds on time-delays

$$\hat{\tau} = 0.8 \ , \ \check{\tau} = 0.3$$

and having mode-switching generator

$$\Im = \begin{bmatrix} -3 & 3 \\ 2 & -2 \end{bmatrix}$$

The associate date are given by:

Mode 1

$$\begin{aligned}
A_o(1) &= \begin{bmatrix} -2 & 0 \\ 1 & -3 \end{bmatrix}, \; A_d(1) = \begin{bmatrix} -1 & 0 \\ -0.8 & -1 \end{bmatrix} \\
B_o(1) &= \begin{bmatrix} 0.2 & 0 \\ 0 & 0.2 \end{bmatrix}, \; D_d(1) = \begin{bmatrix} 0.1 & 1 \end{bmatrix} \\
C_o(1) &= \begin{bmatrix} 1 & 0 \end{bmatrix}, \; \Gamma(1) = \begin{bmatrix} 0.8 & 0.5 \end{bmatrix} \\
L(1) &= \begin{bmatrix} 0.5 & 0.4 \end{bmatrix} \\
M_a(1) &= \begin{bmatrix} 0.1 & 0.05 \\ -0.02 & 0.1 \end{bmatrix}, \; N_a(1) = \begin{bmatrix} 0.8 & 0 \\ 0 & 0.8 \end{bmatrix} \\
M_c(1) &= \begin{bmatrix} -0.2 & 0.8 \end{bmatrix}, \; N_b(1) = \begin{bmatrix} 0.02 & 0.01 \\ 0.2 & 0.5 \end{bmatrix} \\
B_d(1) &= \begin{bmatrix} 0.5 & 0.8 \end{bmatrix}, \; N_d(1) = \begin{bmatrix} 0.7 & 0 \\ 0 & 0.6 \end{bmatrix}
\end{aligned}$$

Mode 2

$$\begin{aligned}
A_o(2) &= \begin{bmatrix} -1 & 1 \\ 0 & -4 \end{bmatrix}, \; A_d(2) = \begin{bmatrix} -0.8 & 0 \\ -0.7 & -0.8 \end{bmatrix} \\
B_o(2) &= \begin{bmatrix} 0.3 & 0 \\ 0 & 0.3 \end{bmatrix}, \; D_d(2) = \begin{bmatrix} 0.3 & 0.7 \end{bmatrix} \\
C_o(2) &= \begin{bmatrix} 0.5 & 0.5 \end{bmatrix}, \; \Gamma(2) = \begin{bmatrix} 0.5 & 0.8 \end{bmatrix} \\
L(2) &= \begin{bmatrix} 0.4 & 0.4 \end{bmatrix} \\
M_a(2) &= \begin{bmatrix} 0.2 & 0.03 \\ -0.02 & 0.3 \end{bmatrix}, \; N_a(2) = \begin{bmatrix} 0.7 & 0 \\ 0 & 0.7 \end{bmatrix} \\
M_c(2) &= \begin{bmatrix} -0.2 & 0.8 \end{bmatrix}, \; N_b(2) = \begin{bmatrix} 0.05 & 0.05 \\ 0.\text{[illegible]} & 0.4 \end{bmatrix} \\
B_d(2) &= \begin{bmatrix} 0.8 & 0.5 \end{bmatrix}, \; N_d(2) = \begin{bmatrix} 0.6 & 0 \\ 0 & 0.7 \end{bmatrix}
\end{aligned}$$

Application of **theorem** 6.11 and solving LMIs (6.133)-(6.135) with

$$\gamma = 1.15 \quad , \quad R = \begin{bmatrix} 1.8 & 0 \\ 0 & 3.3 \end{bmatrix}$$

the state-space matrices of the Markovian filter (6.66)are given by:

$$\begin{aligned} A_f(1) &= \begin{bmatrix} 1.1922 & 0.1958 \\ 2.4809 & -3.8907 \end{bmatrix} \quad , \quad K_f(1) = \begin{bmatrix} -0.3376 \\ -1.5528 \end{bmatrix} \\ A_f(2) &= \begin{bmatrix} 1.1092 & -0.7988 \\ 2.1064 & -8.2876 \end{bmatrix} \quad , \quad K_f(2) = \begin{bmatrix} -0.6107 \\ -3.2087 \end{bmatrix} \end{aligned}$$

6.6 Notes and References

This chapter have developed a linear matrix inequality based methodology to study the problems of robust filtering for a class of linear systems subject to uncertain parameters and Markovian jump parameters and in which the delay factor is wither mode-dependent, weakly-dependent or strong-dependent. The results can generally be extended in various directions by dualizing the results of Chapter 4.

Chapter 7

Neutral Jumping Systems

7.1 Introduction

An integral part of functional differential systems (FDS) [62] is the class of neutral-type systems which can be found in several applications including, but not limited to, chemical reactor, rolling mill, indeed grinding, lossless transmission lines and hydraulic systems. Stability analysis and feedback stabilization for neutral FDS have been studied in [82, 137, 156] and other related work can be found in [98]. Recently $\mathcal{H}_\infty$ state-feedback control has been developed in [99] for a class of linear neutral systems with parametric uncertainties and preliminary results have been reported in [123, 124] .

This Chapter represents a basic departure from the main stream followed by the past Chapters. It is entirely devoted to study a general class of uncertain neutral jumping systems (NJS) with different lengths of distributed state-delays and multiple state-derivative delays. As in previous chapters, the jumping parameters are treated as continuous-time, discrete-state Markov process and the uncertainties are parametric and norm-bounded. We intend to provide a comprehensive treatment of NJS which, by and large, will generalize available results in the literature in the sense that the developed results are new and will encompass the previously published works as special cases.

The Chapter will be divided into two major parts and each part consists of consecutive sections. In turn, each section will deal with a topic from the subject matter. After presenting the model description, the first part starts by providing LMI-based sufficient conditions for the robust stochastic stability. Then robust stabilization is established using memoryless state feedback and distributed feedback. Afterwards, the results are extended to the nominal neutral jumping and the uncertain retarded systems. Robust $\mathcal{H}_\infty$ output feedback control is finally examined to design observer-based controllers which render the combined neutral system and the proposed controller asymptotically stable with a guaranteed performance measure for all admissible uncertainties.

In the second part of the Chapter, the observer problem for a class of uncertain NJS is addressed. Both robust observation and robust $\mathcal{H}_\infty$ observation methods are developed using linear state-delayed observers. In case of robust observation, we establish sufficient conditions for asymptotic stability using a linear matrix inequality approach . The results are then extended to robust $\mathcal{H}_\infty$ observation which renders the augmented system asymptotically stable independent of delay with a guaranteed performance measure. Then we designed memoryless state-estimate feedback to stabilize the closed-loop neutral system. In all cases, the gain matrices are determined by solving linear matrix inequalities. All the technical results throughout the Chapter are cast into LMI-format. The analytical developments throughout the Chapter are organized into theorems whereby the results are presented in a systematic and gradual build-up. Appropriate remarks are inserted to evaluate our results with respect to others.

7.2 Model Description

A general class of uncertain neutral systems with Markovian jump parameters over the space $(\Omega, \mathcal{F}, \mathbf{P})$ has the form:

$$\dot{x}(t) = \mathbb{F}\Big(t, \eta_t, x(t-\tau_1), \cdots, x(t-\tau_q), \dot{x}(t-\pi_1), \cdots, \dot{x}(t-\pi_p)\Big) \tag{7.1}$$

where the factors $\{\tau_1 > 0, \cdots \tau_q > 0\ ,\ \pi_1 > 0, \cdots \pi_p > 0\}$ are constant scalars representing the amount of time-lags in the respective states and their derivatives. For the purpose of stability and control studies, a convenient linearizable representation of (7.1) is described by for $\eta_t = i \in \mathcal{S}$:

$$\begin{aligned}
(\Sigma_{\Delta n})\ \mathcal{M}(\dot{x}_t) &\stackrel{\Delta}{=} \dot{x}(t) - \sum_{j=1}^{p} B_j(\eta_t)\dot{x}(t-\pi_j) = A_{\Delta o}(t, \eta_t)x(t) \\
&+ \sum_{k=1}^{q} A_{\Delta d_k}(t, \eta_t)x(t-\tau_k) \\
&+ F_{\Delta o}(t, i)u(t) + N(i)w(t) && (7.2) \\
x(\vartheta) &= \phi(\vartheta) \in \mathbb{C}([-\max\{\tau_k, k \in \mathcal{Z}_q\}, 0], \mathbb{R}^n) && (7.3) \\
y(t) &= C_{\Delta o}(t, i)x(t) + \sum_{k=1}^{q} C_{\Delta d_k}(t, \eta_t)x(t-\tau_k) \\
&+ M(\eta_t)w(t) && (7.4) \\
z(t) &= L(\eta_t)x(t) && (7.5)
\end{aligned}$$

where

$$\begin{bmatrix} A_{\Delta o}(t,\eta_t) & A_{\Delta d_1}(t,\eta_t) \cdots A_{\Delta d_q}(t,\eta_t) \\ C_{\Delta o}(t,\eta_t) & C_{\Delta d_1}(t,\eta_t) \cdots C_{\Delta d_q}(t,\eta_t) \end{bmatrix} =$$
$$\begin{bmatrix} A_o(t,\eta_t) & A_{d_1}(t,\eta_t) \cdots A_{d_q} \\ C_o(t,\eta_t) & C_{d_1}(t,\eta_t) \cdots C_{d_q} \end{bmatrix} +$$
$$\begin{bmatrix} \Delta A_o(t,\eta_t) & \Delta A_{d_1}(t,\eta_t) \cdots \Delta A_{d_q}(t,\eta_t) \\ \Delta C_o(t,\eta_t) & \Delta C_{d_1}(t,\eta_t) \cdots \Delta C_{d_q}(t,\eta_t) \end{bmatrix}$$
$$\begin{bmatrix} \Delta A_o(t,\eta_t) & \Delta A_{d_1}(t,\eta_t) \cdots \Delta A_{d_q}(t,\eta_t) \\ \Delta C_o(t,\eta_t) & \Delta C_{d_1}(t,\eta_t) \cdots \Delta C_{d_q}(t,\eta_t) \end{bmatrix} =$$

$$\begin{bmatrix} M_a(\eta_t) \\ M_c(\eta_t) \end{bmatrix} \Delta(t,\eta_t)\,[N_a(\eta_t) \quad N_{d_1}(\eta_t)\cdots N_{d_q}(\eta_t)]$$

$$\forall \Delta^t(t,\eta_t)\Delta(t,\eta_t) \le I\,, \qquad \forall t \tag{7.6}$$

$$\begin{aligned} F_{\Delta o}(t,\eta_t) &= F_o(\eta_t) + \Delta F_o(t,\eta_t) \\ &= F_o(\eta_t) + M_a(\eta_t)\Delta(t)N_f(\eta_t) \end{aligned} \tag{7.7}$$

where $x(t) \in \mathbb{R}^n$ is the system state , $u(t) \in \mathbb{R}^v$ is the control input , $y(t) \in \mathbb{R}^m$ is the measurement output , $z(t) \in \mathbb{R}^h$ is the controlled output , $w(t) \in \mathcal{L}_2[0,\infty)$ is the disturbance input and $z(t) \in \mathbb{R}^r$ is the controlled output which belongs to $\mathcal{L}_2\big[(\Omega,\mathcal{F},\mathbf{P}),[0,\infty)\big]$. The matrix $\Delta(t,\eta_t) \in \mathbb{R}^{\alpha\times\beta}$ is unknown with Lebesgue measurable elements. In the sequel, we let

$$\mathcal{Z}_r \triangleq \{1,\cdots,r\}$$

be the set of the r first positive integers and introduce

$$\tau_M = -\max\{\tau_k, k \in \mathcal{Z}_q\}$$

$$\delta_M = -\max\{\delta_j, j \in \mathcal{Z}_p$$

. The initial condition is specified as $\beta_o \triangleq \langle x(0), x(s)\rangle = \langle x_o, \phi(s)\rangle$, where $\phi(\cdot) \in \mathcal{L}_2[-\max\{\tau_k, k \in \mathcal{Z}_q\}, 0]$. Frequently the term

$$\mathcal{M}(x_t) : \mathbb{C}[-\tau,0] \to \mathbb{R}^n \triangleq x(t) - \sum_{j=1}^{p} B_j(\eta_t)x(t-\pi_j)$$

is called the *difference operator* and it offers a fundamental role in the analytical development throughout this chapter.

In the absence of uncertainties ($\Delta(.) \equiv 0$), we obtain the following nominal neutral system

$$(\Sigma_n) \quad \mathcal{M}(\dot{x}_t) \triangleq \dot{x}(t) - \sum_{j=1}^{p} B_j(\eta_t)\dot{x}(t-\pi_j) = A_o(t,\eta_t)x(t)$$

$$
\begin{aligned}
&+ \quad \sum_{k=1}^{q} A_{d_k}(\eta_t)x(t-\tau_k) \\
&+ \quad F_o(t,\eta_t)u(t) + N(\eta_t)w(t) && (7.8) \\
x(\vartheta) &= \phi(\vartheta) \in \mathbb{C}([-\max\{\tau_k, k \in \mathcal{Z}_q\}, 0], \mathbb{R}^n) && (7.9) \\
y(t) &= C_o(t,\eta_t)x(t) + \sum_{k=1}^{q} C_{d_k}(\eta_t)x(t-\tau_k) \\
&+ \quad M(\eta_t)w(t) && (7.10) \\
z(t) &= L(\eta_t)x(t) && (7.11)
\end{aligned}
$$

For each possible value $\eta_t = i \in \mathcal{S}$, we will denote the system matrices of $(\Sigma_{\Delta n})$ associated with mode i by

$$
\begin{aligned}
A_o(\eta_t) &\triangleq A_o(i) \ , \ A_{d_1}(\eta_t) \triangleq A_{d_1}(i), \cdots, A_{d_k}(\eta_t) \triangleq A_{d_q}(i) \\
N(\eta_t) &\triangleq N(i) \ , \ C_o(\eta_t) \triangleq C_o(i) \\
C_{d_1}(\eta_t) &\triangleq C_{d_1}(i), \cdots, C_{d_k}(\eta_t) \triangleq C_{d_q}(i) \ , \ M(\eta_t) \triangleq M(i) \\
B_1(\eta_t) &\triangleq B_1(i), \cdots, B_p(\eta_t) \triangleq B_p(i) \ , \ M_a(\eta_t) \triangleq M_a(i) \\
N_f(\eta_t) &\triangleq N_f(i) \ , \ N_{d_1}(\eta_t) \triangleq N_{d_1}(i) \cdots N_{d_q}(\eta_t) \triangleq N_{d_q}(i) \\
M_c(\eta_t) &\triangleq M_c(i) \ , \ N_a(\eta_t) \triangleq N_a(i) && (7.12)
\end{aligned}
$$

where $A_o(i) \in \mathbb{R}^{n\times n}$, $A_{d_k}(i) \in \mathbb{R}^{n\times n}; k \in \mathcal{Z}_q$, $B_j(i) \in \mathbb{R}^{n\times n}; j \in \mathcal{Z}_p$, $C_o(i) \in \mathbb{R}^{m\times n}$, $C_{d_k}(i) \in \mathbb{R}^{m\times n}; k \in \mathcal{Z}_q$, $F_o(i) \in \mathbb{R}^{n\times v}$, $N(i) \in \mathbb{R}^{n\times r}$, $M(i) \in \mathbb{R}^{m\times r}$ and $L(i) \in \mathbb{R}^{h\times n}$ are known real constant matrices which describe the nominal system of $(\Sigma_{\Delta n})$. Also, $M_a(i) \in \mathbb{R}^{n\times\alpha}$, $M_c(i) \in \mathbb{R}^{p\times\alpha}$, $N_N(i) \in \mathbb{R}^{\beta\times n}$, $N_d(i) \in \mathbb{R}^{\beta\times n}$ and $E_f(i) \in \mathbb{R}^{\beta\times v}$.

The following assumptions on systems $(\Sigma_{\Delta n})$ and (Σ_n) are recalled:

Assumption 7.1 : $\lambda[A_o(i)] \subset \mathbb{C}^-, \ i \in \mathcal{S}$

Assumption 7.2 : $\sum_{j=1}^{p} ||B_j(i)|| \ < \ 1 \ , \ det[B_j(i)] \neq 0, \ i \in \mathcal{S} \ j \in \mathcal{Z}_p$

Remark 7.1 *: We should note that system (7.1)-(7.4) is a hybrid system in which one state $x(t)$ takes values continuously, and another "state" $\eta_t(t)$ takes values discretely. Being continuous in time and represents a wide class of physical systems thus* **Assumption** *(7.1) is quite standard. On the other hand,* **Assumption** *(7.2) provides a sufficient condition on the eigen spectrum in the discrete space and its major role will be clarified in the sequel. An alternative interpretation of* **Assumption** *(7.2) is that the difference operator $\mathcal{M}(x_t)$ is delay-independently stable. The kind of systems (7.2)-(7.5) can be used to represent many important physical systems subject to random failures and structure changes, such as electric power systems [155], control systems of a solar thermal central receiver, communications systems, aircraft flight control, and manufacturing systems [42, 47, 76, 133]*

Our primary objective in this part of the Chapter is to derive LMI-based sufficient conditions characterizing robust stochastic stability and stabilization and designing robust $\mathcal{H}_\infty$ controllers for the neutral system $(\Sigma_{\Delta n})$ Then extend these designs to the neutral system (Σ_n). Towards our goal, we Let $\mathsf{X}(t, \beta_o, \eta_o)$ denote the trajectory of the state $x(t)$ from the initial state (β_o, η_o) and recall the following definition:

Definition 7.1 *System $\Sigma_{\Delta n}$ is said to be* **robustly stochastically stable weakly delay dependent (RSSWDD)** *given $\tau_{kM} = \max\{\tau_k, k \in \mathcal{Z}_q\}$ if for all finite initial vector function $\phi(\cdot) \in \mathcal{L}_2[-\max\{\tau_k, k \in \mathcal{Z}_q\}, 0]$ defined on the interval $[-\max\{\tau_k, k \in \mathcal{Z}_q\}, 0]$ and initial mode $\eta_o \in \mathcal{S}$*

$$\left\{ \int_0^\infty \mathbb{E}\left\{ ||\mathsf{X}(t, \beta_o, \eta_o)||^2 \right\} dt \right\} < +\infty$$

for all admissible uncertainties satisfying (7.6).

7.2.1 Stability Analysis

The following theorems establish that the stability behavior of system $(\Sigma_{\Delta n})$ or (Σ_n) is related to the existence of a positive definite solution of linear matrix inequalities thereby providing a clear key to designing the feedback controllers. In the sequel, for given matrices $0 < \mathbb{L}_k(i) = \mathbb{L}_k^t(i) \in \mathbb{R}^{n\times n}$, $k \in \mathcal{Z}_q$, $0 < \mathbb{N}_j(i) = \mathbb{N}_j^t(i) \in \mathbb{R}^{n\times n}$, $j \in \mathcal{Z}_p$, $i \in \mathcal{S}$, we introduce for $i \in \mathcal{S}$

$$\begin{aligned}
\hat{\mathbb{L}}_k(i) &= \mathbb{L}_k(i) + \xi_{1_k}(i) \sum_{m=1}^{s} \alpha_{im} \mathbb{L}_{1_k}(m) \\
\bar{\mathbb{L}}_k(i) &= \mathbb{L}_k(i) - \xi_{1_k}^{-1}(i) \sum_{m=1}^{s} \alpha_{im} \mathbb{L}_{1_k}(m) \\
\hat{\mathbb{N}}_j(i) &= \mathbb{N}_j(i) + \xi_{2_j}(i) \sum_{m=1}^{s} \alpha_{im} \mathbb{N}_{2_j}(m) \\
\bar{\mathbb{N}}_j(i) &= \mathbb{N}_j(i) - \xi_{2_j}^{-1}(i) \sum_{m=1}^{s} \alpha_{im} \mathbb{N}_j(m)
\end{aligned} \tag{7.13}$$

such that for some scalars $\xi_{1_k}(i) > 0,\ \xi_{2_j}(i) > 0,\ i \in \mathcal{S},\ k \in \mathcal{Z}_q,\ j \in \mathcal{Z}_p$ we guarantee that $\bar{\mathbb{L}}_k(i) > 0,\ \bar{\mathbb{N}}_j(i) > 0,\ i \in \mathcal{S},\ k \in \mathcal{Z}_q,\ j \in \mathcal{Z}_p$.

Next, for $0 < \mathbb{P}(i) = \mathbb{P}^t(i) \in \mathbb{R}^{n\times n}$, $i \in \mathcal{S}$ define the matrix expressions:

$$\begin{aligned}
\Omega_\Delta(t,i) &= \mathbb{P}(i) A_{\Delta o}(t,i) + \sum_{k=1}^{q} \hat{\mathbb{L}}_k(i) + \sum_{j=1}^{p} \hat{\mathbb{N}}_j(i) \\
\Upsilon_\Delta(t,i) &= \Omega_\Delta(i) + A_{\Delta o}^t(t,i)\mathbb{P}(i) + \sum_{m=1}^{s} \alpha_{im} \mathbb{P}(m) \\
\Xi(i) &\triangleq \sum_{j=1}^{p} \left\{ \Xi_j(i) \right\} \\
&= \sum_{j=1}^{p} \left\{ \bar{\mathbb{N}}_j(i) - B_j^t(i) \left[\hat{\mathbb{N}}_j(i) + \sum_{k=1}^{q} \hat{\mathbb{L}}_k(i) \right] B_j(i) \right\} \\
\mathbb{D}(i) &= [A_{d_1}(i), \cdots, A_{d_q}(i)] \\
\Phi(i) &= diag\left[\bar{\mathbb{L}}_k(i) - \upsilon(i) E_{d_k}^t(i) E_{d_k}(i) \right]
\end{aligned}$$

$$\mathbb{B}(i) = [B_1(i), \cdots, B_p(i)] \tag{7.14}$$

Theorem 7.1 *Subject to* **Assumptions** *(7.1)-(7.2), the neutral system* $(\Sigma_{\Delta n})$ *with* $w \equiv 0$ *and* $u \equiv 0$ *is* **(RSSWDD)** *given* τ_M & δ_M *if for given scalars* $\xi_{1_k}(i) > 0$, $\xi_{2_j}(i) > 0$, $i \in \mathcal{S}$, $k \in \mathcal{Z}_q$, $j \in \mathcal{Z}_p$ *and matrices* $0 < \mathbb{L}(i) = \mathbb{L}^t(i) \in \mathbb{R}^{n\times n}$, $i \in \mathcal{S}$ *and* $0 < \mathbb{N}(i) = \mathbb{N}^t(i) \in \mathbb{R}^{n\times n}$, $i \in \mathcal{S}$ *satisfying (7.13), there exist matrices* $0 < \mathbb{P}(i) = \mathbb{P}^t(i) \in \mathbb{R}^{n\times n}$, $i \in \mathcal{S}$ *and scalars* $\varepsilon(i) > 0$, $\varrho(i) > 0$, $\upsilon(i) > 0$, $i \in \mathcal{S}$ *satisfying the following LMIs for all* $i \in \mathcal{S}$:

$$\begin{bmatrix} \Upsilon_t & \Lambda_a(i) & \Lambda_n(i) & \mathbb{P}(i)\mathbb{D}(i) \\ \Lambda_a^t(i) & -\Psi_a(i) & 0 & 0 \\ \Lambda_n^t(i) & 0 & \begin{array}{c}-\Xi(i)+ \\ \varepsilon(i)N_a^t(i)N_a(i)\end{array} & 0 \\ \mathbb{D}^t(i)\mathbb{P}(i) & 0 & 0 & -\Phi(i) \end{bmatrix} < 0 \tag{7.15}$$

where

$$\begin{aligned} \Upsilon_t(i) &= \mathbb{P}(i)A_o(i) + A_o^t(i)\mathbb{P}(i) + \sum_{k=1}^{q} \hat{\mathbb{L}}_k(i) \\ &+ \sum_{j=1}^{p} \hat{\mathbb{N}}_j(i) + \sum_{m=1}^{s} \alpha_{im}\mathbb{P}(m) + \varrho(i)E_a^t(i)E_a(i) \\ \Lambda_a(i) &= [\mathbb{P}(i)H_a(i) \quad \mathbb{P}(i)H_a(i) \quad \mathbb{P}(i)H_a(i)] \\ \Lambda_n(i) &= \left[\mathbb{P}(i)A_o(i) + \sum_{k=1}^{q} \hat{\mathbb{L}}_k(i) + \sum_{j=1}^{p} \hat{\mathbb{N}}_j(i)\right]\mathbb{B}(i) \\ \Psi_a(i) &= diag[\varrho(i)I \quad \varepsilon(i)I \quad \upsilon(i)I] \end{aligned} \tag{7.16}$$

for all admissible uncertainties satisfying (7.6).

Proof: Let $\mathbf{x}_s(t) \triangleq x(s+t)$, $t - \tau_{\eta_t} \leq s \leq t$ and define the process $\{(\mathbf{x}(t), \eta_t),\ t \geq 0\}$ over the state space $\bar{\mathcal{C}}$. It should be observed that $\{(\mathbf{x}(t), \eta_t),\ t \geq 0\}$ is strong Markovian [78]. For $\eta_t = i \in \mathcal{S}$ and given $0 < \mathbb{L}_k(i) = \mathbb{L}_k^t(i)$, $k \in \mathcal{Z}_q$, $0 < \mathbb{N}_j(i) = \mathbb{N}_j^t(i)$, $k \in \mathcal{Z}_p$, let the Lyapunov functional $V(\cdot) : \mathbb{R}^n \times \mathbb{R}_+ \times \mathcal{S} \to \mathbb{R}_+$ be selected as

$$V(t, x, \eta_t = i) \triangleq V(t, x, i) = \mathcal{M}(x_t)^t\mathbb{P}(i)\mathcal{M}(x_t)$$

$$
\begin{aligned}
&+ \sum_{k=1}^{q} \int_{-\tau_k}^{0} x^t(t+\theta)\mathbb{L}_k(i)x(t+\theta)\, d\theta \\
&+ \sum_{j=1}^{p} \int_{-\pi_j}^{0} x^t(t+\theta)\mathbb{N}_j(i)x(t+\theta)\, d\theta \qquad (7.17)
\end{aligned}
$$

The weak infinitesimal operator $\Im_a^x[\cdot]$ of the process $\{x(t), \eta_t, t \geq 0\}$ for system (7.2) at the point $\{t, x, i\}$ is given by [78]:

$$
\Im_a^x[V] = \partial V/\partial t + \dot{\mathcal{M}}^t(x_t)(t)\, \partial V/\partial x\, |_{\eta_t=i} + \sum_{m=1}^{s} \alpha_{im} V(t, x, i, m) \qquad (7.18)
$$

Using (7.2) into (7.17)-(7.18) and manipulating the terms we get:

$$
\begin{aligned}
\Im_a^\zeta[V] &= \mathcal{M}^t(x_t)\mathbb{P}(i)\Big[A_{\Delta o}(t,i)x(t) + \sum_{k=1}^{q} A_{\Delta d_k}(t,i)x(t-\tau_k)\Big] \\
&+ \Big[A_{\Delta o}(t,i)x(t) + \sum_{k=1}^{q} A_{d_k}(t,i)x(t-\tau_k)\Big]^t \mathbb{P}\mathcal{M}(x_t) \\
&+ \mathcal{M}^t(x_t) \sum_{m=1}^{s} \alpha_{im}\mathbb{P}(m)\mathcal{M}(x_t) \\
&+ x^t(t)\Big(\sum_{k=1}^{q} \mathbb{L}_k(i) + \sum_{j=1}^{p} \mathbb{N}_j(i)\Big)x(t) \\
&- \sum_{k=1}^{q} x^t(t-\tau_k)\mathbb{L}_k(i)x(t-\tau_k) - \sum_{j=1}^{p} x^t(t-\pi_j)\mathbb{N}_j(i)x(t-\pi_j) \\
&+ \sum_{m=1}^{s} \alpha_{im}\Big[\sum_{k=1}^{q} \int_{-\tau_k}^{0} x^t(t+\theta)\mathbb{L}_k(m)x(t+\theta)d\theta\Big] \\
&+ \sum_{m=1}^{s} \alpha_{im}\Big[\sum_{j=1}^{p} \int_{-\pi_j}^{0} x^t(t+\theta)\mathbb{N}_j(m)x(t+\theta)d\theta\Big] \\
&\leq \mathcal{M}^t(x_t)\mathbb{P}(i)\Big[A_{\Delta o}(t,i)x(t) + \sum_{k=1}^{q} A_{d_k}(i)x(t-\tau_k)\Big] \\
&+ \Big[x^t(t)A_{\Delta o}^t(t,i) + \sum_{k=1}^{q} x^t(t-\tau_k)A_{d_k}^t(i)\Big]\mathbb{P}\mathcal{M}(x_t)
\end{aligned}
$$

$$
\begin{aligned}
&+ \quad \mathcal{M}^t(x_t) \sum_{m=1}^{s} \alpha_{im} \mathbb{P}(m) \mathcal{M}(x_t) \\
&+ \quad x^t(t) \Big(\sum_{k=1}^{q} \hat{\mathbb{L}}_k(i) + \sum_{j=1}^{p} \hat{\mathbb{N}}_j(i) \Big) x(t) \\
&- \quad \sum_{k=1}^{q} x^t(t-\tau_k) \bar{\mathbb{L}}_k(i) x(t-\tau_k) \\
&- \quad \sum_{j=1}^{p} x^t(t-\pi_j) \bar{\mathbb{N}}_j(i) x(t-\pi_j) \qquad (7.19)
\end{aligned}
$$

Further matrix manipulations of (7.19) using (7.14) yield:

$$
\begin{aligned}
\Im_a^x[V] \quad \leq & \quad \mathcal{M}^t(x_t) \Upsilon_\Delta(t,i) \mathcal{M}(x_t) \\
+ & \quad \mathcal{M}^t(x_t) \Omega_\Delta(t,i) \sum_{j=1}^{p} B_j(i) x(t-\pi_j) \\
+ & \quad \sum_{j=1}^{p} x^t(t-\pi_j) B_j^t(i) \Omega_\Delta^t(t,i) \mathcal{M}(x_t) \\
+ & \quad \mathcal{M}^t(x_t) \mathbb{P}(i) \sum_{k=1}^{q} A_{\Delta d_k}(t,i) x(t-\tau_k) \\
+ & \quad \sum_{k=1}^{q} x^t(t-\tau_k) A_{\Delta d_k}^t(t,i) \mathbb{P} \mathcal{M}(x_t) \\
- & \quad \sum_{k=1}^{q} x^t(t-\tau_k) \bar{\mathbb{L}}_k(i) x(t-\tau_k) \\
- & \quad \sum_{j=1}^{p} x^t(t-\pi_j) \Xi_j(i) \sum_{j=1}^{p} x(t-\pi_j) \\
\leq & \quad \mathcal{M}^t(x_t) \Big\{ \Upsilon_\Delta(t,i) + \Omega_\Delta(t,i) \mathbb{B}(i) \Xi^{-1}(i) \mathbb{B}^t(i) \Omega_\Delta^t(t,i) \\
+ & \quad \mathbb{P}(i) \sum_{k=1}^{q} A_{\Delta d_k}(t,i) \bar{\mathbb{L}}_k^{-1}(i) A_{\Delta d_k}^t(t,i) \mathbb{P}(i) \Big\} \mathcal{M}(x_t) \qquad (7.20)
\end{aligned}
$$

Using **Facts 1-2** and (7.16), it follows from (7.20) for some scalars $\varepsilon(i) >$

$0,\ \upsilon(i) > 0,\ \varrho(i) > 0,\ i \in \mathcal{S}$ that:

$$\begin{aligned}
\Im_a^x[V] &\leq \mathcal{M}^t(x_t)\Big\{\Upsilon_t + \varrho^{-1}(i)\mathbb{P}(i)M_a(i)M_a^t(i)\mathbb{P}(i) \\
&+ [\varepsilon^{-1}(i) + \upsilon^{-1}(i)]\mathbb{P}(i)M_a(i)M_a^t(i)\mathbb{P}(i) \\
&+ \Lambda_n(i)\Big[\Xi(i) - \varepsilon(i)N_a^t(i)N_a(i)\Big]^{-1}\Lambda_n^t(i) \\
&+ \mathbb{P}(i)\mathbb{D}(i)\Phi^{-1}(i)\mathbb{D}(i)\mathbb{P}(i)\Big\}\mathcal{M}(x_t) \\
&\triangleq \mathcal{M}^t(x_t)\Pi(i)\mathcal{M}(x_t)
\end{aligned} \tag{7.21}$$

By the Schur complements, inequality (7.21) is equivalent to LMIs (7.14)-(7.15) from which we conclude that for all admissible uncertainties satisfying (7.6)

$$\Im_a^x[V] < 0 \ \ \forall \ \ x \neq 0 \quad \& \quad \Im_a^x[V] \ \leq \ 0 \ \ \forall \ x$$

Since

$$||x(t+\beta)|| \ \leq \varphi||x(t)|, \quad \forall \beta \in [-\tau_M, 0]$$

and some $\varphi > 0$ [85], it follows from (7.17) that

$$\begin{aligned}
V(t,x,i) &\leq \mathcal{M}^t(x_t)\mathbb{P}(i)\mathcal{M}(x_t) + \mu||x||^2 \\
\mu &= \varphi\Big(\max_i \lambda_M[\mathbb{P}(i)] + \sum_{k=1}^{q} \tau_k \lambda_M[\mathbb{L}_k(i)] \\
&+ \sum_{j=1}^{p} \pi_j \lambda_M[\mathbb{N}_j(i)]\Big)
\end{aligned}$$

Therefore, for all $x \neq 0$, we have

$$\begin{aligned}
\frac{\Im_a^x[V]}{V(t,x,i)} &\leq \frac{\mathcal{M}^t(x_t)\Pi(i)\mathcal{M}(x_t)}{\mathcal{M}^t(x_t)\mathbb{P}(i)\mathcal{M}(x_t) + \mu||x||^2} \leq -\sigma \\
&\triangleq -\min_{i \in \mathcal{S}}\left\{\frac{\lambda_m[-\Pi(i)](1 + \varphi\sum_{j=1}^{p}||B_j(i)||)^2}{\lambda_M[\mathbb{P}(i)](1 + \varphi\sum_{j=1}^{p}||B_j(i)||)^2 + \mu}\right\}
\end{aligned} \tag{7.22}$$

It is readily seen from (7.22) that $\sigma > 0$ and hence we get

$$\Im_a^{\zeta}[V] \ \leq -\sigma\, V(t,x,i)$$

Then, it follows from [78], by using the Gronwall-Bellman lemma [98] and letting $x(t=0,\phi,i)=x_o$, that

$$\mathbb{E}[V(t,x,i)|\phi,\eta_o] \;\leq\; e^{-\sigma\, t}\, V(t,x_o,\eta_o) \tag{7.23}$$

Therefore

$$\begin{aligned}
\mathbb{E}[V(t,x,i)|\phi,\eta_o] &= \mathbb{E}\Big\{\mathcal{M}^t(x_t)\mathbb{P}(i)\mathcal{M}(x_t) \\
&+ \sum_{k=1}^{q}\int_{-\tau_k}^{0} x^t(t+\theta)\mathbb{L}_k(i)x(t+\theta)d\theta|\phi,\eta_o \\
&+ \sum_{j=1}^{p}\int_{-\pi_j}^{0} x^t(t+\theta)\mathbb{N}_j(i)x(t+\theta)d\theta|\phi,\eta_o\Big\} \\
&= \mathbb{E}\Big\{\mathcal{M}^t(x_t)\mathbb{P}(i)\mathcal{M}(x_t)|\phi,\eta_o\Big\} \\
&+ \mathbb{E}\Big\{\sum_{k=1}^{q}\int_{-\tau_k}^{0} x^t(t+\theta)\mathbb{L}_k(i)x(t+\theta)d\theta|\phi,\eta_o\Big\} \\
&+ \mathbb{E}\Big\{\sum_{j=1}^{p}\int_{-\pi_j}^{0} x^t(t+\theta)\mathbb{N}_j(i)x(t+\theta)d\theta|\phi,\eta_o\Big\} \\
&\leq e^{-\sigma\, t}\, V(t,x_o,\eta_o)
\end{aligned} \tag{7.24}$$

Since

$$\mathbb{E}\sum_{k=1}^{q}\Big\{\int_{-\tau_k}^{0} x^t(t+\theta)\mathbb{L}_k(i)x(t+\theta)d\theta|\phi,\eta_o\Big\} \geq 0$$

and

$$\mathbb{E}\sum_{j=1}^{p}\Big\{\int_{-\pi_j}^{0} x^t(t+\theta)\mathbb{N}_j(i)x(t+\theta)d\theta|\phi,\eta_o\Big\} \geq 0$$

then some algebraic manipulation of (7.24) yields:

$$\begin{aligned}
&\mathbb{E}\Big\{\mathcal{M}^t(x_t)\mathbb{P}(i)\mathcal{M}(x_t)|\phi,\eta_o\Big\} \\
\leq\;& e^{-\sigma\, t}V(t,x_o,\eta_o) \Longrightarrow
\end{aligned}$$

$$\begin{aligned}
& \mathbb{E}\Big\{ \int_0^{\mathcal{T}} \mathcal{M}^t(x_t)\mathbb{P}(i)\mathcal{M}(x_t)dt|\phi,\eta_o \Big\} \\
\leq\; & \Big[\int_0^{\mathcal{T}} e^{-\sigma\, t}dt\Big] V(x_o,\eta_o) = \frac{1}{\sigma}[e^{-x\,\mathcal{T}} \;-\; 1]\; V(t,x_o,\eta_o) \Longrightarrow \\
& \lim_{\mathcal{T}\to\infty} \;\mathbb{E}\Big\{ \int_0^{\mathcal{T}} \mathcal{M}^t(x_t)\mathbb{P}(i)\mathcal{M}(x_t)dt|\phi,\eta_o \Big\} \\
\leq\; & \frac{1}{\sigma}x_o^t\mathbb{P}(i)x_o + \frac{1}{\sigma}\Big(\sum_{k=1}^{q} \tau_k \mathbb{L}_k(i)||x(t+\theta)||_*^2 \\
+\; & \sum_{j=1}^{p} \pi_j \mathbb{N}_j(i)||x(t+\delta)||_*^2 \Big) \\
& \forall\theta \in [-\tau_M, 0] \;\;,\;\; \forall\delta \in [-\delta_M, 0]
\end{aligned} \tag{7.25}$$

Now, let

$$\bar{\mathbb{P}}(i) \;=\; \max_{i\in\mathcal{S}} \left\{ \frac{[\mathbb{P}(i)]||x_o||^2 + \varphi\big(\sum_{k=1}^{q} \tau_k[\mathbb{L}_k(i)] + \sum_{j=1}^{p} \pi_j[\mathbb{N}_j(i)]\big)||x(t)||^2}{\sigma[\mathbb{P}(i)]||x_o||^2} \right\}$$

it follows from (7.25) for $i \in \mathcal{S}$ that

$$\begin{aligned}
& \lim_{\mathcal{T}\to\infty} \;\mathbb{E}\Big\{ \int_0^{\mathcal{T}} x^t(t)x(t)dt|\phi,\eta_o \Big\} \\
\leq\; & x_o^t\lambda_M(\bar{\mathbb{P}}(i))x_o < +\infty
\end{aligned}$$

which, in the light of **Definition** 7.1, shows that system $(\Sigma_{\Delta n})$ is **RSSWDD**. $\nabla\nabla\nabla$

Theorem 7.2 *Subject to* **Assumptions** *(7.1)-(7.2), the neutral system* (Σ_n) *with* $w \equiv 0$ *and* $u \equiv 0$ *is* **stochastically stable and weakly delay dependent (SSWDD)** *given* τ_M & δ_M *if for given scalars* $\xi_{1_k}(i) > 0,\; \xi_{2_j}(i) > 0,\; i \in \mathcal{S}\;,\; k \in \mathcal{Z}_q,\; j \in \mathcal{Z}_p$ *and matrices* $0 < \mathbb{L}(i) = \mathbb{L}^t(i) \in \mathbb{R}^{n\times n},\; i \in \mathcal{S}$ *and* $0 < \mathbb{N}(i) = \mathbb{N}^t(i) \in \mathbb{R}^{n\times n},\; i \in \mathcal{S}$ *satisfying (7.13), there exist matrices* $0 < \mathbb{P}(i) = \mathbb{P}^t(i) \in \mathbb{R}^{n\times n},\; i \in \mathcal{S}$ *satisfying the following LMIs for all* $i \in \mathcal{S}$*:*

$$\begin{bmatrix} \Upsilon_{to} & \Lambda_n(i) & \mathbb{P}(i)\mathbb{D}(i) \\ \Lambda_n^t(i) & -\Xi(i) & 0 \\ \mathbb{D}^t(i)\mathbb{P}(i) & 0 & -\Phi_o(i) \end{bmatrix} < 0 \tag{7.26}$$

$$
\begin{aligned}
\Upsilon_{to} &= I\!P(i)A_o(i) + A_o^t(i)I\!P(i) + \sum_{k=1}^{q} \hat{I\!L}_k(i) \\
&+ \sum_{j=1}^{p} \hat{I\!N}_j(i) + \sum_{m=1}^{s} \alpha_{im} I\!P(m) \\
\Phi_o &= \sum_{k=1}^{q} \bar{I\!L}_k
\end{aligned} \tag{7.27}
$$

Proof: Follows from **Theorem** (7.1) by setting $N_a \equiv 0,\; N_{d_k} \equiv 0, M_a \equiv 0$. $\nabla\nabla\nabla$

Remark 7.2 : *The need for* **Assumption** *(7.2) is clearly evident from (7.14)-(7.15) and (7.26) in which case the conditions $\Xi_j(i) > 0,\ i \in \mathcal{S}$ are required and the corresponding result reveals a discrete Lyapunov inequality.*

Remark 7.3 *It should be remarked that both* **Theorems** *(7.1) and (7.2) offer new analytical developments for the class of neutral-type dynamical systems under consideration. The results are conveniently cast into LMI format for which the* **MATLAB-LMI** *software is readily available [57]. The generality of these results is readily evident as they encompass several available published work. Specifically, in the special case $B(i) \equiv 0 \Rightarrow \mathcal{M}(.) = x_t$, systems $(\Sigma_{\Delta A})$ and (Σ_A) become of retarded-type for which* **Theorems** *7.1 and 7.2 retrieve the results of [114, 149]. More importantly,* **Theorem** *3.2 with $p = 1, q = 1, i = 1$ recovers the result of [161] and* **Theorems** *3.1-3.2 with $p = 1, q = 1, i = 1$ and $\tau \equiv \pi$ corresponds to [99]. Other special cases could be readily derived as well.*

Definition 7.2 *System $(\Sigma_{\Delta n})$ is said to be* **RSSWDD with a disturbance attenuation** *γ if for zero initial vector function $\phi \equiv 0$ and initial mode $\eta_o \in \mathcal{S}$*

$$
||z(t)||_{E_2} \;:=\; I\!E\left[\int_0^\infty z^t(t)z(t)dt\right]^{1/2} \;<\; \gamma\, ||w(t)||_2
$$

for all $0 \neq w(t) \in \mathcal{L}_2[0,\infty)$ and for all admissible uncertainties satisfying (7.6).

Theorem 7.3 *Subject to* **Assumptions** *(7.1)-(7.2), the neutral system* $(\Sigma_{\Delta n})$ *with* $u \equiv 0$ *is* **RSSWDD with a disturbance attenuation** γ *given* τ_M & δ_M *if for given scalars* $\xi_{1_k}(i) > 0,\ \xi_{2_j}(i) > 0,\ i \in \mathcal{S}\ ,\ k \in \mathcal{Z}_q,\ j \in \mathcal{Z}_p$ *and matrices* $0 < \mathbb{L}(i) = \mathbb{L}^t(i) \in \mathbb{R}^{n \times n},\ i \in \mathcal{S}$ *and* $0 < \mathbb{N}(i) = \mathbb{N}^t(i) \in \mathbb{R}^{n \times n},\ i \in \mathcal{S}$ *satisfying (7.13), there exist matrices* $\mathbb{P}(i) = \mathbb{P}^t(i) > 0,\ i \in \mathcal{S}$ *and scalars* $\varepsilon(i) > 0,\ \upsilon(i) > 0,\ \varrho(i) > 0,\ i \in \mathcal{S}$*, satisfying the following LMIs for all* $i \in \mathcal{S}$*:*

$$\begin{bmatrix} \Upsilon_{tw} & \Lambda_a(i) & \Lambda_n(i) & \Lambda_{nd} \\ \Lambda_a^t(i) & -\Psi_a(i) & 0 & 0 \\ \Lambda_n^t(i) & 0 & \begin{array}{c} -\Xi(i)+ \\ \varepsilon(i) N_a^t(i) N_a(i) \end{array} & 0 \\ \Lambda_{nd}^t & 0 & 0 & -\Lambda_{dd} \end{bmatrix} < 0 \qquad (7.28)$$

where

$$\begin{aligned} \Omega_{\Delta w}(t,i) &= \Omega_{\Delta}(t,i) + L^t(i)L(i) \\ \Gamma_j(i) &= \Xi_j(i) - B_j^t(i)L^t(i)L(i)B_j(i) \\ \Gamma(i) &= \sum_{j=1}^{p} \Gamma_j(i) \\ \Upsilon_{\Delta w}(t,i) &= \Upsilon_{\Delta}(t,i) + L^t(i)L(i) \\ \Upsilon_{tw}(i) &= \Upsilon_t(i) + L^t(i)L(i) \\ \Lambda_{nd} &= [\mathbb{P}(i)\mathbb{D}(i) \quad \mathbb{P}(i)N(i)] \\ \Lambda_{dd} &= diag[-\Phi(i) \quad -\gamma^2 I] \end{aligned} \qquad (7.29)$$

for all admissible uncertainties satisfying (7.6).

Proof: The stochastic stability follows from **Theorem** 7.1. We need to show now that system (Σ_n) has a disturbance attenuation γ. Let the Lyapunov functional $V(t, x, \eta_t)$, for $\eta_t = i$ be given by (7.17). By evaluating the weak infinitesimal operator $\Im^x[\cdot]$ of the process $\{x(t), \eta_t, t \geq 0\}$ for system (7.1)-(7.5) at the point $\{t, x, \eta_t\}$ using (7.19) and manipulating we get

$$\Im^x[V] \;\leq\; \Im_a^x[V]$$

$$
\begin{aligned}
&+ \quad \mathcal{M}^t(x_t)\mathbb{P}(i)N(i)w(t) \; + \; w^t(t)N^t(i)\mathbb{P}(i)\mathcal{M}(x_t) \\
&\triangleq \quad \Im_b^x[V]
\end{aligned} \tag{7.30}
$$

Now, we introduce the performance measure

$$
\mathcal{J}(w) = \mathbb{E}\left\{ \int_0^\infty [z^t(t)z(t) \; - \; \gamma^2 w^t(t)w(t) \;]dt \right\}
$$

By Dynkin's formula [78], one has

$$
\mathbb{E}\left\{ \int_0^\infty \Im_b^x[V]dt \right\} = \mathbb{E}\{V(t,x,\eta_t)|_{t=\infty}\} - V(t,x,\eta_t)|_{t=0} \geq 0
$$

On using (7.5) and (7.14), we obtain:

$$
\begin{aligned}
\mathcal{J}(x) \;\; &= \;\; \mathbb{E}\left\{ \int_0^\infty [z^t(t)z(t) \; - \; \gamma^2 w^t(t)w(t) \; + \Im_b^x[V] - \Im_b^x[V]]dt \right\} \\
&\leq \;\; \mathbb{E}\left\{ \int_0^\infty [x^t(t)L^t(i)L(i)x(t) \; - \; \gamma^2 w^t(t)w(t) \; + \Im_b^x[V]]dt \right\} \\
&< \;\; \mathbb{E}\left\{ \int_0^\infty \Big[\mathcal{M}^t(x_t)\Upsilon_{\Delta w}(t,i)\mathcal{M}(x_t) \right. \\
&+ \;\; \mathcal{M}^t(x_t)\Omega_{\Delta w}(t,i)\sum_{j=1}^{p} B_j(i)x(t-\pi_j) \\
&+ \;\; \sum_{j=1}^{p} x^t(t-\pi_j)B_j^t(i)\Omega_{\Delta w}^t(t,i)\mathcal{M}(x_t) \\
&+ \;\; \mathcal{M}^t(x_t)\mathbb{P}(i)\sum_{k=1}^{q} A_{\Delta d_k}(t,i)x(t-\tau_k) \\
&+ \;\; \sum_{k=1}^{q} x^t(t-\tau_k)A_{\Delta d_k}^t(t,i)\mathbb{P}\mathcal{M}(x_t) \\
&- \;\; \sum_{k=1}^{q} x^t(t-\tau_k)\bar{\mathbb{L}}_k(i)x(t-\tau_k) \\
&+ \;\; \mathcal{M}^t(x_t)\mathbb{P}(i)N(i)w(t) \; + \; w^t(t)N^t(i)\mathbb{P}(i)\mathcal{M}(x_t) + \\
&- \;\; \left. \sum_{j=1}^{p} x^t(t-\pi_j)\Gamma_j(i)\sum_{j=1}^{p} x(t-\pi_j) - \gamma^2 w^t(t)w(t) \Big] dt \right\}
\end{aligned}
$$

$$
\begin{aligned}
&\leq \mathbb{E}\Big\{ \int_0^\infty \Big[\mathcal{M}^t(x_t) \Big(\Upsilon_{\Delta w}(t,i) + \gamma^{-2} \mathbb{P}(i) N(i) N^t(i) \mathbb{P}(i) \\
&+ \mathbb{P}(i) \sum_{k=1}^{q} A_{\Delta d_k}(t,i) \bar{\mathbb{L}}_k^{-1}(i) A_{\Delta d_k}^t(t,i) \mathbb{P}(i) \Big) \mathcal{M}(x_t) \Big] dt \Big\} \\
&+ \Omega_{\Delta w}(t,i) \mathbb{B}(i) \Gamma^{-1}(i) \mathbb{B}^t(i) \Omega_{\Delta w}^t(i) \qquad (7.31)
\end{aligned}
$$

Using **Facts 1-2**, (7.16) and (7.29), it follows from (7.31) for some scalars $\varepsilon(i) > 0,\ \upsilon(i) > 0,\ \varrho(i) > 0,\ i \in \mathcal{S}$ that:

$$
\begin{aligned}
\mathcal{J}(x) &\leq \mathbb{E}\Big\{ \int_0^\infty \Big[\mathcal{M}^t(x_t) \Big(\Upsilon_{tw} + \gamma^{-2} \mathbb{P}(i) N(i) N^t(i) \mathbb{P}(i) \\
&+ \varrho^{-1}(i) \mathbb{P}(i) M_a(i) M_a^t(i) \mathbb{P}(i) \\
&+ \Lambda_n(i) \Big[\Gamma(i) - \varepsilon(i) N_a^t(i) N_a(i) \Big]^{-1} \Lambda_n^t(i) \\
&+ [\varepsilon^{-1}(i) + \upsilon^{-1}(i)] \mathbb{P}(i) M_a(i) M_a^t(i) \mathbb{P}(i) \\
&+ \mathbb{P}(i) \mathbb{D}(i) \Phi^{-1}(i) \mathbb{D}^t(i) \mathbb{P}(i) \Big) \mathcal{M}(x_t) \Big] dt \Big\} \qquad (7.32)
\end{aligned}
$$

By using (7.28) and the results of **Theorem** 7.1, it follows from inequality (7.32) that $\mathcal{J}(x) < 0$ and by **Definition** (7.2), the proof is completed. ∇∇∇

Theorem 7.4 *Subject to* **Assumptions** *(7.1)-(7.2), the neutral system* (Σ_n) *with* $u \equiv 0$ *is* **RSSWDD with a disturbance attenuation** γ *given* τ_M & δ_M *if for given scalars* $\xi_{1_k}(i) > 0,\ \xi_{2_j}(i) > 0,\ i \in \mathcal{S}\ ,\ k \in \mathcal{Z}_q,\ j \in \mathcal{Z}_p$ *and matrices* $0 < \mathbb{L}(i) = \mathbb{L}^t(i) \in \mathbb{R}^{n \times n},\ i \in \mathcal{S}$ *and* $0 < \mathbb{N}(i) = \mathbb{N}^t(i) \in \mathbb{R}^{n \times n},\ i \in \mathcal{S}$ *satisfying (7.13), there exist matrices* $\mathbb{P}(i) = \mathbb{P}^t(i) > 0,\ i \in \mathcal{S}$ *satisfying the following LMIs for all* $i \in \mathcal{S}$

$$
\begin{bmatrix} \Upsilon_{to} + L^t(i)L(i) & \Lambda_n(i) & \mathbb{P}(i)\mathbb{D}(i) & \mathbb{P}(i)N(i) \\ \Lambda_n^t(i) & -\Xi(i) & 0 & 0 \\ \mathbb{D}^t(i)\mathbb{P}(i) & 0 & -\Phi_o(i) & 0 \\ N^t(i)\mathbb{P}(i) & 0 & 0 & -\gamma^2 I \end{bmatrix} < 0 \qquad (7.33)
$$

Proof: Follows from **Theorem** 7.3 by setting $N_a \equiv 0,\ N_{d_k} \equiv 0, M_a \equiv 0$. ∇∇∇

7.3 Robust Stabilization

In this section, we consider the problem of robust stabilization of systems $(\Sigma_{\Delta n})$ and (Σ_n) using a linear state measurements such that the resulting closed-loop system enjoys some desirable stability properties. Two distinct cases will be analyzed: The first case is memoryless feedback and the second is distributed delayed feedback.

7.3.1 Memoryless Feedback

In this case, the state measurements are instantaneous leading to a control law of the form

$$u(t) \; = \; K_s(i)\, x(t) \tag{7.34}$$

By applying controller (7.34) to system (7.1)-(7.5), we obtain the closed-loop system for $\eta_t = i \in \mathcal{S}$

$$\begin{aligned}
(\Sigma_{\Delta nc}): \; \mathcal{M}(\dot{x}_t) &\triangleq \dot{x}(t) - \sum_{j=1}^{p} B_j(i)\dot{x}(t-\pi_j) \\
&= A_{\Delta co}(t,i)x(t) + \sum_{k=1}^{q} A_{\Delta d_k}(t,i)x(t-\tau_k) \\
&+ \; N(i)w(t) &\quad (7.35) \\
x(\vartheta) &= \phi(\vartheta) \in \mathbb{C}([-\max\{\tau_k, k \in \mathcal{Z}_q\}, 0], \mathbb{R}^n), \\
&\quad \forall\, \vartheta \in [-\max\{\tau_k, k \in \mathcal{Z}_q\}, 0] &\quad (7.36) \\
z(t) &= L(i)x(t) &\quad (7.37)
\end{aligned}$$

Alternatively, by combining (7.8)-(7.11) and (7.34), we obtain the nominal closed-loop system

$$(\Sigma_{nc}): \; \mathcal{M}(\dot{x}_t) \; \triangleq \; \dot{x}(t) - \sum_{j=1}^{p} B_j(i)\dot{x}(t-\pi_j)$$

$$
\begin{aligned}
&= A_{co}(t,i)x(t) + \sum_{k=1}^{q} A_{d_k}(i)x(t-\tau_k) \\
&+ N(i)w(t) \qquad (7.38)
\end{aligned}
$$

$$
\begin{aligned}
x(\vartheta) &= \phi(\vartheta) \in \mathbb{C}([-\max\{\tau_k, k \in \mathcal{Z}_q\}, 0], \mathbb{R}^n), \\
& \quad \forall\, \vartheta \in [-\max\{\tau_k, k \in \mathcal{Z}_q\}, 0] \qquad (7.39) \\
z(t) &= L(i)x(t) \qquad (7.40)
\end{aligned}
$$

where for $\eta_t = i \in \mathcal{S}$

$$
\begin{aligned}
A_{\Delta co}(t,i) &= A_{co}(i) + \Delta A_{co}(t,i) \\
A_{co}(i) &= A_o(i) + F_o(i)K_s(i) \\
\Delta A_{co}(t,i) &= M_a(i)\Delta(t,i)[N_a(i) + N_f(i)K_s(i)] \qquad (7.41)
\end{aligned}
$$

First, we focus on system $(\Sigma_{\Delta nc})$.

It follows from **Theorem** 7.3 that this system is **RSSWDD with a disturbance attenuation** γ if the inequality

$$
\begin{aligned}
&\Upsilon_{\Delta wc}(t,i) + \gamma^{-2}\mathbb{P}(i)N(i)N^t(i)\mathbb{P}(i) + \Omega_{\Delta wc}(t,i)\mathbb{B}(i)\Gamma^{-1}(i)\mathbb{B}^t(i)\Omega^t_{\Delta wc}(i) + \\
&\mathbb{P}(i)\sum_{k=1}^{q} A_{\Delta d_k}(t,i)\bar{\mathbb{L}}_k^{-1}(i)A^t_{\Delta d_k}(t,i)\mathbb{P}(i)\ , < 0 \qquad (7.42)
\end{aligned}
$$

holds for all $i \in \mathcal{S}$ and for all uncertainties satisfying (7.6), where

$$
\begin{aligned}
\Omega_{\Delta wc}(t,i) &= \mathbb{P}(i)A_{\Delta co}(t,i) + \sum_{k=1}^{q} \hat{\mathbb{L}}_k(i) \\
&+ \sum_{j=1}^{p} \hat{\mathbb{N}}_j(i) \\
\Upsilon_{\Delta wc}(t,i) &= \Omega_{\Delta wc}(i) + A^t_{\Delta co}(t,i)\mathbb{P} \\
&+ \sum_{m=1}^{s} \alpha_{im}\mathbb{P}(m) \qquad (7.43)
\end{aligned}
$$

Using **Facts 1-2**, it follows from (7.42) for some scalars $\varepsilon(i) > 0,\ \upsilon(i) > 0,\ \varrho(i) > 0,\ i \in \mathcal{S}$ that:

$$\begin{aligned}
& \Upsilon_{twc} + \gamma^{-2}\mathbb{P}(i)N(i)N^t(i)\mathbb{P}(i) \\
+ \ & \varrho^{-1}(i)\mathbb{P}(i)M_a(i)M_a^t(i)\mathbb{P}(i) + \mathbb{P}(i)\mathbb{D}(i)\Phi^{-1}(i)\mathbb{D}^t(i)\mathbb{P}(i) \\
+ \ & \Lambda_{nc}(i)\left[\Gamma(i) - \varepsilon(i)[N_a(i) + N_f(i)K_s(i)]^t[N_a(i) + N_f(i)K_s(i)]\right]^{-1}\Lambda_{nc}^t(i) \\
+ \ & [\varepsilon^{-1}(i) + \upsilon^{-1}(i)]\mathbb{P}(i)M_a(i)M_a^t(i)\mathbb{P}(i) < 0
\end{aligned} \tag{7.44}$$

where

$$\begin{aligned}
\Upsilon_{twc} &= \mathbb{P}(i)A_{co}(i) + A_{co}^t(i)\mathbb{P}(i) + \sum_{k=1}^{q}\hat{\mathbb{L}}_k(i) \\
&+ \sum_{j=1}^{p}\hat{\mathbb{N}}_j(i) + \sum_{m=1}^{s}\alpha_{im}\mathbb{P}(m) \\
&+ \varrho(i)[N_a(i) + N_f(i)K_s(i)]^t[N_a(i) + N_f(i)K_s(i)] \\
&+ L^t(i)L(i) \\
\Lambda_{nc}(i) &= \left[\mathbb{P}(i)A_{co}(i) + \sum_{k=1}^{q}\hat{\mathbb{L}}_k(i) + \sum_{j=1}^{p}\hat{\mathbb{N}}_j(i)\right]\mathbb{B}(i)
\end{aligned} \tag{7.45}$$

The following theorems establish the main results:

Theorem 7.5 *Subject to* **Assumptions** *(7.1)-(7.2), the neutral system* $(\Sigma_{\Delta nc})$ *is* **RSSWDD via memoryless state-feedback** $u(t) = K_s(i)x(t)$ **with a disturbance attenuation** γ *given* τ_M & δ_M *if for given scalars* $\xi_{1_k}(i) > 0,\ \xi_{2_j}(i) > 0,\ i \in \mathcal{S}\ ,\ k \in \mathcal{Z}_q,\ j \in \mathcal{Z}_p$ *and matrices* $0 < \mathbb{L}(i) = \mathbb{L}^t(i) \in \mathbb{R}^{n\times n},\ i \in \mathcal{S}$ *and* $0 < \mathbb{N}(i) = \mathbb{N}^t(i) \in \mathbb{R}^{n\times n},\ i \in \mathcal{S}$ *satisfying (7.13), there exist matrices* $\mathcal{Y}(i) = \mathcal{Y}^t(i) > 0 \in \mathbb{R}^{n\times n},\ ,\mathcal{Z}(i) = \mathcal{Z}^t(i) > 0 \in \mathbb{R}^{n\times n},\ \mathcal{X}(i) \in \mathbb{R}^{v\times n},\ i \in \mathcal{S}$ *and scalars* $\varepsilon(i) > 0,\ \upsilon(i) > 0,\ \varrho(i) > 0,\ \mu(i) > 0,\ i \in \mathcal{S},$

satisfying the following LMIs for all $i \in \mathcal{S}$*:*

$$\begin{bmatrix} \Pi_a(\mathcal{Y},\mathcal{Z},\mathcal{X},i) & \Lambda_h(i) & \Lambda_g(\mathcal{Z},\mathcal{X},i) & \mathbb{D}(i) & \Pi_b(\mathcal{Y},\mathcal{X},i) \\ \Lambda_h^t(i) & -\Psi_h(i) & 0 & 0 & 0 \\ \Lambda_g^t(\mathcal{Z},\mathcal{X},i) & 0 & -\Psi_g(\mathcal{Z},\mathcal{X},i) & 0 & 0 \\ \mathbb{D}^t(i) & 0 & 0 & -\Phi(i) & 0 \\ \Pi_b^t(\mathcal{Y},\mathcal{X},i) & 0 & 0 & 0 & -\mu(i)I \end{bmatrix} < 0$$

$$\begin{bmatrix} \varrho(i) & 1 \\ 1 & \mu(i) \end{bmatrix} \geq 1 \quad , \quad \begin{bmatrix} \mathcal{Y}(i) & I \\ I & \mathcal{Z}(i) \end{bmatrix} \geq I \tag{7.46}$$

for all admissible uncertainties satisfying (7.6). Moreover, the feedback gain is given by

$$K_s(i) = \mathcal{X}(i)\mathcal{Z}(i) \tag{7.47}$$

where

$$\begin{aligned} \Pi_a(\mathcal{Y},\mathcal{Z},\mathcal{X},i) &= A_o(i)\mathcal{Y}(i) + \mathcal{Y}(i)A_o^t(i) + F_o(i)\mathcal{X}(i) + \mathcal{X}^t(i)F_o^t(i) \\ &+ \mathcal{Y}(i)\Bigg[\sum_{k=1}^{q} \hat{\mathbb{L}}_k(i) + \sum_{j=1}^{p} \hat{\mathbb{N}}_j(i) \\ &+ L^t(i)L(i) + \sum_{m=1}^{s} \alpha_{im}\mathcal{Z}(m)\Bigg]\mathcal{Y}(i) \\ \Lambda_h(i) &= [M_a(i) \quad M_a(i) \quad M_a(i) \quad N(i)] \\ \Pi_b(\mathcal{Y},\mathcal{X},i) &= \mathcal{Y}(i)N_a^t(i) + \mathcal{X}^t(i)N_f^t(i) \\ \Lambda_g(\mathcal{Z},\mathcal{X},i) &= \Bigg[A_o(i) + F_o(i)\mathcal{Z}(i)\mathcal{X}(i) + \mathcal{Y}(i)\Bigg(\sum_{k=1}^{q} \hat{\mathbb{L}}_k(i) + \sum_{j=1}^{p} \hat{\mathbb{N}}_j(i)\Bigg)\Bigg]\mathbb{B}(i) \\ \Psi_g(\mathcal{Z},\mathcal{X},i) &= \Gamma(i) - \varepsilon(i)[N_a(i) + N_f(i)\mathcal{X}(i)\mathcal{Z}(i)]^t \\ &\cdot [N_a(i) + N_f(i)\mathcal{X}(i)\mathcal{Z}(i)] \\ \Psi_h(i) &= [\varepsilon(i)I \quad \upsilon(i) \quad \varrho(i)I \quad \gamma^2 I] \end{aligned} \tag{7.48}$$

Proof: Pre- and post-multiplying (7.44) by $\mathbb{P}(i)$, $i \in \mathcal{S}$, letting $\mathcal{Y}(i) = \mathbb{P}^{-1}(i)$ and using (7.47) we get

$$A_o(i)\mathcal{Y}(i) + \mathcal{Y}(i)A_o^t(i) + F_o(i)K_s(i)\mathcal{Y}(i) + \mathcal{Y}^t(i)K_s^t(i)F_o^t(i)$$

$$
\begin{aligned}
&+ \quad \mathbb{D}^t(i)\Phi^{-1}(i)\mathbb{D}(i) + \gamma^{-2}N(i)N^t(i) \\
&+ \quad \mathcal{Y}(i)\Big[\sum_{k=1}^{q}\hat{\mathbb{L}}_k(i) + \sum_{j=1}^{p}\hat{\mathbb{N}}_j(i) + L^t(i)L(i) + \sum_{m=1}^{s}\alpha_{im}\mathcal{Z}(m)\Big]\mathcal{Y}(i) \\
&+ \quad \Big[A_o(i) + F_o(i)\mathcal{Z}(i)\mathcal{X}(i) + \sum_{k=1}^{q}\hat{\mathbb{L}}_k(i) + \sum_{j=1}^{p}\hat{\mathbb{N}}_j(i)\Big]\mathbb{B}(i)\cdot \\
&\qquad \Big[\Gamma(i) - \varepsilon(i)[N_a(i) + N_f(i)\mathcal{X}(i)\mathcal{Z}(i)]^t[N_a(i) + N_f(i)\mathcal{X}(i)\mathcal{Z}(i)]\Big]^{-1} \\
&\qquad \mathbb{B}^t(i)\Big[A_o(i) + F_o(i)\mathcal{Z}(i)\mathcal{X}(i) + \sum_{k=1}^{q}\hat{\mathbb{L}}_k(i) + \sum_{j=1}^{p}\hat{\mathbb{N}}_j(i)\Big]^t \\
&+ \quad \varrho(i)\Big[\mathcal{Y}(i)N_a^t(i) + \mathcal{Y}^t(i)K_s^t(i)N_f^t(i)][\mathcal{Y}(i)N_a^t(i) + \mathcal{Y}^t(i)K_s^t(i)N_f^t(i)]^t \\
&+ \quad [\varepsilon^{-1}(i) + \upsilon^{-1}(i) +^{-1} \varrho(i)]M_a^t(i)M_a(i) < 0 \qquad (7.49)
\end{aligned}
$$

By **Fact 1** and using (7.46)-(7.47), inequality (7.49) under the equality constraints $\mu(i)\varrho(i) = 1$, $\mathcal{Y}(i)\mathcal{Z}(i) = I$, can be readily arranged into the LMIs (7.46) which completes the proof. $\nabla\nabla\nabla$

By deleting out the uncertainties, **Theorem** 7.5 specializes into

Theorem 7.6 *Subject to* **Assumptions** *(7.1)-(7.2), the neutral system* (Σ_{nc}) *is* **RSSWDD via memoryless state-feedback** $u(t) = K_s(i)x(t)$ **with a disturbance attenuation** γ *given* τ_M & δ_M *if for given scalars* $\xi_{1_k}(i) > 0$, $\xi_{2_j}(i) > 0$, $i \in \mathcal{S}$, $k \in \mathcal{Z}_q$, $j \in \mathcal{Z}_p$ *and matrices* $0 < \mathbb{L}(i) = \mathbb{L}^t(i) \in \mathbb{R}^{n\times n}$, $i \in \mathcal{S}$ *and* $0 < \mathbb{N}(i) = \mathbb{N}^t(i) \in \mathbb{R}^{n\times n}$, $i \in \mathcal{S}$ *satisfying (7.13), there exist matrices* $\mathcal{Y}(i) = \mathcal{Y}^t(i) > 0 \in \mathbb{R}^{n\times n}$, $,\mathcal{Z}(i) = \mathcal{Z}^t(i) > 0 \in \mathbb{R}^{n\times n}$, $\mathcal{X}(i) \in \mathbb{R}^{v\times n}$, $i \in \mathcal{S}$ *satisfying the following LMIs for all* $i \in \mathcal{S}$

$$
\begin{aligned}
&\begin{bmatrix} \Pi_a(\mathcal{Y},\mathcal{Z},\mathcal{X},i) & \Lambda_g(\mathcal{Z},\mathcal{X},i) & \mathbb{D}(i) & \Pi_b(\mathcal{Y},\mathcal{X},i) \\ \Lambda_g^t(\mathcal{Z},\mathcal{X},i) & -\Gamma(i) & 0 & 0 \\ \mathbb{D}^t(i) & 0 & -\Phi(i) & 0 \\ \Pi_b^t(\mathcal{Y},\mathcal{X},i) & 0 & 0 & -\mu(i)I \end{bmatrix} < 0 \\
&\begin{bmatrix} \mathcal{Y}(i) & I \\ I & \mathcal{Z}(i) \end{bmatrix} \geq I \quad i \in \mathcal{S} \qquad (7.50)
\end{aligned}
$$

Moreover, the feedback gain is given by

$$K_s(i) = \mathcal{X}(i)\mathcal{Z}(i) \tag{7.51}$$

7.3.2 Distributed Feedback

In this case, the state measurements are made up of the delayed states and hence the control law has the form

$$u(t) = \sum_{k=1}^{q} K_k^{+}(i)\, x(t-\tau_k) \tag{7.52}$$

Now, by applying controller (7.52) to system (7.1)-(7.5) we obtain the closed-loop system for $i \in \mathcal{S}$

$$\begin{aligned}
(\Sigma_{\Delta nd})\ \mathcal{M}(\dot{x}_t) &\triangleq \dot{x}(t) - \sum_{j=1}^{p} B_j(i)\dot{x}(t-\pi_j) \\
&= A_{\Delta o}(t,i)x(t) + \sum_{k=1}^{q} A_{\Delta cd_k}(t,i)x(t-\tau_k) \\
&+ N(i)w(t) && (7.53)\\
x(\vartheta) &= \phi(\vartheta) \in \mathbb{C}([-\tau_M, 0], \mathbb{R}^n) && (7.54)\\
z(t) &= L(i)x(t) && (7.55)
\end{aligned}$$

where for $i \in \mathcal{S}$

$$\begin{aligned}
A_{\Delta cd_k}(t,i) &= A_{cd_k}(i) + \Delta A_{cd_k}(t,i)\ ,\quad A_{cd_k}(i) = A_{d_k}(i) + F_o(i)K_k^{+}(i) \\
\Delta A_{cd_k}(t,i) &= M_a(i)\Delta(t,i)[N_d(i) + N_f(i)K_k^{+}(i)] && (7.56)
\end{aligned}$$

On the other hand, by combining (7.8)-(7.11) and (7.52), we obtain the nominal closed-loop system

$$(\Sigma_{nd})\ \mathcal{M}(\dot{x}_t) \triangleq \dot{x}(t) - \sum_{j=1}^{p} B_j(i)\dot{x}(t-\pi_j)$$

$$
\begin{aligned}
&= A_o(t,i)x(t) + \sum_{k=1}^{q} A_{cd_k}(i)x(t-\tau_k) \\
&+ N(i)w(t) \qquad (7.57) \\
x(\vartheta) &= \phi(\vartheta) \in \mathbb{C}([-\tau_M, 0], \mathbb{R}^n) \qquad (7.58) \\
z(t) &= L(i)x(t) \qquad (7.59)
\end{aligned}
$$

One of the direct consequences of **Theorem** 7.3 is that system $(\Sigma_{\Delta nd})$ is **RSSWDD with a disturbance attenuation** γ if the inequality

$$
\Upsilon_{\Delta w}(t,i) + \gamma^{-2}\mathbb{P}(i)N(i)N^t(i)\mathbb{P}(i) + \Omega_{\Delta w}(t,i)\mathbb{B}(i)\Gamma^{-1}(i)\mathbb{B}^t(i)\Omega^t_{\Delta w}(i) + \mathbb{P}(i)\sum_{k=1}^{q} A_{\Delta cd_k}(t,i)\bar{\mathbb{L}}_k^{-1}(i)A^t_{\Delta cd_k}(t,i)\mathbb{P}(i) < 0 \qquad (7.60)
$$

holds for all $i \in \mathcal{S}$ and for all uncertainties satisfying (7.6). Now it follows from (7.60) on using **Facts 1-2** for some scalars $\varepsilon(i) > 0,\ \upsilon(i) > 0,\ \varrho(i) > 0,\ i \in \mathcal{S}$ that:

$$
\begin{aligned}
&\Upsilon_{tw}(i) + \gamma^{-2}\mathbb{P}(i)N(i)N^t(i)\mathbb{P}(i) \\
&+ [\varepsilon^{-1}(i) + \upsilon^{-1}(i) + \varrho^{-1}(i)]\mathbb{P}(i)M_a(i)M_a^t(i)\mathbb{P}(i) \\
&+ \Lambda_n(i)\Big[\Gamma(i) - \varepsilon(i)N_a^t(i)N_a(i)\Big]^{-1}\Lambda_n^t(i) \\
&+ \mathbb{P}(i)\mathbb{D}_d(i)\Phi_d^{-1}(i)\mathbb{D}_d^t(i)\mathbb{P}(i) < 0 \qquad (7.61)
\end{aligned}
$$

where

$$
\begin{aligned}
\mathbb{K}(i) &= \Big[K_1^+(i), \ldots\ldots, K_q^+(i)\Big] \\
\Phi_d(i) &= diag\Big[\bar{\mathbb{L}}_k(i) - \upsilon(i)[N_{d_k}(i) + N_f(i)K_k^+(i)]\Big] \\
\mathbb{D}_d(i) &= \mathbb{D}(i) + F_o(i)\mathbb{K}(i) \qquad (7.62)
\end{aligned}
$$

In the manner of **Theorems** 7.5-7.6, the following two theorems are easily established:

Theorem 7.7 *Subject to* **Assumptions** *(7.1)-(7.2), the neutral system* $(\Sigma_{\Delta nd})$ *is* **RSSWDD via distributed feedback** $u(t) = \sum_{k=1}^{q} K_k^+(i)x(t-\tau_k)$ **with a disturbance attenuation** γ *given* τ_M & δ_M *if for given scalars* $\xi_{1_k}(i) > 0$, $\xi_{2_j}(i) > 0$, $i \in \mathcal{S}$, $k \in \mathcal{Z}_q$, $j \in \mathcal{Z}_p$ *and matrices* $0 < \mathbb{L}(i) = \mathbb{L}^t(i) \in \mathbb{R}^{n\times n}$, $i \in \mathcal{S}$ *and* $0 < \mathbb{N}(i) = \mathbb{N}^t(i) \in \mathbb{R}^{n\times n}$, $i \in \mathcal{S}$ *satisfying (7.13), there exist matrices* $\mathcal{Y}(i) = \mathcal{Y}^t(i) > 0 \in \mathbb{R}^{n\times n}$, $,\mathcal{Z}(i) = \mathcal{Z}^t(i) > 0 \in \mathbb{R}^{n\times n}$, $\mathcal{X}_k(i) \in \mathbb{R}^{v\times n}$, $k \in \mathcal{Z}_q$, $i \in \mathcal{S}$ *and scalars* $\varepsilon(i) > 0$, $v(i) > 0$, $\varrho(i) > 0$, $\mu(i) > 0$, $i \in \mathcal{S}$, *satisfying the following LMIs for all* $i \in \mathcal{S}$

$$\begin{bmatrix} \Pi_c(\mathcal{Y},\mathcal{Z},\mathcal{X},i) & \Lambda_h(i) & \Lambda_d(\mathcal{Z},\mathcal{X},i) & \bar{\mathbb{D}}_d(i) & \mathcal{Y}(i)N_a^t(i) \\ \Lambda_h^t(i) & -\Psi_h(i) & 0 & 0 & 0 \\ \Lambda_d^t(\mathcal{Z},\mathcal{X},i) & 0 & -\Psi_d(i) & 0 & 0 \\ \bar{\mathbb{D}}_d^t(i) & 0 & 0 & -\bar{\Phi}_d(i) & 0 \\ N_a(i)\mathcal{Y}(i) & 0 & 0 & 0 & -\mu(i)I \end{bmatrix} < 0$$

$$\begin{bmatrix} \varrho(i) & 1 \\ 1 & \mu(i) \end{bmatrix} \geq 1 \ , \quad \begin{bmatrix} \mathcal{Y}(i) & I \\ I & \mathcal{Z}(i) \end{bmatrix} \geq I \tag{7.63}$$

for all admissible uncertainties satisfying (7.6). Moreover, the feedback gain is given by

$$K_k^+(i) = \mathcal{X}_k(i)\mathcal{Z}(i) \ , \quad k \in \mathcal{Z}_q \tag{7.64}$$

where

$$\begin{aligned} \Pi_c(\mathcal{Y},\mathcal{Z},\mathcal{X},i) &= A_o(i)\mathcal{Y}(i) + \mathcal{Y}(i)A_o^t(i) \\ &+ \mathcal{Y}(i)\Big[\sum_{k=1}^{q} \hat{\mathbb{L}}_k(i) + \sum_{j=1}^{p} \hat{\mathbb{N}}_j(i) + L^t(i)L(i) \\ &+ \sum_{m=1}^{s} \alpha_{im}\mathcal{Z}(m)\Big]\mathcal{Y}(i) \\ \Psi_d(i) &= \Gamma(i) - \varepsilon(i)N_a^t(i)N_a(i) \\ \Lambda_d(\mathcal{Z},\mathcal{X},i) &= \Big[A_o(i) + \mathcal{Y}(i)\Big(\sum_{k=1}^{q} \hat{\mathbb{L}}_k(i) + \sum_{j=1}^{p} \hat{\mathbb{N}}_j(i)\Big)\Big]\mathbb{B}(i) \end{aligned}$$

$$\begin{aligned}
\bar{\Phi}_d(i) &= diag\Big[\bar{I\!L}_k(i) - \upsilon(i)[E_{d_k}(i) + E_f(i)\mathcal{X}_k(i)\mathcal{Z}(i)]^t \\
&\quad \cdot \; [E_{d_k}(i) + E_f(i)\mathcal{X}_k(i)\mathcal{Z}(i)]\Big] \\
\bar{\mathcal{X}}(i) &= [\mathcal{X}_1(i), ..., \mathcal{X}_q(i)] \\
\bar{I\!D}_d(i) &= I\!D(i) + F_o(i)\bar{\mathcal{X}}(i)\mathcal{Z}(i) \qquad (7.65)
\end{aligned}$$

Theorem 7.8 *Subject to* **Assumptions** *(7.1)-(7.2), the neutral system* (Σ_{nd}) *is* **RSSWDD via distributed feedback** $u(t) = \sum_{k=1}^{q} K_k^+(i)x(t-\tau_k)$ **with a disturbance attenuation** γ *given* τ_M & δ_M *if for given scalars* $\xi_{1_k}(i) > 0,\ \xi_{2_j}(i) > 0,\ i \in \mathcal{S}\ ,\ k \in \mathcal{Z}_q,\ j \in \mathcal{Z}_p$ *and matrices* $0 < I\!L(i) = I\!L^t(i) \in I\!R^{n\times n},\ i \in \mathcal{S}$ *and* $0 < I\!N(i) = I\!N^t(i) \in I\!R^{n\times n},\ i \in \mathcal{S}$ *satisfying (7.13), there exist matrices* $\mathcal{Y}(i) = \mathcal{Y}^t(i) > 0 \in I\!R^{n\times n},\ ,\mathcal{Z}(i) = \mathcal{Z}^t(i) > 0 \in I\!R^{n\times n},\ \mathcal{X}_k(i) \in I\!R^{v\times n},\ k \in \mathcal{Z}_q,\ i \in \mathcal{S}$ *satisfying the following LMIs for all* $i \in \mathcal{S}$*:*

$$\begin{bmatrix} \Pi_c(\mathcal{Y},\mathcal{Z},\mathcal{X},i) & \Lambda_d(i) & \bar{I\!D}_d(i) \\ \Lambda_d^t(i) & -\Gamma(i) & 0 \\ \bar{I\!D}_d^t(i) & 0 & -\bar{\Phi}_d(i) \end{bmatrix} < 0$$

$$\begin{bmatrix} \varrho(i) & 1 \\ 1 & \mu(i) \end{bmatrix} \geq 1 \ , \quad \begin{bmatrix} \mathcal{Y}(i) & I \\ I & \mathcal{Z}(i) \end{bmatrix} \geq I \qquad (7.66)$$

for all admissible uncertainties satisfying (7.6). Moreover, the feedback gain is given by

$$K_k^+(i) = \mathcal{X}_k(i)\mathcal{Z}(i)\ , \quad k \in \mathcal{Z}_q \qquad (7.67)$$

Remark 7.4 *To shed more light on the novelty of the developed results, it is interesting to note that* **Theorems** *(7.7)-(7.8) offer new LMI-based sufficient stability conditions for the class of neutral systems under consideration for which several special cases could be easily derived. This includes the case of equal time-delays* $(q = p,\ \tau_k = \pi_j)$ *and single state-derivative delay* $(p = 1)$*. To further illuminate the generality of these results, we consider the case with* $B_j(i) \equiv$

$0,\ j \in \mathcal{Z}_p$ *corresponding to the class of distributed time-delay systems of retarded type with Markovian jump parameters:*

$$
\begin{aligned}
(\Sigma_{\Delta r}):\ \dot{x}(t) &= A_{\Delta o}(t,\eta_t)x(t) + \sum_{k=1}^{q} A_{\Delta d_k}(t,\eta_t)x(t-\tau_k) + F_{\Delta o}(t,\eta_t)u(t) \\
&+ N(\eta_t)w(t) \\
x(\vartheta) &= \phi(\vartheta) \in \mathbb{C}([-\max\{\tau_k, k \in \mathcal{Z}_q\},0], \mathbb{R}^n), \\
& \quad \forall\, \vartheta \in [-\max\{\tau_k, k \in \mathcal{Z}_q\},0] \\
z(t) &= L(\eta_t)x(t)
\end{aligned}
$$

This system is **RSSWDD via distributed feedback** $u(t) = \sum_{k=1}^{q} K_k^+(i)x(t-\tau_k)$ **with a disturbance attenuation** γ *given* τ_M & δ_M *if for given scalars* $\xi_1(i) > 0,\ \xi_2(i) > 0,\ i \in \mathcal{S}$ *and matrices* $0 < \mathbb{L}(i) = \mathbb{L}^t(i) \in \mathbb{R}^{n \times n},\ i \in \mathcal{S}$ *and* $0 < \mathbb{N}(i) = \mathbb{N}^t(i) \in \mathbb{R}^{n\times n},\ i \in \mathcal{S}$ *satisfying (7.13), there exist matrices* $\mathcal{Y}(i) = \mathcal{Y}^t(i) > 0 \in \mathbb{R}^{n\times n},\ ,\mathcal{Z}(i) = \mathcal{Z}^t(i) > 0 \in \mathbb{R}^{n\times n},\ \mathcal{X}_k(i) \in \mathbb{R}^{v \times n},\ k \in \mathcal{Z}_q,\ i \in \mathcal{S}$ *and scalars* $\varepsilon(i) > 0,\ v(i) > 0,\ \varrho(i) > 0,\ \mu(i) > 0,\ i \in \mathcal{S}$*, satisfying the following LMIs for all* $i \in \mathcal{S}$*:*

$$
\begin{bmatrix}
\Pi_c(\mathcal{Y},\mathcal{Z},\mathcal{X},i) & \Lambda_h(i) & \bar{\mathbb{D}}_d(i) & \mathcal{Y}(i)N_a^t(i) \\
\Lambda_h^t(i) & -\Psi_h(i) & 0 & 0 \\
\bar{\mathbb{D}}_d^t(i) & 0 & -\bar{\Phi}_d(i) & 0 \\
N_a(i)\mathcal{Y}(i) & 0 & 0 & -\mu(i)I
\end{bmatrix} < 0
$$

$$
\begin{bmatrix} \varrho(i) & 1 \\ 1 & \mu(i) \end{bmatrix} \geq 1 \ , \quad \begin{bmatrix} \mathcal{Y}(i) & I \\ I & \mathcal{Z}(i) \end{bmatrix} \geq I
$$

for all admissible uncertainties satisfying (7.6). Moreover, the feedback gain is given by $K_k^+(i) = \mathcal{X}_k(i)\mathcal{Z}(i)\ ,\ k \in \mathcal{Z}_q$ *and the different matrices are as stated earlier. In the context of time-delay systems, this result, in the mannar of Chapter 3, is a new contribution to robust stabilization of distributed state-delay systems. Indeed, similar results could be equally provided for the nominal retarded jumping system.*

7.4 $\mathcal{H}_\infty$-Output Feedback

Having developed results pertaining to stochastic stability and stabilization of uncertain NJS, We are now in a position to proceed one more step and consider the design of an $\mathcal{H}_\infty$-output feedback controller for these systems as given by (7.2)-(7.5). For simplicity in exposition, we treat here the case with $\Delta F_o(t,i) \equiv 0$.

We consider an observer-based output feedback control scheme for $i \in \mathcal{S}$ in the following form:

$$
\begin{aligned}
(\Sigma_f): \quad \mathcal{M}(\dot{\beta}_t) &\triangleq \dot{\beta}(t) - \sum_{j=1}^{p} B_j(\eta_t)\dot{\beta}(t-\pi_j) \\
&= A_O(i)\beta(t) + \sum_{k=1}^{q} A_{d_k}(i)\beta(t-\tau_k) \\
&+ K_O(i)\left[y(t) - C_o(i)\beta(t)\right], \quad \beta(0) = 0 \\
\hat{z}(t) &= L(i)\beta(t) \\
u(t) &= G_O(i)\beta(t)
\end{aligned}
\tag{7.68}
$$

where $A_O(i) \in \mathbb{R}^{n\times n}$, $K_O(i) \in \mathbb{R}^{n\times m}$, $G_O(i) \in \mathbb{R}^{v\times m}$, $i \in \mathcal{S}$ are the gains of the observer-based controller to be designed such that the closed-loop system achieves desirable stability properties for all admissible uncertainties satisfying (7.6).

7.4.1 The Closed-Loop System

A state-space augmented model of the output error, $\tilde{z}(t) = z(t) - \hat{z}(t)$, can be constructed in terms of the augmented state vector and the extended matrix $\mathbb{B}(i)$ for each possible value $i \in \mathcal{S}$

$$
\zeta(t) \triangleq \begin{bmatrix} \beta(t) \\ x(t) \end{bmatrix}, \quad \mathcal{B}_j(i) = \begin{bmatrix} B_j(i) & 0 \\ 0 & B_j(i) \end{bmatrix} \tag{7.69}
$$

From (7.2)-(7.5) and (7.68)-(7.69), the augmented dynamics can be represented by

$$
\begin{aligned}
(\Sigma_{\Delta A}): \qquad \mathcal{M}(\dot{\zeta}_t) &\triangleq \dot{\zeta}(t) - \sum_{j=1}^{p} \mathcal{B}_j(i)\dot{\zeta}(t-\pi_j) \\
&= \mathcal{A}_{\Delta o}(t,i)\zeta(t) + \sum_{k=1}^{q} \mathcal{D}_{\Delta d_k}(t,i)\zeta(t-\tau_k) \\
&+ \mathsf{W}(i)w(t) && (7.70) \\
\tilde{z}(t) &= L_f(i)\zeta(t) && (7.71)
\end{aligned}
$$

where

$$
\begin{aligned}
\mathcal{A}_{\Delta o}(t,i) &= \left[\mathsf{A}_o(i) + \bar{H}_A(i)\Delta(t,i)\bar{E}_A(i)\right] \\
\mathcal{D}_{\Delta d_k}(t,i) &= \left[\mathsf{D}_k(i) + \bar{H}_A(i)\Delta(t,i)\bar{E}_{D_k}(i)\right] \\
\mathsf{A}_o(i) &= \begin{bmatrix} A_O(i) - K_O(i)C_o(i) & K_O(i)C_o(i) \\ F_o(i)G_O(i) & A_o(i) \end{bmatrix} \\
\zeta(t-\tau) &= \begin{bmatrix} \beta(t-\tau) \\ x(t-\tau) \end{bmatrix} \\
\bar{E}_{D_k}(i) &= [0 \quad N_{d_k}(i)]\,,\ \bar{E}_A(i) = [0 \quad N_a(i)] && (7.72) \\
\mathsf{D}_k(i) &= \begin{bmatrix} A_{d_k}(i) & K_O(i)C_{d_k} \\ 0 & A_{d_k}(i) \end{bmatrix},\ \mathsf{W}(i) = \begin{bmatrix} K_O(i)M(i) \\ N(i) \end{bmatrix} && (7.73) \\
\bar{H}_A(i) &= \begin{bmatrix} K_O(i)M_c(i) \\ M_a(i) \end{bmatrix},\ L_f(i) = \left[L(i) \quad 0\right] && (7.74)
\end{aligned}
$$

To facilitate further developments, we introduce the following matrix expressions:

$$
\begin{aligned}
\mathcal{P}(i) &= \begin{bmatrix} \mathcal{P}_f(i) & 0 \\ 0 & \mathcal{P}_s(i) \end{bmatrix},\ \bar{\mathbb{B}}(i) = \begin{bmatrix} \mathbb{B}(i) & 0 \\ 0 & \mathbb{B}(i) \end{bmatrix} \\
\mathbb{L}_{t_k}(i) &= \begin{bmatrix} \mathbb{L}_{f_k}(i) & 0 \\ 0 & \mathbb{L}_{s_k}(i) \end{bmatrix},\ \mathbb{N}_{t_j}(i) = \begin{bmatrix} \mathbb{N}_{f_j}(i) & 0 \\ 0 & \mathbb{N}_{s_j}(i) \end{bmatrix} && (7.75) \\
\Omega_{\Delta A}(t,i) &= \mathcal{P}(i)\mathcal{A}_{\Delta o}(t,i) + \sum_{k=1}^{q} \hat{\mathbb{L}}_{t_k}(i)
\end{aligned}
$$

$$
\begin{aligned}
&+ \sum_{j=1}^{p} \hat{\mathbb{N}}_{t_j}(i) + L_f^t(i) L_f(i) \\
\Upsilon_{\Delta A}(t,i) &= \Omega_{\Delta A}(i) + \mathcal{A}_{\Delta o}^t(t,i)\mathcal{P}(i) \\
&+ \sum_{m=1}^{s} \alpha_{im}\mathcal{P}(m)
\end{aligned} \tag{7.76}
$$

where

$$
\begin{aligned}
&0 < \mathcal{P}_f(i) = \mathcal{P}_f^t(i) \in \mathbb{R}^{n\times n} \quad , \quad 0 < \mathcal{P}_s(i) = \mathcal{P}_s^t(i) \in \mathbb{R}^{n\times n} \\
&0 < \mathbb{L}_{f_k}(i) = \mathbb{L}_{f_k}^t(i) \in \mathbb{R}^{n\times n} \quad , \quad 0 < \mathbb{L}_{s_k}(i) = \mathbb{L}_{s_k}^t(i) \in \mathbb{R}^{n\times n} \\
&0 < \mathbb{N}_{f_j}(i) = \mathbb{N}_{f_j}^t(i) \in \mathbb{R}^{n\times n} \quad , \quad 0 < \mathbb{N}_{s_j}(i) = \mathbb{N}_{s_j}^t(i) \in \mathbb{R}^{n\times n}
\end{aligned} \tag{7.77}
$$

such that

$$
\begin{aligned}
\hat{\mathbb{L}}_{t_k}(i) &= \mathbb{L}_{t_k}(i) + \xi_{1_k}(i) \sum_{m=1}^{s} \alpha_{im}\mathbb{L}_{t_k}(m) \\
\bar{\mathbb{L}}_{t_k}(i) &= \mathbb{L}_{t_k}(i) - \xi_{1_k}^{-1}(i) \sum_{m=1}^{s} \alpha_{im}\mathbb{L}_{t_k}(m) \\
\hat{\mathbb{N}}_{t_j}(i) &= \mathbb{N}_{t_j}(i) + \xi_{2_j}(i) \sum_{m=1}^{s} \alpha_{im}\mathbb{N}_{t_j}(m) \\
\bar{\mathbb{N}}_{t_j}(i) &= \mathbb{N}_{j}(i) - \xi_{2_j}^{-1}(i) \sum_{m=1}^{s} \alpha_{im}\mathbb{N}_{t_j}(m)
\end{aligned} \tag{7.78}
$$

for some scalars $\xi_{1_k}(i) > 0,\ \xi_{2_j}(i) > 0,\ i \in \mathcal{S},\ k \in \mathcal{Z}_q,\ j \in \mathcal{Z}_p$ guaranteeing that

$$
\bar{\mathbb{L}}_{t_k}(i) > 0,\ \ \bar{\mathbb{N}}_{t_j}(i) > 0,\ \ i \in \mathcal{S},\ k \in \mathcal{Z}_q,\ j \in \mathcal{Z}_p
$$

Based on **Theorem** 7.3, it immediately follows that the robust stochastic stability with disturbance attenuation γ of the closed-loop system $(\Sigma_{\Delta A})$ is guaranteed if the inequality:

$$
\Upsilon_{\Delta A}(t,i) + \gamma^{-2}\mathcal{P}(i)\mathsf{W}(i)\mathsf{W}^t(i)\mathcal{P}(i)
$$

$$
\begin{aligned}
&+ \quad \Omega_{\Delta A}(t,i)\bar{\mathbb{B}}(i)\bar{\Gamma}^{-1}(i)\bar{\mathbb{B}}^t(i)\Omega^t_{\Delta A}(i) \\
&+ \quad \mathcal{P}(i)\sum_{k=1}^{q}\mathcal{D}_{\Delta d_k}(t,i)\bar{\mathbb{L}}^{-1}_{t_k}(i)\mathcal{D}^t_{\Delta d_k}(t,i)\mathcal{P}(i) \quad < \quad 0 \qquad (7.79)
\end{aligned}
$$

holds for all $i \in \mathcal{S}$ and for all uncertainties satisfying (7.6) where

$$
\begin{aligned}
\bar{\Gamma}_j(i) &= \bar{\mathbb{N}}_{t_j}(i) - \bar{\mathbb{B}}^t(i)\left[\hat{\mathbb{N}}_{t_j}(i) + \sum_{k=1}^{q}\hat{\mathbb{L}}_{t_k}(i) + L^t_f(i)L_f(i)\right]\bar{\mathbb{B}}(i) \\
\bar{\Gamma}(i) &= \sum_{j=1}^{p}\bar{\Gamma}_j(i) \qquad (7.80)
\end{aligned}
$$

It follows from (7.79) on using **Facts 1-2** for some scalars $\varepsilon(i) > 0,\ \upsilon(i) > 0,\ \varrho(i) > 0,\ i \in \mathcal{S}$ that:

$$
\begin{aligned}
& \Upsilon_{twt} + \gamma^{-2}\mathcal{P}(i)\mathsf{W}(i)\mathsf{W}^t(i)\mathcal{P}(i) \\
+ \quad & [\varepsilon^{-1}(i) + \upsilon^{-1}(i) + \varrho^{-1}(i)]\mathcal{P}(i)\bar{H}_A(i)\bar{H}^t_A(i)\mathcal{P}(i) \\
+ \quad & \bar{\Lambda}_n(i)\left[\bar{\Gamma}(i) - \upsilon(i)\bar{E}^t_A(i)\bar{E}_A(i)\right]^{-1}\bar{\Lambda}^t_n(i) \\
+ \quad & \mathcal{P}(i)\sum_{k=1}^{q}\mathsf{D}_k(i)\bar{\Phi}^{-1}_d(i)\mathsf{D}^t_k(i)\mathcal{P}(i) \\
\triangleq \quad & \begin{bmatrix} \Sigma_{1,1}(i) & \Sigma_{1,2}(i) \\ \Sigma^t_{1,2}(i) & \Sigma_{2,2}(i) \end{bmatrix} \quad < \quad 0 \qquad (7.81)
\end{aligned}
$$

where for $i \in \mathcal{S}$

$$
\begin{aligned}
\Upsilon_{twt} &= \mathcal{P}(i)\mathsf{A}_o(i) + \mathsf{A}^t_o(i)\mathcal{P}(i) + \sum_{k=1}^{q}\hat{\mathbb{L}}_{t_k}(i) \\
&+ \quad \sum_{j=1}^{p}\hat{\mathbb{N}}_{t_j}(i) + \sum_{m=1}^{s}\alpha_{im}\mathcal{P}(m) \\
&+ \quad \varepsilon(i)\bar{E}^t_A(i)\bar{E}_A(i) + L^t_f(i)L_f(i) \\
\bar{\Lambda}_n(i) &= \left[\mathcal{P}(i)\mathsf{A}_o(i) + \sum_{k=1}^{q}\hat{\mathbb{L}}_{t_k}(i) + \sum_{j=1}^{p}\hat{\mathbb{N}}_{t_j}(i)\right]\bar{\mathbb{B}}(i) \\
\bar{\Phi}_d(i) &= diag\left[\bar{\mathbb{L}}_{t_k}(i) - \varrho(i)E^t_{D_k}(i)E_{D_k}(i)\right] \qquad (7.82)
\end{aligned}
$$

Before proceeding further, we introduce the following matrix expressions:

$$
\begin{aligned}
\bar{\Gamma}_{f_j}(i) &= \bar{\mathbb{N}}_{f_j}(i) - B_j^t(i)\Big[\hat{\mathbb{N}}_{f_j}(i) + \sum_{k=1}^{q}\hat{\mathbb{L}}_{f_k}(i) + L^t(i)L(i)\Big]B_j(i) \\
\bar{\Gamma}_{s_j}(i) &= \bar{\mathbb{N}}_{s_j}(i) - B_j^t(i)\Big[\hat{\mathbb{N}}_{s_j}(i) + \sum_{k=1}^{q}\hat{\mathbb{L}}_{s_k}(i) + \upsilon(i)N_a^t(i)N_a(i)\Big]B_j(i) \\
\bar{\Gamma}_f(i) &= \sum_{j=1}^{p}\bar{\Gamma}_{f_j}(i) \ , \ \ \bar{\Gamma}_s(i) = \sum_{j=1}^{p}\bar{\Gamma}_{s_j}(i) \\
\bar{C}_o(i) &= C_o(i)\Big[I + \mathbb{B}(i)\Gamma_f^{-1}(i)\mathbb{B}^t(i)\mathcal{R}_{f1}^t(i)\Big] \\
\mathcal{C}_d(i) &= [C_{d_1}(i), \cdots, C_{d_q}(i)] \qquad (7.83) \\
\mathcal{R}_1(i) &= \mathcal{P}_f(i)\Big(A_O(i) - K_O(i)C_o(i)\Big) + \sum_{k=1}^{q}\hat{\mathbb{L}}_{f_k}(i) + \sum_{j=1}^{p}\hat{\mathbb{N}}_{f_j}(i) \\
\mathcal{R}_{f1}(i) &= \Big(\mathcal{P}_f(i)A_o(i) + \sum_{k=1}^{q}\hat{\mathbb{L}}_{s_k}(i) + \sum_{j=1}^{p}\hat{\mathbb{N}}_{s_j}(i)\Big) \\
\mathcal{R}_{s1}(i) &= \Big(\mathcal{P}_s(i)A_o(i) + \sum_{k=1}^{q}\hat{\mathbb{L}}_{s_k}(i) + \sum_{j=1}^{p}\hat{\mathbb{N}}_{s_j}(i)\Big) \\
\Phi_f(i) &= diag[\bar{\mathbb{L}}_{f_k}] \ , \ \ \Phi_s(i) = diag[\bar{\mathbb{L}}_{s_k} - \varrho(i)E_{d_k}^t E_{d_k}] \\
\mathcal{R}_2(i) &= C_o(i)\Big(I + \mathbb{B}(i)\Gamma_s^{-1}(i)\mathcal{R}_{s1}^t(i)\Big) \\
&+ \Big([\varepsilon^{-1}(i) + \upsilon^{-1}(i) + \varrho^{-1}(i)]H_c(i)H_a^t(i)\mathcal{P}_s(i) \\
&+ \mathcal{C}_d(i)\Phi_s^{-1}(i)\mathbb{D}^t(i)\mathcal{P}_s(i)\Big) \qquad (7.84) \\
\mathcal{Q}(i) &= \gamma^{-2}M(i)M^t(i) + [\varepsilon^{-1}(i) + \upsilon^{-1}(i) + \varrho^{-1}(i)]H_c(i)H_c^t(i) \\
&+ C_o(i)\mathbb{B}(i)\Big[\bar{\Gamma}_s^{-1}(i) + \Gamma_f^{-1}(i)\Big]\mathbb{B}^t(i)C_o^t(i) + \mathcal{C}_d(i)\Phi_s^{-1}(i)\mathcal{C}_d^t(i) \\
\mathcal{R}_C(i) &= \mathbb{B}(i)\Gamma_f^{-1}(i)\mathbb{B}^t(i)C_o^t(i)\mathcal{R}_2^t(i)\mathcal{Q}^{-1}(i)\bar{C}_o(i) \ - \ \mathcal{R}_{f2}(i) \\
\mathcal{R}_{f2}(i) &= I + \mathcal{R}_{f1}(i)\mathbb{B}(i)\Gamma_f^{-1}(i)\mathbb{B}^t(i) \qquad (7.85)
\end{aligned}
$$

Expansion of (7.81) using (7.73)-(7.76) and (7.82)-(7.85) yields:

$$
\begin{aligned}
\Sigma_{1,1}(i) &= \mathcal{P}_f(i)\Big(A_O(i) - K_O(i)C_o(i)\Big) + \Big(A_O(i) - K_O(i)C_o(i)\Big)^t \mathcal{P}_f(i) \\
&+ \sum_{k=1}^{q} \hat{\mathbb{L}}_{f_k}(i) + \sum_{j=1}^{p} \hat{\mathbb{N}}_{f_j}(i) + \sum_{m=1}^{s} \alpha_{im}\mathcal{P}_f(m) + L^t(i)L(i) \\
&+ \gamma^{-2}\mathcal{P}_f(i)K_O(i)M(i)M^t(i)K_O^t(i)\mathcal{P}_f(i) \\
&+ [\varepsilon^{-1}(i) + \upsilon^{-1}(i) + \varrho^{-1}(i)]\mathcal{P}_f(i)K_O(i)H_c(i)H_c^t(i)K_O^t(i)\mathcal{P}_f(i) \\
&+ \mathcal{P}_f(i)K_O(i)\mathcal{C}_d(i)\mathbb{B}(i)\Phi_s^{-1}(i)\mathbb{B}^t(i)\mathcal{C}_d^t(i)K_O^t(i)\mathcal{P}_f(i) \\
&+ \mathcal{R}_1(i)\mathbb{B}(i)\Gamma_f^{-1}(i)\mathbb{B}^t(i)\mathcal{R}_1^t(i) \\
&+ \mathcal{P}_f(i)\mathbb{D}(i)\Phi_f^{-1}(i)\mathbb{D}^t(i)\mathcal{P}_f(i) + \\
&+ \mathcal{P}_f(i)K_O(i)C_o(i)\Gamma_s^{-1}(i)C_o^t(i)K_O^t(i)\mathcal{P}_f(i) \qquad (7.86)
\end{aligned}
$$

$$
\begin{aligned}
\Sigma_{2,2}(i) &= \mathcal{P}_s(i)A_o(i) + A_o^t(i)\mathcal{P}_s(i) + \sum_{k=1}^{q} \hat{\mathbb{L}}_{s_k}(i) \\
&+ \sum_{j=1}^{p} \hat{\mathbb{N}}_{s_j}(i) + \sum_{m=1}^{s} \alpha_{im}\mathcal{P}_s(m) + \gamma^{-2}\mathcal{P}_s(i)N(i)N^t(i)\mathcal{P}_s(i) \\
&+ [\varepsilon^{-1}(i) + \upsilon^{-1}(i) + \varrho^{-1}(i)]\mathcal{P}_s(i)M_a(i)M_a^t(i)\mathcal{P}_s(i) \\
&+ \mathcal{P}_s(i)F_o(i)G_O(i)\mathbb{B}(i)\Gamma_f^{-1}(i)\mathbb{B}^t(i)G_O^t(i)F_o^t\mathcal{P}_s(i) \\
&+ \mathcal{R}_{s1}(i)\mathbb{B}(i)\Gamma_s^{-1}(i)\mathbb{B}^t(i)\mathcal{R}_{s1}^t(i) \\
&+ \mathcal{P}_s(i)\mathbb{D}(i)\Phi_s^{-1}(i)\mathbb{D}^t(i)\mathcal{P}_s(i) \qquad (7.87)
\end{aligned}
$$

$$
\begin{aligned}
\Sigma_{1,2}(i) &= \mathcal{P}_f(i)K_O(i)C_o(i) + G_O^t(i)F_o^t(i)\mathcal{P}_s(i) \\
&+ \gamma^{-2}\mathcal{P}_f(i)K_O(i)M(i)N^t(i)\mathcal{P}_s(i) \\
&+ [\varepsilon^{-1}(i) + \upsilon^{-1}(i) + \varrho^{-1}(i)]\mathcal{P}_f(i)K_O(i)M_c(i)M_a^t(i)\mathcal{P}_s(i) \\
&+ \mathcal{R}_1(i)\mathbb{B}(i)\Gamma_f^{-1}(i)\mathbb{B}^t(i)G_O^t(i)F_o^t(i)\mathcal{P}_s(i) \\
&+ \mathcal{P}_f(i)K_O(i)\mathcal{C}_d(i)\Phi_s^{-1}(i)\mathbb{D}^t(i)\mathcal{P}_s(i) \\
&+ \mathcal{P}_f(i)K_O(i)C_o(i)\Gamma_s^{-1}(i)\mathcal{R}_{s1}^t(i) \qquad (7.88)
\end{aligned}
$$

The following theorem establishes the main result:

Theorem 7.9 *Subject to* **Assumptions** *(7.1)-(7.2), the augmented neutral system* $(\Sigma_{\Delta A})$ *is* **RSSWDD with a disturbance attenuation** γ *given* τ_M & δ_M *if for given scalars* $\xi_{1_k}(i) > 0,\ \xi_{2_j}(i) > 0,\ i \in \mathcal{S}\ ,\ k \in \mathcal{Z}_q,\ j \in \mathcal{Z}_p$ *and matrices* $0 < I\!\!L_{t_k} = I\!\!L_{t_k}^t \in I\!\!R^{n\times n},\ i \in \mathcal{S}$ *and* $0 < I\!\!N_{t_j} = I\!\!N_{t_j}^t \in I\!\!R^{n\times n},\ i \in \mathcal{S},\ k \in \mathcal{Z}_q,\ j \in \mathcal{Z}_p$ *satisfying (7.78), there exist matrices* $\mathcal{P}_f(i) = \mathcal{P}_f^t(i) > 0,\ \mathcal{P}_s(i) = \mathcal{P}_s^t(i) > 0,\ i \in \mathcal{S}$ *and scalars* $\varepsilon(i) > 0,\ \upsilon(i) > 0,\ \varrho_1(i) > 0,\ \varrho_2(i) > 0,\ \varrho_3(i) > 0,\ i \in \mathcal{S}$*, satisfying the following LMIs for all* $i \in \mathcal{S}$

$$\begin{bmatrix} \Upsilon_{twf} & \mathcal{P}_f(i)I\!\!D(i) & \mathcal{R}_{f1}(i)I\!\!B(i) \\ I\!\!D^t(i)\mathcal{P}_f(i) & -\Phi_f(i) & 0 \\ \mathcal{R}_{f1}^t(i)I\!\!B^t(i) & \bar{\Gamma}_f & 0 \end{bmatrix} < 0 \tag{7.89}$$

$$\Upsilon_{twf}(i) = \mathcal{P}_f(i)A_o(i) + A_o^t(i)\mathcal{P}_f(i) + \sum_{k=1}^{q} \hat{I\!\!L}_{f_k}(i) + \sum_{j=1}^{p} \hat{I\!\!N}_{f_j}(i) + \sum_{m=1}^{s} \alpha_{im}\mathcal{P}_f(m) - \bar{C}_o^t(i)\mathcal{Q}^{-1}(i)\bar{C}_o(i) \tag{7.90}$$

$$\begin{bmatrix} \Upsilon_{tws} & \Lambda_{as}(i) & \mathcal{P}_s(i)I\!\!D(i) & \mathcal{R}_{s1}(i)I\!\!B(i) & \mathcal{P}_s(i)N(i) \\ \Lambda_{as}^t(i) & -\Psi_a(i) & 0 & 0 & 0 \\ I\!\!D^t(i)\mathcal{P}_s(i) & 0 & -\Phi_s(i) & 0 & 0 \\ \mathcal{R}_{s1}^t(i)I\!\!B^t(i) & 0 & 0 & -\bar{\Gamma}_s & 0 \\ N^t(i)\mathcal{P}_s(i) & 0 & 0 & 0 & -\gamma^2 I \end{bmatrix} < 0 \tag{7.91}$$

$$\Upsilon_{tws}(i) = \mathcal{P}_s(i)A_o(i) + A_o^t(i)\mathcal{P}_s(i) + \sum_{k=1}^{q} \hat{I\!\!L}_{s_k}(i) + \sum_{j=1}^{p} \hat{I\!\!N}_{s_j}(i) + \sum_{m=1}^{s} \alpha_{im}\mathcal{P}_s(m) + \varepsilon(i)E_a^t(i)E_a(i) + \mathcal{R}_S(i)$$

$$\mathcal{R}_S(i) = \mathcal{R}_2^t(i)\mathcal{Q}^{-1}(i)\bar{C}_o(i)\mathcal{R}_C^{-1}(i)I\!\!B(i)\bar{\Gamma}_s^{-1}(i)I\!\!B^t(i)\mathcal{R}_C^{-t}(i)\mathcal{Q}^{-1}(i)\bar{C}_o^t(i)\mathcal{R}_2(i)$$

$$\Lambda_{as}(i) = [\mathcal{P}_s(i)H_a(i) \quad \mathcal{P}_s(i)H_a(i) \quad \mathcal{P}_s(i)H_a(i)] \tag{7.92}$$

for all admissible uncertainties satisfying (7.6) where the controller gains are given by

$$
\begin{aligned}
A_O(i) &= A_o(i) \quad , \quad K_O(i) = \mathcal{P}_f^{-1}(i)\bar{C}_o^t(i)\mathcal{Q}^{-1}(i) \\
G_O(i) &= F_o^{\dagger}(i)\mathcal{P}_s^{-1}(i)\mathcal{R}_2^t(i)\mathcal{Q}^{-1}(i)\bar{C}_o(i)\mathcal{R}_C^{-1}(i)
\end{aligned} \tag{7.93}
$$

where $F_o^{\dagger}$ is the pseudo-inverse of F_o.

Proof: We start with $\Sigma_{1,1}(i)$. Substituting $A_O(i), K_O(i)$ from (7.93) into (7.86) using (7.83)-(7.85), rearranging and applying **Fact 1**, we obtain the LMI (7.89). From (7.93), it is readily seen that $\Sigma_{1,2}(i) \equiv 0$. By the Schur complements, we can put $\Sigma_{2,2}(i)$ using (7.93) into the LMI (7.91) which completes the proof. $\nabla\nabla\nabla$

Had we combined the nominal neutral system (7.8)-(7.11) and the controller (7.68), we would have obtained the nominal augmented system

$$
\begin{aligned}
(\Sigma_A): \quad \mathcal{M}(\dot{\zeta}_t) &\stackrel{\Delta}{=} \dot{\zeta}(t) - \sum_{j=1}^{p} \mathcal{B}_j(i)\dot{\zeta}(t-\pi_j) \\
&= \mathsf{A}_o(t,i)\zeta(t) + \sum_{k=1}^{q} \mathsf{D}_{d_k}(t,i)\zeta(t-\tau_k) \\
&+ \mathsf{W}(i)w(t) && (7.94) \\
\tilde{z}(t) &= L_f(i)\zeta(t) && (7.95)
\end{aligned}
$$

for which the following theorem can be readily proven.

Theorem 7.10 *Subject to* **Assumptions** *(7.1)-(7.2), the augmented neutral system* (Σ_A) *is* **RSSWDD with a disturbance attenuation** γ τ_M & δ_M *if for given scalars* $\xi_{1_k}(i) > 0,\ \xi_{2_j}(i) > 0,\ i \in \mathcal{S}\ ,\ k \in \mathcal{Z}_q,\ j \in \mathcal{Z}_p$ *and matrices* $0 < I\!\!L_{t_k} = I\!\!L_{t_k}^t \in I\!\!R^{n\times n},\ i \in \mathcal{S}$ *and* $0 < I\!\!N_{t_j} = I\!\!N_{t_j}^t \in I\!\!R^{n\times n},\ i \in \mathcal{S},\ k \in \mathcal{Z}_q,\ j \in \mathcal{Z}_p$ *satisfying (7.78), there exist matrices* $\mathcal{P}_f(i) = \mathcal{P}_f^t(i) > 0,\ \mathcal{P}_s(i) = \mathcal{P}_s^t(i) > 0,\ i \in \mathcal{S}$ *satisfying the following LMIs for all* $i \in \mathcal{S}$

$$
\begin{bmatrix}
\Upsilon_{twfo} & \mathcal{P}_f(i)I\!\!D(i) & \mathcal{R}_{f1}(i)I\!\!B(i) \\
I\!\!D^t(i)\mathcal{P}_f(i) & -\Phi_f(i) & 0 \\
\mathcal{R}_{f1}^t(i)I\!\!B^t(i) & 0 & -\bar{\Gamma}_f
\end{bmatrix} < 0 \qquad i \in \mathcal{S} \tag{7.96}
$$

$$\Upsilon_{twf}(i) = \mathcal{P}_f(i)A_o(i) + A_o^t(i)\mathcal{P}_f(i) + \sum_{k=1}^{q} \hat{\mathbb{L}}_{f_k}(i) + \sum_{j=1}^{p} \hat{\mathbb{N}}_{f_j}(i) +$$

$$\sum_{m=1}^{s} \alpha_{im}\mathcal{P}_f(m) - \bar{C}_o^t(i)\mathcal{Q}_o^{-1}(i)\bar{C}_o(i) \qquad (7.97)$$

$$\begin{bmatrix} \Upsilon_{twso} & \mathcal{P}_s(i)\mathbb{D}(i) & \mathcal{R}_{s1}(i)\mathbb{B}(i) & \mathcal{P}_s(i)N(i) \\ \mathbb{D}^t(i)\mathcal{P}_s(i) & -\Phi_s(i) & 0 & 0 \\ \mathcal{R}_{s1}^t(i)\mathbb{B}^t(i) & 0 & -\bar{\Gamma}_{so} & 0 \\ N^t(i)\mathcal{P}_s(i) & 0 & 0 & -\gamma^2 I \end{bmatrix} < 0 \qquad (7.98)$$

$$\mathcal{R}_{So}(i) = \mathcal{R}_2^t(i)\mathcal{Q}_o^{-1}(i)\bar{C}_o(i)\mathcal{R}_C^{-1}(i)\mathbb{B}(i)\bar{\Gamma}_{so}^{-1}(i)\mathbb{B}^t(i)\mathcal{R}_C^{-t}(i)\mathcal{Q}_o^{-1}(i)\bar{C}_o^t(i)\mathcal{R}_2(i)$$

$$\Upsilon_{twso}(i) = \mathcal{P}_s(i)A_o(i) + A_o^t(i)\mathcal{P}_s(i) + \sum_{k=1}^{q} \hat{\mathbb{L}}_{s_k}(i) + \sum_{j=1}^{p} \hat{\mathbb{N}}_{s_j}(i)$$

$$+ \quad \sum_{m=1}^{s} \alpha_{im}\mathcal{P}_s(m) + \mathcal{R}_{So}(i) \qquad (7.99)$$

for all admissible uncertainties satisfying (7.6) where the controller gains are given by

$$\begin{aligned} A_O(i) &= A_o(i) \quad , \quad K_{OO}(i) = \mathcal{P}_f^{-1}(i)\bar{C}_o^t(i)\mathcal{Q}_o^{-1}(i) \\ G_{OO}(i) &= F_o^{\dagger}(i)\mathcal{P}_s^{-1}(i)\mathcal{R}_2^t(i)\mathcal{Q}_o^{-1}(i)\bar{C}_o(i)\mathcal{R}_C^{-1}(i) \end{aligned} \qquad (7.100)$$

where

$$\begin{aligned} \bar{\Gamma}_{so_j}(i) &= \bar{\mathbb{N}}_{s_j}(i) - B_j^t(i)\left[\hat{\mathbb{N}}_{s_j}(i) + \sum_{k=1}^{q} \hat{\mathbb{L}}_{s_k}(i)\right]B_j(i) \\ \bar{\Gamma}_{so}(i) &= \sum_{j=1}^{p} \bar{\Gamma}_{s_j}(i) \\ \Phi_{so}(i) &= diag[\bar{\mathbb{L}}_{s_k}] \\ \mathcal{R}_{2o}(i) &= C_o(i)\left(I + \mathbb{B}(i)\bar{\Gamma}_{so}^{-1}(i)\mathcal{R}_{s1}^t(i)\right) \\ &+ \left(\mathcal{C}_d(i)\Phi_{so}^{-1}(i)\mathbb{D}^t(i)\mathcal{P}_s(i)\right) \end{aligned}$$

$$
\begin{aligned}
\mathcal{Q}_o(i) &= \gamma^{-2} M(i) M^t(i) \\
&+ C_o(i) \mathbb{B}(i) \left[\bar{\Gamma}_{so}^{-1}(i) + \Gamma_f^{-1}(i) \right] \mathbb{B}^t(i) C_o^t(i) \\
&+ \mathcal{C}_d(i) \Phi_{so}^{-1}(i) \mathcal{C}_d^t(i)
\end{aligned} \tag{7.101}
$$

7.4.2 Example 7.1

We consider a pilot-scale single-reach water quality system which can fall into the type (7.2)-(7.5) with

$$\tau_1 = 0.75 \ , \ \tau_2 = 0.65 \ , \ \pi_1 = 0.45 \ , \ \pi_2 = 0.55$$

. Let the Markov process governing the mode switching has generator

$$\Im = \begin{bmatrix} -2 & 2 \\ 5 & -5 \end{bmatrix}$$

For the two operating conditions (modes), the associated date are:

Mode 1:

$$
\begin{aligned}
A_o(1) &= \begin{bmatrix} -2.5 & -0.5 \\ 0 & -3 \end{bmatrix}, \ A_{d_1}(1) = \begin{bmatrix} -0.1 & -0.05 \\ 0.03 & -0.1 \end{bmatrix} \\
A_{d_2}(1) &= \begin{bmatrix} 0.1 & 0 \\ 0.1 & 0.1 \end{bmatrix}, \ B_1(1) = \begin{bmatrix} 0.2 & 0 \\ 0 & 0.1 \end{bmatrix} \\
B_2(1) &= \begin{bmatrix} 0.1 & 0 \\ 00 & 0.2 \end{bmatrix}, \ C_o(1) = \begin{bmatrix} 2 & 0 \\ 0 & 1 \end{bmatrix} \\
C_{d_1}(1) &= \begin{bmatrix} 0.2 & 0 \\ 0 & 0.1 \end{bmatrix}, \ C_{d_2}(1) = \begin{bmatrix} 0.1 & 0 \\ 0 & 0.2 \end{bmatrix} \\
F_o(1) &= \begin{bmatrix} 0.5 & 0 \\ 0 & 0.7 \end{bmatrix}, \ N(1) = \begin{bmatrix} 0.1 \\ 0.3 \end{bmatrix}, \ M(1) = \begin{bmatrix} 0.2 \\ 0.2 \end{bmatrix} \\
L(1) &= [0.3 \ \ 0.3] \ , \ N_a(1) = [0.2 \ \ 0.4] \\
N_{d_1}(1) &= [0.3 \ \ 0] \ , \ M_a(1) = [0.5 \ \ 1] \\
N_{d_2}(1) &= [0.3 \ \ 0] \ , \ M_c(1) = [0.3 \ \ 0]
\end{aligned}
$$

Mode 2:

$$A_o(1) = \begin{bmatrix} -2 & 1 \\ 0 & -4 \end{bmatrix}, \ A_{d_1}(1) = \begin{bmatrix} 0.1 & -0.05 \\ 0.03 & 0.1 \end{bmatrix}$$

$$
\begin{aligned}
A_{d_2}(1) &= \begin{bmatrix} -0.1 & 0.04 \\ 0.05 & -0.1 \end{bmatrix}, \; B_1(1) = \begin{bmatrix} 0.3 & 0 \\ 0 & 0.3 \end{bmatrix} \\
B_2(1) &= \begin{bmatrix} 0.4 & 0 \\ 00 & 0.4 \end{bmatrix}, \; C_o(1) = \begin{bmatrix} 1 & 0 \\ 0 & 2 \end{bmatrix} \\
C_{d_1}(1) &= \begin{bmatrix} 0.25 & 0 \\ 0 & 0.15 \end{bmatrix}, \; C_{d_2}(1) = \begin{bmatrix} 0.15 & 0 \\ 0 & 0.25 \end{bmatrix} \\
F_o(1) &= \begin{bmatrix} 1 & 0 \\ 0 & 0.4 \end{bmatrix}, \; N(1) = \begin{bmatrix} 0.3 \\ 0.1 \end{bmatrix}, \; M(1) = \begin{bmatrix} 0.2 \\ 0.2 \end{bmatrix} \\
L(1) &= [0.3 \;\; 0.3], \; N_a(1) = [0.4 \;\; 0.2] \\
N_{d_1}(1) &= [0 \;\; 0.3], \; M_a(1) = [0.1 \;\; 5] \\
N_{d_2}(1) &= [0 \;\; 0.3], \; M_c(1) = [0 \;\; 0.3]
\end{aligned}
$$

First we note that **Assumptions** (7.1)-(7.2) are met for both modes. The initial data for $i = 1, 2$ are :

$$
\begin{aligned}
\mathbb{L}_e(1) &= \begin{bmatrix} 5 & 0 \\ 0 & 5 \end{bmatrix}, \; \mathbb{L}_x(1) = \begin{bmatrix} 5 & 0 \\ 0 & 5 \end{bmatrix}, \; \xi(1) = 15 \\
\mathbb{L}_e(2) &= \begin{bmatrix} 4 & 0 \\ 0 & 4 \end{bmatrix}, \; \mathbb{L}_x(2) = \begin{bmatrix} 4 & 0 \\ 0 & 4 \end{bmatrix}, \; \xi(2) = 20
\end{aligned}
$$

which ensures that $\bar{\mathbb{L}}_e(1) > 0$, $\hat{\mathbb{L}}_e(1) > 0$, $\bar{\mathbb{L}}_x(2) > 0$, $\hat{\mathbb{L}}_x(2) > 0$. Invoking the software environment [57], we solve the LMIs (7.50) and compute the feedback gain (7.51). The feasible solutions are given by:

$$
\begin{aligned}
\mathcal{Y}(1) &= \begin{bmatrix} 2.4886 & 0.0007 \\ 0.0007 & 0.7145 \end{bmatrix}, \; \mathcal{Z}(1) = \begin{bmatrix} 0.7413 & 0.5638 \\ 0.0029 & 0.9746 \end{bmatrix} \\
\varrho(1) &= 2.7694, \; \varepsilon(1) = 6.1246, \; \upsilon(1) = 3.7412, \; \gamma = 1.1045 \\
K_s(1) &= \begin{bmatrix} 0.2977 & 0.7888 \\ 0.0008 & 1.3640 \end{bmatrix} \\
\mathcal{Y}(2) &= \begin{bmatrix} 4.0161 & 0.0302 \\ 0.0302 & 2.1105 \end{bmatrix}, \; \mathcal{Z}(2) = \begin{bmatrix} 1.1041 & 0.0204 \\ 0.9683 & 0.7869 \end{bmatrix} \\
\varrho(2) &= 3.9554, \; \varepsilon(2) = 5.1286, \; \upsilon(2) = 4.7572 \\
K_s(2) &= \begin{bmatrix} 0.2688 & 0.0058 \\ 0.2331 & 0.3695 \end{bmatrix}
\end{aligned}
$$

In case of distributed feedback delay, we solve the LMIs (7.63) and compute the feedback gains (7.64). The feasible solutions are given by:

$$\begin{aligned}
\mathcal{X}_1(1) &= \begin{bmatrix} 2.2507 & 0.0132 \\ 0.0132 & 1.9754 \end{bmatrix}, \ \mathcal{Z}(1) = \begin{bmatrix} 0.4556 & 0.3638 \\ 0.2184 & 1.4376 \end{bmatrix} \\
\mathcal{X}_2(1) &= \begin{bmatrix} 2.7148 & 0.2321 \\ 0.2321 & 1.7648 \end{bmatrix} \\
\varrho(1) &= 2.7694 \ , \ \varepsilon(1) = 6.1246 \\
\upsilon(1) &= 3.7412 \ , \ \gamma = 1.2005 \ , \ \mu(1) = 2.6543 \\
K_1^+(1) &= \begin{bmatrix} 0.2014 & 0.1828 \\ 0.0928 & 0.7271 \end{bmatrix}, \ K_2^+(1) = \begin{bmatrix} 0.1519 & 0.1862 \\ 0.0109 & 0.8132 \end{bmatrix} \\
\mathcal{X}_1(2) &= \begin{bmatrix} 3.1119 & 0.0045 \\ 0.0045 & 2.4002 \end{bmatrix}, \ \mathcal{Z}(2) = \begin{bmatrix} 0.5058 & 0.3368 \\ 0.1528 & 2.8435 \end{bmatrix} \\
\mathcal{X}_2(2) &= \begin{bmatrix} 2.4257 & 0.1602 \\ 0.1602 & 2.6543 \end{bmatrix} \\
\varrho(2) &= 2.4557 \ , \ \varepsilon(2) = 7.1203 \\
\upsilon(2) &= 3.1227 \ , \ \mu(2) = 2.4653 \\
K_1^+(2) &= \begin{bmatrix} 0.1623 & 0.1400 \\ 0.0474 & 1.1846 \end{bmatrix} \\
K_2^+(2) &= \begin{bmatrix} 0.2009 & 0.1148 \\ -0.0078 & 1.0718 \end{bmatrix}
\end{aligned}$$

7.5 Robust Observers

This is the second part of the Chapter in which we consider the state observation and stabilization problems for a class of linear neutral jumping systems with norm-bounded uncertainties. Initially, we address both problems of robust state observation and robust $\mathcal{H}_\infty$ observation and employ a new linear state-delayed observer such that the asymptotic stability of the combined neutral system and the proposed observer is guaranteed for all admissible uncertainties. The main tool for solving the foregoing problems is the linear matrix inequality approach. In this regard, it will be shown that the solution of robust is expressed in terms of two LMIs involving scaling parameters. Looked at in this light, the developed methods provide new results which in some sense are the dual of [99]. Then, the robust stabilization problem is considered by designing memoryless state-estimate feedback such the asymptotic stability of the closed-loop stability is guaranteed.

We consider a class of stochastic uncertain neutral systems with Markovian jump parameters described over the space $(\Omega, \mathcal{F}, \mathbf{P})$ by:

$$
\begin{aligned}
(\Sigma_{\Delta n})\ \mathcal{M}(\dot{x}_t) &\stackrel{\Delta}{=} \dot{x}(t) - B(\eta_t)\dot{x}(t-\tau) = [A_o(\eta_t) + \Delta A_o(t,\eta_t)]x(t) \\
&+ [A_d(\eta_t) + \Delta A_d(t,\eta_t)]x(t-\tau) \\
&+ N(\eta_t)w(t) \\
&= A_{\Delta o}(t,\eta_t)x(t) + A_{\Delta d}(t,\eta_t)x(t-\tau) \\
&+ N(\eta_t)w(t) \qquad (7.102)\\
x(\eta) &= \phi(\eta) \in \mathbb{C}([-\tau,0], \mathbb{R}^n), \quad \forall\, \eta \in [-\tau, 0] \qquad (7.103)\\
y(t) &= \Big[C_o(\eta_t) + \Delta C_o(t,\eta_t)\Big]x(t) \\
&+ \Big[C_d(\eta_t) + \Delta C_d(t,\eta_t)\Big]x(t-\tau) + M(\eta_t)w(t) \\
&= C_{\Delta o}(t,\eta_t)x(t) + C_{\Delta d}(t,\eta_t)x(t-\tau) \\
&+ M(\eta_t)w(t) \qquad (7.104)\\
z(t) &= L(\eta_t)x(t) \qquad (7.105)
\end{aligned}
$$

where $x(t) \in \mathbb{R}^n$ is the system state , $y(t) \in \mathbb{R}^m$ is the measurement output , $z(t) \in \mathbb{R}^p$ is the controlled output , $w(t) \in \mathcal{L}_2[0,\infty)$ is the disturbance input, $z(t) \in \mathbb{R}^r$ is the controlled output which belongs to $\mathcal{L}_2\big[(\Omega,\mathcal{F},\mathbf{P}),[0,\infty)\big]$ and the factor $\tau > 0$ is a constant scalar representing the amount of time-lag in the state.

Two points are in order. First the model (7.102)-(7.105) is a special case of the model (7.2)-(7.5) by setting $p = 1$ and our selection this way is to simplify the analysis to follow. Second as mentioned before, the term $\mathcal{M}(x_t) : \mathbb{C}[-\tau, 0] \rightarrow \mathbb{R}^n \stackrel{\Delta}{=} x(t) - Bx(t-\tau)$ is called the *difference operator.*

For each possible value $\eta_t = i \in \mathcal{S}$, we will denote the system matrices of $(\Sigma_{\Delta n})$ associated with mode i by

$$
\begin{aligned}
A_o(\eta_t) &\stackrel{\Delta}{=} A_o(i) \;,\; A_d(\eta_t) \stackrel{\Delta}{=} A_d(i) \;,\; N(\eta_t) \stackrel{\Delta}{=} N(i) \\
C_o(\eta_t) &\stackrel{\Delta}{=} C_o(i) \;,\; C_d(\eta_t) \stackrel{\Delta}{=} C_d(i) \;,\; M(\eta_t) \stackrel{\Delta}{=} M(i) \\
B(\eta_t) &\stackrel{\Delta}{=} B(i)
\end{aligned}
\tag{7.106}
$$

where $A_o(i) \in \mathbb{R}^{n \times n}$, $A_d(i) \in \mathbb{R}^{n \times n}$, $B(i) \in \mathbb{R}^{n \times n}$, $C_o(i) \in \mathbb{R}^{m \times n}$, $C_d(i) \in \mathbb{R}^{m \times n}$, $N(i) \in \mathbb{R}^{n \times r}$, $M(i) \in \mathbb{R}^{m \times r}$ and $L(i) \in \mathbb{R}^{p \times n}$ are known real constant matrices. $A_o(i), A_d(i), C_o(i), C_d(i), B(i), N(i)$ and $M(i)$ are known real constant matrices of appropriate dimensions which describe the nominal system of $(\Sigma_{\Delta n})$. The matrices $\Delta A_o(t, \eta_t)$, $\Delta A_d(t, \eta_t)$ $\Delta C_o(t, \eta_t)$, and $\Delta C_d(t, \eta_t)$ are real, time-varying matrix functions representing the norm-bounded parameter uncertainties. For $\eta_t = i$, the admissible uncertainties are represented by:

$$
\begin{aligned}
\begin{bmatrix} \Delta A_o(t,i) & \Delta A_d(t,i) \\ \Delta C_o(t,i) & \Delta C_d(t,i) \end{bmatrix} &= \begin{bmatrix} M_a(i) \\ M_c(i) \end{bmatrix} \Delta(t) \, [N_a(i) \quad N_d(i)] \\
\forall \Delta^t(t) \Delta(t) &\leq I \;,\; \forall t
\end{aligned}
\tag{7.107}
$$

where $M_a(i) \in \mathbb{R}^{n \times \alpha}$, $M_c(i) \in \mathbb{R}^{p \times \alpha}$, $N_a(i) \in \mathbb{R}^{\beta \times n}$ and $N_d(i) \in \mathbb{R}^{\beta \times n}$, are known real constant matrices and $\Delta(t) \in \mathbb{R}^{\alpha \times \beta}$ is an unknown matrix with Lebesgue measurable elements. The initial condition is specified as $\beta_o \stackrel{\Delta}{=} \langle x(0), x(s) \rangle = \langle x_o, \phi(s) \rangle$, where $\phi(\cdot) \in \mathcal{L}_2[-\tau, 0]$.

In the absence of uncertainties ($\Delta(.) \equiv 0$) and for each possible value $\eta_t = i$, $i \in \mathcal{S}$, we obtain the nominal neutral system

$$
(\Sigma_n) \;\; \mathcal{M}(\dot{x}_t) \stackrel{\Delta}{=} \dot{x}(t) - B(i)\dot{x}(t-\tau) = A_o(i)x(t)
$$

$$
\begin{aligned}
& + \quad A_d(i)x(t-\tau) + N(i)w(t) && (7.108)\\
x(\eta) &= \phi(\eta) \in \mathbb{C}([-\tau,0],\mathbb{R}^n), \quad \forall\, \eta \in [-\tau,0] && (7.109)\\
y(t) &= C_o(i)x(t) + C_d(i)x(t-\tau) + M(i)w(t) && (7.110)\\
z(t) &= L(i)x(t) && (7.111)
\end{aligned}
$$

Assumptions (7.1)-(7.2) are recalled here on systems $(\Sigma_{\Delta n})$ and (Σ_n) as well.

Our primary objective here is to design robust state and robust $\mathcal{H}_\infty$ observers for the neutral system $(\Sigma_{\Delta n})$ with some desirable stability behavior and then extend these designs to the neutral system (Σ_n). Towards our goal, we Let $\mathsf{X}(t,\beta_o,\eta_o)$ denote the trajectory of the state $x(t)$ from the initial state (β_o,η_o) and recall **Definition** 7.1 concerning the robust stochastic stability with weak delay dependence

7.5.1 Structure of Observer

In the sequel, to derive the state estimate $\hat{x} \in \mathbb{R}^n$ we will utilize the following linear Markovian state-delayed observer for each possible value $\eta_t = i,\ i \in \mathcal{S}$

$$
\begin{aligned}
(\Sigma_e) \quad \mathcal{M}(\dot{\hat{x}}_t) &\stackrel{\Delta}{=} \dot{\hat{x}}(t) - B\dot{\hat{x}}(t-\tau) = A_f(i)\hat{x}(t)\\
&+ \quad A_d(i)\hat{x}(t-\tau) + K_f(i)\,y(t)\,, \quad \hat{x}(0) = 0\\
\hat{z}(t) &= L(i)\hat{x}(t) \qquad\qquad (7.112)
\end{aligned}
$$

where $A_f(i) \in \mathbb{R}^{n\times n}$, $K_f(i) \in \mathbb{R}^{n\times m}, i \in \mathcal{S}$ are the observer matrix gains to be designed such that $\hat{x}$ reproduce x asymptotically for all admissible uncertainties satisfying (7.107).

Let the state error be

$$
\tilde{x} \stackrel{\Delta}{=} x(t) \;-\; \hat{x}(t) \qquad (7.113)
$$

From (7.102)-(7.104) and (7.112)-(7.113), the state error dynamics can be represented by

$$
\begin{aligned}
(\Sigma_{\Delta e}) \quad \mathcal{M}(\dot{\tilde{x}}_t) &\triangleq \dot{\tilde{x}}(t) - B(i)\dot{\tilde{x}}(t-\tau) \\
&= A_f(i)\tilde{x}(t) + A_d(i)\tilde{x}(t-\tau) \\
&+ \big[A_o(i) - K_f(i)C_o(i) - A_f(i)\big]x(t) \\
&+ \big[\Delta A_o(t,i) - K_f(i)\Delta C_o(t,i)\big]x(t) \\
&- K_f(i)[C_d(i) + \Delta C_d(t,i)]x(t-\tau) \\
&+ \big[N(i) - K_f(i)M(i)\big]w(t) \qquad (7.114)
\end{aligned}
$$

A state-space augmented model of the observation error

$$
\tilde{z}(t) = z(t) - \hat{z}(t)
$$

can then be constructed in terms of the augmented state vector and the extended matrix $\mathbb{B}(i)$ for each possible value $\eta_t = i \in \mathcal{S}$

$$
\zeta(t) \triangleq \begin{bmatrix} \tilde{x}(t) \\ x(t) \end{bmatrix} , \quad \mathbb{B}(i) = \begin{bmatrix} B(i) & 0 \\ 0 & B(i) \end{bmatrix} \qquad (7.115)
$$

by using (7.102)-(7.105) and (7.114)-(7.115) as follows:

$$
\begin{aligned}
(\Sigma_{\Delta A}): \quad \mathcal{M}(\dot{\zeta}_t) &\triangleq \dot{\zeta}(t) - \mathbb{B}(i)\dot{\zeta}(t-\tau) \\
&= \mathcal{A}_{\Delta o}(t,i)\zeta(t) + \mathcal{D}_{\Delta o}(t,i)\zeta(t-\tau) \\
&+ \mathsf{B}(i)w(t) \qquad (7.116) \\
\tilde{z}(t) &= L_f(i)\zeta(t) \qquad (7.117)
\end{aligned}
$$

where

$$
\begin{aligned}
\mathcal{A}_{\Delta o}(t,i) &= \Big[\mathsf{A}(i) + \bar{H}_A(i)\Delta(t,i)\bar{E}_A(i)\Big] \\
\mathcal{D}_{\Delta o}(t,i) &= \Big[\mathsf{D}(i) + \bar{H}_D(i)\Delta(t,i)\bar{E}_D(i)\Big]
\end{aligned}
$$

$$
\begin{aligned}
L_f(i) &= \begin{bmatrix} L(i) & 0 \end{bmatrix} \\
\mathsf{A}(i) &= \begin{bmatrix} A_f(i) & A_o(i) - K_f(i)C_o(i) - A_f(i) \\ 0 & A_o(i) \end{bmatrix} \\
\zeta(t-\tau) &= \begin{bmatrix} \tilde{x}(t-\tau) \\ x(t-\tau) \end{bmatrix}, \; \bar{E}_A(i) = [0 \quad N_a(i)] \\
\bar{E}_D(i) &= [0 \quad N_d(i)] \qquad (7.118) \\
\mathsf{D}(i) &= \begin{bmatrix} A_d(i) & -K_f(i)C_d(i) \\ 0 & A_d(i) \end{bmatrix} \\
\mathsf{B}(i) &= \begin{bmatrix} N(i) - K_f(i)M(i) \\ N(i) \end{bmatrix} \qquad (7.119) \\
\bar{H}_A &= \begin{bmatrix} M_a(i) - K_f(i)H_c(i) \\ M_a(i) \end{bmatrix} \\
\bar{H}_D(i) &= \begin{bmatrix} -K_f(i)H_c(i) \\ H_c(i) \end{bmatrix} \qquad (7.120)
\end{aligned}
$$

Had we followed another route and combined systems (Σ_n) and (Σ_e), we would have obtained the nominal augmented system

$$
\begin{aligned}
(\Sigma_A): \quad \mathcal{M}(\dot{\zeta}_t) &\triangleq \dot{\zeta}(t) - \mathbb{B}(i)\dot{\zeta}(t-\tau) \\
&= \mathsf{A}(i)\zeta(t) + \mathsf{D}(i)\zeta(t-\tau) + \mathsf{B}(i)w(t) \qquad (7.121) \\
\tilde{z}(t) &= L_f(i)\zeta(t) \qquad (7.122)
\end{aligned}
$$

Remark 7.5 : *It should be stressed that system $(\Sigma_{\Delta A})$ describes a linear uncertain jumping system of neutral-type the nominal version of which is represented by systems (Σ_A). The matrices of both systems depend on the gains $A_f(i)$, $K_f(i)$, $i \in \mathcal{S}$.*

Following the results developed in section 7.2, the following theorems establish the stability behavior of system $(\Sigma_{\Delta A})$ or (Σ_A).

Theorem 7.11 *Given gain matrices $A_f(i), K_f(i), i \in \mathcal{S}$ and subject to* **Assumptions** *(7.1)-(7.2), the neutral system $(\Sigma_{\Delta A})$ with $w \equiv 0$ is* **RSSWDD** *if* *for given matrices* $0 < \mathbb{L}(i) = \mathbb{L}^t(i) \in \mathbb{R}^{2n\times 2n}$, $i \in \mathcal{S}$, *and letting*

$$\begin{aligned} \bar{\mathbb{L}}(i) &= \mathbb{L}(i) - \xi^{-1}(i) \sum_{m=1}^{s} \alpha_{im} \mathbb{L}(m) \\ \hat{\mathbb{L}}(i) &= \mathbb{L}(i) + \xi(i) \sum_{m=1}^{s} \alpha_{im} \mathbb{L}(m), \qquad i \in \mathcal{S} \end{aligned}$$

for some scalars $\xi(i) > 0,\ i \in \mathcal{S}$, *there exist matrices* $0 < \mathbb{P}(i) = \mathbb{P}^t(i) \in \mathbb{R}^{2n\times 2n}$, $i \in \mathcal{S}$ *and scalars* $\varepsilon(i) > 0,\ \varrho(i) > 0,\ i \in \mathcal{S}$ *satisfying the following LMIs for all* $i \in \mathcal{S}$:

$$\begin{bmatrix} \Upsilon_t & \Lambda_a(i) & \Lambda_n(i) \\ \Lambda_a^t(i) & -\Phi_a(i) & 0 \\ \Lambda_n^t(i) & 0 & -\Theta_a(i) \end{bmatrix} < 0 \tag{7.123}$$

$$\begin{bmatrix} -\bar{\mathbb{L}}(i) & \mathbb{B}^t(i)\hat{\mathbb{L}}(i) & \bar{E}_D^t(i) & \varepsilon(i)\mathbb{B}^t(i)\bar{E}_A^t(i) \\ \hat{\mathbb{L}}(i)\mathbb{B}(i) & \hat{\mathbb{L}}(i) & 0 & 0 \\ \bar{E}_D(i) & 0 & -I & 0 \\ \varepsilon(i)\bar{E}_A(i)\mathbb{B}(i) & 0 & 0 & -\varepsilon(i)I \end{bmatrix} < 0 \tag{7.124}$$

where

$$\begin{aligned} \Upsilon_t &= \mathbb{P}(i)\mathsf{A}(i) + \mathsf{A}^t(i)\mathbb{P}(i) + \hat{\mathbb{L}}(i) \\ &+ \sum_{m=1}^{s} \alpha_{im}\mathbb{P}(m) + \varrho(i)\bar{E}_A^t(i)\bar{E}_A(i) \\ \Lambda_a &= [\mathbb{P}(i)\bar{H}_A(i) \quad \mathbb{P}(i)\bar{H}_A(i) \quad \mathbb{P}(i)\bar{H}_D(i)] \\ \Phi_a &= diag[\varrho(i)I \quad \varepsilon(i)I \quad \varepsilon(i)I] \\ \Theta_a &= \bar{\mathbb{L}}(i) - \mathbb{B}^t(i)\hat{\mathbb{L}}(i)\mathbb{B}(i) \\ &- \varepsilon(i)\big(\mathbb{B}^t(i)\bar{E}_A^t(i)\bar{E}_A(i)\mathbb{B}(i) + \bar{E}_D^t(i)\bar{E}_D(i)\big) \\ \Lambda_n(i) &= \mathbb{P}(i)[\mathsf{A}(i)\mathbb{B}(i) + \mathsf{D}(i)] + \hat{\mathbb{L}}(i)\mathbb{B}(i) \end{aligned} \tag{7.125}$$

for all admissible uncertainties satisfying (7.107)

Proof: Follows from parallel development to **Theorem** 7.1. ∇∇∇

Theorem 7.12 *Given gain matrices $A_f(i), K_f(i), i \in \mathcal{S}$ and subject to* **Assumptions** *(7.1)-(7.2), the neutral system (Σ_A) with $w \equiv 0$ is* **stochastically stable with weak delay dependence (SSWDD)** *if for given matrices $0 < I\!L(i) = I\!L^t(i) \in I\!R^{2n \times 2n}$, $i \in \mathcal{S}$, and letting*

$$\begin{aligned} \bar{I\!L}(i) &= I\!L(i) - \xi^{-1}(i) \sum_{m=1}^{s} \alpha_{im} I\!L(m) \\ \hat{I\!L}(i) &= I\!L(i) + \xi(i) \sum_{m=1}^{s} \alpha_{im} I\!L(m) \end{aligned}$$

for some scalars $\xi(i) > 0$, $i \in \mathcal{S}$, there exist matrices $0 < I\!P(i) = I\!P^t(i) \in I\!R^{2n \times 2n}$, $i \in \mathcal{S}$ satisfying the following LMIs for all $i \in \mathcal{S}$:

$$\begin{bmatrix} \Upsilon_{to} & \Lambda_n(i) \\ \Lambda_n^t(i) & -\Theta_{ao}(i) \end{bmatrix} < 0$$

$$\begin{bmatrix} -\bar{I\!L}(i) & I\!B^t(i)\hat{I\!L}(i) \\ \hat{I\!L}(i) I\!B(i) & \hat{I\!L}(i) \end{bmatrix} < 0 \tag{7.126}$$

where

$$\begin{aligned} \Upsilon_{to} &= I\!P(i)\mathsf{A}(i) + \mathsf{A}^t(i) I\!P(i) + \hat{I\!L}(i) + \sum_{m=1}^{s} \alpha_{im} I\!P(m) \\ \Theta_{ao} &= \bar{I\!L}(i) - I\!B^t(i)\hat{I\!L}(i) I\!B(i) \end{aligned} \tag{7.127}$$

Proof: Follows from **Theorem** (7.11) by setting $\bar{E}_A^t \equiv 0$, $\bar{E}_D^t \equiv 0, \bar{H}_A^t \equiv 0$. $\nabla\nabla\nabla$

Remark 7.6 *It should be remarked that both* **Theorems** *(7.11) and (7.12) offer new analytical results for the class of neutral-type dynamical systems under consideration. The results are cast in LMI format for which the* **MATLAB-LMI** *software is readily available [57]. More importantly, in the case $B(i) \equiv 0 \Rightarrow \mathcal{M}(.) = x_t$, systems $(\Sigma_{\Delta A})$ and (Σ_A) become of retarded-type for which* **Theorems** *(7.11) and (7.12) retrieve the results of [114, 149].*

Remark 7.7 *The need for* **Assumption** *(7.2) is quite evident from (7.125) and (7.127) in which case the conditions $\Theta_a > 0, \Theta_{ao} > 0$ are required, respectively. In both cases, the result reveals a discrete Lyapunov inequality.*

7.5.2 Design Procedure

Proceeding further, we now provide expressions for the gain matrices of the observer (7.112) when applied to the neural systems $(\Sigma_{\Delta A})$ and (Σ_A). To facilitate further development, we introduce the following matrix expressions for some scalars $\varepsilon(i) > 0,\ \varrho(i) > 0,\ i \in \mathcal{S}$:

$$\mathbb{P}(i) = \begin{bmatrix} \mathbb{P}_e(i) & 0 \\ 0 & \mathbb{P}_x(i) \end{bmatrix}, \quad \mathbb{L}(i) = \begin{bmatrix} \mathbb{L}_e(i) & 0 \\ 0 & \mathbb{L}_x(i) \end{bmatrix} \tag{7.128}$$

$$\begin{aligned} \Xi_a(i) &= \bar{\mathbb{L}}_e(i) - B^t(i)\hat{\mathbb{L}}_e(i)B(i) \\ \Xi_b(i) &= \bar{\mathbb{L}}_x(i) - B^t(i)\hat{\mathbb{L}}_x(i)B(i) \\ &- \varepsilon(i)\big(N_a^t(i)N_a(i) + N_d^t(i)N_d(i)\big) \\ \Xi_c(i) &= B(i)\Xi_b^{-1}(i)\Big[B^t(i)\hat{\mathbb{L}}_x(i) \\ &+ [A_d^t(i) + B^t(i)A_o^t(i)]\mathbb{P}_x(i)\Big] \end{aligned} \tag{7.129}$$

$$\begin{aligned} \hat{A}_o(i) &= A_o(i) \\ &+ \varrho^{-1}(i)M_a(i)M_a^t(i)\mathbb{P}_x(i)\big(I + \Xi_c(i)\big)^{-1} \\ \hat{C}_o(i) &= C_o(i) \\ &+ \Big[C_d(i)B^{-1}(i)\Xi_c(i) + \big(\varepsilon^{-1}(i) + \varrho^{-1}(i)\big)M_c(i)M_a^t(i)\mathbb{P}_x(i) \\ &+ \varepsilon^{-1}(i)M_c(i)M_c^t(i)\mathbb{P}_x(i)\Big]\big(I + \Xi_c(i)\big)^{-1} \end{aligned} \tag{7.130}$$

$$\begin{aligned} \Xi_k(i) &= \Big(2\,\varepsilon^{-1}(i) + \varrho^{-1}(i)\Big)M_c(i)M_c^t(i) \\ &+ \hat{C}_o(i)B(i)\Xi_a^{-1}(i)B^t(i)\hat{C}_o^t(i) \\ &+ \Big[[\hat{C}_o(i) + C_o(i)]B(i) + C_d(i)\Big]\Xi_b^{-1}(i) \end{aligned}$$

$$
\cdot \quad \left[[\hat{C}_o(i) + C_o(i)]B(i) + C_d(i) \right]^t \tag{7.131}
$$

$$
\begin{aligned}
\Xi_g(i) &= \left(\varepsilon^{-1}(i) + \varrho^{-1}(i) \right) M_a(i) M_c^t(i) \\
&+ \left(\hat{A}_o(i) + A_d(i) + \mathbb{L}_e(i)B(i) \right) \Xi_a^{-1}(i) B^t(i) \hat{C}_o(i) \\
&+ \left(\hat{A}_o(i) - A_o(i) + \mathbb{L}_e(i)B(i) \right) \Xi_b^{-1}(i) \\
&\cdot \left[[\hat{C}_o(i) + C_o(i)]B(i) + C_d(i) \right]^t
\end{aligned} \tag{7.132}
$$

The main results are summarized by the following theorems:

Theorem 7.13 *The neutral augmented system* $\Sigma_{\Delta A}$ *is* **RSSWDD**, *if given matrices* $0 < I\!\!L_e(i) = I\!\!L_e^t(i) \in I\!\!R^{n \times n}$, $0 < I\!\!L_x(i) = I\!\!L_x^t(i) \in I\!\!R^{n \times n}$ *and scalars* $\xi(i) > 0$ *such that for all* $i \in \mathcal{S}$

$$
\begin{aligned}
\bar{I\!\!L}_x(i) &= I\!\!L_x(i) - \xi^{-1}(i) \sum_{m=1}^{s} \alpha_{im} I\!\!L_x(m) \\
\hat{I\!\!L}_x(i) &= I\!\!L_x(i) + \xi(i) \sum_{m=1}^{s} \alpha_{im} I\!\!L_x(m) \\
\bar{I\!\!L}_e(i) &= I\!\!L_e(i) - \xi^{-1}(i) \sum_{m=1}^{s} \alpha_{im} I\!\!L_e(m) \\
\hat{I\!\!L}_e(i) &= I\!\!L_x(i) + \xi(i) \sum_{m=1}^{s} \alpha_{im} I\!\!L_e(m)
\end{aligned}
$$

there exist matrices $0 < I\!\!P_e(i) = I\!\!P_e^t(i) \in I\!\!R^{n \times n}$, $0 < I\!\!P_x(i) = I\!\!P_x^t(i) \in I\!\!R^{n \times n}$, $i \in \mathcal{S}$ *and scalars* $\varepsilon(i) > 0$, $\varrho(i) > 0$ *satisfying the LMIs for all* $i \in \mathcal{S}$

$$
\begin{bmatrix}
\Upsilon_e(i) & \Lambda_b(i) & \Lambda_d(i) & \Lambda_f(i) \\
\Lambda_b^t(i) & -\Phi_b(i) & 0 & 0 \\
\Lambda_d^t(i) & 0 & -\Xi_a(i) & 0 \\
\Lambda_f^t(i) & 0 & 0 & -\Xi_b(i)
\end{bmatrix} < 0
$$

$$
\begin{bmatrix}
\Upsilon_x(i) & \Lambda_c(i) & \Lambda_h(i) \\
\Lambda_c^t(i) & -\Phi_c(i) & 0 \\
\Lambda_h^t(i) & 0 & -\Xi_b(i)
\end{bmatrix} < 0 \tag{7.133}
$$

where

$$
\begin{aligned}
\Upsilon_e(i) &= I\!P_e(i)\hat{A}_o(i) + \hat{A}_o^t(i) I\!P_e(i) + \hat{I\!L}_e(i) \\
&+ \sum_{m=1}^{s} \alpha_{im} I\!P_e(m) - I\!P_e \Xi_g^t \Xi_k^{-1} \Xi_g I\!P_e \\
\Lambda_b(i) &= [I\!P_e(i) M_a(i) \quad I\!P_e(i) M_a(i)] \\
\Phi_b(i) &= diag[\varepsilon(i) I \quad \varrho(i) I] \\
\Lambda_d(i) &= I\!P_e(i)[\hat{A}_o(i) + A_d(i)] + \hat{I\!L}_e(i) B(i) \\
\Lambda_f(i) &= I\!P_e(i)[\hat{A}_o(i) - A_o(i)] + \hat{I\!L}_e(i) B(i) \qquad (7.134) \\
\Upsilon_x(i) &= I\!P_x(i) A_o(i) + A_o^t(i) I\!P_x(i) + \hat{I\!L}_x(i) \\
&+ \varrho N_a^t(i) N_a(i) + \sum_{m=1}^{s} \alpha_{im} I\!P_x(m) \\
\Lambda_h(i) &= I\!P_x(i) A_o(i) B(i) + I\!P_x(i) A_d(i) + \hat{I\!L}_x(i) B(i) \\
\Lambda_c(i) &= [I\!P_e(i) M_a(i) \quad I\!P_e(i) M_a(i) \quad I\!P_x(i) M_c(i)] \\
\Phi_c(i) &= diag[\varepsilon(i) I \quad \varrho(i) I \quad \varepsilon(i) I] \qquad (7.135)
\end{aligned}
$$

for all admissible uncertainties satisfying (7.107). Moreover, the estimator gains are given by:

$$
\begin{aligned}
K_f(i) &= \Xi_g(i)\, \Xi_k^{-1}(i) \\
A_f(i) &= \hat{A}_o(i) - K_f(i) \hat{C}_o(i) \qquad (7.136)
\end{aligned}
$$

Proof: Extending on **Theorem** 7.12 by using (7.118)-(7.120) and (7.128)-(7.132) into (7.126) and expanding terms we express the result into the block form

$$
\begin{bmatrix} \Sigma_{11}(i) & \Sigma_{12}(i) \\ \Sigma_{12}^t(i) & \Sigma_{22}(i) \end{bmatrix}
$$

where

$$
\Sigma_{11}(i) = \mathbb{P}_e(i) A_f(i) + A_f^t(i) \mathbb{P}_e(i) + \hat{\mathbb{L}}_e(i) + \sum_{m=1}^{s} \alpha_{im} \mathbb{P}_e(m)
$$

$$
\begin{aligned}
&+ \; \big(\varepsilon^{-1}(i) + \varrho^{-1}(i)\big)\mathbb{P}_e(i)\big[M_a(i) - K_f(i)M_c(i)\big] \\
&\cdot \; \big[M_a(i) - K_f(i)M_c(i)\big]^t\mathbb{P}_e(i) \\
&+ \; \varepsilon^{-1}(i)\mathbb{P}_e(i)K_f(i)M_c(i)M_c^t(i)K_f^t(i)\mathbb{P}_e(i) \\
&+ \; \Big[\mathbb{P}_e(i)[A_o(i) - A_f(i) - K_f(i)C_o(i)]B(i) - \mathbb{P}_e(i)K_f(i)C_d(i)\Big]\Xi_b^{-1}(i) \\
&\cdot \; \Big[\mathbb{P}_e(i)[A_o(i) - A_f(i) - K_f(i)C_o(i)]B(i) - \mathbb{P}_e(i)K_f(i)C_d(i)\Big]^t \\
&+ \; \Big[\mathbb{P}_e(i)[A_f(i)B(i) + A_d(i)] + \hat{\mathbb{L}}_e(i)B(i)\Big]\Xi_a^{-1}(i) \\
&\cdot \; \Big[\mathbb{P}_e(i)[A_f(i)B(i) + A_d(i)] + \hat{\mathbb{L}}_e(i)B(i)\Big]^t
\end{aligned} \tag{7.137}
$$

$$
\begin{aligned}
\Sigma_{22}(i) \; &= \; \mathbb{P}_x(i)A_o(i) + A_o^t(i)\mathbb{P}_x(i) + \hat{\mathbb{L}}_x(i) \\
&+ \; \big(\varepsilon^{-1}(i) + \varrho^{-1}(i)\big)\mathbb{P}_x(i)M_a(i)M_a^t(i)\mathbb{P}_x(i) \\
&+ \; \varepsilon^{-1}(i)\mathbb{P}_x(i)M_c(i)M_c^t(i)\mathbb{P}_x(i) \\
&+ \; \Big[\mathbb{P}_x(i)[A_o(i)B(i) + A_d(i)] + \hat{\mathbb{L}}_x(i)B(i)\Big]\Xi_b^{-1}(i) \\
&\cdot \; \Big[\mathbb{P}_x(i)[A_o(i)B(i) + A_d(i)] + \hat{\mathbb{L}}_x(i)B(i)\Big]^t \\
&+ \; \sum_{m=1}^{s} \alpha_{im}\mathbb{P}_x(m) + \varrho(i)E_a^t(i)E_a(i)
\end{aligned} \tag{7.138}
$$

$$
\begin{aligned}
\Sigma_{12}(i) \; &= \; \Big[\mathbb{P}_e(i)[A_o(i) - A_f(i) - K_f(i)C_o(i)]B(i) - \mathbb{P}_e(i)K_f(i)C_d(i)\Big]\Xi_b^{-1}(i) \\
& \; \Big[\mathbb{P}_x(i)[A_o(i)B(i) + A_d(i) + \hat{\mathbb{L}}_x(i)B(i)\Big]^t \\
&- \; \varepsilon^{-1}\mathbb{P}_e(i)K_f(i)M_c(i)M_c^t(i)K_f^t(i)\mathbb{P}_x(i) \\
&+ \; \mathbb{P}_e(i)[A_o(i) - A_f(i) - K_f(i)C_o(i)] \\
&+ \; \big(\varepsilon^{-1}(i) + \varrho^{-1}(i)\big)\mathbb{P}_e(i)\big[M_a(i) - K_f(i)H_c(i)\big]H_a^t(i)\mathbb{P}_x(i)
\end{aligned} \tag{7.139}
$$

Applying **Fact 1** to the matrix block Σ_{22}, we can readily obtain one of the LMIs (7.133). The substitution of (7.130)-(7.132) into (7.139) renders $\Sigma_{12} \equiv 0$.

Using (7.130)-(7.132) and (7.136) into (7.137) with some matrix manipulations and applying the Schur complements we get the other LMI (7.133). $\nabla\nabla\nabla$

Theorem 7.14 *The neutral augmented system* Σ_A *is* **(SSWDD)**, *if given matrices* $0 < \mathbb{L}_e(i) = \mathbb{L}_e^t(i) \in \mathbb{R}^{n\times n}$, $0 < \mathbb{L}_x(i) = \mathbb{L}_x^t(i) \in \mathbb{R}^{n\times n}$ *and scalars* $\xi(i) > 0$ *such that*

$$\begin{aligned}
\bar{\mathbb{L}}_x(i) &= \mathbb{L}_x(i) - \xi^{-1}(i)\sum_{m=1}^{s}\alpha_{im}\mathbb{L}_x(m) \\
\hat{\mathbb{L}}_x(i) &= \mathbb{L}_x(i) + \xi(i)\sum_{m=1}^{s}\alpha_{im}\mathbb{L}_x(m) \\
\bar{\mathbb{L}}_e(i) &= \mathbb{L}_e(i) - \xi^{-1}(i)\sum_{m=1}^{s}\alpha_{im}\mathbb{L}_e(m) \\
\hat{\mathbb{L}}_e(i) &= \mathbb{L}_x(i) + \xi(i)\sum_{m=1}^{s}\alpha_{im}\mathbb{L}_e(m)
\end{aligned}$$

there exist matrices $0 < \mathbb{P}_e(i) = \mathbb{P}_e^t(i) \in \mathbb{R}^{n\times n}$, $0 < \mathbb{P}_x(i) = \mathbb{P}_x^t(i) \in \mathbb{R}^{n\times n}$, $i \in \mathcal{S}$ *satisfying the LMIs for all* $i \in \mathcal{S}$

$$\begin{aligned}
&\begin{bmatrix} \Upsilon_{eo}(i) & \Lambda_{do}(i) & \hat{\mathbb{L}}_e(i)B(i) \\ \Lambda_{do}^t(i) & -\Xi_a(i) & 0 \\ B^t(i)\hat{\mathbb{L}}_e^t(i) & 0 & -\Xi_{bo}(i) \end{bmatrix} < 0 \\
&\begin{bmatrix} \Upsilon_{xo}(i) & \Lambda_h(i) \\ \Lambda_h^t(i) & -\Xi_{bo}(i) \end{bmatrix} < 0
\end{aligned} \qquad (7.140)$$

where

$$\begin{aligned}
\Upsilon_{eo}(i) &= \mathbb{P}_e(i)A_o(i) + A_o^t(i)\mathbb{P}_e(i) + \hat{\mathbb{L}}_e(i) \\
&+ \sum_{m=1}^{s}\alpha_{im}\mathbb{P}_e(m) - \mathbb{P}_e\Xi_{go}^t\Xi_{ko}^{-1}\Xi_{go}\mathbb{P}_e \\
\Lambda_{do}(i) &= \mathbb{P}_e(i)[A_o(i) + A_d(i)] + \hat{\mathbb{L}}_e(i)B(i) \\
\Upsilon_{xo}(i) &= \mathbb{P}_x(i)A_o(i) + A_o^t(i)\mathbb{P}_x(i) + \hat{\mathbb{L}}_x(i) + \sum_{m=1}^{s}\alpha_{im}\mathbb{P}_x(m) \\
\Lambda_h(i) &= \mathbb{P}_x(i)A_o(i)B(i) + \mathbb{P}_x(i)A_d(i) + \hat{\mathbb{L}}_x(i)B(i)
\end{aligned} \qquad (7.141)$$

Moreover, the estimator gains are given by:

$$\begin{aligned} K_f(i) &= \Xi_{go}(i)\,\Xi_{ko}^{-1}(i) \\ A_f(i) &= A_o(i) - K_f(i)\bar{C}_o(i) \end{aligned} \tag{7.142}$$

Proof: Define

$$\begin{aligned} \Xi_{bo}(i) &= \bar{\mathbb{L}}_x(i) - B^t(i)\hat{\mathbb{L}}_x(i)B(i) \\ \Xi_{co}(i) &= B(i)\Xi_{bo}^{-1}(i)\left[B^t(i)\hat{\mathbb{L}}_x(i) + [A_d^t(i) + B^t(i)A_o^t(i)]\mathbb{P}_x(i)\right] \\ \bar{C}_o(i) &= C_o(i) + C_d(i)B^{-1}(i)\Xi_{co}(i)\big(I + \Xi_{co}(i)\big)^{-1} \\ \Xi_{ko}(i) &= \bar{C}_o(i)B(i)\Xi_a^{-1}(i)B^t(i)\bar{C}_o^t(i) \\ &+ \left[[\bar{C}_o(i) + C_o(i)]B(i) + C_d(i)\right]\Xi_{bo}^{-1}(i) \\ &\cdot \left[[\bar{C}_o(i) + C_o(i)]B(i) + C_d(i)\right]^t \\ \Xi_{go}(i) &= \left(A_o(i) + A_d(i) + \hat{\mathbb{L}}_e(i)B(i)\right)\Xi_a^{-1}(i)B^t(i)\bar{C}_o(i) \\ &+ \hat{\mathbb{L}}_e(i)B(i)\Xi_{co}^{-1}(i)\left[[\bar{C}_o(i) + C_o(i)]B(i) + C_d(i)\right]^t \end{aligned} \tag{7.143}$$

By setting $\varrho \equiv 0, \varepsilon \equiv 0, M_a \equiv 0, N_a \equiv 0, M_c \equiv 0, N_d \equiv 0$ while using (7.143) and following similar technique to the one employed in **Theorem** 7.13, the desired result is achieved. $\nabla\nabla\nabla$

7.5.3 $\mathcal{H}_\infty$ Performance

In order to improve the foregoing robust observer results further, one would direct the design effort on robust observation in an $\mathcal{H}_\infty$ setting. Therefore, our immediate objective is to design robust observers for the neutral system $(\Sigma_{\Delta n})$ with some desirable stability behavior and guaranteed $\mathcal{H}_\infty$ performance and then extend this design to the neutral system (Σ_n).

Based thereon, the following theorems could be established in a straightforward manner:

Theorem 7.15 *Given gain matrices* $A_f(i), K_f(i), i \in \mathcal{S}$ *and subject to* **Assumptions** *(7.1)-(7.2), the neutral system* $(\Sigma_{\Delta A})$ *is* **RSSWDD with a disturbance attenuation** γ *if given matrices* $0 < \mathbb{L}(i) = \mathbb{L}^t(i) \in \mathbb{R}^{2n \times 2n}$ *and scalars* $\xi(i) > 0$ *such that for all* $i \in \mathcal{S}$

$$\begin{aligned} \bar{\mathbb{L}}(i) &= \mathbb{L}(i) - \xi^{-1}(i) \sum_{m=1}^{s} \alpha_{im} \mathbb{L}_x(m) \\ \hat{\mathbb{L}}(i) &= \mathbb{L}(i) + \xi(i) \sum_{m=1}^{s} \alpha_{im} \mathbb{L}(m) \end{aligned}$$

there exist matrices $0 < \mathbb{P}(i) = \mathbb{P}^t(i) \in \mathbb{R}^{2n \times 2n}$, $i \in \mathcal{S}$ *and scalars* $\gamma > 0$, $\varepsilon(i) > 0$, $\varrho(i) > 0$ *satisfying the LMIs for all* $i \in \mathcal{S}$

$$\begin{bmatrix} \Upsilon_w(i) & \Lambda_a(i) & \Gamma_a(i) & \mathbb{P}(i)\mathsf{B}(i) \\ \Lambda_a^t(i) & -\Phi_a(i) & 0 & 0 \\ \Gamma_a^t(i) & 0 & -\Theta_b(i) & 0 \\ \mathsf{B}^t(i)\mathbb{P}(i) & 0 & 0 & \begin{matrix} -\gamma^2 I + \\ L_f^t(i) L_f(i) \end{matrix} \end{bmatrix} < 0 \tag{7.144}$$

$$\begin{bmatrix} -\bar{\mathbb{L}}(i) & \mathbb{B}^t(i) L_f^t(i) & \mathbb{B}^t(i)\hat{\mathbb{L}}(i) & \bar{E}_D^t(i) & \varepsilon \mathbb{B}^t(i) \bar{E}_A^t(i) \\ L_f(i)\mathbb{B}(i) & -I & 0 & 0 & 0 \\ \hat{\mathbb{L}}(i)\mathbb{B}(i) & 0 & -\hat{\mathbb{L}}(i) & 0 & 0 \\ \bar{E}_D(i) & 0 & 0. & -I & 0 \\ \varepsilon \bar{E}_A \mathbb{B} & 0 & 0 & 0 & -\varepsilon I \end{bmatrix} < 0 \tag{7.145}$$

where

$$\begin{aligned} \Upsilon_w(i) &= \Upsilon_t(i) + L_f^t(i) L_f(i) \\ \Gamma_a(i) &= \Lambda_n(i) + L_f^t(i) L_f(i) \\ \Theta_b(i) &= \Theta_a(i) + L_f^t(i) L_f(i) \end{aligned} \tag{7.146}$$

for all admissible uncertainties satisfying (7.107)

Theorem 7.16 *Given gain matrices $A_f(i), K_f(i), i \in \mathcal{S}$ and subject to* **Assumptions** *(7.1)-(7.2), the neutral system (Σ_A) is* **SSWDD with a disturbance attenuation** γ *if given matrices $0 < \mathbb{L}(i) = \mathbb{L}^t(i) \in \mathbb{R}^{2n\times 2n}$, $i \in \mathcal{S}$ and scalars $\xi(i) > 0$ such that for all $i \in \mathcal{S}$*

$$\begin{aligned} \bar{\mathbb{L}}(i) &= \mathbb{L}(i) - \xi^{-1}(i)\sum_{m=1}^{s}\alpha_{im}\mathbb{L}(m) \\ \hat{\mathbb{L}}(i) &= \mathbb{L}(i) + \xi(i)\sum_{m=1}^{s}\alpha_{im}\mathbb{L}(m) \end{aligned}$$

there exist matrices $0 < \mathbb{P}(i) = \mathbb{P}^t(i) \in \mathbb{R}^{2n\times 2n}$, $i \in \mathcal{S}$ satisfying the following LMIs for all $i \in \mathcal{S}$

$$\begin{bmatrix} \Upsilon_w(i) & \Gamma_a(i) & \mathbb{P}(i)\mathsf{B}(i) \\ \Gamma_a^t(i) & -\Theta_b(i) & 0 \\ \mathsf{B}^t(i)\mathbb{P}(i) & 0 & -\gamma^2 I + L_f^t(i)L_f(i) \end{bmatrix} < 0 \qquad (7.147)$$

$$\begin{bmatrix} -\bar{\mathbb{L}}(i) & \mathbb{B}^t(i)L_f^t(i) & \mathbb{B}^t(i)\hat{\mathbb{L}}(i) \\ L_f(i)\mathbb{B}(i) & -I & 0 \\ \hat{\mathbb{L}}(i)\mathbb{B}(i) & 0 & -\hat{\mathbb{L}}(i) \end{bmatrix} < 0 \qquad (7.148)$$

Proof: Followed from **Theorem** 7.15 by setting $\bar{E}_A^t \equiv 0,\ \bar{E}_D^t \equiv 0, \bar{H}_A^t \equiv 0$. $\nabla\nabla\nabla$

Having developed the basic analytical results in **Theorems** 7.15-7.16, we provide in the sequel expressions for the gain matrices of the observer (7.112) when applied to the neutral systems $(\Sigma_{\Delta A})$ and (Σ_A) while guaranteeing $\mathcal{H}_\infty$ performance in the light of **Definition** (7.2).

For simplicity in exposition, we introduce the following matrix expressions for some scalars $\varepsilon > 0,\ \varrho > 0$:

$$\begin{aligned} \bar{A}_o(i) &= \hat{A}_o(i) + \gamma^{-2}N(i)N^t(i)\mathbb{P}_x(i)\big(I + \Xi_c(i)\big)^{-1} \\ \bar{C}_o(i) &= \hat{C}_o(i) + M(i)[\gamma^2 I - L^t(i)L(i)]^{-1}N^t(i)\mathbb{P}_x(i)\big(I + \Xi_c(i)\big)^{-1} \qquad (7.149) \\ \Xi_{kk}(i) &= \Big(\varepsilon^{-1}(i) + \varrho^{-1}(i)\Big)H_c(i)H_c^t(i) + \bar{C}_o(i)B(i)\Xi_{aa}^{-1}(i)B^t(i)\hat{C}_o^t(i) \end{aligned}$$

$$
\begin{aligned}
&+ \left[[\bar{C}_o(i) + C_o(i)]B(i) + C_d(i)\right]\Xi_b^{-1}(i)\left[[\bar{C}_o(i) + C_o(i)B(i) + C_d(i)\right]^t \\
\Xi_{gg}(i) &= \left(\bar{A}_o(i) + A_d(i) + \hat{\mathbb{L}}_e(i)B(i)\right)\Xi_a^{-1}(i)B^t(i)\bar{C}_o(i) \\
&+ \left(\bar{A}_o(i) - A_o(i) + \hat{\mathbb{L}}_e(i)B(i)\right)\Xi_b^{-1}(i)\left[[\bar{C}_o(i) + C_o(i)B(i) + C_d(i)\right]^t \\
&+ \left(\varepsilon^{-1}(i) + \varrho^{-1}(i)\right)M_a(i)M_c^t(i) \qquad (7.150)
\end{aligned}
$$

The main results are summarized by the following theorems:

Theorem 7.17 *The neutral augmented system* $\Sigma_{\Delta A}$ *is* **RSSWDD with a disturbance attenuation** γ, *if given matrices* $0 < \mathbb{L}(i) = \mathbb{L}^t(i) \in \mathbb{R}^{2n\times 2n}$ *and scalars* $\xi(i) > 0$ *such that for all* $i \in \mathcal{S}$

$$
\begin{aligned}
\bar{\mathbb{L}}(i) &= \mathbb{L}(i) - \xi^{-1}(i)\sum_{m=1}^{s}\alpha_{im}\mathbb{L}_x(m) \\
\hat{\mathbb{L}}(i) &= \mathbb{L}(i) + \xi(i)\sum_{m=1}^{s}\alpha_{im}\mathbb{L}(m)
\end{aligned}
$$

there exist matrices $0 < \mathbb{P}_e(i) = \mathbb{P}_e^t(i) \in \mathbb{R}^{n\times n}$, $0 < \mathbb{P}_x(i) = \mathbb{P}_x^t(i) \in \mathbb{R}^{n\times n}$, $i \in \mathcal{S}$ *and scalars* $\varepsilon(i) > 0$, $\varrho(i) > 0$, $i \in \mathcal{S}$ *satisfying the LMIs for all* $i \in \mathcal{S}$

$$
\begin{bmatrix}
\Upsilon_e(i) & \Lambda_b(i) & \Lambda_d(i) & \Lambda_f(i) & \mathbb{P}_x(i)N(i) \\
\Lambda_b^t(i) & -\Phi_b(i) & 0 & 0 & 0 \\
\Lambda_d^t(i) & 0 & -\Xi_a(i) & 0 & 0 \\
\Lambda_f^t(i) & 0 & 0 & -\Xi_b(i) & 0 \\
N^t(i)\mathbb{P}_x(i) & 0 & 0 & 0 & -\gamma^{-2}I + L^t(i)L(i)
\end{bmatrix} < 0 \qquad (7.151)
$$

$$
\begin{bmatrix}
\Upsilon_x(i) & \Lambda_c(i) & \Lambda_h(i) & \mathbb{P}_x(i)N(i) \\
\Lambda_c^t(i) & -\Phi_c(i) & 0 & 0 \\
\Lambda_h^t(i) & 0 & -\Xi_b(i) & 0 \\
N^t(i)\mathbb{P}_x(i) & 0 & 0 & -\gamma^{-2}I
\end{bmatrix} < 0 \qquad (7.152)
$$

for all admissible uncertainties satisfying (7.107). Moreover, the estimator gains are given by:

$$\begin{aligned} K_f(i) &= \Xi_{gg}(i)\, \Xi_{kk}^{-1}(i) \\ A_f(i) &= \bar{A}_o(i) - K_f(i)\bar{C}_o(i) \end{aligned} \tag{7.153}$$

Proof: Proceeding like **Theorem** 7.13, we express the expansion of (7.147) using (7.126) into the block form

$$\begin{bmatrix} \bar{\Sigma}_{11} & \bar{\Sigma}_{12} \\ \bar{\Sigma}_{12}^t & \bar{\Sigma}_{22} \end{bmatrix}$$

where

$$\begin{aligned} \bar{\Sigma}_{11}(i) &= \mathbb{P}_e(i)A_f(i) + A_f^t(i)\mathbb{P}_e(i) + \hat{\mathbb{L}}_e(i) \\ &+ \sum_{m=1}^{s} \alpha_{im}\mathbb{P}_e(m) + L^t(i)L(i) \\ &+ \left(\varepsilon^{-1}(i) + \varrho^{-1}(i)\right)\mathbb{P}_e(i)\left[M_a(i) - K_f(i)H_c(i)\right] \\ &\cdot \left[M_a(i) - K_f(i)H_c(i)\right]^t\mathbb{P}_e(i) \\ &+ \left[\mathbb{P}_e(i)[A_f(i)B(i) + A_d(i)] + \hat{\mathbb{L}}_e(i)B(i)\right]\Xi_a^{-1}(i) \\ &\quad \left[\mathbb{P}_e(i)[A_f(i)B(i) + A_d(i)] + \hat{\mathbb{L}}_e(i)B(i)\right]^t \\ &+ \left[\mathbb{P}_e(i)[A_o(i) - A_f(i) - K_f(i)C_o(i)]B(i) - \mathbb{P}_e(i)K_f(i)C_d(i)\right]\Xi_b^{-1}(i) \\ &\quad \left[\mathbb{P}_e(i)[A_o(i) - A_f(i) - K_f(i)C_o(i)]B(i) - \mathbb{P}_e(i)K_f(i)C_d(i)\right]^t \\ &+ \mathbb{P}_e(i)\left[N(i) - K_f(i)M(i)\right][\gamma^2 I - L^t(i)L(i)]^{-1} \\ &\cdot \left[N(i) - K_f(i)M(i)\right]^t(i)\mathbb{P}_x(i) \\ &+ \varepsilon^{-1}(i)\mathbb{P}_e(i)K_f(i)H_c(i)H_c^t(i)K_f^t(i)\mathbb{P}_e(i) \end{aligned} \tag{7.154}$$

$$\begin{aligned} \bar{\Sigma}_{12}(i) &= \Sigma_{12}(i) \\ &+ \mathbb{P}_e(i)\left[N(i) - K_f(i)M(i)\right][\gamma^2 I - L^t(i)L(i)]^{-1}N^t(i)\mathbb{P}_x(i) \end{aligned} \tag{7.155}$$

$$\bar{\Sigma}_{22}(i) = \Sigma_{22}(i) + \gamma^{-2}\mathbb{P}_x(i)N(i)N^t(i)\mathbb{P}_x(i) \tag{7.156}$$

Applying **Fact 1** to the matrix block $\bar{\Sigma}_{22}$, we can readily obtain the LMI (7.152). The substitution of (7.149)-(7.150) into (7.155) renders $\bar{\Sigma}_{12} \equiv 0$. Using (7.149)-(7.150) and (7.153) into (7.154) with some matrix manipulations and applying the Schur complements we get the LMI (7.151). $\nabla\nabla\nabla$

Theorem 7.18 *The neutral augmented system Σ_A is* **SSWDD with a disturbance attenuation** γ, *if given matrices $0 < \mathbb{L}_e(i) = \mathbb{L}_e^t(i) \in \mathbb{R}^{n\times n}$, $0 < \mathbb{L}_x(i) = \mathbb{L}_x^t(i) \in \mathbb{R}^{n\times n}$ and scalars $\xi(i) > 0$ such that for all $i \in \mathcal{S}$*

$$\begin{aligned}
\bar{\mathbb{L}}_x(i) &= \mathbb{L}_x(i) - \xi^{-1}(i)\sum_{m=1}^{s}\alpha_{im}\mathbb{L}_x(m) \\
\hat{\mathbb{L}}_x(i) &= \mathbb{L}_x(i) + \xi(i)\sum_{m=1}^{s}\alpha_{im}\mathbb{L}_x(m) \\
\bar{\mathbb{L}}_e(i) &= \mathbb{L}_e(i) - \xi^{-1}(i)\sum_{m=1}^{s}\alpha_{im}\mathbb{L}_e(m) \\
\hat{\mathbb{L}}_e(i) &= \mathbb{L}_x(i) + \xi(i)\sum_{m=1}^{s}\alpha_{im}\mathbb{L}_e(m)
\end{aligned}$$

there exist matrices $0 < \mathbb{P}_e(i) = \mathbb{P}_e^t(i) \in \mathbb{R}^{n\times n}$, $0 < \mathbb{P}_x(i) = \mathbb{P}_x^t(i) \in \mathbb{R}^{n\times n}$, $i \in \mathcal{S}$ satisfying the LMIs for all $i \in \mathcal{S}$

$$\begin{bmatrix} \Upsilon_{eo}(i) & \Lambda_{do}(i) & \Lambda_{fo}(i) & \mathbb{P}_x(i)N(i) \\ \Lambda_{do}^t(i) & -\Xi_a(i) & 0 & 0 \\ \Lambda_{fo}^t(i) & 0 & -\Xi_{bo}(i) & 0 \\ N^t(i)\mathbb{P}_x(i) & 0 & 0 & -\gamma^{-2}I + L^t(i)L(i) \end{bmatrix} < 0 \tag{7.157}$$

$$\begin{bmatrix} \Upsilon_{xo}(i) & \Lambda_h(i) & \mathbb{P}_x(i)N(i) \\ \Lambda_h^t(i) & -\Xi_{bo}(i) & 0 \\ N^t(i)\mathbb{P}_x(i) & 0 & -\gamma^{-2}I \end{bmatrix} < 0 \tag{7.158}$$

Moreover, the estimator gains are given by:

$$\begin{aligned}
K_f(i) &= \Xi_{ggo}(i)\,\Xi_{kko}^{-1}(i) \\
A_f &= \bar{A}_{oo} - K_f\bar{C}_{oo}
\end{aligned} \tag{7.159}$$

Proof: By introducing

$$
\begin{aligned}
\bar{A}_{oo}(i) &= A_o(i) + \gamma^{-2} N(i) N^t(i) \mathbb{P}_x(i) \big(I + \Xi_{co}(i)\big)^{-1} \\
\bar{C}_{oo}(i) &= \bar{C}_o + M(i)[\gamma^2 I - L^t(i) L(i)]^{-1} N^t(i) \mathbb{P}_x \big(I + \Xi_{co}\big)^{-1} \\
\Xi_{kko} &= \Big(\bar{C}_o B \Xi_a^{-1} B^t(i) \bar{C}_o^t(i) + \Big[[\bar{C}_o(i) + C_o(i)] B(i) + C_d(i) \Big] \Xi_{bo}^{-1}(i) \\
& \Big[[\bar{C}_o(i) + C_o(i)] B(i) + C_d(i) \Big]^t \\
\Xi_{ggo}(i) &= \Big\{ \Big(\bar{A}_{oo}(i) + A_d(i) + \hat{\mathbb{L}}_e(i) B(i) \Big) \Xi_a^{-1}(i) B^t(i) \bar{C}_o(i) \\
&+ \Big(\bar{A}_{oo}(i) - A_o(i) + \hat{\mathbb{L}}_e(i) B(i) \Big) \Xi_{bo}^{-1}(i) \Big[[\bar{C}_o(i) + C_o(i)] B(i) + C_d(i) \Big]^t \\
\Lambda_{fo}(i) &= \hat{\mathbb{L}}_e(i) B(i)
\end{aligned}
\tag{7.160}
$$

and applying similar procedure to **Theorem** (7.13) while setting $\varrho \equiv 0, \varepsilon \equiv 0, M_a \equiv 0, N_a \equiv 0, M_c \equiv 0, N_d \equiv 0$ and using (7.154)-(7.156), the proof is completed . $\nabla\nabla\nabla$

7.5.4 Robust Stabilization

The foregoing theorems provided ways to produce a good replica of the state of the neutral system. Quite naturally, the attractive step would be to derive a robust state-estimate feedback control. For this purpose, we consider the following linear uncertain model:

$$
\begin{aligned}
(\Sigma_{\Delta nc})\ \mathcal{M}(\dot{x}_t) &\triangleq \dot{x}(t) - B(i)\dot{x}(t-\tau) \\
&= A_{\Delta o}(t,i)x(t) + A_{\Delta d}(t,i)x(t-\tau) \\
&+ F_{\Delta o}(t,i)u(t) + Nw(t) && (7.161) \\
x(\eta) &= \phi(\eta) \in \mathbb{C}([-\tau,0], \mathbb{R}^n), \quad \forall\, \eta \in [-\tau, 0] && (7.162) \\
y(t) &= C_{\Delta o}(t,i)x(t) + C_{\Delta d}(t,i)x(t-\tau)
\end{aligned}
$$

$$
\begin{aligned}
& + \quad M(i)w(t) && (7.163)\\
z(t) &= L(i)x(t) && (7.164)
\end{aligned}
$$

where $u(t) \in \mathbb{R}^q$ is the control input and

$$
\begin{aligned}
F_{\Delta o}(t,i) &= F_o(i) + \Delta F_o(t,i) \\
&= F_o(i) + M_a(i)\Delta(t)N_b(i) && (7.165)
\end{aligned}
$$

where $F_o(i) \in \mathbb{R}^{n\times q}$ and $N_b(i) \in \mathbb{R}^{\beta\times q}$ are known real matrices. The remaining matrices are as in section 7.2.

In the absence of uncertainties ($\Delta(.) \equiv 0$), we obtain the following nominal neutral system

$$
\begin{aligned}
(\Sigma_{nc})\ \mathcal{M}(\dot{x}_t) &\triangleq \dot{x}(t) - B(i)\dot{x}(t-\tau) \\
&= A_o(i)x(t) + A_d(i)x(t-\tau) \\
&+ F_o(i)u(t) + N(i)w(t) && (7.166)\\
x(\eta) &= \phi(\eta) \in \mathbb{C}([-\tau,0],\mathbb{R}^n), \quad \forall\, \eta \in [-\tau,0] && (7.167)\\
y(t) &= C_o(i)x(t) + C_d(i)x(t-\tau) \\
&+ Mw(t) && (7.168)\\
z(t) &= L(i)x(t) && (7.169)
\end{aligned}
$$

In the sequel, we consider the problems of stabilization of the neutral system (Σ_{nc}) and robust stabilization of the uncertain neutral system $(\Sigma_{\Delta nc})$, using a linear memoryless state-estimate feedback control

$$
u(t) = K_s(i)\hat{x}(t),\ i \in \mathcal{S}
$$

where $\hat{x}(t)$ is generated by (7.112).

It can be readily shown that the closed-loop system of $(\Sigma_{\Delta nc})$ and (Σ_e) takes the form:

$$
(\Sigma_{\Delta AC}):\ \mathcal{M}(\dot{\zeta}_t) \triangleq \dot{\zeta}(t) - \mathbb{B}(i)\dot{\zeta}(t-\tau)
$$

$$
\begin{aligned}
&= \mathcal{A}_{\Delta c}(t,i)\zeta(t) + \mathcal{D}_{\Delta o}(t,i)\zeta(t-\tau) \\
&+ \mathsf{B}(i)w(t) \qquad (7.170) \\
\tilde{z}(t) &= L_f(i)\zeta(t) \qquad (7.171)
\end{aligned}
$$

where

$$
\begin{aligned}
\mathcal{A}_{\Delta c}(t,i) &= \left[\mathsf{A}_c(i) + \bar{H}_A(i)\Delta(t)\bar{E}_{AC}(i)\right] \\
\bar{E}_{AC}(i) &= [0 \quad N_a(i) + N_b(i)K_s(i)] \qquad (7.172) \\
\mathsf{A}_c(i) &= \begin{bmatrix} A_f(i) - F_o(i)K_s(i) & \bar{A}_{afk} \\ 0 & A_o(i) \end{bmatrix} \\
\bar{A}_{afk} &= A_o(i) + F_o(i)K_s(i) - K_f(i)C_o(i) - A_f(i) \qquad (7.173)
\end{aligned}
$$

The remaining matrices are given by (7.118)-(7.120).

It follows from **Theorem** 7.11 that $(\Sigma_{\Delta AC})$ is **RSSWDD** if for given matrices $0 < \mathbb{L}(i) = \mathbb{L}^t(i) \in \mathbb{R}^{2n\times 2n}$, $i \in \mathcal{S}$, there exist matrices $0 < \mathbb{P}(i) = \mathbb{P}^t(i) \in \mathbb{R}^{2n\times 2n}$, $i \in \mathcal{S}$ and scalars $\xi(i) > 0$, $\varepsilon(i) > 0$, $\varrho(i) > 0$, $i \in \mathcal{S}$ such that the following inequality holds :

$$
\begin{aligned}
& \mathbb{P}(i)\mathsf{A}_c(i) + \mathsf{A}_c^t(i)\mathbb{P}(i) + \hat{\mathbb{L}}(i) + \sum_{m=1}^{s} \alpha_{im}\mathbb{P}(m) + \varrho^{-1}(i)\mathbb{P}(i)\bar{H}_A(i)\bar{H}_A^t(i)\mathbb{P}(i) \\
+ \;& \varrho(i)\bar{E}_{AC}^t(i)\bar{E}_{AC}(i) + \varepsilon^{-1}(i)\mathbb{P}(i)\left[\bar{H}_A(i)\bar{H}_A^t(i) + \bar{H}_D(i)\bar{H}_D^t(i)\right]\mathbb{P}(i) \\
+ \;& \Big(\mathbb{P}(i)[\mathsf{A}_c(i)\mathbb{B}(i) + \mathsf{D}(i)] + \hat{\mathbb{L}}(i)\mathbb{B}(i)\Big) \\
& \left[\mathbb{L} - \mathbb{B}^t(i)\hat{\mathbb{L}}(i)\mathbb{B}(i) - \varepsilon(i)\Big(\bar{E}_A^t(i)\bar{E}_A(i) + \bar{E}_D^t(i)\bar{E}_D(i)\Big)\right]^{-1} \\
& \Big(\mathbb{P}(i)[\mathsf{A}_c(i)\mathbb{B}(i) + \mathsf{D}(i)] + \hat{\mathbb{L}}(i)\mathbb{B}(i)\Big)^t \qquad (7.174)
\end{aligned}
$$

for all $i \in \mathcal{S}$. Taking into account (7.118)-(7.120), (7.128)-(7.130) and (7.173), we express (7.174) into the block form

$$
\begin{bmatrix} \Omega_{11}(i) & \Omega_{12}(i) \\ \Omega_{12}^t(i) & \Omega_{22}(i) \end{bmatrix}
$$

where

$$
\begin{aligned}
\Omega_{11}(i) &= \mathbb{P}_e(i)[A_f(i) - F_o(i)K_s(i)] + [A_f(i) - F_o(i)K_s(i)]^t\mathbb{P}_e(i) + \hat{\mathbb{L}}_e(i) \\
&+ \varepsilon^{-1}(i)\mathbb{P}_e(i)K_f(i)H_c(i)H_c^t(i)K_f^t(i)\mathbb{P}_e(i) + \sum_{m=1}^{s} \alpha_{im}\mathbb{P}_e(m) \\
&+ \Big(\varepsilon^{-1}(i) + \varrho^{-1}(i)\Big)\mathbb{P}_e(i)\big[M_a(i) - K_f(i)H_c(i)\big] \\
&\cdot \big[M_a(i) - K_f(i)M_c(i)\big]^t\mathbb{P}_e(i) \\
&+ \Big[\mathbb{P}_e(i)[A_o(i) + F_o(i)K_s(i) - A_f(i) - K_f(i)C_o(i)]B(i) \\
&- \mathbb{P}_e(i)K_f(i)C_d(i)\Big]\Xi_w^{-1}(i) \\
&\Big[\mathbb{P}_e(i)[A_o(i) + F_o(i)K_s(i) - A_f(i) - K_f(i)C_o(i)]B(i) \\
&- \mathbb{P}_e(i)K_f(i)C_d(i)\Big]^t \\
&+ \Big[\mathbb{P}_e(i)[A_f(i)B(i) + A_d(i)] + \hat{\mathbb{L}}_e(i)B(i)\Big]\Xi_a^{-1}(i) \\
&\Big[\mathbb{P}_e(i)[A_f(i)B(i) + A_d(i)] + \hat{\mathbb{L}}_e(i)B(i)\Big]^t \qquad (7.175) \\
\Omega_{22}(i) &= \mathbb{P}_x(i)A_o(i) + A_o^t(i)\mathbb{P}_x(i) + \hat{\mathbb{L}}_x(i) \\
&+ \sum_{m=1}^{s} \alpha_{im}\mathbb{P}_x(m) + \varrho(i)[N_a(i) + N_b(i)K_s(i)]^t(i) \\
&\cdot [N_a(i) + N_b(i)K_s(i)] \\
&+ \big(\varepsilon^{-1}(i) + \varrho^{-1}(i)\big)\mathbb{P}_x(i)M_a(i)M_a^t(i)\mathbb{P}_x(i) \\
&+ \varepsilon^{-1}(i)\mathbb{P}_x(i)M_c(i)M_c^t(i)\mathbb{P}_x(i) \\
&+ \Big[\mathbb{P}_x(i)[A_o(i)B(i) + A_d(i)] + \hat{\mathbb{L}}_x(i)B(i)\Big]\Xi_b^{-1}(i) \\
&\Big[\mathbb{P}_x(i)[A_o(i)B(i) + A_d(i)] + \hat{\mathbb{L}}_x(i)B(i)\Big]^t \qquad (7.176) \\
\Omega_{12}(i) &= \Big[\mathbb{P}_e(i)[A_o(i) + F_o(i)K_s(i) - A_f(i) - K_f(i)C_o(i)]B(i)
\end{aligned}
$$

$$
\begin{aligned}
&- \quad \mathbb{P}_e(i)K_f(i)C_d(i)\Big]\Xi_b^{-1}(i) \\
&\quad \Big[\mathbb{P}_x(i)[A_o(i)B(i) + A_d(i)] + \hat{\mathbb{L}}_x(i)B(i)\Big]^t \\
&- \quad \varepsilon^{-1}\mathbb{P}_e(i)K_f(i)M_c(i)H_c^t(i)K_f^t(i)\mathbb{P}_x(i) \\
&+ \quad \Big(\varepsilon^{-1}(i) + \varrho^{-1}(i)\Big)\mathbb{P}_e(i)\big[M_a(i) - K_f(i)M_c(i)\big]M_a^t(i)\mathbb{P}_x(i) \\
&+ \quad \mathbb{P}_e(i)[A_o(i) + F_o(i)K_s(i) - A_f(i) - K_f(i)C_o(i)] \qquad (7.177)
\end{aligned}
$$

where

$$
\begin{aligned}
\Xi_w(i) &= \bar{\mathbb{L}}_x(i) - \varepsilon(i)\Big([N_a(i) + N_b(i)K_s(i)]^t[N_a(i) + N_b(i)K_s(i)] \\
&+ \quad N_d^t(i)N_d(i)\Big) \\
&- \quad B^t(i)\hat{\mathbb{L}}_x(i)B(i) \qquad (7.178)
\end{aligned}
$$

The main robust stabilization result is now summarized by the following theorem

Theorem 7.19 *The closed-loop neutral system* $(\Sigma_{\Delta AC})$ *is* **RSSWDD via memoryless state-feedback** $u(t) = K_s(i)x(t),\ i \in \mathcal{S}$, *if given matrices* $0 < \mathbb{L}_e(i) = \mathbb{L}_e^t(i) \in \mathbb{R}^{n\times n}$, $0 < \mathbb{L}_x(i) = \mathbb{L}_x^t(i) \in \mathbb{R}^{n\times n}$ *and scalars* $\xi(i) > 0$ *such that for all* $i \in \mathcal{S}$

$$
\begin{aligned}
\bar{\mathbb{L}}_x(i) &= \mathbb{L}_x(i) - \xi^{-1}(i)\sum_{m=1}^{s} \alpha_{im}\mathbb{L}_x(m) \\
\hat{\mathbb{L}}_x(i) &= \mathbb{L}_x(i) + \xi(i)\sum_{m=1}^{s} \alpha_{im}\mathbb{L}_x(m) \\
\bar{\mathbb{L}}_e(i) &= \mathbb{L}_e(i) - \xi^{-1}(i)\sum_{m=1}^{s} \alpha_{im}\mathbb{L}_e(m) \\
\hat{\mathbb{L}}_e(i) &= \mathbb{L}_x(i) + \xi(i)\sum_{m=1}^{s} \alpha_{im}\mathbb{L}_e(m)
\end{aligned}
$$

there exist matrices $0 < Y(i) = Y^t(i) \in \mathbb{R}^{n\times n}$, $0 < X(i) = X^t(i) \in \mathbb{R}^{n\times n}$, $0 < Z(i) = Z^t(i) \in \mathbb{R}^{n\times n}$, $W(i) \in \mathbb{R}^{q\times n}$, $i \in \mathcal{S}$ *and scalars* $\xi(i) > 0$, $\varepsilon(i) >$

0, $\varrho(i) > 0$, $i \in \mathcal{S}$ *satisfying the LMIs for all* $i \in \mathcal{S}$

$$\begin{bmatrix} \Gamma_{ek}(i) & \Lambda_b(i) & \Lambda_m(i) & \Lambda_q(i) \\ \Lambda_b^t(i) & -\Phi_b(i) & 0 & 0 \\ \Lambda_m^t(i) & 0 & -\Xi_a(i) & 0 \\ \Lambda_q^t(i) & 0 & 0 & -\Xi_b(i) \end{bmatrix} < 0$$

$$\begin{bmatrix} \Upsilon_{xk}(i) & \Lambda_c(i) & \Lambda_{vx}(i) \\ \Lambda_c^t(i) & -\Phi_c(i) & 0 \\ \Lambda_{vx}^t(i) & 0 & -\Xi_b(i) \end{bmatrix} < 0 \tag{7.179}$$

$$\begin{bmatrix} Y(i) & I \\ I & Z(i) \end{bmatrix} \geq 0 \tag{7.180}$$

where

$$\begin{aligned}
\Xi_{gs}(i) &= \Xi_g(i) + F_o(i)W(i)Z(i)\Big\{\Xi_a^{-1}(i)B^t(i)\hat{C}_o(i) + \Xi_b^{-1}(i) \\
&\quad \Big[[\hat{C}_o(i) + C_o(i)]B(i) + C_d(i)\Big]^t\Big\} \\
\Gamma_{ek}(i) &= \hat{A}_o(i)Y(i) + Y(i)\hat{A}_o^t(i) + Y(i)\hat{I\!L}_e(i)Y(i) - \Xi_{gs}^t(i)\Xi_k^{-1}(i)\Xi_{gs}(i) \\
&+ F_o(i)W(i) + W^t(i)F_o^t(i) \\
\Lambda_m(i) &= \hat{A}_o(i) + A_d(i) + Y(i)\hat{I\!L}_e(i)B(i) + F_o(i)W(i)Z(i) + Z(i)W^t(i)F_o^t(i) \\
\Lambda_q(i) &= \hat{A}_o(i) - A_o(i) + Y(i)\hat{I\!L}_e(i)B(i) + F_o(i)W(i)Z(i) + Z(i)W^t(i)F_o^t(i) \\
\Upsilon_{xk}(i) &= X(i)A_o(i) + A_o^t(i)X(i) + \hat{I\!L}_x(i) \\
&+ \varrho(i)[N_a(i) + N_b(i)W(i)Z(i)]^t[N_a(i) + N_b(i)W(i)Z(i)] \\
\Xi_s(i) &= B(i)\Xi_b^{-1}(i)\Big[B^t(i)\hat{I\!L}_x(i) + [A_d^t(i) + B^t(i)A_o^t(i)]X(i)\Big] \\
\Lambda_{vx}(i) &= X(i)A_o(i)B(i) + X(i)A_d(i) + \hat{I\!L}_x(i)B(i) \\
\tilde{A}_o(i) &= A_o(i) + M_a(i)M_a^t(i)X(i)\big(I + \Xi_s(i)\big)^{-1}
\end{aligned} \tag{7.181}$$

for all admissible uncertainties satisfying (7.107). Moreover, the associated gains are given by:

$$K_f(i) = \Xi_{gs}(i)\,\Xi_k^{-1}(i)$$

$$
\begin{aligned}
A_f(i) &= \tilde{A}_o(i) + F_o(i)W(i)Z(i) - K_f(i)\hat{C}_o(i) \\
K_s(i) &= W(i)Z(i)
\end{aligned}
\tag{7.182}
$$

Proof: By defining

$$
\mathbb{P}_x(i) = X(i) \ , \ \mathbb{P}_e^{-1}(i) = Y(i), K_s(i) = W(i)Z(i)
$$

and applying the technique of **Theorem** 7.13 using (7.134)-(7.135) and (7.182), the desired result is readily obtained. $\nabla\nabla\nabla$

On the other hand, by combining the nominal system of (Σ_{nc}) and (Σ_e) we obtain the closed-loop system:

$$
\begin{aligned}
(\Sigma_{AC}): \ \mathcal{M}(\dot{\zeta}_t) &\triangleq \dot{\zeta}(t) - \mathbb{B}\dot{\zeta}(t-\tau) \\
&= \mathsf{A}_c\zeta(t) + \mathsf{D}\zeta(t-\tau) + \mathsf{B}w(t) && (7.183) \\
\tilde{z}(t) &= L_f\zeta(t) && (7.184)
\end{aligned}
$$

and for which we prove the following theorem

Theorem 7.20 *The closed-loop neutral system* (Σ_{AC}) *is* **SSWDD via memoryless state-feedback** $u(t) = K_s(i)x(t), \ i \in \mathcal{S}$, *if there exist matrices* $0 < \mathbb{L}_e(i) \in \mathbb{R}^{n\times n}$, $0 < \mathbb{L}_x(i) \in \mathbb{R}^{n\times n}$, $0 < Y(i) = Y^t(i) \in \mathbb{R}^{n\times n}$, $0 < X(i) = X^t(i) \in \mathbb{R}^{n\times n}$, $0 < Z(i) = Z^t(i) \in \mathbb{R}^{n\times n}$, $W(i) \in \mathbb{R}^{q\times n}$, $i \in \mathcal{S}$ *satisfying the LMIs for all* $i \in \mathcal{S}$

$$
\begin{bmatrix} \Gamma_{eko}(i) & \Lambda_{mo}(i) & \Lambda_{qo}(i) \\ \Lambda_{mo}^t(i) & -\Xi_a(i) & 0 \\ \Lambda_{qo}^t(i) & 0 & -\Xi_{bo}(i) \end{bmatrix} < 0
$$

$$
\begin{bmatrix} \Upsilon_{xko}(i) & \Lambda_{vx}(i) \\ \Lambda_{vx}^t(i) & -\Xi_{bo}(i) \end{bmatrix} < 0 \tag{7.185}
$$

$$
\begin{bmatrix} Y(i) & I \\ I & Z(i) \end{bmatrix} \geq 0 \tag{7.186}
$$

where

$$
\begin{aligned}
\Xi_{gso} &= \Xi_{go}(i) + F_o(i)W(i)Z(i)\Big\{\Xi_a^{-1}(i)B^t(i)\bar{C}_o(i) + \Xi_b^{-1}(i) \\
& \quad \Big[[\bar{C}_o(i) + C_o(i)]B(i) + C_d(i)\Big]^t\Big\} \\
\Gamma_{eko}(i) &= A_o(i)Y(i) + Y(i)A_o^t(i) + Y(i)\hat{I\!\!L}_e(i)Y(i) - \Xi_{gso}^t(i)\Xi_{ko}^{-1}(i)\Xi_{gso}(i) \\
&+ F_o(i)W(i) + W^t(i)F_o^t(i) \\
\Lambda_{mo}(i) &= A_o(i) + A_d(i) + Y(i)\hat{I\!\!L}_e(i)B(i) + F_o(i)W(i)Z(i) + Z(i)W^t(i)F_o^t(i) \\
\Lambda_{qo}(i) &= Y(i)\hat{I\!\!L}_e(i)B(i) + F_o(i)W(i)Z(i) + Z(i)W^t(i)F_o^t(i) \\
\Upsilon_{xko}(i) &= X(i)A_o(i) + A_o^t(i)X(i) + \hat{I\!\!L}_x(i) \\
\Xi_s(i) &= B(i)\Xi_b^{-1}(i)\Big[B^t(i)\hat{I\!\!L}_x(i) + [A_d^t(i) + B^t(i)A_o^t(i)]X(i)\Big] \\
\Lambda_{vx}(i) &= X(i)A_o(i)B(i) + X(i)A_d(i) + \hat{I\!\!L}_x(i)B(i) \\
\tilde{A}_o(i) &= A_o(i) \qquad (7.187)
\end{aligned}
$$

Moreover, the associated gains are given by:

$$
\begin{aligned}
K_f(i) &= \Xi_{gso}(i)\,\Xi_{ko}^{-1}(i) \\
A_f(i) &= A_o(i) + F_o(i)W(i)Z(i) - K_f(i)\bar{C}_o(i) \\
K_s(i) &= W(i)Z(i) \qquad (7.188)
\end{aligned}
$$

Proof: Follows from **Theorem** 7.19 in the manner **Theorem** 7.18 by suppressing the uncertain terms. ∇∇∇

7.6 Examples

In order to illustrate the theoretical results of this paper, we provide some numerical examples.

7.6.1 Example 7.2

We consider a pilot-scale single-reach water quality system which can fall into the type (7.102)-(7.104) with $\tau = 0.75$. Let the Markov process governing the mode switching has generator

$$\Im = \begin{bmatrix} -4 & 4 \\ 3 & -3 \end{bmatrix}$$

For the two operating conditions (modes), the associated date are:

Mode 1:

$$\begin{aligned}
A_o(1) &= \begin{bmatrix} 0 & 2 \\ -3 & -6 \end{bmatrix}, \; A_d(1) = \begin{bmatrix} -0.1 & 0 \\ 0.2 & -0.3 \end{bmatrix} \\
C_o(1) &= \begin{bmatrix} 1 & 0 \\ 0 & 2 \end{bmatrix}, \; C_d(1) = \begin{bmatrix} 0.1 & 0 \\ 0 & 0.1 \end{bmatrix} \\
B(1) &= \begin{bmatrix} 0.15 & 0 \\ 0 & 0.15 \end{bmatrix}, \; N(1) = \begin{bmatrix} 0.1 \\ 0.2 \end{bmatrix} \\
M(1) &= \begin{bmatrix} 0.2 \\ 0.1 \end{bmatrix}, \; M_a(1) = \begin{bmatrix} 0.1 \\ 0 \end{bmatrix} \\
L(1) &= [0.1 \;\; 0.1], \; N_a(1) = [0.1 \;\; 0.2], \; M_d(1) = [0.1 \;\; 0]
\end{aligned}$$

Mode 2:

$$\begin{aligned}
A_o(2) &= \begin{bmatrix} 0 & 2 \\ -3 & -4 \end{bmatrix}, \; A_d(2) = \begin{bmatrix} -0.1 & 0 \\ 0.2 & -0.4 \end{bmatrix} \\
C_o(2) &= \begin{bmatrix} 1 & 0 \\ 0 & 1 \end{bmatrix}, \; C_d(2) = \begin{bmatrix} 0.2 & 0 \\ 0 & 0.2 \end{bmatrix} \\
B(2) &= \begin{bmatrix} 0.1 & 0 \\ 0 & 0.1 \end{bmatrix}, \; N(2) = \begin{bmatrix} 0.2 \\ 0.1 \end{bmatrix} \\
M(2) &= \begin{bmatrix} 0.1 \\ 0.2 \end{bmatrix}, \; M_a(2) = \begin{bmatrix} 0 \\ 0.2 \end{bmatrix} \\
L(2) &= [0.1 \;\; 0.1], \; N_a(2) = [0.2 \;\; 0.1], \; N_d(2) = [0 \;\; 0.3]
\end{aligned}$$

First we note that **Assumptions** (7.1)-(7.2) are met for both modes. Invoking the software environment [57], we proceed to solve the LMIs (7.133) using

(7.129)-(7.135) and the initial data for $i = 1, 2$:

$$\begin{aligned} \mathbb{L}_e(1) &= \begin{bmatrix} 5 & 0 \\ 0 & 5 \end{bmatrix}, \ \mathbb{L}_x(1) = \begin{bmatrix} 5 & 0 \\ 0 & 5 \end{bmatrix}, \ \xi(1) = 20 \\ \mathbb{L}_e(2) &= \begin{bmatrix} 4 & 0 \\ 0 & 4 \end{bmatrix}, \ \mathbb{L}_x(2) = \begin{bmatrix} 4 & 0 \\ 0 & 4 \end{bmatrix}, \ \xi(2) = 25 \end{aligned}$$

which ensures that $\bar{\mathbb{L}}_e(1) > 0$, $\hat{\mathbb{L}}_e(1) > 0$, $\bar{\mathbb{L}}_x(2) > 0$, $\hat{\mathbb{L}}_x(2) > 0$. The feasible solutions are given by:

$$\begin{aligned} \mathbb{P}_e(1) &= \begin{bmatrix} 1.8010 & 0.0577 \\ 0.0577 & 1.3405 \end{bmatrix}, \ \mathbb{P}_x(1) = \begin{bmatrix} 1.0134 & 0.0106 \\ 0.0106 & 0.8245 \end{bmatrix} \\ \varrho(1) &= 2.7694 \ , \ \varepsilon(1) = 6.1246 \ , \ \gamma = 1.1045 \\ K_f(1) &= \begin{bmatrix} 0.0404 & -2.6255 \\ 1.0095 & -3.1423 \end{bmatrix}, \ A_f(1) = \begin{bmatrix} 0.0322 & 1.3146 \\ -2.3011 & -5.8945 \end{bmatrix} \\ \mathbb{P}_e(2) &= \begin{bmatrix} 2.9627 & 0.1527 \\ 0.1527 & 0.8405 \end{bmatrix}, \ \mathbb{P}_x(2) = \begin{bmatrix} 0.9467 & 0.0105 \\ 0.0105 & 1.18542 \end{bmatrix} \\ \varrho(2) &= 3.2658 \ , \ \varepsilon(2) = 4.8453 \\ K_f(2) &= \begin{bmatrix} 0.1107 & -2.9167 \\ 1.1945 & -4.4627 \end{bmatrix}, \ A_f(2) = \begin{bmatrix} 0.0176 & 1.5031 \\ -3.0111 & -3.9213 \end{bmatrix} \end{aligned}$$

This verifies **Theorem** 7.17 and in turn confirms the robust stochastic stability independent of delay and with disturbance attenuation $\gamma = 1.1045$ of the water quality model.

7.6.2 Example 7.3

To illustrate **Theorem** 7.18, we consider the numerical data of **Example 7.2** in addition to

$$\begin{aligned} F_o(1) &= \begin{bmatrix} 1 & 0 \\ 0 & 2 \end{bmatrix}, \ F_o(2) = \begin{bmatrix} 2 & 0 \\ 0 & 1 \end{bmatrix} \\ N_b(1) &= [0.1 \ \ 0.1] \ , \ N_b(2) = [0.2 \ \ 0.2] \end{aligned}$$

and rely again on the software package [57]. Here, we solve the LMIs (7.179)-(7.180) using (7.181) for $i = 1, 2$ to produce the feasible solutions:

$$X(1) = \begin{bmatrix} 2.3506 & 0.9457 \\ 0.9457 & 2.5294 \end{bmatrix}, \ Y(1) = \begin{bmatrix} 1.5980 & 0.2654 \\ 0.2654 & 1.4533 \end{bmatrix}$$

$$
\begin{aligned}
Z(1) &= \begin{bmatrix} 1.5512 & -0.2055 \\ -0.2055 & 0.8733 \end{bmatrix}, \; W(1) = \begin{bmatrix} -0.7524 & -0.1445 \\ 0.1281 & 1.2206 \end{bmatrix} \\
\varrho(1) &= 3.1338 \,, \; \varepsilon(1) = 8.1457 \,, \; \gamma = 1.2671 \\
K_f(1) &= \begin{bmatrix} 0.0404 & -2.6255 \\ 1.0095 & -3.1423 \end{bmatrix}, \; A_f(1) = \begin{bmatrix} 0.0322 & 1.3146 \\ -2.3011 & -5.8945 \end{bmatrix} \\
K_s(1) &= \begin{bmatrix} -1.1374 & 0.284 \\ -0.0521 & 1.0396 \end{bmatrix} \\
Y(2) &= \begin{bmatrix} 2.9627 & 0.1527 \\ 0.1527 & 0.8405 \end{bmatrix}, \; X(2) = \begin{bmatrix} 1.1174 & 0.0109 \\ 0.0109 & 1.1657 \end{bmatrix} \\
Z(2) &= \begin{bmatrix} 1.1054 & -0.2104 \\ -0.2104 & 1.2132 \end{bmatrix}, \; W(2) = \begin{bmatrix} 0.9450 & -0.5117 \\ 0.4402 & -1.4204 \end{bmatrix} \\
\varrho(2) &= 3.4116 \,, \; \varepsilon(2) = 5.5514 \\
K_f(2) &= \begin{bmatrix} 0.1107 & -2.9167 \\ 1.1945 & -4.4627 \end{bmatrix}, \; A_f(2) = \begin{bmatrix} 0.0176 & 1.5031 \\ -3.0111 & -3.9213 \end{bmatrix} \\
K_s(2) &= \begin{bmatrix} 1.0578 & -0.8196 \\ 0.7414 & -1.8158 \end{bmatrix}
\end{aligned}
$$

7.7 Notes and References

We have established complete results on the robust stability analysis, robust stabilization, robust observation , robust $\mathcal{H}_\infty$ observation and the design of robust $\mathcal{H}_\infty$ output feedback controllers for a general class of uncertain neutral jumping systems with different lengths of distributed state-delays and multiple state-derivative delays. The jumping parameters are treated as continuous-time, discrete-state Markov process and the uncertainties are parametric and norm-bounded.

In all cases, the gain matrices are determined by solving linear matrix inequalities with scaling parameters. It has been shown that our results encompass almost all of the previously published works.

We have designed

1) a linear state-delayed estimator

2) a memoryless state-feedback stabilizer

3) a distributed state-feedback stabilizer

such that the resulting closed-loop system achieve desirable stability properties with weak delay dependence.

There are several avenues to elaborate on the results for strong-delay dependence, mode-dependent delays and using transformation methods

Chapter 8

Interconnected Systems

8.1 Introduction

Problems of decentralized control and stabilization of interconnected systems are receiving considerable interests [8, 9, 10, 58] where most of the effort are focused on dealing with the interaction patterns. Quadratic stabilization of classes of interconnected systems have been presented in [93] where the closed-loop feedback subsystems are cast into $\mathcal{H}_\infty$ control problems. When the interconnected system involves delays, only few studies are available. In [64, 80] the focus have been on delays in the interaction patterns with the subsystem dynamics being known completely. In [87], a class of uncertain systems is considered where the delays occur within the subsystems.

The problems of decentralized robust stabilization and robust $\mathcal{H}_\infty$ performance for a class of uncertain interconnected time-delay systems with jumping parameters seem to have been overlooked in the literature. The only exception to this is [121, 125] where preliminary results have been addressed.

Therefore the objective of this Chapter is to provide an overview of the subject and present pertinent coherent results in line with the foregoing Chapters. We examine closely the problems of stochastic stability and stabilization for a class of interconnected systems with Markovian jump parameters with or without a prescribed $\mathcal{H}_\infty$-performance. The jumping parameters are treated as continuous-time, discrete-state Markov process. In the model setup, the delays are time-varying in the state of each subsystem as well as in the interconnections among the subsystems. All the design effort is to be undertaken at the subsystem level. It is shown that the robust control design problems can be solved with the aid of two algebraic Riccati inequalities. A class of decentralized feedback controllers (state-feedback and observer-based feedback) is developed to render the closed-loop interconnected system stochastically stable. The robust $\mathcal{H}_\infty$-control problem with uncertain jumping rates has been also examined. Since Lyapunov theory is the main vehicle in stability analysis, the resulting conditions are only sufficient. In the course of the development, some parameters have been introduced as manipulative factors to reduce the degree of conservativeness.

8.2 Problem Statement

Given a probability space $(\Omega, \mathcal{F}, \mathbf{P})$ where Ω is the sample space, $\mathcal{F}$ is the algebra of events and $\mathbf{P}$ is the probability measure defined on $\mathcal{F}$, as detailed in Chapter 1. We consider a class $\mathbf{\Sigma_{n_s}}$ of continuous-time systems with Markovian jump parameters described for $\eta_t = i \in \mathcal{S}$ by:

$$\begin{aligned} \Sigma_{n_s}: \ \dot{x}(t) &= A(t,\eta_t)x(t) + B(t,\eta_t)u(t) + \Gamma(t,\eta_t)w(t) \\ x_o &= \phi \ , \ \eta_o \ = \ i \ , \ t \ \in \ [0, \mathcal{T}] \\ z(t) &= G(t,\eta_t)x(t) + F(t,\eta_t)u(t) \end{aligned} \tag{8.1}$$

which we will recognize in the sequel as an interconnection of n_s coupled subsystems Σ_j described over the space $(\Omega, \mathcal{F}, \mathbf{P})$ by:

$$\begin{aligned}
\Sigma_j: \quad \dot{x}_j(t) &= A_j(t,\eta_t)x_j(t) + B_j(t,\eta_t)u_j(t) + \Gamma_j(\eta_t)w_j(t) \\
&+ E_j(t,\eta_t)x_j(t-\tau_j) + \sum_{k=1}^{n_s} A_{jk}(t,\eta_t)x_k(t-\psi_{jk}) \qquad (8.2) \\
y_j(t) &= C_j(t,\eta_t)x_j(t) + D_j(t,\eta_t)u_j(t) + \Pi_j(\eta_t)w_j(t) \qquad (8.3) \\
z_j(t) &= G_j(\eta_t)x_j(t) + F_j(\eta_t)u_j(t) \qquad (8.4)
\end{aligned}$$

where for (8.1) x, u, w and z satisfy $x = (x_1^t, ..., x_{n_s}^t)^t$, $u = (u_1^t, ..., u_{n_s}^t)^t$, $w = (w_1^t, ..., w_{n_s}^t)^t$, $z = (z_1^t, ..., z_{n_s}^t)^t$. For (8.2)-(8.4) with $j \in \{1, ..., n_s\}$, $x_j(t) \in \mathbb{R}^{n_j}$ is the state vector; $u_j(t) \in \mathbb{R}^{m_j}$ is the control input; $w_j(t) \in \mathbb{R}^{q_j}$ is the disturbance input which belongs to $\mathcal{L}_2[0, \mathcal{T}]$; $y_j(t) \in \mathbb{R}^{p_j}$ is the measured output and $z_j(t) \in \mathbb{R}^{r_j}$ is the controlled output which belongs to $\mathcal{L}_2\big[(\Omega, \mathcal{F}, \mathbf{P}), [0, \mathcal{T}]\big]$ and ψ_{jk}, τ_j are unknown time-delays within known ranges such that

$$\begin{aligned}
0 \leq \psi_{jk} \leq \psi^+ \ , \ \dot{\psi}_{jk} \leq \psi^* < 1 \ , \forall j, k \in \{1, ..., n_s\} \\
0 \leq \tau_j \leq \tau^+ \ , \ \dot{\tau}_j \leq \tau^* < 1 \ , \ \forall j, k \in \{1, ..., n_s\}
\end{aligned}$$

in order to guarantee smooth growth of the state trajectories.

From now onwards, the notations **Lss** and **Css** refer, respectively, to the original large-scale system (8.1) and composite subsystem representation (8.2)-(8.4). An important identity that links both representations are expressed as [83]:

$$\sum_{j=1}^{n_s} \left\{ \sum_{k=1}^{n_s} A_{jk}(t,\eta_t)x_k(t) \right\} = \sum_{k=1}^{n_s} \left\{ \sum_{j=1}^{n_s} A_{kj}(t,\eta_t)x_j(t) \right\} \qquad (8.5)$$

The main difference in the underlying treatment of both representations is the explicit modeling of interconnections among subsystems as represented by the vector $g_j(t) \in \Re^{n_j}$ which in effect designates an interaction input to the jth subsystem. For various technical and operational factors, it is considered convenient to deal with **Css** instead of **Lss** and hence, in the remaining part of this work, we will base the analysis and design on the subsystem level. This implies that we will closely examine the role of interactions on the system behavior and therefore our work will eventually be a departure from the available results [32, 42, 47, 45] and the references therein.

For each possible value $\eta_t = i \in \mathcal{S}$, we will denote the system matrices of Σ_j) associated with mode i by

$$\begin{aligned} A_j(t,\eta_t) &\triangleq A_j(t,i)\ ,\ B_o(t,\eta_t) \triangleq B_j(t,i)\ ,\ \Gamma(\eta_t) \triangleq \Gamma(i) \\ E_j(t,\eta_t) &\triangleq E_j(t,i)\ ,\ C_j(t,\eta_t) \triangleq C(t,i)\ ,\ D_j(t,\eta_t) \triangleq D_j(t,i) \\ A_{jk}(t,\eta_t) &\triangleq A_{jk}(t,i)\ ,\ \Pi(\eta_t) \triangleq \Pi(i)\ ,\ G_j(\eta_t) \triangleq G_j(i) \\ F_j(\eta_t) &\triangleq F_j(i) \end{aligned} \tag{8.6}$$

where $\Pi_j(i), G_j(i), F_j(i), \Gamma_j(i)$ are known real constant matrices of appropriate dimensions and the matrices $A_j(t,i)$, $A_{jk}(t,i)$, $E_j(t,i)$, $C_j(t,i)$, $D_j(t,i)$, and $B_j(t,i)$ are real, time-varying matrix functions, possibly fast time varying, and representing the parameter uncertainties. Here, we follow the linear fractional representation method in which the uncertainties are expressed as:

$$\begin{aligned} \begin{bmatrix} A_j(t,i) & B_j(t,i) \\ C_j(t,i) & D_j(t,i) \end{bmatrix} &= \begin{bmatrix} A_{oj}(i) & B_{oj}(i) \\ C_{oj}(i) & D_{oj}(i) \end{bmatrix} \\ &+ \begin{bmatrix} R_j(i) \\ L_j(i) \end{bmatrix} H_j(t,i) \begin{bmatrix} M_j(i) & N_j(i) \end{bmatrix} \end{aligned} \tag{8.7}$$

$$\begin{aligned} E_j(t,i) &= E_{oj}(i) + M_{ej}(i) H_{ej}(t,i) N_{ej}(i) \\ A_{jk}(t,i) &= A_{jko}(i) + M_{jk}(i) H_{jk}(t,i) N_{jk}(i) \\ H_j^t(t,i) H_j(t,i) &\le I\ ;\ H_{jk}(t,i) H_{jk}^t(t,i) \le I \\ H_{ej}(t,i) H_{ej}^t(t,i) &\le I \quad \forall t, \forall j,\, k \in \{1, ..., n_s\} \end{aligned} \tag{8.8}$$

where $H_j(t,i) \in \Re^{\phi_j \times \varphi_j}$, $H_{jk}(t,i) \in \Re^{\nu_j \times \epsilon_k}$, $H_{ej}(t,i) \in \Re^{\upsilon_j \times \varsigma_j}$ are unknown time-varying matrices whose elements are Lebesgue measurable; $R_j(i)$, $L_j(i)$, $M_j(i)$, $N_j(i)$, $M_{jk}(i)$, $N_{jk}(i)$, $M_{ej}(i)$, $N_{ej}(i)$ are real, known and constant matrices with appropriate dimensions and $A_{oj}(i)$, $B_{oj}(i)$, $C_{oj}(i)$, $D_{oj}(i)$, and $E_{oj}(i)$ are real constant and known matrices representing the nominal decoupled system (without uncertainties and interactions):

$$\begin{aligned} \Sigma_j^{oo}: \quad \dot{x}_j(t) &= A_{oj}(i)x_j(t) + B_{oj}(i)u_j(t) + \Gamma_j(i)w_j(t) \\ y_j(t) &= C_{oj}(i)x_j(t) + D_{oj}(i)u_j(t) + \Pi_j(i)w_j(t) \\ z_j(t) &= G_j(i)x_j(t) + F_j(i)u_j(t) \end{aligned} \tag{8.9}$$

In the sequel, we assume $\forall j \in \{l, ..., n_s\}$ that the n_s-pairs $(A_{oj}(i), B_{oj}(i))$ and $(A_{oj}(i), C_{oj}(i))$ are stabilizable and detectable, respectively. The initial condition is specified as $\langle x_j(0), x_j(s)\rangle = \langle x_{jo}, \kappa_j(s)\rangle$, where $\kappa_j(.) \in \mathcal{L}_2[-d_j, 0]$, $d_j = \max_j\{-\tau_j, \psi_{jk}\}$, $j, k \in \{1, .., n_s\}$. Let $\mathsf{X}(t, \kappa)$ denotes the state trajectory of the interconnected system (8.2)-(8.4) from the initial condition $\{x_o, \kappa\}$, $x_o \equiv [x_{1o}^t, ..., x_{n_so}^t]^t$, $\kappa \equiv [\kappa_1^t, ..., \kappa_{n_s}^t]^t$.

Remark 8.1 *In the literature on state-space models containing parametric uncertainties, there has been different methods to characterize the uncertainty. In one method, the uncertainty is assumed to satisfy the so-called matching condition [8]. Loosely speaking, this condition implies that the uncertainties cannot enter arbitrarily into the system dynamics but are rather restricted to lie in the range space of the input matrix. By a second method, the uncertainty is represented by rank-1 decomposition [147]. It is well-known that both methods are quite restrictive in practice. This limitation can be overcome by using the generalized matching conditions [52] through an iterative procedure of constructing stabilizing controllers. There is a fourth method in which the dynamic model is cast into the polytopic format [132] which implies that the systems of the associated state-space model depend on a single parameter vector. The interest in* *the uncertainty characterization (8.6-8.7) is supported by the fact that quadratic stabilizability of feedback systems with norm-bounded uncertainties is equivalent to the standard H_∞ control problem [144]*

In the spirit of Chapter 3 with regard to the stochastic stability concepts and related issues, the following definitions pertaining to either **Css** or **Lss** are provided

Definition 8.1 *The subsystem (8.2)-(8.4) is said to be* **robustly stochastically stable with weak-delay dependence (RSSWDD)** *given $\tau^+, \tau^*, \psi^+, \psi^*$ if there exists a constant $\mathcal{Q}(\eta_o, \phi) \geq 0$ such that for all finite initial vector function $\phi \in I\!\!R^n$ defined on the interval $[-d_j, 0]$, $d_j = \max_j\{\tau_j, \psi_{jk}\}$, $j, k \in \{1, .., n_s\}$, for all interaction inputs $g_j(t) \in I\!\!R^{n_j}$, $j \in \{1, .., n_s\}$ and initial mode $\eta_o = i \in \mathcal{S}$*

$$I\!\!E\left[\int_0^\infty \{||\mathsf{X}(t, x_o, \kappa)||^2\}\, dt \,\middle|\, \eta_o, \kappa(s), s \in [-d, 0]\right] \quad \leq \mathcal{Q}(\eta_o, \kappa)$$

for all admissible uncertainties satisfying (8.7)-(8.8)

Definition 8.2 *The subsystem (8.2)-(8.4) is said to be* **RSSWDD with a disturbance attenuation** γ *given $\tau^+, \tau^*, \psi^+, \psi^*$ if for zero initial vector function $\kappa \equiv 0$ and initial mode $\eta_o = i \in \mathcal{S}$ the following inequality holds*

$$||z(t)||_{E_2} \;\triangleq\; I\!\!E\left[\int_0^\infty z^t(t) z(t) dt\right]^{1/2} \quad < \gamma\, ||w(t)||_2$$

for all $0 \neq w(t) \in \mathcal{L}_2[0, \infty)$, where $\gamma > 0$ is a prescribed level of disturbance attenuation for all admissible uncertainties satisfying (8.7)-(8.8.

Obviously in the absence of uncertainties, we have the following definitions

Definition 8.3 *The subsystem (8.2)-(8.4) is said to be* **stochastically stable with weak-delay dependence (SSWDD)** *given $\tau^+, \tau^*, \psi^+, \psi^*$ if there exists*

a constant $\mathcal{Q}(\eta_o, \phi) \geq 0$ *such that for all finite initial vector function* $\phi \in \mathbb{R}^n$ *defined on the interval* $[-d_j, 0]$, $d_j = \max_j\{\tau_j, \psi_{jk}\}$, $j, k \in \{1, .., n_s\}$, *for all interaction inputs* $g_j(t) \in \mathbb{R}^{n_j}$, $j \in \{1, .., n_s\}$ *and initial mode* $\eta_o = i \in \mathcal{S}$

$$\mathbb{E}\left[\int_0^\infty \{||\mathsf{X}(t, x_o, \kappa)||^2\}\, dt \middle| \eta_o, \kappa(s), s \in [-d, 0]\right] \quad \leq \mathcal{Q}(\eta_o, \kappa)$$

Definition 8.4 *The subsystem (8.2)-(8.4) is said to be* **SSWDD with a disturbance attenuation** γ *given* $\tau^+, \tau^*, \psi^+, \psi^*$ *if for zero initial vector function* $\kappa \equiv 0$ *and initial mode* $\eta_o = i \in \mathcal{S}$ *the following inequality holds*

$$||z(t)||_{E_2} \triangleq \mathbb{E}\left[\int_0^\infty z^t(t) z(t) dt\right]^{1/2} < \gamma\, ||w(t)||_2$$

for all $0 \neq w(t) \in \mathcal{L}_2[0, \infty)$, *where* $\gamma > 0$ *is a prescribed level of disturbance attenuation.*

Remark 8.2 *It should be emphasized that the stability definitions posed above have two particular features: 1) They are presented to the individual subsystem which comes in line with the prevailing trends in large-scale systems, and 2) They treat the interconnection variables* $g_j(t)$ *as external signal to preserve autonomy of the subsystems. In some sense, these definitions bear the decentralized properties which are fundamental to the analysis and design of interconnected systems*

8.3 Nominal Analysis and Design

We initially focus on the nominal case in which the uncerainties are not present and for simplicity we will suppress the interconnection delays ψ_{jk}.

8.3.1 Stability Results

The corresponding time-invariant model for $\eta_t = i \in \mathcal{S}$ is given by:

$$\Sigma_j^o : \quad \dot{x}_j(t) \quad = \quad A_j(i) x_j(t) + B_j(i) u_j(t) + \Gamma_j(i) w_j(t)$$

$$
\begin{aligned}
& + \quad E_j(i)x_j(t-\tau_j) + g_j(t)\ , \\
x_o &= \phi\ , \quad \eta_o = i\ , \quad t \in [0, \mathcal{T}]\ , j = \{1, ..., n_s\} && (8.10) \\
g_j(t) &= \sum_{k=1, j\neq k}^{n_s} A_{jk}(i)x_k(t) && (8.11) \\
y_j(t) &= x_j(t) && (8.12) \\
z_j(t) &= G_j(i)x_j(t) + F_j(i)u_j(t) && (8.13)
\end{aligned}
$$

for which we have the following stability result:

Theorem 8.1 *Consider the subsystem Σ_j^o with $u_j(t) \equiv 0,\ w_j(t) \equiv 0$. For any matrix sequence $Q_j(i) = Q_j^t(i) > 0,\ i \in \mathcal{S}$ satisfying*

$$
\sum_{m=1}^{n_s} \alpha_{im}\, Q_j(m) \ \leq \ 0 \tag{8.14}
$$

if there exist matrices $P_j(i) = P_j^t(i) > 0,\ i \in \mathcal{S}\ ,\ j\ \in \{1, ..., n_s\}$ satisfying the system of LMIs for all $i \in \mathcal{S}$

$$
\Xi_{jt} = \begin{bmatrix} \Upsilon_{ao}(i) & \Upsilon_b(i) \\ \Upsilon_b^t(i) & -\Upsilon_c(i) \end{bmatrix} < 0 \tag{8.15}
$$

where

$$
\begin{aligned}
\Upsilon_{ao}(i) &= A_j^t(i)P_j(i) + P_j(i)A_j(i) \\
&+ \sum_{m=1}^{n_s} \alpha_{im} P_j(m) + Q_j(i) + (n_s - 1)I \\
\Upsilon_b(i) &= [P_j(i)E_j(i) \ \ P_j(i) \sum_{k=1, j\neq k}^{n_s} A_{jk}(i)] \\
\Upsilon_c(i) &= [(1 - \tau_j^+)Q_j(i) \ \ I] && (8.16)
\end{aligned}
$$

then the subsystem Σ_j^o is **SSWDD**.

Proof: Let $\mathbf{x_j}_s(t) \stackrel{\Delta}{=} x_j(s+t), t - \tau \leq s \leq t,\ j \in \{1, \cdots, n_s\}$ and define the processes $\{(\mathbf{x}(t), \eta_t),\ t\ \geq\ 0\}$ and $\{(\mathbf{x}_j(t), \eta_t),\ t\ \geq\ 0\}$ associated with the **Lss**

and **Css** representations at the point $\{t, x, \eta_t\}$ over the state space $\bar{\mathcal{C}}$. It should be observed that both processes $\{(\mathbf{x}(t), \eta_t),\ t \geq 0\}$ and $\{(\mathbf{x}_j(t), \eta_t),\ t \geq 0\}$ are strong Markovian [78] and let $\Im^x[\cdot]$ and $\Im^x_j[\cdot]$ be the infinitesimal operators, respectively, of these processes. For $\eta_t = i \in \mathcal{S}$ and given $Q(i) = Q^t(i) > 0$, let the Lyapunov functional $V_j(\cdot) = V_j(x, \eta_t = i) : \Re^{n_j} \times \Re_+ \times \mathcal{S} \rightarrow \Re_+$ be selected as

$$V_j(x,i) \quad = \quad x_j^t(t)P_j(i)x(t) \; + \; \int_{-\tau}^{0} x_j^t(t+\theta)Q_j(i)x_j(t+\theta)d\theta \tag{8.17}$$

The weak infinitesimal operators $\Im^x[\cdot]$ and $\Im^x_j[\cdot]$ are given by:

$$\begin{aligned} \Im^x_j[V_j] \quad &= \quad \frac{\partial V_j}{\partial t} \; + \; \dot{x}^t(t)\frac{\partial V_j}{\partial x}|_{\eta_t=i} \; + \; \sum_{k=1}^{n_s} \alpha_{ik} V_j(t,x,k,i) && (8.18) \\ V(t,x,\eta_t) \quad &= \quad \sum_{j=1}^{n_s} \; V_j(t,x,\eta_t) \\ \Im^x[V(t,x,\eta_t)] \quad &= \quad \sum_{j=1}^{n_s} \; \Im^x_j[V_j(t,x,\eta_t)] && (8.19) \end{aligned}$$

For $\eta_t = i \in \mathcal{S}$, we get from (8.27)-(8.11) and (8.18)-(8.19):

$$\begin{aligned} \Im^x[V] \;&\triangleq\; \sum_{j=1}^{n_s} \Im^x_j[V_j] \\ &= \; \sum_{j=1}^{n_s} \Big\{ x_j^t(t)\; [A_j^t(i)P_j(i) + P_j(i)A_j(i)]\; x_j(t) \\ &+ \; x_j^t(t)P_j(i) \sum_{m=1, j\neq m}^{n_s} A_{jm}(\eta_t)x_m(t) \\ &+ \; \sum_{m=1, j\neq m}^{n_s} x_m^t(t)A_{jm}^t(i)\; P_j(i)x_j^t(t) \\ &+ \; x_j^t(t)\Big[\sum_{m=1}^{n_s} \alpha_{im}x_j^t(t)P_j(m)x_j(t) + Q_j(i)\Big]\; x_j(t) \end{aligned}$$

$$
\begin{aligned}
&+ \sum_{m=1}^{s} \alpha_{im} \int_{-\tau_j}^{0} x_j^t(t+\theta) Q_j(m) x_j(t+\theta)\, d\theta \\
&+ x_j^t(t-\tau_j) E_j^t(i) P_j(i) x_j(t) + x_j^t(t) P_j(i) E_j(i) x_j(t-\tau_j) \\
&- (1-\dot{\tau}_j) x_j^t(t-\tau_j) Q_j(i) x_j(t-\tau_j) \Big\} \qquad (8.20)
\end{aligned}
$$

In view of (8.14), we have

$$
\begin{aligned}
& \sum_{m=1}^{n_s} \alpha_{im} \int_{-\tau_j}^{0} x_j^t(t+\theta) Q_j(m) x_j(t+\theta)\, d\theta \\
&= \int_{-\tau_j}^{0} x_j^t(t+\theta) \left[\sum_{m=1}^{n_s} \alpha_{im}\, Q_j(m) \right] x_j(t+\theta)\, d\theta \\
&\leq 0 \qquad (8.21)
\end{aligned}
$$

Applying **Fact 1** and rearranging using (8.21) it follows from (8.20) that

$$
\begin{aligned}
&\Im^x[V] \\
&\leq \sum_{j=1}^{n_s} \left\{ x_j^t(t) \left[A_j^t(i) P_j(i) + P_j(i) A_j(i) + \sum_{m=1}^{n_s} \alpha_{im} P_j(m) \right.\right. \\
&+ Q_j(i) + P_j(i) E_j(i) \Big((1-\tau_j^+) Q_j(i) \Big)^{-1} j(i) E_j^t(i) P_j(i) \Big) \\
&+ \left.\left. P_j(i) [\sum_{k=1, j\neq k}^{n_s} A_{jk}(i) A_{jk}^t(i)] P_j(i) + \sum_{k=1, j\neq k}^{n_s} x_k^t(t) x_k(t) \right] x_j(t) \right\} \\
&\leq \sum_{j=1}^{n_s} \left\{ x_j^t(t) \left[A_j^t(i) P_j(i) + P_j(i) A_j(i) + \sum_{m=1}^{n_s} \alpha_{im} P_j(m) \right.\right. \\
&+ Q_j(i) + P_j(i) E_j(i) \Big((1-\tau_j^+) Q_j(i) \Big) j(i) E_j^t(i) P_j(i) \Big) \\
&+ \left.\left. P_j(i) [\sum_{k=1, j\neq k}^{n_s} A_{jk}(i) A_{jk}^t(i)] P_j(i) + (n_s - 1) I \right] x_j(t) \right\} \\
&\triangleq \sum_{j=1}^{n_s} \Im_j^x[V_j] \qquad (8.22)
\end{aligned}
$$

By using the Schur complements, it is readily seen from (8.21) and (8.22) that

$$\Im_j^x[V_j] \; < \; 0$$

The remaining part of the proof follows from **Theorem** 3.1 of Chapter 3. $\nabla\nabla\nabla$

Remark 8.3 *In [45], it has been established that, for subsystem (8.2)-(8.4) with $j = 1$, the terms " stochastically stable ", " exponentially mean-square stable ", and "asymptotically mean-square stable ", are equivalent, and any of them can imply " almost surely asymptotically stable ". Extending on these results, we have introduced* **Definition** *8.1 to suit* **Lss** *and* **Css** *representations. Accordingly, we can pose the equivalent terms " stochastically decentrally stable ", " exponentially mean-square decentrally stable ", and "asymptotically mean-square decentrally stable ", interchangeably for subsystem (8.2)-(8.4) with $j \in \{1, ..., n_s\}$.*

Following parallel development, the next theorem can be easily established

Theorem 8.2 *Consider the subsystem Σ_j^o with $u_j(t) \equiv 0$ and a scalar $\gamma > 0$. For any matrix sequence $Q_j(i) = Q_j^t(i) > 0,\ i \in \mathcal{S}$ satisfying*

$$\sum_{m=1}^{n_s} \alpha_{im}\, Q_j(m) \; \leq \; 0 \tag{8.23}$$

if there exist matrices $P_j(i) = P_j^t(i) > 0,\ i \in \mathcal{S}\ ,\ j\ \in \{1, ..., n_s\}$ satisfying the system of LMIs for all $i \in \mathcal{S}$

$$\Xi_{jt} = \begin{bmatrix} \Upsilon_{go}(i) & \Upsilon_b(i) & P_j(i)\Gamma_j(i) \\ \Upsilon_b^t(i) & -\Upsilon_c(i) & 0 \\ \Gamma_j^t(i)P_j(i) & 0 & -\gamma^2(i) \end{bmatrix} \; < \; 0 \tag{8.24}$$

where

$$\Upsilon_{go}(i) \;\; = \;\; \Upsilon_{ao}(i) + G^t(i)G(i) \tag{8.25}$$

then the subsystem Σ_j^o is **SSWDD with a disturbance attenuation** *γ.*

8.3.2 Stabilization Results

Introducing the decentralized control law for $\eta_j = i \in \mathcal{S}$

$$u_j(t) = -K_j^*(i)x_j(t) \ , \ j \in \{1, ..., n_s\} \ , \ ||K_j^*(i)|| < \infty \tag{8.26}$$

to subsystem Σ_j^o, $j = \{1, ..., n_s\}$, we obtain the closed-loop subsystem:

$$\begin{aligned} \Sigma_j^c : \ \dot{x}_j(t) &= \bar{A}_j(i)x_j(t) + \Gamma_j(i)w_j(t) \\ &+ E_j(i)x_j(t - \tau_j) + g_j(t) \ , \\ x_o &= \phi \ , \ \eta_o = i \ , \ t \in [0, \mathcal{T}] \end{aligned} \tag{8.27}$$

$$g_j(t) = \sum_{k=1, j \neq k}^{n_s} A_{jk}(i)x_k(t) \tag{8.28}$$

$$y_j(t) = x_j(t) \tag{8.29}$$

$$z_j(t) = \bar{G}_j(i)x_j(t) \tag{8.30}$$

$$\begin{aligned} \bar{A}_j(i) &= A_j(i) - B_j(i)K_j^*(i) \\ \bar{G}_j(i) &= G_j(i) - F_j(i)K_j^*(i) \end{aligned} \tag{8.31}$$

Extending on **Theorems** 8.1-8.2, we have the following results for the stochastic decentralized stabilizability of subsystem Σ_j^c.

Theorem 8.3 *Consider the subsystem Σ_j^c with $u_j(t) \equiv 0$, $w_j(t) \equiv 0$. For any matrix sequence $Q_j(i) = Q_j^t(i) > 0$, $i \in \mathcal{S}$ satisfying*

$$\sum_{m=1}^{n_s} \alpha_{im} \, Q_j(m) \ \leq \ 0 \tag{8.32}$$

if there exist matrices $P_j(i) = P_j^t(i) > 0$, $i \in \mathcal{S}$, $j \in \{1, ..., n_s\}$ satisfying the system of LMIs for all $i \in \mathcal{S}$

$$\Xi_{js} = \begin{bmatrix} \Upsilon_{ao}(i) & \Upsilon_b(i) \\ \Upsilon_b^t(i) & -\Upsilon_c(i) \end{bmatrix} < 0 \tag{8.33}$$

where

$$\begin{aligned}\bar{\Upsilon}_{ao}(i) &= \bar{A^t}_j(i)P_j(i) + P_j(i)\bar{A}_j(i) \\ &+ \sum_{m=1}^{n_s} \alpha_{im} P_j(m)Q_j(i) + (n_s - 1)I\end{aligned} \tag{8.34}$$

then the closed-loop subsystem Σ_j^o *is* **SSWDD**.

Theorem 8.4 *Consider the subsystem* Σ_j^c *with* $u_j(t) \equiv 0$ *and a scalar* $\gamma > 0$. *For any matrix sequence* $Q_j(i) = Q_j^t(i) > 0,\ i \in \mathcal{S}$ *satisfying*

$$\sum_{m=1}^{n_s} \alpha_{im}\, Q_j(m) \ \leq \ 0 \tag{8.35}$$

if there exist matrices $P_j(i) = P_j^t(i) > 0,\ i \in \mathcal{S}\ ,\ j\ \in \{1, ..., n_s\}$ *satisfying the system of LMIs for all* $i \in \mathcal{S}$

$$\Xi_{jh} = \begin{bmatrix} \bar{\Upsilon}_{go}(i) & \Upsilon_b(i) & P_j(i)\Gamma_j(i) \\ \Upsilon_b^t(i) & -\Upsilon_c(i) & 0 \\ \Gamma_j^t(i)P_j(i) & 0 & -\gamma^2(i) \end{bmatrix} < 0 \tag{8.36}$$

where

$$\bar{\Upsilon}_{go}(i) \ = \ \bar{\Upsilon}_{ao}(i) + \bar{G}^t(i)\bar{G}(i) \tag{8.37}$$

then the subsystem Σ_j^c *is* **SSWDD with a disturbance attenuation** γ.

Remark 8.4 *All the foregoing theorems show that the stochastic stability and stabilizability of every nominal jump subsystem is related to the existence of positive-definite solutions to a set of* $\mathbf{2\ n_s \times s}$ *coupled linear matrix inequalities. Equivalently stated, the stochastic stabilizability of the interconnected nominal jump system amounts to the existence of positive-definite solutions to a coupled set of* $\mathbf{2\ n_s^2 \times s}$ *LMIs.*

8.3.3 $\mathcal{H}_\infty$-State Feedback Control

In this section, associated with Css (8.2)-(8.4) with $j \in \{1, ..., n_s\}$, we consider the feedback design problem of Lss (8.1) with $\mathcal{H}_\infty$ performance using decentralized state-feedback controllers of the type (8.26) under the assumption that state information is available for feedback. The objective is to design a decentralized feedback controller $\mathcal{G}_j(t, \eta_t)$ such that, for all nonzero $w(t) \in \mathcal{L}_2$

$$||z_j(t)||_{E_2} \triangleq \mathbb{E}\left[\int_0^{\mathcal{T}} z_j^t(t) z_j(t) dt\right]^{1/2} < \gamma \, ||w_j(t)||_2 \ , \ j \in \{1, ..., n_s\} \quad (8.38)$$

where $\gamma > 0$ is a prescribed level of disturbance attenuation. When system (8.2)-(8.4) under the action of the controller $\mathcal{G}_j(t, i)$ satisfies condition (8.38), the interconnected controlled system is said to have an $\mathcal{H}_\infty$-performance over the horizon $[0, \mathcal{T}]$.

Two distinct cases arise:

(1) The finite-horizon case in which the system (8.2)-(8.4) with $j \in \{1, ..., n_s\}$ under the decentralized feedback controller $\mathcal{G}_j(t, i)$, $\eta_t = i \in \mathcal{S}$ has performance (8.38) over a given horizon $[0, \mathcal{T}]$;

(2) The infinite-horizon case in which the system (8.2)-(8.4) with $j \in \{1, ..., n_s\}$ under the decentralized feedback controller $\mathcal{G}_j(t, i)$, $\eta_t = i \in \mathcal{S}$ is stochastically decentrally stable and has performance (8.38) over the horizon $[0, \infty)$.

To treat both cases, we consider the following time-varying jumping time-delay system for $\eta_t = i \in \mathcal{S}$

$$\begin{aligned} \Sigma_j^t : \ \dot{x}_j(t) &= A_j(t, i) x_j(t) + B_j(t, i) u_j(t) \\ &+ E_j(i) x_j(t - \tau_j) + g_j(t) + \Gamma_j(t, i) w_j(t) \\ &+ \sum_{k=1, j \neq k}^{n_s} A_{jk}(t, i) x_k(t) \ , \ x_o = \phi \ , \ t \in [0, \mathcal{T}] \quad (8.39) \\ g_j(t) &= \sum_{k=1, j \neq k}^{n_s} A_{jk}(t, \eta_t) x_k(t) \quad (8.40) \end{aligned}$$

$$y_j(t) = x_j(t) \tag{8.41}$$

$$z_j(t) = G_j(t,i)x_j(t) + F_j(t,i)u_j(t) \tag{8.42}$$

and make the following assumptions:

Assumption 8.1 *For all $i \in \mathcal{S}$ on $[0, \mathcal{T}]$ and for all $j \in \{1, ..., n_s\}$,*

$$F_j^t(t,i)\ F_j(t,i) = R(t,i),\ \ R_j(t,i) = R^t(t,i) > 0.$$

Assumption 8.2 *For all $i \in \mathcal{S}$ and for all $j \in \{1, ..., n_s\}$,*

(1) $\{A_j(i), B_j(i)\}$ is stochastically decentrally stabilizable;

(2) $\{C_j(i), A_j(i)\}$ is stochastically decentrally observable.

Remark 8.5 Assumption *8.1 ensures that the $\mathcal{H}_\infty$-control problem for system (8.39)-(8.42) is nonsingular and corresponds to the standard assumption in $\mathcal{H}_\infty$-control theory for linear systems without jump parameters.* **Assumption** *8.2 guarantees the existence of a decentralized stabilizing controller for system (8.39)-(8.42) over the probability space $(\Omega, \mathcal{F}, \mathbf{P})$. The term "decentrally" is used to emphasize that the underlying condition is satisfied on the subsystem level.*

Finite Horizon

Now, we consider the design of a decentralized $\mathcal{H}_\infty$-state feedback controller for (8.39)-(8.42) subject to the condition of the probability space $(\Omega, \mathcal{F}, \mathbf{P})$ over a finite horizon.

Theorem 8.5 *Consider subsystem Σ_j^t over the probability space $(\Omega, \mathcal{F}, \mathbf{P})$ with $j \in \{1, ..., n_s\}$. Then for a given $\gamma > 0$, there exists a decentralized state-feedback controller $u_j(t)$ of the type (8.26) satisfying (8.38) for all nonzero $w(t) \in \mathcal{L}_2[0, \mathcal{T}]$ if the following set of $n_s \times s$ coupled differential Riccati equations*

(DREs) for all $i \in \mathcal{S}$*:*

$$\begin{aligned}
& \dot{P}_j(t,i) \; + \; P_j(t,i)A_j(t,i) + A_j^t(t,i)P_j(t,i) + (n_s - 1)I \\
+ \; & G_j^t(t,i)G_j(t,i) + \sum_{m=1, i \neq m}^{n_s} \alpha_{im} P_j(t,m) \\
+ \; & P_j(i)E_j(i)\Big((1-\tau_j^+)Q_j(i)\Big)^{-1} E_j^t(i)P_j(i) \\
+ \; & P_j(t,i)\Big\{ \sum_{k=1, j \neq k}^{n_s} A_{jk}(t,i)A_{t,jk}^t(t,i) + \gamma^{-2}\Gamma_j(i)\Gamma_j^t(t,i) \\
- \; & B_j(t,i)R_j^{-1}(t,i)B_j^t(t,i)\Big\}P_j(t,i) \; = \; 0, \\
& P_j(\mathcal{T}) \; = \; 0, \; , \;\; t \in [0, \mathcal{T}], \;\; j \in \{1, ..., n_s\}
\end{aligned} \tag{8.43}$$

has a solution $P_j(t,i) = P_j^t(t,i) > 0, \; i \in \mathcal{S}, \; j \in \{1, ..., n_s\}$ *on* $[0, \mathcal{T}]$*. Moreover, the decentralized controller is given by:*

$$\begin{aligned}
u_j(t) \;\; &= \;\; -K_j^*(t,i)x_j(t) \\
K^*(t,\eta_t) \;\; &= \;\; R_j^{-1}(t,i)\Big[B_j^t(t,i)P_j(t,i) + F_j^t(t,i)G_j(t,i)\Big] \\
& \qquad t \in [0, \mathcal{T}], \;\; j \; \in \{1, ..., n_s\}
\end{aligned} \tag{8.44}$$

Proof: Denote

$$\mathcal{J}(x_j) \triangleq \mathbb{E}\Big\{ \int_0^{\mathcal{T}} z_j^t(t)z_j(t) \; - \; \gamma^2 w_j^t(t)w_j(t) \, dt \Big\} \tag{8.45}$$

and let $\mathbf{x_j}_s(t) \triangleq x_j(s+t), t-\tau \leq s \leq t, \; j \in \{1, \cdots, n_s\}$ and define the processes $\{(\mathbf{x}(t), \eta_t), \; t \; \geq \; 0\}$ and $\{(\mathbf{x}_j(t), \eta_t), \; t \; \geq \; 0\}$ associated with the **Lss** and **Css** representations at the point $\{t, x, \eta_t\}$ over the state space $\bar{\mathcal{C}}$. It should be observed that both processes $\{(\mathbf{x}(t), \eta_t), \; t \; \geq \; 0\}$ and $\{(\mathbf{x}_j(t), \eta_t), \; t \; \geq \; 0\}$ are strong Markovian [78] and let $\Im^x[\cdot]$ and $\Im_j^x[\cdot]$ be the infinitesimal operators, respectively, of these processes. For $\eta_t \; = \; i \; \in \; \mathcal{S}$ and given $Q(i) \; = \; Q^t(i) \; > \; 0$, let the Lyapunov functional $V_j(\cdot) = V_j(x, \eta_t = i)$ be given by ((refInt7). In this case, the weak infinitesimal operators $\Im^x[\cdot]$ and $\Im_j^x[\cdot]$ as defined by (8.18)-(8.19) can be obtained as:

$$\begin{aligned}
\Im^x[V] &\triangleq \sum_{j=1}^{n_s} \Im_j^x[V_j] \\
&= \sum_{j=1}^{n_s} \Big\{ \dot{x}_j^t(t) P_j(t,i) x_j(t) \; + \; x_j^t(t) \dot{P}_j(t,i) x_j(t) \; + \; x_j^t(t) P_j(t,i) \dot{x}_j(t) \\
&+ \sum_{m=1}^{n_s} \alpha_{im} x_j^t(t) P_j(t,m) x_j(t) \Big\} \\
&+ \sum_{m=1}^{s} \alpha_{im} \int_{-\tau_j}^{0} x_j^t(t+\theta) Q_j(m) x_j(t+\theta)\, d\theta \Big\} \\
&= \sum_{j=1}^{n_s} \Big\{ x_j^t(t) \Big[\dot{P}_j(t,i) \; + \; A_j^t(t,i) P_j(t,i) + P_j(t,i) A_j(t,i) \Big] x_j(t) \; + \\
&+ \Big[x_j^t(t) P_j(t,i) B_j(t,i) u_j(t) + u_j^t(t) B_j^t(t,i) P_j(t,i) x_j(t) \Big] \\
&+ \Big[x_j^t(t) P_j(t,i) \Gamma_j(t,i) w_j(t) + w_j^t(t) \Gamma_j^t(t,i) P_j(t,i) x_j(t) \Big] \\
&+ x_j^t(t) P_j(t,i) \sum_{k=1, j\neq k}^{n_s} A_{jk}(t,\eta_t) x_k(t) \\
&+ \sum_{k=1, j\neq k}^{n_s} x_k^t(t) A_{jk}^t(t,\eta_t) \; P_j(t,i) x_j^t(t) \\
&+ \sum_{m=1}^{n_s} \alpha_{im} x_j^t(t) P_j(t,m) x_j(t) + Q_j(i) \\
&+ x_j^t(t-\tau_j) E_j^t(i) P_j(i) x_j(t) + x_j^t(t) P_j(i) E_j(i) x_j(t-\tau_j) \\
&+ \sum_{m=1}^{s} \alpha_{im} \int_{-\tau_j}^{0} x_j^t(t+\theta) Q_j(m) x_j(t+\theta)\, d\theta \\
&- (1-\dot{\tau}_j) x_j^t(t-\tau_j) Q_j(i) x_j(t-\tau_j) \Big\}
\end{aligned} \tag{8.46}$$

Standard matrix manipulations of (8.46) using (8.45) yields:

$$\begin{aligned}
\mathcal{J}(x) &\triangleq \sum_{j=1}^{n_s} \mathcal{J}(x_j) \\
&= \sum_{j=1}^{n_s} \left\{ \mathbb{E} \int_0^{\mathcal{T}} \left[z_j^t(t) z_j(t) \right.\right. \\
&+ x_j^t(t) \{ \dot{P}_j(t,i) \; + \; A_j^t(t,i) P_j(t,i) + P_j(t,i) A_j(t,i) \} x_j(t) \\
&+ \{ x_j^t(t) P_j(t,i) \Gamma_j(t,i) w_j(t) + w_j^t(t) \Gamma_j^t(t,i) P_j(t,i) x_j(t) \} \\
&+ \{ x_j^t(t) P_j(t,i) B_j(t,i) u_j(t) + u_j^t(t) B_j^t(t,i) P_j(t,i) x_j(t) \} \\
&+ x_j^t(t) P_j(t,i) \sum_{k=1, j \neq k}^{n_s} A_{jk}(t, \eta_t) x_k(t) \\
&+ \sum_{k=1, j \neq k}^{n_s} x_k^t(t) A_{jk}^t(t, \eta_t) \; P_j(t,i) x_j^t(t) \\
&+ \sum_{m=1}^{n_s} \alpha_{im} x_j^t(t) P_j(t,m) x_j(t) + Q_j(i) - \; \gamma^2 w_j^t(t) w_j(t) \\
&+ x_j^t(t - \tau_j) E_j^t(i) P_j(i) x_j(t) + x_j^t(t) P_j(i) E_j(i) x_j(t - \tau_j) \\
&+ \sum_{m=1}^{s} \alpha_{im} \int_{-\tau_j}^{0} x_j^t(t+\theta) Q_j(m) x_j(t+\theta) \, d\theta \\
&- \left.\left. (1 - \dot{\tau}_j) x_j^t(t - \tau_j) Q_j(i) x_j(t - \tau_j) \right] dt \right\} \\
&- \mathbb{E} \left\{ \int_0^{\mathcal{T}} \Im^x[V] dt \right\}
\end{aligned} \tag{8.47}$$

The substitution of (8.21) and (8.41) into (8.47) with standard manipulations yields:

$$\begin{aligned}
\mathcal{J}(x) \; \leq \; & \sum_{j=1}^{n_s} \left\{ \mathbb{E} \int_0^{\mathcal{T}} \left[x_j^t(t) \Big(\dot{P}_j(t,i) + A_j^t(t,i) P_j(t,i) + P_j(t,i) A_j(t,i) \right.\right. \\
+ \; & \sum_{m=1}^{n_s} \alpha_{im} P_j(t,m) + P_j(t,i) \Big[\gamma^{-2} \Gamma_j(t,i) \Gamma_j^t(t,i)
\end{aligned}$$

$$
\begin{aligned}
&- \quad B_j(t,i)R_j^{-1}(t,i)B_j^t(t,i)\Big]P_j(t,i) + G_j^t(t,i)G_j(t,i)\Big)x_j(t) \\
&+ \quad \Big[u_j^t(t) + x_j^t(t)[F_j^t(t,i)G_j(t,i) + P_j(t,i)B_j(t,i)]R_j^{-1}(t,i)\Big]R_j(t,i) \\
&\cdot \quad \Big[u_j(t) + x_j^t(t)[G_j^t(t,i)F_j(t,i) + B_j^t(t,i)P_j(t,i)]R_j^{-1}(t,i)\Big] \\
&- \quad \gamma^2\Big[w_j^t(t) - \gamma^{-2}x_j^t(t)P_j(t,i)\Gamma_j^t(t,i)\Big] \\
&\cdot \quad \Big[w_j(t) - \Gamma_j^t(t,i)P_j(t,i)x_j(t,i)\Big] \\
&+ \quad x_j^t(t)P_j(t,i)\sum_{k=1,j\neq k}^{n_s} A_{jk}(t,\eta_t)x_k(t) \\
&+ \quad \sum_{k=1,j\neq k}^{n_s} x_k^t(t)A_{jk}^t(t,\eta_t)\; P_j(t,i)x_j^t(t)\;\Big] \\
&+ \quad x_j^t(t-\tau_j)E_j^t(i)P_j(i)x_j(t) + x_j^t(t)P_j(i)E_j(i)x_j(t-\tau_j) \\
&- \quad (1-\dot{\tau}_j)x_j^t(t-\tau_j)Q_j(i)x_j(t-\tau_j)\;\Big]\;dt\Big\} \\
&- \quad \mathbb{E}\Big\{\int_0^{\mathcal{T}} \Im^x[V]dt\Big\} \qquad (8.48)
\end{aligned}
$$

In veiw of (8.26), (8.48) reduces to

$$
\begin{aligned}
\mathcal{J}(x) \;&\leq\; \sum_{j=1}^{n_s}\Big\{E\int_0^{\mathcal{T}}\Big\{x_j^t(t)[\dot{P}_j(t,i) + A_j^t(t,i)P_j(t,i) + P_j(t,i)A_j(t,i) \\
&+ \quad \sum_{m=1}^{n_s}\alpha_{im}P_j(t,m) + P_j(t,i)\big[\gamma^{-2}\Gamma_j(t,i)\Gamma_j^t(t,i) \\
&- \quad B_j(t,i)R_j^{-1}(t,i)B_j^t(t,i)\big]P_j(t,i) + G_j^t(t,i)G_j(t,i) \\
&+ \quad P_j(t,i)\Big[\sum_{k=1,j\neq k}^{n_s} A_{jk}(t,\eta_t)A_{jk}^t(t,\eta_t)\Big]P_j(t,i) + (n_s-1)I\Big]x_j(t) \\
&+ \quad P_j(i)E_j(i)\Big((1-\tau_j^+)Q_j(i)\Big)^{-1}E_j^t(i)P_j(i)
\end{aligned}
$$

$$
\begin{aligned}
&+ \quad \Big[u_j^t(t) + x_j^t(t)[F_j^t(t,i)G_j(t,i) + P_j(t,i)B_j(t,i)]R_j^{-1}(t,i)\Big] R_j(t,i) \\
&\cdot \quad \Big[u_j(t) + x_j^t(t)[G_j^t(t,i)F_j(t,i) + B_j^t(t,i)P_j(t,i)]R_j^{-1}(t,i)\Big] \\
&- \quad \gamma^2 \Big[w_j^t(t) - \gamma^{-2} x_j^t(t)P_j(t,i)\Gamma_j^t(t,i)\Big] \\
&\cdot \quad \Big[w_j(t) - \Gamma_j^t(t,i)P_j(t,i)x_j(t,i)\Big] - x_j^t(t)x_j(t)\Big\} \, dt \Big\} \\
&- \quad \mathbb{E}\int_0^{\mathcal{T}} \Im^x[V]dt\Big\}
\end{aligned}
\tag{8.49}
$$

Without loss any generality, we assume that the initial state value $x_j(0) = 0$. By using Dynkin's formula [78], one has

$$
\begin{aligned}
&\sum_{j=1}^{n_s} \Big\{\mathbb{E}\int_0^{\mathcal{T}} \Im^x[V]dt\Big\} \\
= \quad &\sum_{j=1}^{n_s} \Big\{\mathbb{E}[x_j^t(\mathcal{T})P_j(\mathcal{T},i)x_j(\mathcal{T})] - \mathbb{E}[x_j^t(0)P_j(0,i)x_j(0)]\Big\}
\end{aligned}
$$

together with the facts that $x_j(0) = 0$ and $P_j(\mathcal{T},i) = 0$, we now choose the decentralized controller $u_j(t)$ as that of (8.44) with $P_j(t,i)$ satisfying (8.43) and hence we get from (8.49) the inequality

$$
\begin{aligned}
\mathcal{J}(x) \quad \le \quad &\sum_{j=1}^{n_s} \Big\{ -\gamma^2\, \mathbb{E}\Big[\Big[\int_0^{\mathcal{T}} [w_j^t(t) - \gamma^{-2} x_j^t(t)P_j(t,i)\Gamma_j^t(t,i)] \\
\cdot \quad &[w_j(t) - \Gamma_j^t(t,i)P_j(t,i)x_j(t,i)] \\
+ \quad &\gamma^{-2} x_j^t(t)x_j(t)\, dt\Big]\Big]\Big\} \\
< \quad &0
\end{aligned}
\tag{8.50}
$$

and the proof is completed. $\nabla\nabla\nabla$

Infinite Horizon

For the infinite-horizon case, the main result is established by the following theorem.

Theorem 8.6 *Consider subsystem Σ_j^t over the probability space $(\Omega, \mathcal{F}, \mathbf{P})$ with $j \in \{1, ..., n_s\}$. Then, for a given $\gamma > 0$, there exists a decentralized state-feedback controller $u_j(t)$ such that the interconnected closed-loop system is stochastically decentrally stable and*

$$||z_j(t)||_{E_2} \; < \; \gamma \, ||w_j(t)||_2$$

for all nonzero $w(t) \in \mathcal{L}_2[0, \infty]$, if the following set of $\mathbf{n_s} \times \mathbf{s}$ coupled algebraic Riccati equations (AREs) for $i \in \mathcal{S}$:

$$\begin{aligned}
& P_j(i)A_j(i) + A_j^t(i)P_j(i) + (n_s - 1)I \\
+ \;& G_j^t(i)G_j(i) + \sum_{m=1, i \neq m}^{n_s} \alpha_{im} P_j(m) \\
+ \;& P_j(i)E_j(i)\Big((1 - \tau_j^+)Q_j(i)\Big)^{-1} E_j^t(i)P_j(i) \\
+ \;& P_j(i)\Big\{ \sum_{k=1, j \neq k}^{n_s} A_{jk}(i)A_{jk}^t(i) + \gamma^{-2}\Gamma_j(i)\Gamma_j^t(i) \\
- \;& B_j(i)R_j^{-1}(i)B_j^t(i)\Big\}P_j(i) \; = \; 0, \\
& t \in [0, \infty), \quad j \in \{1, ..., n_s\}
\end{aligned} \tag{8.51}$$

has a solution $P_j(i) = P_j^t(i)$, $i \in \mathcal{S}$, $j \in \{1, ..., n_s\}$. Moreover, the decentralized controller is given by:

$$\begin{aligned}
u_j(t) \;&=\; -K_j^*(\eta_t)x_j(t) \\
K^*(\eta_t) \;&=\; R_j^{-1}(\eta_t)[B_j^t(\eta_t)P_j(\eta_t) + F_j^t(\eta_t)G_j(\eta_t)] \\
& \quad t \in [0, \infty], \; j \in \{1, ..., n_s\}
\end{aligned} \tag{8.52}$$

Proof: In terms of the closed-loop system matrix

$$\begin{aligned}\bar{A}_j(i) &= A_j(i) - B_j(i)K_j^*(i) \\ &= A_j(i) - B_j(i)R_j^{-1}(i)[B_j^t(i)P_j(i) + F_j^t(i)G_j(i)] \end{aligned} \tag{8.53}$$

we rewrite (8.52) as

$$\begin{aligned} & P_j(i)\bar{A}_j(i) + \bar{A}_j^t(i)P_j(i) + (n_s - 1)I \\ + \; & G_j^t(i)G_j(i) + \sum_{m=1, i\neq m}^{n_s} \alpha_{im}P_j(m) \\ + \; & P_j(i)E_j(i)\Big((1-\tau_j^+)Q_j(i)\Big)^{-1}E_j^t(i)P_j(i) \\ + \; & P_j(i)\Big\{\sum_{k=1, j\neq k}^{n_s} A_{jk}(i)A_{jk}^t(i) + \gamma^{-2}\Gamma_j(i)\Gamma_j^t(i) \\ - \; & B_j(i)R_j^{-1}(i)B_j^t(i)\Big\}P_j(i) \;=\; 0, \\ & i \in \mathcal{S}, \quad j \in \{1, ..., n_s\} \end{aligned} \tag{8.54}$$

Since $P_j(i) = P_j^t(i) > 0,\ i \in \mathcal{S},\ j \in \{1, ..., n_s\}$ and the n_s-pairs $\{C_j(i), A_j(i)\},\ i \in \mathcal{S},\ j \in \{1, ..., n_s\}$ are decentrally observable, the stochastic stability of the interconnected closed-loop systems follows from the results of [42]. The $\mathcal{H}_\infty$-performance $||z_j(t)||_{E_2} < \gamma\ ||w_j(t)||_2$ for all nonzero $w_j(t) \in \mathcal{L}_2[0, \infty]$, can be readily obtained in the manner of **Theorem** 8.5. $\nabla\nabla\nabla$

Remark 8.6 Theorems *8.5-8.6 establish sufficient solvability conditions for the $\mathcal{H}_\infty$-control problem of the interconnected system (8.2)-(8.4) over the finite-horizon and infinite-horizon cases, respectively. The resulting conditions are expressed in terms $n_s \times s$ coupled differential and algebraic Riccati equations, respectively. It should be noted that when $\eta_t = 1$ and $A_{jk} \equiv 0$,* **Theorems** *8.5-8.6 recover the standard results of $\mathcal{H}_\infty$-control problems of single linear systems, see for example, [36].*

8.4 Robust Analysis and Design

In this section, we consider the design of a decentralized robust $\mathcal{H}_\infty$ feedback controller for the interconnected system (8.1) with uncertain parameters. In this case, the state-space model is given by (8.2)-(8.4) with $D_j(i) \equiv 0,\ \Pi_j(i) \equiv 0$. Specifically, we consider the problem of robust state-feedback control of the uncertain, interconnected Markovian jumping system Σ_j with $D_j(i) \equiv 0,\ \Pi_j(i) \equiv 0,\ j = \{1, ..., n_s\}$. Our purpose is to design a decentralized feedback controller $\mathcal{G}_j(t,i)$

$$u_j(t) \;=\; -K_j(t,i)x_j(t) \ , \quad j, \in \{1, ..., n_s\} \tag{8.55}$$

where $||K_j(t,\eta_t)|| < \infty$ such that, for all nonzero $w_j(t) \in \mathcal{L}_2[0,\infty)$ and for all parametric uncertainties satisfying (8.7)-(8.8) ang guaranteeing

$$||z_j(t)||_{E_2} \;<\; \gamma\, ||w_j(t)||_2 \ , \qquad j \in \{1, ..., n_s\} \tag{8.56}$$

where $\gamma > 0$ is a prescribed level of disturbance attenuation.

8.4.1 Robust Analysis

When system Σ_j with $D_j(i) \equiv 0,\ \Pi_j(i) \equiv 0,\ j = \{1, ..., n_s\}$ under the action of the controller (8.55) satisfies condition (8.56), the interconnected controlled system is said to have an $\mathcal{H}_\infty$-performance over the horizon $[0, \mathcal{T}]$.

We now establish some stochastic stability properties based on **Definitions** 8.1-8.2 and extending on **Theorems** 8.1-8.2.

Theorem 8.7 *Consider the subsystem Σ_j with $D_j(i) \equiv 0,\ \Pi_j(i) \equiv 0$. Then, the following statements are equivalent:*

(a) the subsystem Σ_j is **RSSWDD***;*

(b) for any matrix $\Phi_j(i) = \Phi_j^t(i) > 0,\ i \in \mathcal{S}$ and a scalar $\mu_j(i) > 0,\ ,i \in \mathcal{S}$ there exist matrices $\Omega_j(i) = \Omega_j^t(i) > 0,\ i \in \mathcal{S}$ for all $j\ \in \{1, ..., n_s\}$, satisfying

the LMIs for all $i \in \mathcal{S}$

$$\Xi_{jt} = \begin{bmatrix} \Upsilon_{as}(i) & \Upsilon_b(i) & \Omega_j(i)M_j(i) \\ \Upsilon_b^t(i) & -\Upsilon_c(i) & 0 \\ M_j^t(i)\Omega_j(i) & 0 & -\mu_j(i)I \end{bmatrix} < 0 \qquad (8.57)$$

where

$$\begin{aligned} \Upsilon_{as}(i) &= A_j^t(i)\Omega_j(i) + \Omega_j(i)A_j(i) \\ &+ \sum_{m=1}^{n_s} \alpha_{im}\Omega_j(m) + \Phi_j(i) + (n_s - 1)I \\ &+ \mu_j(i)N_j^t(i)N_j(i) \end{aligned} \qquad (8.58)$$

Proof: Let 8.61 have a feasible solution $\Omega_j(i) = \Omega_j^t(i)^t > 0,\ i \in \mathcal{S}$ and $j \in \{1, ..., n_s\}$. For the class of admissible uncertainties $\Delta_j(t,i)$ satisfying (8.7)-(8.8) and for $i \in \mathcal{S}$, we get from **Fact 1**

$$\begin{aligned} &\mu_j^{-1}(i)\Omega_j(i)M_j(i)M_j^t(i)\Omega_j(i) + \mu_j(i)N_j^t(i)N_j(i) \\ &\geq N_j^t(i)\Delta_j^t(i)M_j^t(i)\Pi_j(i) + \Pi_j(i)M_j(i)\Delta_j(i)N_j(i) \end{aligned} \qquad (8.59)$$

It follows from (8.61)-(8.63) with the aid of **Fact 3** that

$$\begin{aligned} &\Omega_j(i)[A_j(i) + M_j(i)\Delta_j(i)N_j(i)] + [A_j(i) + M_j(i)\Delta_j(i)N_j(i)]^t\Omega_j(i) \\ +\ &\Omega_j(i)\{\sum_{m=1,j\neq m}^{n_s} A_{jm}(i)A_{jm}^t\}\Omega_j(i) + (n_s - 1)I + \sum_{m=1,i\neq m}^{n_s} \alpha_{im}\Pi_j(m) \\ +\ &\Phi_j(i) + \Omega_j(i)E_j(i)\left[(1-\tau_j^+)Q_j(i)\right]E_j^t(i)\Omega_j(i) \leq 0 \end{aligned} \qquad (8.60)$$

holds for all admissible uncertainties $\Delta_j(t,i)$ satisfying (8.7)-(8.8). The LMIs (8.61) is readily obtained from application of **Theorem** 8.1. $\nabla\nabla\nabla$

Theorem 8.8 *Consider the subsystem Σ_j with $D_j(i) \equiv 0$, $\Pi_j(i) \equiv 0$. Then, the following statements are equivalent:*

(a) the subsystem Σ_j is **RSSWDD** *by a decentralized control law of the type (8.26)*

(b) for any matrix $\Phi_j(i) = \Phi_j^t(i) > 0,\ i \in \mathcal{S}$ and a scalar $\mu_j(i) > 0,\ ,i \in \mathcal{S}$ there exist matrices $\Omega_j(i) = \Omega_j^t(i) > 0,\ i \in \mathcal{S}$ for all $j\ \in \{1, ..., n_s\}$, satisfying the LMIs for all $i \in \mathcal{S}$

$$\Xi_{jt} = \begin{bmatrix} \bar{\Upsilon}_{as}(i) & \Upsilon_b(i) & \Omega_j(i)M_j(i) \\ \Upsilon_b^t(i) & -\Upsilon_c(i) & 0 \\ M_j^t(i)\Omega_j(i) & 0 & -\mu_j(i)I \end{bmatrix} < 0 \tag{8.61}$$

where

$$\begin{aligned} \Upsilon_{as}(i) &= \tilde{A}_j^t(i)\Omega_j(i) + \Omega_j(i)\tilde{A}_j(i) \\ &+ \sum_{m=1}^{n_s} \alpha_{im}\Omega_j(m) + \Phi_j(i) + (n_s - 1)I \\ &+ \mu_j(i)N_j^t(i)N_j(i) \\ \tilde{A}_j(i) &= A_j(i) - B_j(i)K_j^*(i) \end{aligned} \tag{8.62}$$

Proof: It follows by parallel development to **Theorem** 8.7 and using **Theorem** 8.6.

8.4.2 Robust Design

We now focus attention on the controller design. More specifically, the objective is to design a robust decentralized state-feedback controller $\mathcal{G}_j(t, \eta_t)$ such that:

(1) In the finite-horizon case, system Σ_j with $j \in \{1, ..., n_s\}$ under the decentralized feedback controller $\mathcal{G}_j(t, \eta_t)$ has performance (8.56) over a given horizon $[0, \mathcal{T}]$;

(2) In the infinite-horizon case in which system Σ_j with $j \in \{1, ..., n_s\}$ under the decentralized feedback controller $\mathcal{G}_j(t, \eta_t)$ is stochastically decentrally stable and has performance (8.56) over a given horizon $[0, \infty]$.

The main results are established by the following theorems for the cases of finite-horizon and infinite-horizon cases, respectively

Theorem 8.9 *Consider the subsystem Σ_j with $D_j(i) \equiv 0$, $\Pi_j(i) \equiv 0$ and $j \in \{1, ..., n_s\}$. Then, for a given $\gamma > 0$, there exists a decentralized state-feedback controller $u_j(t)$ such that*

$$||z_j(t)||_{E_2} \;<\; \gamma\, ||w_j(t)||_2 \;, \qquad j \in \{1, ..., n_s\}$$

for all nonzero $w(t) \in \mathcal{L}_2[0, \mathcal{T}]$, and for all admissible uncertainties satisfying (8.7)-(8.8) if for a given scalar $\mu_j(i) > 0$, ,$i \in \mathcal{S}$, the following set of $n_s \times s$ coupled DREs for all $i \in \mathcal{S}$

$$\begin{aligned}
&\dot{\Pi}_j(t,i) \;+\; \Pi_j(t,i)A_j(t,i) + A_j^t(t,i)\Pi_j(t,i) + (n_s - 1)I \\
+\;& G_j^t(t,i)G_j(t,i) + \sum_{m=1, i\neq m}^{n_s} \alpha_{im}\Pi_j(t,m) \\
+\;& \Pi_j(t,i)\Big\{ \sum_{k=1, j\neq k}^{n_s} A_{jk}(t,i)A_{t,jk}^t + \gamma^{-2}\Gamma_j(i)\Gamma_j^t(t,i) \\
-\;& B_j(t,i)R_j^{-1}(t,i)B_j^t(t,i)\Big\}\Pi_j(t,i) \\
+\;& \Pi_j(i)E_j(i)\Big((1-\tau_j^+)Q_j(i)\Big)^{-1}E_j^t(i)\Pi_j(i) \\
+\;& \mu_j^{-1}(i)\Pi_j(t,i)M_j(t,i)M_j^t(t,i)\Pi_j(t,i) \\
+\;& \mu_j(i)N_j^t(t,i)N_j(t,i) \;=\; 0, \qquad (8.63) \\
&\Pi_j(\mathcal{T}) \;=\; 0, \quad i \in \mathcal{S}, \quad t \in [0, \mathcal{T}] \qquad (8.64)
\end{aligned}$$

has a solution $\Pi_j(t,i) = \Pi_j^t(t,i) > 0$, $i \in \mathcal{S}$, $j \in \{1, ..., n_s\}$ on $[0, \mathcal{T}]$. Moreover, the decentralized controller is given by:

$$\begin{aligned}
u_j(t) &= -K_j(t,i)x_j(t) \\
K(t,i) &= R_j^{-1}(t,i)\Big[B_j^t(t,i)\Pi_j(t,i) + F_j^t(t,i)G_j(t,i)\Big]
\end{aligned}$$

$$t \in [0, \mathcal{T}], \;\; i \in \mathcal{S}, \;\; j \in \{1, ..., n_s\} \tag{8.65}$$

Proof: Let (8.63) have a solution $\Pi_j(i) = \Pi_j^t(i)^t > 0, \; i \in \mathcal{S}$ and $j \in \{1, ..., n_s\}$. For the class of admissible uncertainties $\Delta_j(t, \eta_t)$ satisfying (8.7)-(8.8) and for $i \in \mathcal{S}$, and proceeding like **Theorem** 8.8, we have:

$$\begin{aligned}
& \dot{\Pi}_j(t,i) \; + \; \Pi_j(t,i)[A_j(t,i) + \Delta A_j(t,i)] + [A_j(t,i) + \Delta A_j(t,i)]^t \Pi_j(t,i) \\
+ \;\; & (n_s - 1)I + G_j^t(t,i)G_j(t,i) + \sum_{m=1, i \neq m}^{n_s} \alpha_{im} \Pi_j(t,m) \\
+ \;\; & \Pi_j(i)E_j(i)\Big((1 - \tau_j^+)Q_j(i)\Big)^{-1} E_j^t(i)\Pi_j(i) \\
+ \;\; & \Pi_j(t,i)\{ \sum_{k=1, j \neq k}^{n_s} A_{jk}(t,i)A_{t,jk}^t + \gamma^{-2}\Gamma_j(i)\Gamma_j^t(t,i) \\
- \;\; & B_j(t,i)R_j^{-1}(t,i)B_j^t(t,i)\}\Pi_j(t,i) \; = \; 0,
\end{aligned} \tag{8.66}$$

$$\Pi_j(\mathcal{T}) \; = \; 0, \quad t \in [0, \mathcal{T}], \quad j \in \{1, ..., n_s\} \tag{8.67}$$

for all admissible uncertainties $\Delta_j(t,i), i \in \mathcal{S}$ satisfying (8.7)-(8.8). It follows from **Theorem** 8.5 that

$$||z_j(t)||_{E_2} \; < \; \gamma \, ||w_j(t)||_2 \; , \quad j \in \{1, ..., n_s\}$$

and the proof is completed. $\nabla\nabla\nabla$

Theorem 8.10 *Consider the subsystem Σ_j with $D_j(i) \equiv 0, \; \Pi_j(i) \equiv 0$ and $j \in \{1, ..., n_s\}$. Then, for a given $\gamma > 0$, there exists a decentralized the interconnected closed-loop system is* **RSSWDD** *and*

$$||z_j(t)||_{E_2} \; < \; \gamma \, ||w_j(t)||_2 \; , \quad j \in \{1, ..., n_s\}$$

for all nonzero $w(t) \in \mathcal{L}_2[0, \infty]$, and for all admissible uncertainties satisfying (8.7)-(8.8)if for a given scalar $\mu_j(i) > 0$, ,$i \in \mathcal{S}$, if the following set of $n_s \times s$

coupled AREs:

$$
\begin{aligned}
& \Pi_j(i)A_j(i) + A_j^t(i)\Pi_j(i) + \sum_{m=1, i\neq m}^{n_s} \alpha_{im}\Pi_j(m) + (n_s - 1)I + G_j^t(i)G_j(i) \\
+ & \; \Pi_j(i)\{ \sum_{k=1, j\neq k}^{n_s} A_{jk}(i)A_{jk}^t + \gamma^{-2}\Gamma_j(i)\Gamma_j^t(i) - B_j(i)R_j^{-1}(i)B_j^t(i)\}\Pi_j(i) \\
+ & \; \mu_j^{-1}(i)\Pi_j(i)M_j(i)M_j^t(i)\Pi_j(i) + \mu_j(i)N_j^t(i)N_j(i) \; = \; 0, \qquad (8.68) \\
& t \in [0, \infty), \quad j \in \{1, ..., n_s\} \qquad (8.69)
\end{aligned}
$$

has a solution $\Pi_j(i) = \Pi_j^t(i) > 0, \; i \in \mathcal{S}, \; j \in \{1, ..., n_s\}$. *Moreover, the decentralized controller is given by:*

$$
\begin{aligned}
u_j(t) &= -K_j(i)x_j(t) \\
K(i) &= R_j^{-1}(\eta_t)\Big[B_j^t(\eta_t)\Pi_j(\eta_t) + F_j^t(\eta_t)G_j(\eta_t)\Big] \\
& \quad t \in [0, \infty), \; \in \{1, ..., n_s\} \qquad (8.70)
\end{aligned}
$$

Proof: It can be established by following a similar procedure to **Theorem** 8.9 with the help of **Theorem** 8.7. ∇∇∇

Remark 8.7 *Using the convex optimization techniques over linear matrix inequalities, the existence of scaling parameters* $\mu_j(i) > 0, \; , i \in \mathcal{S}, \; j \; \in \{1, ..., n_s\}$ *can be conveniently checked out.*

Remark 8.8 *In terms of* $\mathcal{H}_\infty$ *control theory [11], it can be shown from* **Theorem** *8.9 and* **Theorem** *8.10 that the* $n_s \times s$ *DREs (8.66) and the* $n_s \times s$ *AREs (8.68) are the sufficient stochastic stability conditions for the following* $\mathcal{H}_\infty$ *control problem without parametric uncertainties over the finite-horizon and the infinite-horizon, respectively:*

$$
\tilde{\Sigma}_j : \; \dot{x}_j(t) \; = \; A_j(t, \eta_t)x_j(t)
$$

$$
\begin{aligned}
&+ \quad B_j(t,\eta_t)u_j(t) + \left[\Gamma_j(t,\eta_t) \quad \frac{\gamma}{\sqrt{\mu_j(\eta_t)}}M_j(t,\eta_t)\right]\tilde{w}_j(t) \\
&+ \quad \sum_{k=1,j\neq k}^{n_s} A_{jk}(t,\eta_t)x_k(t) \ , \ x_o = 0 \ , \quad t \in [0,\mathcal{T}] \qquad (8.71)
\end{aligned}
$$

$$
\begin{aligned}
\tilde{z}(t) \quad &= \quad \begin{bmatrix} \sqrt{\mu_j(\eta_t)}N_j(t,\eta_t) \\ G_j(t,\eta_t) \end{bmatrix} x_j(t) \\
&+ \quad \begin{bmatrix} 0 \\ F_j(t,\eta_t) \end{bmatrix} u_j(t) \qquad (8.72)
\end{aligned}
$$

where

$$
\begin{aligned}
\tilde{w}_j(t) \quad &= \quad \begin{bmatrix} w_j(t) \\ \gamma^{-1}\sqrt{\mu_j(\eta_t)}\Delta(t,\eta_t)N(t,\eta_t) \end{bmatrix} \\
\tilde{z}_j(t) \quad &= \quad \begin{bmatrix} \sqrt{\mu_j(\eta_t)}N(t,\eta_t)x_j(t) \\ z_j(t) \end{bmatrix} \qquad (8.73)
\end{aligned}
$$

It is readily seen that

$$
||\tilde{z}_j(t)||_{E_2} \quad < \quad ||z_j(t)||_{E_2}
$$

and hence we conclude that if we solve the $\mathcal{H}_\infty$-control problem for system (8.71)-(8.73) with (8.7)-(8.8), then we can also solve the robust $\mathcal{H}_\infty$-control problem for subsystem Σ_j with (8.7)-(8.8) using the same controller.

8.4.3 Uncertain Jumping Rates

Extension of the developed robustness results to the case where the jumping rates are subject to uncertainties. Specifically, we consider the transition probability from mode i at time t to mode j at time $t+\delta$, $i,j \in \mathcal{S}$ to be:

$$
\begin{aligned}
p_{ij} \quad &= \quad Pr(\eta_{t+\delta} = j \mid \eta_t = i) \\
&= \quad \begin{cases} (\alpha_{ij} + \Delta\alpha_{ij})\delta + o(\delta), & ifi \neq j, \\ 1 + (\alpha_{ii} + \Delta\alpha_{ii})\delta + o(\delta), & ifi = j \end{cases} \qquad (8.74)
\end{aligned}
$$

with transition probability rates $(\alpha_{ij} + \Delta\alpha_{ij}) \geq 0$ for $i,j \in \mathcal{S}, i \neq j$ and

$$
\alpha_{ii} + \Delta\alpha_{ii} \quad = \quad - \sum_{m=1,m\neq i}^{s} (\alpha_{im} + \Delta\alpha_{im}) \qquad (8.75)
$$

We assume that the uncertainties $\Delta\alpha_{ij}$ satisfies

$$||\Delta\alpha_{ij}|| \le \beta_{ij} \ , \qquad \forall\, i,j \in \mathcal{S} \tag{8.76}$$

where β_{ij} are known scalars $\forall\, i,j \in \mathcal{S}$.

In line of **Theorem** 8.9 and **Theorem** 8.10, we have the following robustness results.

Theorem 8.11 *Consider the subsystem Σ_j with $D_j(i) \equiv 0$, $\Pi_j(i) \equiv 0$ and $j \in \{1, ..., n_s\}$. Then, for a given $\gamma > 0$, there exists a decentralized state-feedback controller $u_j(t)$ such that*

$$||z_j(t)||_{E_2} \ < \ \gamma\, ||w_j(t)||_2$$

for all nonzero $w(t) \in \mathcal{L}_2[0, \mathcal{T}]$, and for all admissible uncertainties satisfying (8.7)-(8.8) if for a given scalar $\mu_j(i) > 0$, $, i \in \mathcal{S}$, the following set of $n_s \times s$ coupled DREs for all $i \in \mathcal{S}$

$$\begin{aligned}
& \dot{\Pi}_j(t,i) \ + \ \Pi_j(t,i)A_j(t,i) + A_j^t(t,i)\Pi_j(t,i) + (n_s-1)I \\
+ \ & G_j^t(t,i)G_j(t,i) + \sum_{m=1, i\neq m}^{n_s} \Big[\alpha_{im} + \beta_{im}\Big]\Pi_j(t,m) \\
+ \ & \Pi_j(t,i)\Big\{ \sum_{k=1, j\neq k}^{n_s} A_{jk}(t,i)A_{t,jk}^t + \gamma^{-2}\Gamma_j(i)\Gamma_j^t(t,i) \\
- \ & B_j(t,i)R_j^{-1}(t,i)B_j^t(t,i)\Big\}\Pi_j(t,i) \\
+ \ & \Pi_j(i)E_j(i)\Big((1-\tau_j^+)Q_j(i)\Big)^{-1} E_j^t(i)\Pi_j(i) \\
+ \ & \mu_j^{-1}(i)\Pi_j(t,i)M_j(t,i)M_j^t(t,i)\Pi_j(t,i) \\
+ \ & \mu_j(i)N_j^t(t,i)N_j(t,i) \ = \ 0, && (8.77) \\
& \Pi_j(\mathcal{T}) \ = \ 0, \quad i \in \mathcal{S}, \quad t \in [0, \mathcal{T}] && (8.78)
\end{aligned}$$

has a solution $\Pi_j(t,i) = \Pi_j^t(t,i) > 0,\ i \in \mathcal{S},\ j \in \{1,...,n_s\}$ *on* $[0,\mathcal{T}]$. *Moreover, the decentralized controller is given by:*

$$\begin{aligned} u_j(t) &= -K_j(t,i)x_j(t) \\ K(t,i) &= R_j^{-1}(t,i)\Big[B_j^t(t,i)\Pi_j(t,i) + F_j^t(t,i)G_j(t,i)\Big] \\ & \quad t \in [0,\mathcal{T}],\ \ i \in \mathcal{S},\ \ j \in \{1,...,n_s\} \end{aligned} \tag{8.79}$$

Proof: It can be derived by using similar arguments to **Theorem** 8.9. ∇∇∇

Theorem 8.12 *Consider the subsystem* Σ_j *with* $D_j(i) \equiv 0,\ \Pi_j(i) \equiv 0$ *and* $j \in \{1,...,n_s\}$. *Then, for a given* $\gamma > 0$, *there exists a decentralized the interconnected closed-loop system is* **RSSWDD** *and*

$$||z_j(t)||_{E_2} \ < \ \gamma\, ||w_j(t)||_2\ , \qquad j \in \{1,...,n_s\}$$

for all nonzero $w(t) \in \mathcal{L}_2[0,\infty]$, *and for all admissible uncertainties satisfying (8.7)-(8.8)if for a given scalar* $\mu_j(i) > 0,\ ,i \in \mathcal{S}$, *if the following set of* $n_s \times s$ *coupled AREs:*

$$\begin{aligned} & \Pi_j(i)A_j(i) + A_j^t(i)\Pi_j(i) + \sum_{m=1, i\neq m}^{n_s} \Big[\alpha_{im} + \beta_{im}\Big]\Pi_j(m) \\ + \ & (n_s - 1)I + G_j^t(i)G_j(i) \\ + \ & \Pi_j(i)\{\sum_{k=1,j\neq k}^{n_s} A_{jk}(i)A_{jk}^t + \gamma^{-2}\Gamma_j(i)\Gamma_j^t(i) - B_j(i)R_j^{-1}(i)B_j^t(i)\}\Pi_j(i) \\ + \ & \mu_j^{-1}(i)\Pi_j(i)M_j(i)M_j^t(i)\Pi_j(i) + \mu_j(i)N_j^t(i)N_j(i) \ = \ 0, \end{aligned} \tag{8.80}$$

$$t \in [0,\infty), \quad j \in \{1,...,n_s\} \tag{8.81}$$

has a solution $\Pi_j(i) = \Pi_j^t(i) > 0,\ i \in \mathcal{S},\ j \in \{1,...,n_s\}$. *Moreover, the decentralized controller is given by:*

$$u_j(t) \ = \ -K_j(i)x_j(t)$$

$$
\begin{aligned}
K(i) &= R_j^{-1}(\eta_t)\Big[B_j^t(\eta_t)\Pi_j(\eta_t) + F_j^t(\eta_t)G_j(\eta_t)\Big] \\
& \quad t \in [0,\infty), \ \in \{1,...,n_s\}
\end{aligned} \tag{8.82}
$$

Proof: It can be carried out by parallel development to **Theorem** 8.10. ∇∇∇

8.5 Robust Decentralized Dynamic Feedback

In this section, we proceed further beyond what has been accomplished in the previous sections of this Chapter. In the sequel, we will work on the full model (8.2)-(8.4) to generalize the results which, in turn, enables us to derive several special cases of interest. Therefore, we will seek stabilization of system (8.2)-(8.4) in the absence of the external disturbance, $w_j(t) \equiv 0$, by means of a class of observer-based feedback controllers of the form:

$$
\begin{aligned}
\Sigma_j^{dc}: \ \dot{\hat{x}}_j(t) &= [A_{oj}(i) + \delta A_j(i)]\hat{x}_j(t) + \Theta_j(i)[y_j(t) - \hat{y}_j(t)] \\
&+ B_{oj}(i)u_j(t) \\
\hat{y}_j(t) &= [C_{oj}(i) + \delta C_j(i)]\hat{x}_j(t) + D_{oj}(i)u_j(t) \\
u_j(t) &= -K_j(i)\hat{x}_j(t)
\end{aligned} \tag{8.83}
$$

where the matrices introduced in (8.83) will be specified shortly. We now look for a solution to the decentralized robust stabilization problem. For this purpose, consider that for $j = 1,...,n_s$ there exist sequence of matrices $0 < P_j(i) = P_j^t(i) \in \Re^{n_j \times n_j}$, $0 < \Upsilon_p^t(i) = \Upsilon_p(i) \in \Re^{n_j \times n_j}$, $0 < Q_j(i) = Q_j^t(i) \in \Re^{n_j \times n_j}$, $0 < \Upsilon_q^t(i) = \Upsilon_q(i) \in \Re^{n_j \times n_j}$, $i \in \mathcal{S}$ satisfying the family of LMIs for all $i \in \mathcal{S}$:

$$
\begin{bmatrix}
\Xi_{jp}(i) & P_j(i)X_j(i) & \rho_j(i)M_j^t(i) \\
X_j^t(i)P_j(i) & -I_j & 0 \\
\rho_j(i)M_j(i) & 0 & -I_j
\end{bmatrix} < 0 \tag{8.84}
$$

$$\begin{bmatrix} \Xi_{jq}(i) & Q_j(i)X_j(i) & \tilde{C}^t_{oj} \\ X^t_j(i)Q_j(i) & -I_j & 0 \\ \tilde{C}_{oj}(i) & 0 & -\rho^2_j(i)L_j(i)L^t_j(i) \end{bmatrix} < 0 \tag{8.85}$$

$$\begin{bmatrix} -\Upsilon_q(i) & \rho_j(i)J^t_j \\ \rho_j(i)J_j & L^t_j(i)L_j(i) \end{bmatrix} < 0$$

$$\begin{bmatrix} -\Upsilon_p(i) & \tilde{B}^t_{oj} \\ \tilde{B}_{oj} & \rho^2_j(i)N^t_j(i)N_j(i) \end{bmatrix} < 0 \tag{8.86}$$

along with

$$\begin{aligned}
\Xi_{jp}(i) &= P_j(i)A_{oj}(i) + A^t_{oj}(i)P_j(i) \\
&+ \sum_{m=1}^{s} P_j(m) + \Upsilon_p(i) + [n_s + 1]I_j \\
\Xi_{jq}(i) &= Q_j(i)A_{oj}(i) + Q_j(i)X_j(i)X^t_j(i)P_j(i) \\
&+ A^t_{oj}(i)Q_j(i) + P_j(i)X_j(i)X^t_j(i)Q_j(i) \\
&+ \Upsilon_q(i) + \sum_{m=1}^{s} Q_j(m) \\
X_j(i)X^t_j(i) &= \rho^{-2}_j(i)R_j(i)R^t_j(i) + \Omega_{ej}(i)\Omega^t_{ej}(i) + \Omega_{aj}(i)\Omega^t_{aj}(i) \\
\tilde{B}_{oj}(i) &= B^t_{oj}(i)P_j(i) + \rho^2_j(i)N^t_j(i)M_j(i) \\
\Omega_{ej}(i)\Omega^t_{ej}(i) &= E_{oj}(i)[I_j - \alpha^2_j N^t_{ej}(i)N_{ej}(i)]^{-1}E^t_{oj}(i) \\
&+ \alpha^{-2}_j(i)M_{ej}(i)M^t_{ej}(i) \\
\Omega_{aj}(i)\Omega^t_{aj}(i) &= \sum_{k=1}^{n_s} \{A_{jko}(i)[I_j - \beta^2_{jk}N^t_{jk}(i)N_{jk}(i)]^{-1}A^t_{jko}(i) \\
&+ \beta^{-2}_{jk}(i)M_{jk}(i)M^t_{jk}(i)\} \\
J_j(i) &= C_{oj}(i) + \rho^{-2}_j(i)L_j(i)R^t_j(i)\left[P_j(i) + Q_j(i)\right] \\
\tilde{C}^t_{oj}(i) &= C_{oj}(i) + \rho^2_j(i)L_j(i)S^t_j(i)P_j(i)
\end{aligned} \tag{8.87}$$

for some scalars $\rho_j(i) > 0,\ \alpha_j(i) > 0,\ \beta_{jk}(i) > 0,\ i \in \mathcal{S}$, and $I_j \in \Re^{n_j \times n_j}$ is the identity matrix.

Towards our goal, we introduce the error

$$e_j(t) = \hat{x}_j(t) - x_j(t)$$

Then it follows from (8.2)-(8.4) and (8.83) that this error has the dynamics:

$$\begin{aligned}
\dot{e}_j(t) &= \Big[A_{oj}(i) - \Theta_j(i)C_j + \delta A_j(i) + \Theta_j(i)\delta C_j(i) \\
&+ [R_j(i) - \Theta_j(i)L_j(i)]H_j(t,i)N_j(i)K_j(i)\Big] e_j \\
&+ \Big[\delta A_j(i) - \Theta_j(i)\delta C_j(i) \\
&- [R_j(i) - \Theta_j(i)L_j(i)]H_j(t,i)[M_j(i) - N_j(i)K_j(i)]\Big] x_j \\
&- E_j(t,i)x_j(t-\tau_j) - \sum_{k=1}^{n_s} A_{jk}(t,i)x_k(t-\phi_{jk}) \qquad (8.88)
\end{aligned}$$

In terms of $\psi_j^t = [x_j^t \quad e_j^t] \in \Re^{2n_j}$, the dynamics of the augmented system (8.2) and (8.88) take the form

$$\begin{aligned}
\dot{\psi}_j(t) &= \tilde{A}_j(t,i)\psi_j(t) + \tilde{E}_j(t,i)x_j(t-\tau_j) \\
&+ \sum_{k=1}^{n_s} \tilde{A}_{jk}(t,i)x_k(t-\phi_{jk}) \qquad (8.89)
\end{aligned}$$

where

$$\begin{aligned}
\tilde{A}_j(t,i) &= \breve{A}_j(i) + \widehat{R}_j(i)H_j(t,i)\widehat{M}_j(i) \\
\widehat{M}_j(i) &= [M_j(i) - N_j(i)K_j(i) \quad -N_j(i)K_j(i)] \\
\bar{A}_{ocj}(i) &= A_{oj}(i) - \Theta_j(i)C_{oj}(i) + \delta A_j(i) - \Theta_j(i)\delta C_j(i) \\
\breve{A}_j(i) &= \begin{bmatrix} A_{oj}(i) - B_{oj}(i)K_j(i) & -B_{oj}(i)K_j(i) \\ \delta A_j(i) - \Theta_j(i)\delta C_j(i) & \bar{A}_{ocj}(i) \end{bmatrix} \\
\widehat{R}_j(i) &= \begin{bmatrix} R_j(i) \\ -R_j(i) + \Theta_j(i)L_j(i) \end{bmatrix} \\
\tilde{E}_j(t,i) &= \begin{bmatrix} E_{oj}(i) \\ -E_{oj}(i) \end{bmatrix} + \begin{bmatrix} M_{ej}(i) \\ -M_{ej}(i) \end{bmatrix} H_{ej}(t,i)N_{ej}(i)
\end{aligned}$$

$$
\begin{aligned}
&= \widehat{E}_{oj}(i) + \widehat{M}_{ej}(i)H_{ej}(t,i)N_{ej}(i) \\
\tilde{A}_{jk}(t,i) &= \begin{bmatrix} A_{jko}(i) \\ -A_{jko}(i) \end{bmatrix} + \begin{bmatrix} M_{jk}(i) \\ -M_{jk}(i) \end{bmatrix} H_{jk}(t,i)N_{jk}(i) \\
&= \widehat{A}_{jko}(i) + \widehat{M}_{jk}(i)H_{jk}(t,i)N_{jk}(i) \qquad (8.90)
\end{aligned}
$$

A preliminary result is established first:

Lemma 8.1 *Let the matrices δA_j, δC_j, Θ_j and K_j be defined as*

$$
\begin{aligned}
\delta A_j(i) &= X_j(i)X_j^t(i)P_j(i) \\
\delta C_j(i) &= \rho_j^{-2}(i)L_j(i)S_j^t(i)P_j(i) \qquad (8.91) \\
\Theta_j(i) &= \rho_j^2(i)Q_j^{-1}(i)J_j^t(i)[L_j(i)L_j^t(i)]^{-1} \qquad (8.92) \\
K_j(i) &= \rho_j^{-2}(i)[N_j^t(i)N_j(i)]^{-1} \\
&\cdot \left[B_{oj}^t(i)P_j(i) + \rho_j^2(i)N_j^t(i)M_j(i) \right] \qquad (8.93)
\end{aligned}
$$

then, the following inequality holds for all $i \in \mathcal{S}$

$$
\begin{aligned}
& \tilde{A}_j^t(i)W_j(i) + W_j(i)\tilde{A}_j(i) \\
+ \; & W_j(i)\Big(\tilde{E}_j(t,i)\tilde{E}_j^t(t,i) + \sum_{k=1}^{n_s} \tilde{A}_{jk}(t,i)\tilde{A}_{jk}^t(t,i) \Big) W_j(i) \\
+ \; & \sum_{m=1}^{s} \alpha_{jm}W_j(m) + U_s \\
\triangleq \; & \Xi_{jA}(i) \; < \; 0 \qquad (8.94)
\end{aligned}
$$

where $P_j(i)$, $Q_j(i)$, $i \in \mathcal{S}$ satisfy the LMIs (8.84)-(8.86) and

$$
W_j(i) \triangleq \begin{bmatrix} P_j(i) & 0 \\ 0 & Q_j(i) \end{bmatrix}, \quad U_s = \begin{bmatrix} (n_s+1)I_j & 0 \\ 0 & 0 \end{bmatrix} \qquad (8.95)
$$

Proof. Define the augmented matrices:

$$
\Sigma_j(i) = \begin{bmatrix} \Sigma_{1j}(i) & \vdots & \Sigma_{2j}(i) \\ \cdots & \cdot & \cdots \\ \Sigma_{3j}(i) & \vdots & \Sigma_{4j}(i) \end{bmatrix}
$$

$$
\begin{aligned}
&= \breve{A}_j^t(i) W_j(i) + W_j(i) \breve{A}_j(i) + W_j(i) \widehat{X}_j(i) \widehat{X}_j^t(i) W_j(i) \\
&+ \rho_j^2(i) \widehat{M}_j^t(i) \widehat{M}_j(i) + \sum_{m=1}^{s} \alpha_{jm} W_j(m) + U_s
\end{aligned} \tag{8.96}
$$

and the matrix expressions

$$
\begin{aligned}
\widehat{X}_j(i) \widehat{X}_j^t(i) &= \rho_j^{-2}(i) \widehat{R}_j(i) \widehat{R}_j^t(i) + \widehat{\Omega}_{ej}(i) \widehat{\Omega}_{ej}^t(i) \\
&+ \widehat{\Omega}_{aj}(i) \widehat{\Omega}_{aj}^t(i) \\
\widehat{\Omega}_{ej}(i) \widehat{\Omega}_{ej}^t(i) &= \widehat{E}_{oj}(i) \left[I_j - \alpha_j^2(i) N_{ej}^t(i) N_{ej}(i) \right]^{-1} \widehat{E}_{oj}^t(i) \\
&+ \alpha_j^{-2}(i) \widehat{M}_{ej}(i) \widehat{M}_{ej}^t(i) \\
\widehat{\Omega}_{aj}(i) \widehat{\Omega}_{aj}^t(i) &= \sum_{k=1}^{n_s} \left[\widehat{A}_{jko}(i) \left[I_j - \beta_{jk}^2(i) N_{jk}^t(i) N_{jk}(i) \right]^{-1} \widehat{A}_{jko}^t(i) \right. \\
&+ \left. \beta_{jk}^{-2}(i) \widehat{M}_{jk}(i) \widehat{M}_{jk}^t(i) \right]
\end{aligned} \tag{8.97}
$$

Considering (8.96) and after some algebraic manipulations using (8.90) and (8.97), it can be shown that

$$
\begin{aligned}
\Sigma_{1j}(i) &= [A_{oj}^t(i) - K_j^t(i) B_{oj}^t(i)] P_j(i) + P_j(i) [A_{oj}(i) - B_{oj}(i) K_j(i)] \\
&+ P_j(i) E_{oj}(i) \left[I_j - \alpha_j^2(i) N_{ej}^t(i) N_{ej}(i) \right]^{-1} E_{oj}^t(i) P_j(i) \\
&+ \rho_j^{-2}(i) P_j(i) R_j(i) R_j^t(i) P_j(i) + \alpha_j^{-2}(i) P_j(i) M_{ej}(i) M_{ej}^t(i) P_j(i) \\
&\quad + \sum_{k=1}^{n_s} P_j(i) \left[A_{jko} \left[I_j - \beta_{jk}^2(i) N_{jk}^t(i) N_{jk}(i) \right]^{-1} A_{jko}^t(i) \right. \\
&+ \left. \beta_{jk}^{-2}(i) M_{jk}(i) M_{jk}^t(i) \right] P_j(i) \\
&+ \left[M_j^t(i) - K_j^t(i) N_j^t(i) \right] \left[M_j(i) - N_j(i) K_j(i) \right] \\
&+ \sum_{m=1}^{s} \alpha_{jm} P_j(m) + (n_s + 1) I_j
\end{aligned} \tag{8.98}
$$

$$
\begin{aligned}
\Sigma_{2j}(i) &= [\delta A_j^t(i) - \delta C_j^t(i) \Theta_j^t(i)] Q_j(i) - P_j(i) B_{oj}(i) K_j(i) \\
&+ \rho_j^{-2}(i) P_j(i) R_j(i) [-R_j^t(i) + L_j^t(i) \Theta_j(i)] Q_j(i)
\end{aligned}
$$

$$
\begin{aligned}
& -P_j(i)E_{oj}(i)\left[I_j - \alpha_j^2(i)N_{ej}^t(i)N_{ej}(i)\right]^{-1}E_{oj}^t(i)Q_j(i) \\
- \quad & \alpha_j^{-2}(i)P_j(i)M_{ej}(i)M_{ej}^t(i)Q_j(i) \\
& -\sum_{k=1}^{n_s} P_j(i)\Big[A_{jko}(i)\left[I_j - \beta_{jk}^2(i)N_{jk}^t(i)N_{jk}(i)\right]^{-1}A_{jko}^t(i) \\
+ \quad & \beta_{jk}^{-2}(i)M_{jk}(i)M_{jk}^t(i)\Big]Q_j(i) \\
- \quad & [M_j^t(i) - K_j^t(i)N_j^t(i)]N_j(i)K_j(i) \triangleq \Sigma_{3j}^t(i) \qquad (8.99)
\end{aligned}
$$

$$
\begin{aligned}
\Sigma_{4j}(i) \quad = \quad & [A_{oj}^t(i) - C_{oj}^t(i)\Theta_j^t(i) + \delta A_j^t(i) - \delta C_j^t(i)\Theta_j^t(i)]Q_j(i) \\
+ \quad & Q_j(i)[A_{oj}(i) - \Theta_j(i)C_{oj}(i) + \delta A_j(i) - \Theta_j(i)\delta C_j(i)] \\
+ \quad & \rho_j^{-2}(i)Q_j[-R_j(i) + \Theta_j L_j(i)][-R_j^t(i) + L_j^t(i)\Theta_j(i)]Q_j(i) \\
+ \quad & Q_j(i)E_{oj}(i)\left[I_j - \alpha_j^2(i)N_{ej}^t(i)N_{ej}(i)\right]^{-1}E_{oj}^t(i)Q_j(i) \\
+ \quad & \sum_{k=1}^{n_s} Q_j(i)\Big[A_{jko}\left[I_j - \beta_{jk}^2(i)N_{jk}^t(i)N_{jk}(i)\right]^{-1}A_{jko}^t(i) \\
+ \quad & \beta_{jk}^{-2}(i)M_{jk}(i)M_{jk}^t(i)\Big]Q_j(i) \\
- \quad & \alpha_j^{-2}(i)Q_jM_{ej}(i)M_{ej}^t(i)Q_j(i) + K_j^t(i)N_j^t(i)N_j(i)K_j(i) \\
+ \quad & \sum_{m=1}^{s} \alpha_{jm}Q_j(m) \qquad (8.100)
\end{aligned}
$$

Using (8.91)-(8.93) into (8.98)-(8.100) with some lengthy but standard matrix manipulations, we obtain:

$$
\begin{aligned}
\Sigma_{1j}(i) \quad & < \quad 0 \\
\Sigma_{2j}(i) \quad & = \quad \Sigma_{3j}(i) = 0 \\
\Sigma_{4j}(i) \quad & < \quad 0, \qquad j = 1, ..., n_s \qquad (8.101)
\end{aligned}
$$

This in turn implies that for $j = 1, ..., n_s$

$$
\breve{A}_j^t(i)W_j(i) + W_j(i)\breve{A}_j(i) + W_j(i)\widehat{X}_j(i)\widehat{X}_j^t(i)W_j(i)
$$

$$+ \quad \rho_j^2(i)\widehat{M}_j^t(i)\widehat{M}_j(i) + \sum_{m=1}^{s} \alpha_{jm} W_j(m) + U_s < 0 \tag{8.102}$$

Then by applying **Lemma** 8.1 to (8.102) with the help of (8.84), (8.85), (8.90) and (8.101) we get

$$\begin{aligned} &\tilde{A}_j^t W_j + W_j \tilde{A}_j + W_j(\tilde{E}_j \tilde{E}_j^t + \sum_{k=1}^{n_s} \tilde{A}_{jk} \tilde{A}_{jk}^t) W_j \\ + \quad &\sum_{m=1}^{s} \alpha_{jm} W_j(m) + U_s < 0 \end{aligned} \tag{8.103}$$

which corresponds to (8.94) as desired. ∇∇∇

Remark 8.9 *Taking into account the quadratic nature of the term*

$$W_j(\tilde{E}_j \tilde{E}_j^t + \sum_{k=1}^{n_s} \tilde{A}_{jk} \tilde{A}_{jk}^t) W_j$$

it follows from (8.103) that

$$\begin{aligned} &\tilde{A}_j^t W_j + W_j \tilde{A}_j + W_j(\tilde{E}_j \tilde{E}_j^t + \sum_{k=1}^{r} \tilde{A}_{jk} \tilde{A}_{jk}^t) W_j \\ + \quad &\sum_{m=1}^{s} \alpha_{jm} W_j(m) + U_s < 0 \quad \forall r \le n_s \end{aligned} \tag{8.104}$$

Now, the main stability result is established by the following theorem:

Theorem 8.13 *System (8.1) is robustly stabilizable via the decentralized dynamic feedback controller (8.83) if* $\forall\, j,\, k \in \{l, ..., n_s\}$ *there exist positive scalars* $\rho_j(i)$, $\alpha_j(i)$, $\beta_{jk}(i)$, $i \in \mathcal{S}$ *such that the following conditions are met for* $i \in \mathcal{S}$:

1) The matrices

$$L_j(i) L_j^t(i), \quad \left[I_j - \alpha_j^2(i) N_{ej}^t(i) N_{ej}\right](i)$$

$$N_j^t(i) N_j(i), \quad \left[I_j - \beta_{jk}^2(i) N_{jk}^t(i) N_{jk}(i)\right]$$

are invertible

2) There exist matrices $0 < P_j^t(i) = P_j(i) \in \Re^{n_j \times n_j}$, $0 < Q_j^t(i) = Q_j(i) \in \Re^{n_j \times n_j}$

satisfying (8.84) and (??), respectively.

In this case, the matrices of the stabilizing decentralized controller (8.83) are given by (8.91)-(8.93).

Proof: Let $d_j = \max_j\{\tau_j, \phi_{jk}\}$ and let $\mathbf{x}_{sj}(t) \stackrel{\Delta}{=} x_j(s+t),\ t - d_j \ \le\ s\ \le t,\ j \in \{1, ..., n_s\}$ and define the process $\{(\mathbf{x}_j(t), i),\ t\ \ge\ 0\}$ over the state space $\bar{\mathcal{C}}$. It should be observed that $\{(\mathbf{x}(t), \eta_t),\ t\ \ \ge\ \ 0\}$ is strong Markovian [78] and so is $\{(\psi_j(t), \eta_t),\ t\ \ge\ 0\}$. For $\eta_t = i, i \in \mathcal{S}$, let the Lyapunov functional $V(\cdot) : \Re^n \times \Re_+ \times \mathcal{S} \to \Re_+$ be selected as

$$V(\psi, \eta_t = i) \stackrel{\Delta}{=} V(\psi, i) = \sum_{j=1}^{n_s} V_j(\psi_j, i)\ , \qquad V_j(\psi_j) : \Re^{2n_j} \to \Re_+ \tag{8.105}$$

where

$$\begin{aligned} V_j(\psi_j, i) \quad &= \quad \psi_j^t W_j \psi_j + \int_{t-\tau_j}^{t} x_j^t(\alpha) x_j(\alpha) d\alpha \\ &+ \quad \sum_{k=1}^{n_s} \int_{t-\phi_{jk}}^{t} x_k^t(\beta) x_k(\beta) d\beta \end{aligned} \tag{8.106}$$

which takes into account the present as well as the delayed states. Note that $V(\psi) > 0$ for $\psi \neq 0$. The weak infinitesimal operator $\Im^{\psi}[\cdot]$ of the process $\{x(t), \eta_t = i, t \ge 0\}$ for system (8.89) at the point $\{t, \psi, i\}$ is given by [78, 45]:

$$\Im^{\psi}[V] \ = \ \partial V/\partial t + \dot{\psi}^t(t)\ \partial V/\partial \psi \mid_{\eta_t = i} + \sum_{m=1}^{s} \alpha_{im} V(\psi, i, m) \tag{8.107}$$

Using (8.89) into (8.106)-(8.107), manipulating the terms , applying the argument of 'completing the squares' and over-bounding the result using **Fact 1**, we get:

$$
\begin{aligned}
\Im^{\psi}[V] &= \sum_{j=1}^{n_s} \Big\{ \psi_j^t [W_j(i)\tilde{A}_j^t(i) + \tilde{A}_j(i)W_j(i)]\psi_j + \psi_j^t \sum_{m=1}^{s} \alpha_{im} W(m)\psi_j \\
&+ [\tilde{E}_j(i)x_j(t-\tau_j)]^t W_j(i)\psi_j + \psi_j^t W_j(i)[\tilde{E}_j(i)x_j(t-\tau_j)] \\
&+ x_j^t x_j - x_j^t(t-\tau_j)x_j(t-\tau_j) + \sum_{k=1}^{n_s} x_k^t x_k - x_k^t(t-\phi_{jk})x_k(t-\phi_{jk}) \\
&+ [\sum_{j=1}^{n_s} \tilde{A}_{jk}(i)x_k(t-\phi_{jk})]^t W_j(i)\psi_j \\
&+ \psi_j^t W_j(i)[\sum_{j=1}^{n_s} \tilde{A}_{jk}(i)x_k(t-\phi_{jk})] \Big\}
\end{aligned} \tag{8.108}
$$

On observing the identity

$$
\sum_{j=1}^{n_s}\sum_{k=1}^{n_s} x_j^t x_k = \sum_{k=1}^{n_s}\sum_{j=1}^{n_s} x_k^t x_j \tag{8.109}
$$

and defining the extended state-vector

$$
\chi_j^t = [\psi_j^t,\ x_j^t(t-\tau_j),\ x_1^t(t-\phi_{j1}),\ ...,\ x_{n_s}^t(t-\phi_{jn_s})] \tag{8.110}
$$

it then follows from (8.108) that

$$
\Im^{\psi}[V] < \sum_{j=1}^{n_s} \chi_j^t \Upsilon_{jA} \chi_j \tag{8.111}
$$

where

$$
\Upsilon_{jA} =
$$

$$
\begin{bmatrix}
\begin{array}{c} W_j\tilde{A}_j + \tilde{A}_j^t W_j \\ +U_s + \sum_{m=1}^{s} \alpha_{jm} W(m) \end{array} & W_j\tilde{E}_j & W_j\tilde{A}_{j1} & ... & W_j\tilde{A}_{jn_s} \\
\tilde{E}_j^t W_j & -I & 0 & ... & 0 \\
\tilde{A}_{j1}^t W_j & 0 & -I & ... & 0 \\
. & 0 & 0 & . & . \\
. & . & . & . & . \\
\tilde{A}_{jn_s}^t W_j & 0 & 0 & . & -I
\end{bmatrix}
$$

By the Schur complements [98] and in view of (8.94), inequality (8.111) with (8.110) is equivalent to

$$\sum_{j=1}^{n_s} \psi_j^t \Xi_{jA}(i) \psi_j$$

It is directly evident that $\Xi_j < 0 \; \forall \, i \in \mathcal{S}$, we conclude that

$$\Im^{\psi}[V] \; < \; 0 \quad \forall \;\psi \; \neq \; 0 \quad \& \quad \; \leq 0 \; \forall \; \psi$$

Since

$$||\psi_j(t+\beta)|| \; \leq \varphi||\psi_j(t)|, \; \forall \beta \in [-d_j, 0]$$

and some $\varphi > 0$ [85], it follows from (8.106) that

$$V(\psi, i) \leq \sum_{j=1}^{n_s} [\psi_j^t(t) \Xi_j(i) \psi_j(t) + \mu_j ||\psi_j||^2]$$

where

$$\mu_j = \varphi d_j \big(\max_i \lambda_M[W(i)] + I \big)$$

Therefore, for all $\psi \neq 0$, we have

$$\begin{aligned} \frac{\Im^{\psi}[V]}{V(\psi, i)} \quad &\leq \quad \frac{\sum_{j=1}^{n_s} \psi_j^t \Xi_j \psi_j}{\sum_{j=1}^{n_s} [\psi_j^t(t) \Upsilon_j(i) \psi_j(t) + \mu_j ||\psi_j||^2]} \\ &\leq \quad - \xi \\ &\triangleq \quad - \min_{i \in \mathcal{S}} \left\{ \frac{\sum_{j=1}^{n_s} \lambda_m[-\Upsilon_j(i)]}{\sum_{j=1}^{n_s} \lambda_M[W_j(i)] + \mu_j} \right\} \end{aligned} \tag{8.112}$$

It is readily seen from (8.112) that

$$\xi \; > \; 0$$

and hence we get

$$\Im^{\psi}[V] \;\; \leq -\xi \; V(\psi, i)$$

It follows from [78] by using the Gronwall-Bellman lemma [98] and letting $\psi(t = 0, \varphi, \eta_o) = \psi_o$, one has

$$\mathbb{E}[V(\psi, i)|\varphi, \eta_o] \;\leq\; e^{-\xi\, t}\, V(psi_o, i) \tag{8.113}$$

Since

$$\mathbb{E}\Big\{\int_{t-\tau_j}^{t} x_j^t(\alpha)x_j(\alpha)d\alpha + \sum_{k=1}^{n_s}\int_{t-\phi_{jk}}^{t} x_k^t(\beta)x_k(\beta)d\beta\Big\} \geq 0$$

it is easy to see from (8.106) that

$$\begin{aligned}
&\mathbb{E}\Big\{\sum_{j=1}^{n_s}\psi_j^t W_j\psi_j|\phi, \eta_o\Big\} \leq e^{-\xi\, t}V(\psi_o, i) \Longrightarrow \\
&\mathbb{E}\Big\{\int_0^{\mathcal{T}}\psi_j^t W_j\psi_j dt|\phi, \eta_o = i\Big\} \\
&\leq \Big[\int_0^{\mathcal{T}} e^{-\xi\, t}dt\Big]V(\psi_o, i) = \frac{1}{\psi}[e^{-\xi\,\mathcal{T}} \;-\; 1]\; V(\psi_o, i) \Longrightarrow \\
&\lim_{\mathcal{T}\to\infty}\;\mathbb{E}\Big\{\int_0^{\mathcal{T}}\sum_{j=1}^{n_s}\psi_j^t W_j\psi_j dt|\phi, \eta_o = i\Big\} \\
&\leq\; \frac{1}{\xi}\psi_o^t W(\eta_o)\psi_o + \frac{\tau^*}{\xi}||\psi_j(t+\theta)||_*^2, \\
&\forall\theta \in [-d_j, 0]
\end{aligned} \tag{8.114}$$

where

$$||\psi_j(t+\theta)||_*^2 \stackrel{\Delta}{=} \sup_{\theta\in[-d_j,0]}\; ||\psi_j(t+\theta)||_2^2$$

Let

$$\bar{W}(i) \;=\; \max_{i\in\mathcal{S}}\left\{\frac{W_j(\eta_o)||\psi_o||^2 + \tau^*||\psi_j(t+\theta)||_*^2}{\xi[W_j(\eta_o)]||\psi_o||^2}\right\}$$

it follows from (8.114) for $i \in \mathcal{S}$ that

$$\lim_{\mathcal{T}\to\infty}\;\mathbb{E}\Big\{\sum_{j=1}^{n_s}\int_0^{\mathcal{T}}\psi_j^t(t)\psi_j(t)dt\big|\phi, \eta_o = i\Big\} \leq \psi_o^t\lambda_M(\bar{W}(i))\psi_o < +\infty \tag{8.115}$$

which shows in the light of **Definition** 8.1 that the interconnected system (8.2)-(8.4) is **SSWDD** by the decentralized dynamic feedback controller (8.83). ∇∇∇

Remark 8.10 *It is readily evident from the preceding result that the closed-loop system stability is weakly delay dependent. This is a pleasing result in view of what is available in the literature; see [87] and the references cited therein. The real need for bounded delays stems from the requirement that the state trajectories should behave regularly without abrupt changes. Had we followed another approach, we could have obtained delay-dependent stability results [90].*

Remark 8.11 *It is to be noted that the developed conditions of* **Theorem** *8.13 are only sufficient and therefore the results can be generally conservative. In order to reduce this conservativeness, some parameters* $(\alpha_j,\ \beta_{jk},\ \rho_j)$ *are left to be adjusted by the designer.*

8.5.1 Algorithm

Now to utilize **Theorem** 8.13 in system applications, the following computational procedure is recommended:

STEP 1 Read the nominal matrices of subsystem j, $\forall j \in \{1, ..., n_s\}$ as given in model (8.2)-(8.4),

STEP 2 Identify the matrices of the uncertainty structure (8.7)-(8.8),

STEP 3 Select the scalars α_j, β_{jk} $\forall j, k \in \{1, ..., n_s\}$ such that the matrices

$$(I_j - \alpha_j^2 N_{ej}^t N_{ej})^{-1} \quad , \quad (I_j - \beta_{jk}^2 N_{jk}^t N_{jk})^{-1}$$

exist.

STEP 4 Select the scalar ρ_j $\forall j \in \{1, ..., n_s\}$ and solve the LMIs (8.84)-(**??**) for P_j, Q_j, respectively. Change ρ_j whenever necessary to ensure feasible solutions. If no feasible solution exists, update α_j, β_{jk} and go to **STEP 3**. If a solution exists, record the result and **STOP**.

Experience has indicated that proper initial choice of α_j, β_{jk} always guarantees the solvability of LMIs (8.84)-(8.86) for a wide range of the parameter ρ_j.

Corollary 8.1 *Consider system (8.1) without time-delay; that is set*

$$\tau_j = 0, \ \ \phi_{jk} = 0, \ \ \forall j, k \in \{1, ..., n_s\}$$

In this case we get

$$E_j = 0 \Rightarrow \Omega_{ej} \Omega_{ej}^t = 0, \ \ \forall j \in \{1, ..., n_s\}$$

It follows that the LMIs (8.84)-(8.86) reduce to:

$$\begin{bmatrix} \Xi_{jp}(i) & P_j(i)\Omega_{sj}(i) & \rho_j(i)M_j^t(i) \\ X_j^t(i)P_j(i) & -I_j & 0 \\ \rho_j(i)M_j(i) & 0 & -I_j \end{bmatrix} < 0 \qquad (8.116)$$

$$\begin{bmatrix} \Xi_{jqr}(i) & Q_j(i)\Omega_{sj}(i) & \rho_j^{-1}(i)\tilde{C}_{oj}^t \\ \Omega_{sj}^t(i)Q_j(i) & -I_j & 0 \\ \tilde{C}_{oj}(i) & 0 & -\rho_j^2(i)L_j(i)L_j^t(i) \end{bmatrix} < 0 \qquad (8.117)$$

$$\begin{bmatrix} -\Upsilon_{qw}(i) & -\rho_j(i)J_j^t \\ -\rho_j(i)J_j & -L_j^t(i)L_j(i) \end{bmatrix} < 0$$

$$\begin{bmatrix} -\Upsilon_{pw}(i) & -\tilde{B}_{oj}^t \\ -\tilde{B}_{oj} & \rho_j^2(i)N_j^t(i)N_j(i) \end{bmatrix} < 0 \qquad (8.118)$$

where

$$\begin{aligned} \Omega_{sj} &= [\Omega_{aj} \cdots \rho_j^{-1} S_j] \\ \Xi_{jqr}(i) &= Q_j(i)A_{oj}(i) + Q_j(i)\Omega_{sj}\Omega_{sj}^t P_j(i) \\ &+ A_{oj}^t(i)Q_j(i) + P_j(i)\Omega_{sj}\Omega_{sj}^t Q_j(i) \\ &+ \Upsilon_q(i) + \sum_{m=1}^{s} Q_j(m) \end{aligned} \qquad (8.119)$$

Corollary 8.2 *Consider system (8.1) in the case that all state variables are fully measurable and available for feedback. It is readily seen that this system is* **RSSWDD** *via a decentralized state feedback controller structure for all* $j \in \{1, ..., n_s\}$

$$u_j(t) = -K_j(i)x_j(t)$$

$$= \quad -\rho_j^{-2}(i)\left[N_j^t(i)N_j(i)\right]^{-1}\tilde{B}_{oj}(i)x_j(t) \qquad (8.120)$$

if there exist $\rho_j(i) > 0$ *and* $0 < P_j(i) = P_j^t(i) \in \Re^{n_j \times n_j}$, $0 < \Upsilon_{pw}^t(i) = \Upsilon_{pw}(i) \in \Re^{n_j \times n_j}$ *satisfying the LMIs for all* $i \in \mathcal{S}$

$$\begin{bmatrix} \Xi_{jp}(i) & P_j(i)X_j(i) & \rho_j(i)M_j^t(i) \\ X_j^t(i)P_j(i) & -I_j & 0 \\ \rho_j(i)M_j(i) & 0 & -I_j \end{bmatrix} < 0$$

$$\begin{bmatrix} -\Upsilon_{pw}(i) & -\tilde{B}_{oj}^t \\ -\tilde{B}_{oj} & \rho_j^2(i)N_j^t(i)N_j(i) \end{bmatrix} < 0 \qquad (8.121)$$

and the inverse $\left[N_j^t(i)N_j(i)\right]^{-1}$ *exists*

Corollary 8.3 *The standard centralized solution of the* $\mathcal{H}_\infty$ *robust stabilization of single-mode dynamical systems without delay can be readily deduced from our results by simply dropping out the subscripts* j *and* k *and setting* $n_s = 1$ $\&i = 1$. *In addition,*

$$A_{jk} = 0 \Rightarrow A_{jko} = 0,\ N_{jk} = 0,\ M_{jk} = 0 \Rightarrow \Omega_{jk} = 0$$

and

$$E_j = 0 \Rightarrow E_{oj} = 0,\ M_{ej} = 0,\ N_{ej} = 0$$

Accordingly, it follows that there exist matrices $0 < P = P \in I\!R^{n \times n}$, $0 < Q = Q^t \in I\!R^{n \times n}$, $0 < \Upsilon_{po}^t(i) = \Upsilon_{po}(i) \in \Re^{n_j \times n_j}$, $0 < \Upsilon_{qo}^t(i) = \Upsilon_{qo}(i) \in \Re^{n_j \times n_j}$ *and scalar* $\rho > 0$ *satisfying the LMIs*

$$\begin{bmatrix} \Xi_{po} & PR & \rho M^t \\ R^tP & -\rho^2 I & 0 \\ M & 0 & -I \end{bmatrix} < 0 \qquad (8.122)$$

$$\begin{bmatrix} \Xi_{qo}(i) & QR & \begin{matrix} C_o + \\ \rho^2 LR^tP \end{matrix} \\ R^tQ & -\rho^2 I & 0 \\ \begin{matrix} C_o^t + \\ \rho^2 PRL^t \end{matrix} & 0 & -\rho^2(i)LL^t \end{bmatrix} < 0 \qquad (8.123)$$

$$\begin{bmatrix} -\Upsilon_{po} & -PB_o - \rho^2 M^t N \\ -B_o^t P - \rho^2 N^t M & -\rho^2 N^t N \end{bmatrix} < 0$$

$$\begin{bmatrix} -\Upsilon_{qo} & \begin{matrix} -\rho^2 C_o - \\ LR^t(P+Q) \end{matrix} \\ \begin{matrix} -\rho^2 C_o^t - \\ (P+Q)RL^t \end{matrix} & -LL^t \end{bmatrix} < 0 \tag{8.124}$$

where

$$\begin{aligned} \Xi_{po} &= PA_o + A_o^t P + I + \Upsilon_{po} \\ \Xi_{qo} &= Q\left[A_o + \rho^{-2} RR^t P\right] + \left[A_o + \rho^{-2} RR^t P\right]^t Q \\ &+ \Upsilon_{qo} \end{aligned} \tag{8.125}$$

8.6 Decentralized Robust $\mathcal{H}_\infty$ Performance

Now, we move to consider the stabilization of the interconnected system (8.1) by solving the problem of decentralized robust $\mathcal{H}_\infty$ performance. Extending on (8.83), we use the observer-based feedback controller

$$\begin{aligned} \Sigma_j^{cw}: \quad \dot{\hat{x}}_j(t) &= [A_{oj}(i) + \delta A_j(i)]\hat{x}_j(t) + \Theta_j(i)[y_j(t) - \hat{y}_j(t)] \\ &+ B_{oj}(i)u_j(t) + +\Gamma_j(i)\hat{w}_j(t) \\ \hat{y}_j(t) &= [C_{oj}(i) + \delta C_j(i)]\hat{x}_j(t) \\ &+ D_{oj}(i)u_j(t) + \Pi_j(i)\hat{w}_j(t) \\ u_j(t) &= -\tilde{K}_j(i)\hat{x}_j(t) \end{aligned} \tag{8.126}$$

which has the same structure as (8.83) in addition to the auxiliary signal $\hat{w}_j(t)$. This signal affects both the systems dynamics and the measured output and is introduced in order to cope with the external disturbance w_j. The different constant matrices in (8.126) are those of (8.83) in addition to

$$\delta A_j(i) = X_j(i)X_j^t(i)P_j(i) \ ; \ \delta C_j(i) = \rho_j^{-2}(i)L_j R_j^t(i)P_j(i)$$

$$\hat{w}_j(t) = \gamma_j^{-2}\Gamma_j^t(i)P_j(i)\hat{x}_j \tag{8.127}$$

and the gains $\tilde{\Theta}_j(i)$, $\tilde{K}_j(i)$; $\forall j \in \{1, ..., n_s\}$ will be specified shortly.

In terms of the error

$$e_j(t) = \hat{x}_j(t) - x_j(t)$$

we obtain from (8.90), (8.126) and (8.127) the error model:

$$\begin{aligned} \dot{e}_j(t) &= \Big[A_{oj}(i) - \tilde{\Theta}_j(i)C_j + \delta A_j(i) + \tilde{\Theta}_j(i)\delta C_j(i) \\ &+ [R_j(i) - \tilde{\Theta}_j(i)L_j(i)]H_j(t,i)N_j(i)K_j(i) \\ &+ \gamma_j^{-2}\Big(\Gamma_j(i) - \tilde{\Theta}_j(i)\Pi_j(i)\Big)\Gamma_j^t(i)P_j(i)\Big]e_j \\ &+ \Big[\delta A_j(i) - \tilde{\Theta}_j(i)\delta C_j(i) \\ &- [R_j(i) - \tilde{\Theta}_j(i)L_j(i)]H_j(t,i)[M_j(i) - N_j(i)\tilde{K}_j(i)] \\ &+ \gamma_j^{-2}\Big(\Gamma_j(i) - \tilde{\Theta}_j(i)\Pi_j(i)\Big)\Gamma_j^t(i)P_j(i)\Big]x_j \\ &- [\Gamma_j(i) - \tilde{\Theta}_j(i)\Pi_j(i)]w_j - E_j(t,i)x_j(t-\tau_j) \\ &- \sum_{k=1}^{n_s} A_{jk}(t,i)x_k(t-\phi_{jk}) \end{aligned} \tag{8.128}$$

In terms of

$$\psi_j^t = [x_j^t \cdots\cdots e_j^t] \quad \in \Re^{2n_j}$$

the dynamics of the augmented system (8.83) and (8.128) can be put in the compact form:

$$\begin{aligned} \dot{\psi}_j(t) &= \hat{A}_j(t,i)\psi_j(t) + \tilde{\Gamma}_j(i)w_j + \tilde{E}(i)x_j(t-\tau_j) \\ &+ \sum_{k=1}^{n_s} \tilde{A}_{jk}(t)x_k(t-\phi_{jk}) \end{aligned} \tag{8.129}$$

where

$$\hat{A}_j(t) = \widehat{A_j}(i) + \tilde{R}(i)H_j(i)\tilde{M}_j(i)$$

$$
\begin{aligned}
\widehat{A}_j(i) &= \begin{bmatrix} A_{oj}(i) - B_{oj}(i)\tilde{K}_j(i) & \vdots & -B_{oj}(i)\tilde{K}_j(i) \\ \cdots & \cdots & \cdots \\ \widehat{A}_{j1}(i) & \vdots & \widehat{A}_{j2}(i) \end{bmatrix} \\
\tilde{\Gamma}_j &= \begin{bmatrix} \Gamma_j(i) \\ -\left(\Gamma_j(i) - \tilde{\Theta}_j(i)\Pi_j(i)\right) \end{bmatrix} \\
\tilde{R}_j &= \begin{bmatrix} R_j \\ -\left(R_j - \tilde{\Theta}_j(i)L_j(i)\right) \end{bmatrix} \\
\tilde{M}_j^t &= \begin{bmatrix} \left(M_j(i) - N_j(i)\tilde{K}_j(i)\right)^t \\ -\left(N_j(i)\tilde{K}_j(i)\right)^t \end{bmatrix} \\
\widehat{A}_{j1}(i) &= \delta A_j(i) - \tilde{\Theta}_j(i)\delta C_j(i) \\
&+ \gamma_j^{-2}\left(\Gamma_j(i) - \tilde{\Theta}_j(i)\Pi_j(i)\right)\Gamma_j^t(i)P_j(i) \\
\widehat{A}_{j1}(i) &= A_{oj}(i) - \tilde{\Theta}_j(i)C_{oj}(i) + \delta A_j(i) - \tilde{\Theta}_j(i)\delta C_j(i) \\
&+ \gamma_j^{-2}\left(\Gamma_j(i) - \tilde{\Theta}_j(i)\Pi_j(i)\right)\Gamma_j^t(i)P_j(i) \qquad (8.130)
\end{aligned}
$$

and the remaining matrices are given by (8.90) and (8.97). From (8.1) and $e_j(t) = \hat{x}_j(t) - x_j(t)$, the controlled output has the form

$$
\begin{aligned}
z_j &= [(G_j - F_j\tilde{K}_j) \quad - F_j\tilde{K}_j]\psi_j \\
&= \widehat{G}_j\Psi_j \qquad (8.131)
\end{aligned}
$$

Introduce a matrix $0 < W_j^t(i) = W_j(i) \in \Re^{2n_j \times 2n_j}$ such that

$$
W_j(i) = \begin{bmatrix} P_j(i) & 0 \\ 0 & Q_j(i) \end{bmatrix}
$$

where for all $j = \{1, ..., n_s\}$ the matrices $0 < P_j(i) = P_j^t(i) \in \Re^{n_j \times n_j}$, $0 < \Upsilon_{pw}^t(i) = \Upsilon_{pw}(i) \in \Re^{n_j \times n_j}$, $0 < Q_j(i) = Q_j^t(i) \in \Re^{n_j \times n_j}$, $0 < \Upsilon_{qw}^t(i) =$

$\Upsilon_{qw}(i) \in \Re^{n_j \times n_j}$, $i \in \mathcal{S}$ satisfy the LMIs for all $i \in \mathcal{S}$:

$$\begin{bmatrix} \Xi_{jpw}(i) & P_j(i)Y_j(i) & \rho_j(i)M_j^t(i) \\ Y_j^t(i)P_j(i) & -I_j & 0 \\ \rho_j(i)M_j(i) & 0 & -I_j \end{bmatrix} < 0 \tag{8.132}$$

$$\begin{bmatrix} \Xi_{jqw}(i) & Q_j(i)Y_j(i) & \hat{B}_{oj}^t(i) \\ Y_j^t(i)Q_j(i) & -I_j & 0 \\ \hat{B}_{oj}(i) & 0 & -\hat{N}_j^t(i) \end{bmatrix} < 0 \tag{8.133}$$

$$\begin{bmatrix} -\Upsilon_{qw}(i) & \hat{J}_j^t(i) \\ \hat{J}_j(i) & \hat{L}(i) \end{bmatrix} < 0$$

$$\begin{bmatrix} -\Upsilon_{pw}(i) & \hat{B}_{oj}^t(i) \\ \hat{B}_{oj} & \hat{N}_j^t(i) \end{bmatrix} < 0 \tag{8.134}$$

along with

$$\begin{aligned} \Xi_{jpw}(i) &= P_j(i)A_{oj}(i) + A_{oj}^t(i)P_j(i) + G_j^t(i)G_j(i) \\ &+ \sum_{m=1}^{s} P_j(m) + \Upsilon_{pw}(i) + [n_s + 1]I_j \\ \Xi_{jqw}(i) &= Q_j(i)\Big[A_{oj}(i) + Y_j(i)Y_j^t(i)P_j(i)\Big] \\ &+ \Big[A_{oj}^t(i)Q_j(i) + P_j(i)Y_j(i)Y_j^t(i)\Big]Q_j(i) \\ &+ \Upsilon_{qw}(i) + \sum_{m=1}^{s} Q_j(m) \end{aligned} \tag{8.135}$$

$$\begin{aligned} \hat{B}_{oj}(i) &= B_{oj}^t(i)P_j(i) + \rho_j^2(i)N_j^t(i)M_j(i) + F_j^t(i)G_j(i) \\ \hat{N}_j(i) &= \rho_j^2(i)N_j^t(i)N_j(i) + F_j^t(i)F_j(i) \\ \hat{L}_j(i) &= \rho_j^{-2}(i)L_j(i)L_j^t(i) + \gamma_j^{-2}(i)\Pi_j(i)\Pi_j^t(i) \\ Y_j(i) &= [X_j(i) \quad \gamma_j^{-1}\Gamma_j(i)] \\ \hat{J}_j(i) &= J_j(i) + \gamma_j^{-2}\Big(\Gamma_j(i)\Pi_j^t(i)Q_j(i) \\ &+ \Pi_j(i)\Gamma_j^t(i)P_j(i)\Big) \end{aligned} \tag{8.136}$$

Theorem 8.14 *Given the desired levels of disturbance attenuation* $\{\gamma_j > 0,\ j = 1, ..., n_s\}$. *The augmented system (8.129) is* **RSSWDD** *with disturbance attenuation* $\gamma = diag(\gamma_i)$ *via the decentralized observer-based controller (8.126) if there exist scalars* $\rho_j(i) > 0,\ \alpha_j(i) > 0,\ \beta_{jk}(i) > 0,\ i \in \mathcal{S}\ ,\ \forall j = 1, ..., n_s$ *such that the following conditions are met:*

1) The matrices

$$\left[N_j^t(i)N_j(i)\right]\ ,\ \left[L_j(i)L_j^t(i)\right]$$

$$\left[I_j - \alpha_j^2(i)N_{ej}^t(i)N_{ej}(i)\right]$$

$$\left[I_j - \beta_{jk}^2(i)N_{jk}^t(i)N_{jk}(i)\right]$$

$$\left[F_j^t(i)F_j(i)\right]\ ,\ \left[\Pi_j(i)\Pi_j^t(i)\right]$$

are all invertible,

2) There exist $0 < P_j(i) = P_j^t(i) \in \Re^{n_j \times n_j},\ 0 < \Upsilon_{pw}^t(i) = \Upsilon_{pw}(i) \in \Re^{n_j \times n_j},\ 0 < Q_j(i) = Q_j^t(i) \in \Re^{n_j \times n_j},\ 0 < \Upsilon_{qw}^t(i) = \Upsilon_{qw}(i) \in \Re^{n_j \times n_j},\ i \in \mathcal{S}$ *satisfying the LMIs (8.132) - (8.134), respectively.*

The feedback and observer gains are given by

$$\begin{aligned} \tilde{\Theta}_j(i) &= Q_j^{-1}(i)\left[J_j(i) + \gamma_j^{-2}\Gamma_j(i)\Pi_j^t(i)Q_j(i)\right]^t \hat{L}_j^{-1}(i) \\ \tilde{K}_j &= \hat{N}_j^{-1}(i)\hat{B}_{oj}(i) \end{aligned} \qquad (8.137)$$

Proof: First, we have to establish the stochastic stability of the composite system (8.129)-(8.131). Given $0 < W_j^t(i) = W_j(i) \in \Re^{n_j \times n_j}$ such that

$$W_j(i) = \begin{bmatrix} P_j(i) & 0 \\ 0 & Q_j(i) \end{bmatrix}$$

where for all $j = \{1, ..., n_s\}$ the matrices $0 < P_j(i) = P_j^t(i) \in \Re^{n_j \times n_j},\ 0 < \Upsilon_{pw}^t(i) = \Upsilon_{pw}(i) \in \Re^{n_j \times n_j},\ 0 < Q_j(i) = Q_j^t(i) \in \Re^{n_j \times n_j},\ 0 < \Upsilon_{qw}^t(i) =$

$\Upsilon_{qw}(i) \in \Re^{n_j \times n_j}$, $i \in \mathcal{S}$ satisfy the LMIs (8.132)-(8.134) for all $i \in \mathcal{S}$. Define the augmented matrix:

$$\begin{aligned}
\Xi_{Tj}(i) &= \begin{bmatrix} \Xi_{T1j}(i) & \vdots & \Xi_{T2j}(i) \\ \cdots & \cdot & \cdots \\ \Xi_{T3j}(i) & \vdots & \Xi_{T4j}(i) \end{bmatrix} \\
&= \widehat{A}_j^t(i) W_j(i) + W_j(i) \widehat{A}_j(i) \\
&+ W_j(i) \Big(\widehat{X}_j(i) \widehat{X}_j^t(i) + \gamma_j^{-2} \tilde{\Gamma}_j(i) \tilde{\Gamma}_j^t(i) \Big) W_j(i) \\
&+ \rho_j^2(i) \tilde{M}_j^t(i) \tilde{M}_j(i) + \widehat{G}_j^t(i) \widehat{G}_j(i) + U_s
\end{aligned} \tag{8.138}$$

where $\Xi_{Tj}(i) \in \Re^{2n_j \times 2n_j}$, $\Xi_{Tkj}(i) \in \Re^{n_j \times n_j}$, $k = 1, ..., 4$.

Algebraic manipulation of (8.136) using (8.90) and (8.97) in the manner of **Theorem** 8.13 leads for all $j = 1, ..., n_s$ to:

$$\begin{aligned}
\Xi_{1j} &< 0 \ , \ \Xi_{2j} = 0 \\
\Xi_{3j} &= 0 \ , \ \Xi_{4j} < 0
\end{aligned} \tag{8.139}$$

This in turn implies that

$$\begin{aligned}
& \widehat{A}_j^t W_j + W_j \widehat{A}_j + W_j (\widehat{X}_j \widehat{X}_j^t + \gamma_j^{-2} \tilde{\Gamma}_j \tilde{\Gamma}_j^t) W_j \\
+ \ & \rho_j^2 \tilde{M}_j^t \tilde{M}_j + \widehat{G}_j^t \widehat{G}_j + U_s < 0
\end{aligned} \tag{8.140}$$

By applying **Lemma** 8.1 to (8.140) with the help of (8.130), (8.133) and (8.134), we get

$$\begin{aligned}
& \hat{A}_j^t W_j + W_j \hat{A}_j + \widehat{G}_j^t \widehat{G}_j + W_j (\tilde{E}_j \tilde{E}_j^t + \sum_{k=1}^{n_s} \tilde{A}_{jk} \tilde{A}_{jk}^t) W_j \\
+ \ & \gamma_j^{-2} W_j \tilde{\Gamma}_j \tilde{\Gamma}_j^t W_j + U_s < 0
\end{aligned} \tag{8.141}$$

Using similar procedure to **Theorem** 8.13, the weak-delay dependent robust stability of the closed-loop system (8.129) can be easily deduced.

Next, to establish the desired robust $\mathcal{H}_\infty$ performance we introduce the performance measure

$$J_\infty = \sum_{j=1}^{n_s} \int_0^\infty [z_j^t(t) z_j(t) - \gamma_j^2 w_j^t(t) w_j(t)] dt \tag{8.142}$$

which is bounded in view of the asymptotic stability of the closed-loop system (8.129) and the fact that $w_j \in \mathcal{L}_2[0, \infty)$, $\forall j \in \{1, ..., n_s\}$.

Consider the Lyapunov function candidate $V(\psi)$ given by (8.105)-(8.106). In line of **Theorem** 8.13, the weak infinitesimal operator $\Im^\psi[\cdot]$ of the process $\{x(t), \eta_t = i, t \geq 0\}$ for the composite system (8.129)-(8.131) at the point $\{t, \psi, i\}$ is given by:

$$\begin{aligned}
\Im^\psi[V] \;&=\; \sum_{j=1}^{n_s} \Big\{ \psi_j^t \Big[W_j(i) \hat{A}_j^t(i) + \hat{A}_j(i) W_j(i) \Big] \psi_j + \psi_j^t \sum_{m=1}^{s} \alpha_{im} W(m) \psi_j \\
&+\; \Big[\tilde{E}_j x_j(t - \tau_j) \Big]^t W_j(i) \psi_j + \psi_j^t W_j(i) \Big[\tilde{E}_j(i) x_j(t - \tau_j) \Big] \Big\} \\
&+\; \sum_{j=1}^{n_s} \Big\{ \Big[\tilde{\Gamma}_j(i) w_j(t) \Big]^t W_j(i) \psi_j + \psi_j^t W_j(i) \Big[\tilde{\Gamma}_j(i) w_j(t) \Big] \Big\} \\
&+\; \sum_{j=1}^{n_s} \Big\{ x_j^t x_j - x_j^t(t - \tau_j) x_j(t - \tau_j) \Big\} \\
&+\; \sum_{j=1}^{n_s} \sum_{k=1}^{n_s} \Big\{ x_k^t x_k - x_k^t(t - \phi_{jk}) x_k(t - \phi_{jk}) \Big\} \\
&+\; \sum_{j=1}^{n_s} \Big\{ \Big[\sum_{j=1}^{n_s} \tilde{A}_{jk}(i) x_k(t - \phi_{jk}) \Big]^t W_j(i) \psi_j \\
&+\; \psi_j^t W_j(i) \Big[\sum_{j=1}^{n_s} \tilde{A}_{jk} x_k(t - \phi_{jk}) \Big] \Big\}
\end{aligned} \tag{8.143}$$

By considering (8.129), setting the initial conditions of system (8.83) to zero and using argument like those of **Theorem** 8.6, we can rewrite (8.143) in the compact form:

$$J^a_\infty \le \sum_{j=1}^{n_s} \int_0^\infty [\hat{\chi}^t_j \hat{\Upsilon}(i) \hat{\chi}_j] dt \tag{8.144}$$

where

$$\begin{aligned}
\hat{\chi}^t_j &= [\psi^t_j,\, x^t_j(t-\tau_j),\, x^t_1(t-\eta_{j1}),\, ...,\, x^t_{n_s}(t-\eta_{jn_s}),\, w^t_j] \\
\bar{W}_{aj}(i) &= [W_j(i)\tilde{A}_{j1}(i).....W_j(i)\tilde{A}_{jn_s}(i)] \\
E_{n_s} &= diag[I.....I]
\end{aligned} \tag{8.145}$$

and

$$\hat{\Upsilon} = \begin{bmatrix} \begin{matrix} W_j(i)\hat{A}_j(i) + \hat{A}^t_j(i)W_j(i) \\ +\widehat{G}^t_j(i)\widehat{G}_j(i) + U_s \end{matrix} & W_j(i)\tilde{E}_j & \bar{W}^a_j(i) & W_j\tilde{\Gamma}_j(i) \\ \tilde{E}^t_j(i)W_j(i) & -I & 0 & 0 \\ \bar{W}^t_{aj}(i) & 0 & -E_{n_s} & 0 \\ \tilde{\Gamma}^t_j(i)W_j(i) & 0 & 0 & -\gamma^2_j I \end{bmatrix} \tag{8.146}$$

In view of the negative-definiteness of $\hat{\Upsilon}$, it is readily evident from (8.143) that $J^a_\infty < 0$. This in turn implies that condition (2) is satisfied for all admissible uncertainties and for all non zero disturbances $w_j \in L_2[0, \infty)$, $\forall j \in \{1, ..., n_s\}$. $\nabla\nabla\nabla$

Corollary 8.4 *Consider system (8.83) without time-delay; that is*

$$\tau_j = 0, \quad \phi_{jk} = 0 \quad \forall j,\, k \in \{1, ..., n_s\}$$

In this case we set

$$E_j = 0, \Rightarrow \Omega_{ej}\Omega^t_{ej} \quad \forall j \in \{1, ..., n_s\}$$

It follows that the LMIs (8.132) - (8.134) reduce for all $i \in \mathcal{S}$ *to:*

$$\begin{bmatrix} \Xi_{jpw}(i) & P_j(i)\bar{Y}_j(i) & \rho_j(i)M_j^t(i) \\ \bar{Y}_j^t(i)P_j(i) & -I_j & 0 \\ \rho_j(i)M_j(i) & 0 & -I_j \end{bmatrix} < 0 \tag{8.147}$$

$$\begin{bmatrix} \Xi_{jqw}(i) & Q_j(i)\bar{Y}_j(i) & \hat{B}_{oj}^t(i) \\ \bar{Y}_j^t(i)Q_j(i) & -I_j & 0 \\ \hat{B}_{oj}(i) & 0 & -\hat{N}_j^t(i) \end{bmatrix} < 0 \tag{8.148}$$

$$\begin{bmatrix} -\Upsilon_{qw}(i) & \hat{J}_j^t(i) \\ \hat{J}_j(i) & \hat{L}(i) \end{bmatrix} < 0$$

$$\begin{bmatrix} -\Upsilon_{pw}(i) & \hat{B}_{oj}^t(i) \\ \hat{B}_{oj} & \hat{N}_j^t(i) \end{bmatrix} < 0 \tag{8.149}$$

with

$$\bar{Y}_j(i) = [\Omega_{sj} \quad \gamma^{-1}\Gamma_j(i)] \tag{8.150}$$

and the system is **RSSWDD**

Corollary 8.5 *Consider system (8.83) in the case that all state variables are fully measurable and available for feedback. It is readily seen that this system is* **RSSWDD** *via a decentralized state feedback controller structure*

$$\begin{aligned} u_j(t) &= -K_j(i)x_j(t) \\ &= -\rho_j^{-2}(i)\left[N_j^t(i)N_j(i)\right]^{-1}\tilde{B}_{oj}(i)x_j(t)\ ,\ \forall j \in \{1, ..., n_s\} \end{aligned} \tag{8.151}$$

if there exist $\rho_j(i) > 0,\ i \in \mathcal{S}$ *and* $0 < P_j(i) = P_j^t(i) \in \Re^{n_j \times n_j},\ 0 < \Upsilon_{pw}^t(i) = \Upsilon_{pw}(i) \in \Re^{n_j \times n_j},\ i \in \mathcal{S}$ *satisfying the LMIs for all* $i \in \mathcal{S}$

$$\begin{bmatrix} \Xi_{jp}(i) & P_j(i)Y_j(i) & \rho_j(i)M_j^t(i) \\ Y_j^t(i)P_j(i) & -I_j & 0 \\ \rho_j(i)M_j(i) & 0 & -I_j \end{bmatrix} < 0$$

$$\begin{bmatrix} -\Upsilon_{pw}^t(i) & -\tilde{B}_{oj}^t \\ -\tilde{B}_{oj} & \hat{N}_j(i) \end{bmatrix} < 0 \tag{8.152}$$

8.7 Example 8.1

To illustrate the design procedures developed in **Theorems** 8.13-8.14, we consider a representative water pollution model of three consecutive reaches of the River Nile. This linearized model forms an interconnected system of the type (8.1) for $n_s = 3$ and two-operating conditions $\mathcal{S} = 1,\ 2$ along with a mode-switching generator

$$\Im = \begin{bmatrix} -4 & 4 \\ 3 & -3 \end{bmatrix}$$

and the following information:

Mode 1:

nominal subsystem matrices

$$\begin{aligned}
A_{o1}(1) &= \begin{bmatrix} 0.97 & -0.31 \\ 1.2 & 0 \end{bmatrix} ,\ A_{o2}(1) = \begin{bmatrix} 0 & 1 \\ 1.2 & -0.87 \end{bmatrix} \\
A_{o3}(1) &= \begin{bmatrix} 0 & 1 \\ 0.78 & 1.2 \end{bmatrix} ,\ B_{o1}(1) = \begin{bmatrix} 1 \\ 0 \end{bmatrix} \\
B_{o2}(1) &= \begin{bmatrix} 0 \\ 1 \end{bmatrix} ,\ B_{o3}(1) = \begin{bmatrix} 0 \\ 1 \end{bmatrix} \\
C_{o1}(1) &= \begin{bmatrix} 1 & -1 \end{bmatrix} ,\ C_{o2}(1) = \begin{bmatrix} 1 & 1 \end{bmatrix} \\
C_{o3}(1) &= \begin{bmatrix} -1 & -1 \end{bmatrix}
\end{aligned}$$

delay and disturbance parameters

$$\begin{aligned}
\Gamma_1(1) &= \begin{bmatrix} 0.1 \\ 0.1 \end{bmatrix} ,\ \Gamma_2(1) = \begin{bmatrix} 0 \\ 0.3 \end{bmatrix} ,\ \Gamma_3(1) = \begin{bmatrix} 0 \\ 0.2 \end{bmatrix} \\
E_{o1}(1) &= 0.2 ,\ E_{o2}(1) = 0.1 ,\ E_{o3}(1) = 0.3 \\
\Pi_1(1) &= 0.08 ,\ \Pi_2(1) = 0.15 ,\ \Pi_3(1) = 0.06 \\
D_{o1}(1) &= 0.02 ,\ D_{o2}(1) = 0.03 ,\ D_{o3}(1) = 0.01 \\
G_1(1) &= [0.1 \quad 0.1 \quad 0.2] \\
G_2(1) &= [0.2 \quad 0.1 \quad 0.2]
\end{aligned}$$

$$
\begin{aligned}
G_3(1) &= [0.2 \quad 0.1 \quad 0.1] \\
F_1(1) &= 0.1\ ,\ F_2(1) = 0.1\ ,\ F_3(1) = 0.1
\end{aligned}
$$

coupling matrices

$$
\begin{aligned}
A_{120}(1) &= \begin{bmatrix} -0.5 & 0 \\ 0 & 1 \end{bmatrix} \ ,\ A_{130}(1) = \begin{bmatrix} -0.3 & 0 \\ 0 & 0.4 \end{bmatrix} \\
A_{210}(1) &= \begin{bmatrix} -0.5 & 0 \\ 0 & 1 \end{bmatrix} \ ,\ A_{230}(1) = \begin{bmatrix} 0.5 & 0 \\ 0 & 1 \end{bmatrix} \\
A_{310}(1) &= \begin{bmatrix} -0.5 & 0 \\ 0 & 0.5 \end{bmatrix} \ ,\ A_{320}(1) = \begin{bmatrix} 0.3 & 0 \\ 0 & 0.4 \end{bmatrix} \\
M_{12}(1) &= \begin{bmatrix} 0.2 \\ 0.1 \end{bmatrix} \ ,\ M_{13}(1) = \begin{bmatrix} -0.4 \\ 0.4 \end{bmatrix} \ ,\ M_{21}(1) = \begin{bmatrix} 0.5 \\ -0.3 \end{bmatrix} \\
M_{23}(1) &= \begin{bmatrix} 0.4 \\ 0.5 \end{bmatrix} \ ,\ M_{31}(1) = \begin{bmatrix} -0.6 \\ 0.5 \end{bmatrix} \ ,\ M_{32}(1) = \begin{bmatrix} 0.2 \\ -0.1 \end{bmatrix} \\
N_{12}^t(1) &= \begin{bmatrix} 0.3 \\ 0.1 \end{bmatrix} \ ,\ N_{13}^t(1) = \begin{bmatrix} 0.3 \\ -0.3 \end{bmatrix} \ ,\ N_{21}^t(1) = \begin{bmatrix} -0.1 \\ 0.6 \end{bmatrix} \\
N_{23}^t(1) &= \begin{bmatrix} 0.4 \\ 0.5 \end{bmatrix} \ ,\ N_{31}^t(1) = \begin{bmatrix} 0.3 \\ -0.3 \end{bmatrix} \ ,\ N_{32}^t(1) = \begin{bmatrix} 0.4 \\ -0.3 \end{bmatrix}
\end{aligned}
$$

In terms of the uncertainty structure (8.4), the following data is made available:

$$
\begin{aligned}
R_1(1) &= \begin{bmatrix} 0.1 \\ 0.2 \end{bmatrix} \ ,\ R_2(1) = \begin{bmatrix} 0.2 \\ 0.1 \end{bmatrix} \ ,\ R_3(1) = \begin{bmatrix} 0.2 \\ 0.2 \end{bmatrix} \\
M_1(1) &= [0.8 \quad 0.3]\ ,\ M_2(1) = [0.6 \quad 0.4] \\
M_3(1) &= [0.7 \quad 0.2]\ ,\ N_1(1) = 0.4\ ,\ N_2(1) = -0.5\ ,\ N_3(1) = -0.3 \\
L_1(1) &= 4\ ,\ L_2(1) = 3\ ,\ L_3(1) = 5 \\
M_{e1}(1) &= 0.3\ ,\ M_{e2}(1) = 0.2\ ,\ M_{e3}(1) = 0.4 \\
N_{e1}(1) &= 0.1\ ,\ N_{e2}(1) = 0.2\ ,\ N_{e3}(1) = 0.3
\end{aligned}
$$

Mode 2:

nominal subsystem matrices

$$
A_{o1}(2) = \begin{bmatrix} 1.05 & -0.42 \\ 1.1 & 0 \end{bmatrix} \ ,\ A_{o2}(2) = \begin{bmatrix} 0 & 1 \\ 1.1 & -0.65 \end{bmatrix}
$$

$$
\begin{aligned}
A_{o3}(2) &= \begin{bmatrix} 0 & 1 \\ 0.66 & 1.3 \end{bmatrix} , \; B_{o1}(2) = \begin{bmatrix} 0 \\ 1 \end{bmatrix} \\
B_{o2}(2) &= \begin{bmatrix} 1 \\ 0 \end{bmatrix} , \; B_{o3}(2) = \begin{bmatrix} 0 \\ 1 \end{bmatrix} \\
C_{o1}(2) &= \begin{bmatrix} -1 & 1 \end{bmatrix} , \; C_{o2}(2) = \begin{bmatrix} 1 & 1 \end{bmatrix} \\
C_{o3}(2) &= \begin{bmatrix} -1 & 1 \end{bmatrix}
\end{aligned}
$$

delay and disturbance parameters

$$
\begin{aligned}
\Gamma_1(2) &= \begin{bmatrix} 0.2 \\ 0.3 \end{bmatrix} , \; \Gamma_2(2) = \begin{bmatrix} 0 \\ 0.2 \end{bmatrix} , \; \Gamma_3(2) = \begin{bmatrix} 0 \\ 0.3 \end{bmatrix} \\
E_{o1}(2) &= 0.3 \, , \; E_{o2}(2) = 0.05 \, , \; E_{o3}(2) = 0.4 \\
\Pi_1(2) &= 0.07 \, , \; \Pi_2(2) = 0.2 \, , \; \Pi_3(2) = 0.06 \\
D_{o1}(2) &= 0.02 \, , \; D_{o2}(2) = 0.03 \, , \; D_{o3}(2) = 0.01 \\
G_1(2) &= [0.2 \quad 0.1 \quad 0.2] \\
G_2(2) &= [0.1 \quad 0.2 \quad 0.1] \\
G_3(2) &= [0.3 \quad 0.2 \quad 0.3] \\
F_1(2) &= 0.1 \, , \; F_2(2) = 0.1 \, , \; F_3(2) = 0.1
\end{aligned}
$$

coupling matrices

$$
\begin{aligned}
A_{120}(2) &= \begin{bmatrix} 1 & 0 \\ 0 & -0.5 \end{bmatrix} , \; A_{130}(2) = \begin{bmatrix} 0.4 & 0 \\ 0 & -0.3 \end{bmatrix} \\
A_{210}(1) &= \begin{bmatrix} -0.5 & 0 \\ 0 & 1 \end{bmatrix} , \; A_{230}(2) = \begin{bmatrix} 0.5 & 0 \\ 0 & 0.6 \end{bmatrix} \\
A_{310}(2) &= \begin{bmatrix} -0.7 & 0 \\ 0 & 0.4 \end{bmatrix} , \; A_{320}(2) = \begin{bmatrix} 0.6 & 0 \\ 0 & 0.6 \end{bmatrix} \\
M_{12}(2) &= \begin{bmatrix} 0.3 \\ 0.2 \end{bmatrix} , \; M_{13}(2) = \begin{bmatrix} 0.5 \\ -0.5 \end{bmatrix} , \; M_{21}(2) = \begin{bmatrix} -0.5 \\ 0.3 \end{bmatrix} \\
M_{23}(2) &= \begin{bmatrix} 0.4 \\ 0.5 \end{bmatrix} , \; M_{31}(2) = \begin{bmatrix} 0.5 \\ -0.6 \end{bmatrix} , \; M_{32}(2) = \begin{bmatrix} -0.1 \\ 0.2 \end{bmatrix} \\
N_{12}^t(2) &= \begin{bmatrix} 0.1 \\ 0.3 \end{bmatrix} , \; N_{13}^t(2) = \begin{bmatrix} 0.3 \\ -0.3 \end{bmatrix} , \; N_{21}^t(2) = \begin{bmatrix} 0.1 \\ -0.6 \end{bmatrix}
\end{aligned}
$$

$$N_{23}^t(2) = \begin{bmatrix} 0.3 \\ 0.6 \end{bmatrix} , \; N_{31}^t(2) = \begin{bmatrix} -0.3 \\ 0.3 \end{bmatrix} , \; N_{32}^t(2) = \begin{bmatrix} -0.4 \\ 0.3 \end{bmatrix}$$

In terms of the uncertainty structure (8.4), the following data is made available:

$$\begin{aligned}
R_1(2) &= \begin{bmatrix} 0.1 \\ 0.2 \end{bmatrix} , \; R_2(2) = \begin{bmatrix} 0.1 \\ 0.3 \end{bmatrix} , \; R_3(2) = \begin{bmatrix} 0.3 \\ 0.2 \end{bmatrix} \\
M_1(2) &= [0.7 \;\; 0.4] , \; M_2(2) = [0.5 \;\; 0.5] \\
M_3(2) &= [0.6 \;\; 0.3] , \; N_1(2) = 0.5 , \; N_2(2) = -0.7 , \; N_3(2) = -0.4 \\
L_1(2) &= 3 , \; L_2(2) = 2 , \; L_3(2) = 6 \\
M_{e1}(2) &= 0.3 , \; M_{e2}(2) = 0.3 , \; M_{e3}(2) = 0.3 \\
N_{e1}(2) &= 0.1 , \; N_{e2}(2) = 0.2 , \; N_{e3}(2) = 0.4
\end{aligned}$$

By selecting

$$\alpha_1(1) = 0.5 , \; \alpha_2(1) = 0.5 , \; \alpha_3(1) = 0.5$$

$$\beta_1(1) = 0.5 , \; \beta_2(1) = 0.5 , \; \beta_3(1) = 0.5$$

and

$$\alpha_1(2) = 0.6 , \; \alpha_2(2) = 0.6 , \; \alpha_3(2) = 0.6$$

$$\beta_1(2) = 0.6 , \; \beta_2(2) = 0.6 , \; \beta_3(2) = 0.6$$

as initial guess, it is found by applying the computational procedure set forth that

$$\rho_1(1) = 0.2 , \; \rho_2(2) = 0.3 , \; \rho_3(2) = 0.4$$

and

$$\rho_1(2) = 0.4 , \; \rho_2(2) = 0.2 , \; \rho_3(2) = 0.3$$

can be taken as a successful (first) choice. Using the foregoing nominal data and invoking the MATLAB software, we obtain

$$P_1(1) = \begin{bmatrix} 0.1331 & 0.0364 \\ 0.0364 & 0.9109 \end{bmatrix} , \; P_2(1) = \begin{bmatrix} 0.1445 & 0.0782 \\ 0.0782 & 0.7923 \end{bmatrix}$$

$$P_3(1) = \begin{bmatrix} 0.1557 & 0.1421 \\ 0.1421 & 0.1460 \end{bmatrix}, \quad Q_1(1) = \begin{bmatrix} 0.9070 & -0.5361 \\ -0.5361 & 2.3821 \end{bmatrix}$$

$$Q_2(1) = \begin{bmatrix} 2.2218 & 0.8047 \\ 0.8047 & 1.9533 \end{bmatrix}, \quad Q_3(1) = \begin{bmatrix} 2.6863 & 0.5944 \\ 0.5944 & 2.4828 \end{bmatrix}$$

$$P_1(2) = \begin{bmatrix} 0.1187 & 0.0445 \\ 0.0445 & 1.0563 \end{bmatrix}, \quad P_2(2) = \begin{bmatrix} 0.1385 & 0.0864 \\ 0.0864 & 0.9853 \end{bmatrix}$$

$$P_3(2) = \begin{bmatrix} 0.1673 & 0.1214 \\ 0.1214 & 0.1609 \end{bmatrix}, \quad Q_1(2) = \begin{bmatrix} 0.8886 & -0.5136 \\ -0.5136 & 2.4415 \end{bmatrix}$$

$$Q_2(2) = \begin{bmatrix} 2.2192 & 0.9041 \\ 0.9041 & 1.8563 \end{bmatrix}, \quad Q_3(2) = \begin{bmatrix} 2.7143 & 0.5854 \\ 0.5854 & 2.6627 \end{bmatrix}$$

as feasible solutions of the LMIs (8.84)-(8.85). These give the following gain matrices

$$K_1(1) = [22.8166 \quad 6.4302] \quad , \quad K_2(1) = [2.2767 \quad 34.4125]$$

$$K_3(1) = [7.5354 \quad 78.9143]$$

$$K_1(2) = [19.7425 \quad 9.7462] \quad , \quad K_2(2) = [4.7555 \quad 41.1124]$$

$$K_3(2) = [11.6657 \quad 87.0146]$$

$$\Theta_1(1) = \begin{bmatrix} 0.0223 \\ 0.0235 \end{bmatrix}, \quad \Theta_2(1) = \begin{bmatrix} 0.0027 \\ 0.0202 \end{bmatrix}$$

$$\Theta_3(1) = \begin{bmatrix} -0.0021 \\ 0.0187 \end{bmatrix}$$

$$\Theta_1(2) = \begin{bmatrix} 0.0322 \\ 0.0511 \end{bmatrix}, \quad \Theta_2(2) = \begin{bmatrix} 0.0031 \\ 0.0197 \end{bmatrix}$$

$$\Theta_3(2) = \begin{bmatrix} -0.0019 \\ 0.0202 \end{bmatrix}$$

8.8 Notes and References

This Chapter has just launched initial results for the stochastic stability, stabilization and feedback control design of a class of continuous-time interconnected

JTDS. In view of its significant importance in practice, we expect much more research work in this direction. Preliminary results are available in [121].

Chapter 9

Appendix

9.1 Standard Facts

In this section, we present some mathematical inequalities which are milestones in the literature about uncertain jumping time-delay systems. Their common use has been to overbound certain expressions in stochastic stability studies. Throughout this section, we let

$$\Sigma_1(i)\ ,\ \Sigma_2(i)\ ,\ \Sigma_3(i)\ ; \qquad i \in \mathcal{S}$$

be real constant matrices of compatible dimensions at the mode $\eta_t = i \in \mathcal{S}$ and $\Delta(t,i)$ be a real time-varying matrix function satisfying

$$\Delta(t,i)\ \Delta(t,i)\ \leq\ I; \qquad i \in \mathcal{S}$$

Fact 1:

Let $\eta_t = i \in \mathcal{S}$ then the following inequality holds

$$\begin{aligned} & \Sigma_1(i)^t \Sigma_2(i) + \Sigma_2^t(i)\Sigma_1(i) \\ \leq\ & \zeta(i)\Sigma_1^t(i)\Sigma_1(i) + \zeta^{-1}(i)\Sigma_2^t(i)\Sigma_2(i), \quad \zeta(i) > 0 \end{aligned} \tag{9.1}$$

Proof:

Since $\Sigma^t(i)\Sigma(i) \geq 0\ , \quad i \in \mathcal{S}$ for any real matrix $\Sigma(i)$, then it follows that

$$\left[\zeta^{1/2}(i)\Sigma_1(i) - \zeta^{-1/2}(i)\Sigma_2(i)\right]^t \cdot \left[\zeta^{1/2}(i)\Sigma_1(i) - \zeta^{-1/2}(i)\Sigma_2(i)\right] \geq 0 \tag{9.2}$$

Expansion of (9.2) yields

$$\zeta(i)\Sigma_1^t(i)\Sigma_1(i) - \Sigma_1^t(i)\Sigma_2(i) - \Sigma_2^t(i)\Sigma_1(i) + \zeta^{-1}(i)\Sigma_2^t(i)\Sigma_2(i) \geq 0 \tag{9.3}$$

Rearranging (9.3), it gives

$$\Sigma_1(i)^t\Sigma_2(i) + \Sigma_2^t(i)\Sigma_1(i) \leq \zeta(i)\Sigma_1^t(i)\Sigma_1(i) + \zeta^{-1}(i)\Sigma_2^t(i)\Sigma_2(i), \quad \zeta(i) > 0 \tag{9.4}$$

$\nabla\nabla\nabla$

Fact 2:

Let $\eta_t = i \in \mathcal{S}$ then the following inequality holds

$$\Sigma_1(i)\Delta(t,i)\Sigma_2(i) + \Sigma_2^t(i)\Delta^t(t,i)\Sigma_1^t(i) \leq \rho^{-2}(i)\Sigma_1(i)\Sigma_1^t(i) + \rho^2(i)\Sigma_2^t(i)\Sigma_2(i), \quad \rho(i) > 0 \tag{9.5}$$

Proof:

Consider the matrix function for all $\eta_t = i \in \mathcal{S}$

$$\left[\rho^{-1}(i)\Sigma_1^t(i) - \rho(i)\Delta(t,i)\Sigma_2(i)\right]^t \cdot \left[\rho^{-1}(i)\Sigma_1^t(i) - \rho(i)\Delta(t,i)\Sigma_2(i)\right] \geq 0 \tag{9.6}$$

On expanding (9.5), we get

$$\rho^{-2}(i)\Sigma_1(i)\Sigma_1^t(i) - \Sigma_1(i)\Delta(t,i)\Sigma_2(i)$$

$$
\begin{aligned}
- & \quad \Sigma_2^t(i)\Delta^t(t,i)\Sigma_1^t(i) + \rho^2(i)\Sigma_2^t\Delta^t(t,i)\Delta(t,i)\Sigma_2(i) \\
\cdot & \quad \geq 0
\end{aligned}
\tag{9.7}
$$

and by rearranging the terms using $\Delta(t,i)\ \Delta(t,i) \ \leq \ I; \ \ i \in \mathcal{S}$, we obtain

$$
\begin{aligned}
& \Sigma_1(i)\Delta(t,i)\Sigma_2(i) + \Sigma_2^t(i)\Delta^t(t,i)\Sigma_1^t(i) \\
\cdot & \leq \rho^{-2}(i)\Sigma_1(i)\Sigma_1^t(i) + \rho^2(i)\Sigma_2^t\Sigma_2(i), \quad \rho > 0
\end{aligned}
\tag{9.8}
$$

$\nabla\nabla\nabla$

Fact 3 (Schur Complement):

Given constant matrices $\eta_t = i \in \mathcal{S}\ \Omega_1(i),\ \Omega_2(i),\ \Omega_3(i)$ where

$$
\Omega_1(i) = \Omega_1^t(i) \quad , i \in \mathcal{S}
$$

and

$$
0 < \Omega_2(i) = \Omega_2^t(i)
$$

then

$$
\begin{aligned}
& \Omega_1(i) + \Omega_3^t(i)\Omega_2^{-1}(i)\Omega_3(i) < 0 \\
\Longleftrightarrow & \\
& \begin{bmatrix} \Omega_1(i) & \Omega_3^t(i) \\ \Omega_3(i) & -\Omega_2(i) \end{bmatrix} < 0 \\
or & \\
& \begin{bmatrix} -\Omega_2(i) & \Omega_3(i) \\ \Omega_3^t(i) & \Omega_1(i) \end{bmatrix} < 0
\end{aligned}
\tag{9.9}
$$

Fact 4:

Let $\eta_t = i \in \mathcal{S}$ then $\forall \rho(i) > 0$ such that

$$
\rho^2(i) \quad \Sigma_2^t(i) \quad \Sigma_2(i) \quad < \quad I
$$

the following inequality holds

$$
\Big[\Sigma_3(i) + \Sigma_1(i)\Delta(t,i)\Sigma_2(i)\Big]
$$

$$
\begin{aligned}
& \cdot \quad \left[\Sigma_3^t(i) + \Sigma_2^t(i)\Delta^t(t,i)\Sigma_1^t(i)\right] \\
& \leq \quad \rho^{-2}(i)\Sigma_1(i)\Sigma_1^t(i) \\
& + \quad \Sigma_3(i)\left[I - \rho^2(i)\Sigma_2^t(i)\Sigma_2(i)\right]^{-1}\Sigma_3^t(i)
\end{aligned} \tag{9.10}
$$

Proof:

Consider the matrix function for all $\eta_t = i \in \mathcal{S}$

$$
\begin{aligned}
\Sigma(t,i) \quad = \quad & \left[\rho^{-2}(i)I - \Sigma_2(i)\Sigma_2^t(i)\right]^{-1/2}\Sigma_2(i)\Sigma_3^t(i) \\
- \quad & \left[\rho^{-2}(i)I - \Sigma_2(i)\Sigma_2^t\right]^{1/2}\Delta^t(t,i)(t)\Sigma_1^t
\end{aligned} \tag{9.11}
$$

Expanding

$$
\Sigma(t,i) \;\; \Sigma(t,i) \; \geq 0
$$

using the fact that

$$
\rho^2(i)\Sigma_2^t(i)\Sigma_2(i) < I \;, \;\; i \in \mathcal{S}
$$

we get

$$
\begin{aligned}
& \quad \Sigma_3(i)\Sigma_2^t(i)\Delta^t(t,i)\Sigma_1^t(i) + \Sigma_1(i)\Delta(t,i)\Sigma_2(i)\Sigma_3^t(i) \\
& + \quad \Sigma_1(i)\Delta(t,i)\Sigma_2(i)\Sigma_2^t\Delta^t(t,i)\Sigma_1^t(t,1) \\
& \leq \quad \Sigma_3(i)\Sigma_2^t(i)\left[\rho^{-2}(i)I - \Sigma_2(i)\Sigma_2^2(i)\right]^{-1}\Sigma_2(i)\Sigma_3^t(i) \\
& + \quad \rho^{-2}(i)\Sigma_1(i)\Delta(t,i)\Delta^t(t,i)(t)\Sigma_1^t(i) \\
& + \quad \left(\Sigma_3(i)\Sigma_1(i)\Delta(t,i)\Sigma_2(i)\right)\left(\Sigma_3(i) + \Sigma_1(i)\Delta(t,i)\Sigma_2(i)\right)^t \\
& - \quad \Sigma_3(i)\Sigma_3^t(i) \\
& \leq \quad \Sigma_3(i)\Sigma_2^t(i)[\rho^{-2}(i)I - \Sigma_2(i)\Sigma_2^t(i)]^{-1}\Sigma_2(i)\Sigma_3^t(i) \\
& + \quad \rho^{-2}(i)\Sigma_1(i)\Delta^t(t)\Sigma_1^t(i) \\
& + \quad \left(\Sigma_3(i) + \Sigma_1(i)\Delta(t,i)\Sigma_2(i)\right)
\end{aligned}
$$

$$
\begin{aligned}
& \cdot \quad \Big(\Sigma_3(i) + \Sigma_1(i)\Delta(t,i)(i)\Sigma_2(i)\Delta(t,i)^t \\
\leq \quad & \Sigma_3(i)\Big[I + \Sigma_2^t(i)[\rho^{-2}(i)I - \Sigma_2(i)\Sigma_2^t(i)]^{-1}\Sigma_2(i) \Big] \Sigma_3^t(i) \\
+ \quad & \rho^{-2}(i)\Sigma_1(i)\Delta^t(t,i)\Sigma_1^t(i)
\end{aligned} \tag{9.12}
$$

Since

$$
\begin{aligned}
& \Big[I \; - \; \rho^2(i)\Sigma_2^t(i)\Sigma_2(i) \Big] \\
= \quad & \Big[I + \rho^2(i)\Sigma_2^t(i) \Big(I - \rho^2\Sigma_2^t(i)\Sigma_2(i) \Big)^{-1} \Sigma_2(i) \Big]
\end{aligned} \tag{9.13}
$$

and

$$
\Delta^t(t,i)\Delta(t,i) \leq I \Longrightarrow \Delta(t,i)\Delta^t(t,i) \leq I \tag{9.14}
$$

then

$$
\begin{aligned}
& \Big[\Sigma_3(i) + \Sigma_1(i)\Delta(t,i)\Sigma_2(i) \Big] \Big[\Sigma_3^t(i) + \Sigma_2^t\Delta^t(t,i)\Sigma_1^t \Big] \\
\leq \quad & \rho^{-2}(i)\Sigma_1(i)\Sigma_1^t(i) \\
+ \quad & \Sigma_3(i) \Big[I - \rho^2(i)\Sigma_2^t(i)\Sigma_2(i) \Big]^{-1} \Sigma_3^t
\end{aligned} \tag{9.15}
$$

$\nabla\nabla\nabla$

Fact 5:

Let $\eta_t = i \in \mathcal{S}$ and given matrices $\Sigma_1(i)$, $\Sigma_2(i)$, $\Sigma_3(i)$ such that

$$
\Sigma_1^t(i) \;\; = \;\; \Sigma_1(i)
$$

then the inequality

$$
\begin{aligned}
& \Sigma_1(i) + \Sigma_3(i)\Delta(t,i)\Sigma_2(i) \; + \; \Sigma_2^t(i)\Delta^t(t,i)\Sigma_3^t(i) \; < 0 \\
& \forall \, \Delta^t(t,i)\Delta(t,i) \; \leq \; I
\end{aligned} \tag{9.16}
$$

holds if and only if for some $\varepsilon(i) > 0, \; i \in \mathcal{S}$

$$\Sigma_1(i) + \varepsilon^{-1}(i)\; \Sigma_3(i)\Sigma_3^t(i) + \varepsilon(i)\; \Sigma_2^t(i)\Sigma_2(i) \; < 0 \qquad (9.17)$$

Proof:

Can be proved in line of **Fact 4**. $\nabla\nabla\nabla$

Fact 6:

For any quantities $u(i)$ and $v(i)$ of equal dimensions and for all $\eta_t = i \in \mathcal{S}$, it follows that the following inequality holds

$$||u(i) \; + \; v(i)||^2 \; \leq \; [1 + \beta^{-1}(i)] \; ||u(i)||^2 \; + \; [1 + \beta(i)]||v(i)||^2 \qquad (9.18)$$

for any scalar $\beta(i) > 0, \quad i \in \mathcal{S}$

Proof:

Since

$$\begin{aligned} &[u(i) \; + \; v(i)]^t \; [u(i) \; + \; v(i)] = \\ &u^t(i)\; u(i) \; + \; v^t(i)\; v(i) \; + \; 2\; u^t(i)\; v(i) \end{aligned} \qquad (9.19)$$

It follows by taking norm of both sides of (9.19) for all $i \in \mathcal{S}$ that

$$||u(i) \; + \; v(i)||^2 \; \leq \; ||u(i)||^2 \; + \; ||v(i)||^2 \; + \; 2\; ||u^t(i)\; v(i)|| \qquad (9.20)$$

We know from the triangle inequality that

$$2\; ||u^t(i)\; v(i)|| \; \leq \; \beta^{-1}(i)\; ||u(i)||^2 \; + \; \beta(i)\; ||v(i)||^2 \qquad (9.21)$$

On substituting (9.21) into (9.20), it yields (9.18). $\nabla\nabla\nabla$

9.2 Some Common Lyapunov Functionals

In this section, we provide some Lyapunov functionals and their time-derivatives which would be of common use in deriving the weak infinitesimal operator.

Let $\eta_t = i \in \mathcal{S}$ and define

$$V_1(x,i) = x^t P(i) x + \int_{-\tau}^{0} x^t(t+\theta) Q(i) x(t+\theta)\, d\theta \tag{9.22}$$

$$V_2(x,i) = \int_{-\tau}^{0} \left[\int_{t+\theta}^{t} x^t(\alpha) R(i) x(\alpha)\, d\alpha \right] d\theta \tag{9.23}$$

$$V_3(x,i) = \sum_{j=1}^{r} \int_{-\tau-\beta_j}^{-\beta_j} \left[\int_{t+\theta}^{t} x^t(\alpha) L(i) x(\alpha)\, d\alpha \right] d\theta \tag{9.24}$$

$$V_4(x,i) = \int_{-\tau}^{0} \left[\int_{t+\theta}^{t} \dot{x}^t(\alpha) W(i) \dot{x}(\alpha)\, d\alpha \right] d\theta \tag{9.25}$$

where x is the state vector, τ is a constant delay factor and the matrices $0 < P^t(i) = P(i)$, $0 < Q^t(i) = Q(i)$, $0 < R^t(i) = R(i)$, $0 < L^t(i) = L(i)$, $0 < W^t(i) = W(i)$ are appropriate weighting factors for all $\eta_t = i \in \mathcal{S}$.

Standard matrix manipulations lead to

$$\begin{aligned} \dot{V}_1(x,i) &= \dot{x}^t P(i) x + x^t P(i) \dot{x} \\ &+ x^t(t) Q(i) x(t) - x^t(t-\tau) Q(i) x(t-\tau) \end{aligned} \tag{9.26}$$

$$\begin{aligned} \dot{V}_2(x,i) &= \int_{-\tau}^{0} \Big[x^t(t) R(i) x(t) \\ &- x^t(t+\alpha) R(i) x(t+\alpha) \Big]\, d\,\theta \\ &= \tau\, x^t(t) R(i) x(t) \\ &- \int_{-\tau}^{0} x^t(t+\theta) R(i) x(t+\theta) \Big]\, d\,\theta \end{aligned} \tag{9.27}$$

$$\begin{aligned} \dot{V}_3(x,i) &= \sum_{j=1}^{r} \int_{-\tau-\beta_j}^{-\beta_j} \Big[x^t(t) L(i) x(i) \\ &- x^t(t+\alpha) L(i) x(t+\alpha) \Big] d\alpha \\ &= \sum_{j=1}^{r} \tau\, x^t(t) L(i) x(i) \end{aligned}$$

$$- \sum_{j=1}^{r} \int_{-\tau}^{0} x^t(t+\alpha-\beta_j)L(i)x(t+\alpha-\beta_j)\, d\alpha \qquad (9.28)$$

$$\dot{V}_4(x,i) = \tau\, \dot{x}^t(t)W(i)x(t) - \int_{t-\tau}^{t} \dot{x}^t(\alpha)W(i)\dot{x}(\alpha)\, d\alpha \qquad (9.29)$$

Chapter 10

Bibliography

Bibliography

[1] Abbad, M. and J. A. Filar, "Perturbation and Stability Theory for Markov Control Problems," *IEEE Trans. Automatic Control*, vol. 37, pp. 1415–1420, 1992.

[2] Abou-Kandil, H., G. Freiling and G. Jank, "Solution and Asymptotic Behavior of Coupled Riccati Equations in Jump Linear Systems," *IEEE Trans. Automatic Control*, vol. 39, pp. 1631-16636, 1994.

[3] Abou-Kandil, H., G. Freiling and G. Jank, "On the Solution of Discrete-Time Markovian Jump Linear Quadratic Control Problems," *Automatica*, vol. 31, pp. 765-768, 1995.

[4] Akella, R. and P. R. Kumar, "Optimal Control of Production Rate in a Failure Prone Manufacturing System," *IEEE Trans. Automat. Control*, vol. 31, pp. 116–126, 1986.

[5] Al-Muthairi, N. F. , M. Zribi and M. S. Mahmoud, "Adaptive Identification and Control of Interconnected Systems", *Int. J. Systems Analysis and Modeling Simulation*, Vol. 42, pp. 851-877, 2002.

[6] Anderson, B. D. O. and J. B. Moore, "*Optimal Filtering*", Prentice Hall, New York, 1979.

[7] Apkarian, P., and P. Gahinet, "A Convex Characterization of Gain-Scheduled $\mathcal{H}_\infty$ Controllers," *IEEE Trans. Automatic Control*, vol. 40, pp. 853–864, 1995.

[8] Bahnasawi, A. A., and M. S. Mahmoud, "*Control of Partially-Known Dynamical Systems*", Springer-Verlag, Berlin, 1989.

[9] Bahnasawi, A. A., A.S. Al-Fuhaid and M.S. Mahmoud, "Decentralized and Hierarchical Control of Interconnected Uncertain Systems", *Proc. IEE Part D*, vol. 137, pp. 311-321, 1990.

[10] Bahnasawi,A.A., A.S. Al-Fuhaid and M.S. Mahmoud," A New Hierarchical Control Structure for a Class of Uncertain Discrete Systems", *Control Theory and Advanced Technology*, vol. 6, pp.1-21, 1990.

[11] Basar, T., and P. Bernhard, "$\mathcal{H}_\infty$ *Optimal Control and Related Minimax Problems*", Birkhauser, Berlin, 1990.

[12] Basar, T. and A. Haurie, "Feedback Equilbria in Differential Games with Structural and Modal Uncertainties, *Advances in Large Scale Systems*, J. B. Cruz, Jr (Ed.), JAI Press Inc., pp. 163-201, 1984.

[13] Bernard, F., F. Dufour and P. Bertrand, "On the JLQ Problem with Uncertainty," *IEEE Trans. Automatic Control*, vol. 42, pp. 869-872, 1997.

[14] Bernstein, D. S. and W. A. Haddad, "LQG Control with an $\mathcal{H}_\infty$ Performance Bound: A Riccati Equation Approach," *IEEE Trans. Automatic Control*, vol. 34, pp. 293-305, 1989.

[15] Bernstein, D. S. and W. M. Haddad, "Steady-State Kalman Filtering with an H_∞ Error Bound", *Systems and Control Letters*, vol. 16, pp. 309-317, 1991.

[16] Bielecki, T. R. and J. A. Filar, "Singularly Perturbed Markov Control Problems," *IEEE Trans. Annals of Operations Research*, vol. 25, pp. 153–168, 1991.

[17] Bingulac, S. and H. F. VanLandingham, "*Algorithms for Computer-Aided Design of Multivariable Control Systems*", Marcel-Dekker, New York, 1993.

[18] Boukas, K. and A. Haurie, "Manufacturing Flow Control and Preventive Maintenance: A Stochastic Control Approach," *IEEE Trans. Automatic Control*, vol. 35, pp. 1024–1031, 1990.

[19] Boukas, E. K., "Control of Systems with Controlled Jump Markov Disturbances", *Control Theory and Advanced Technology*, Vol 9, pp. 577-595, 1993.

[20] Boukas, E. K. and P. Shi, "Stochastic stability and guaranteed cost control of discrete-time uncertain systems with Markovian jumping parameters", *Int. J. Robust & Nonlinear Control*, vol.8, no.13, pp. 1155-1167, 1998.

[21] Boukas, E. K. and H. Yang, "Exponential Stabilizability of Stochastic Systems with Markovian Jumping Parameters", *Automatica*, vol. 35, pp. 1437–1441, 1999.

[22] Boukas, E. K., Z. K. Liu, and G. X. Liu, "Delay-Dependent Robust Stability and $\mathcal{H}_\infty$ Control of Jump Linear Systems with Time-Delay", *Int. J. Control*, vol. 74, no. 4, pp. 329-340, 2001.

[23] Boukas, E. K., and Z. K. Liu, "Production and Maintenance Control for Manufacturing Systems," *IEEE Trans. Automatic Control*, vol. 46, pp. 1455–1460, 2001.

[24] Boukas, E. K. and Y. Liu, *Deterministic and Stochastic Time-Delay Systems*", Birkuhauther, Boston, 2002.

[25] Boyd, S. , L. El Ghaoui, E. Feron and V. Balakrishnan, "*Linear Matrix Inequalities in Control*", SIAM Studies in Applied Mathematics, Philadelphia, 1994.

[26] Busenberg, S. and M. Martelli (Editor), "*Delay Differential Equations and Dynamical Systems*", Springer-Verlag, New York, 1991.

[27] Cao Y. Y. and J. Lam, "Robust H_∞ Control of Uncertain Markovian Jump Systems with Time-Delay", *IEEE Trans. Automatic Control*, vol. 45, no. 1, pp. 77-83, 2002.

[28] Chizeck, H. J., A. S. Willsky and D. Castanon, "Discrete-Time Markovian Jump Linear Quadratic Control," *Int. J. Control*, vol. 43, pp. 219-234, 1986.

[29] Costa, O. L. V. and M. D. Fragoso, "Stability results for discrete-time linear systems with Markovian jumping parameters," *J. Math. Anal. Appl.*, vol. 179, no. 2, pp. 154–178, 1993.

[30] Crocco, L., "Aspects of Combustion Stability in Liquid Propellant Rocket Motors: Part I: Fundamentals-Low Frequency Instability with Monopropellants", *J. American Rocket Society*, vol. 21, pp. 163–178, 1951.

[31] Datko, R., "Neutral Autonomous Functional Equations with Quadratic Cost," *SIAM J. Control*, vol. 12, pp. 70-82, 1974.

[32] Davis, M., "*Markov Models and Optimization*", Chapman and Hall, London, 1992.

[33] de Souza, C. E. and M. D. Fragoso, "$\mathcal{H}_\infty$ Control for Linear Systems with Markovian Jumping Parameters," *Control-Theory and Advanced Technology*, vol. 9, no. 2, pp. 457–466, 1993.

[34] de Souza, C. E. and M. D. Fragoso, "Robust $\mathcal{H}_\infty$ Filtering for Uncertain Markovian Jump Linear Systems," *Int. J. Robust & Nonlinear Control*, vol. 12, pp. 435–446, 2002.

[35] Doob, J. L., "*Stochastic Processes*", Wiley & Sons, New York, 1990.

[36] Dorato, P. and R. K. Yedavalli, "*Recent Advances in Robust Control*", IEEE Press, New York, 1990.

[37] Doyle, J. C., K. Glover, P. P. Khargonekar and B. A. Francis, "State Space Solutions to Standard $\mathcal{H}_2$ and $\mathcal{H}_\infty$ Control Problems," *IEEE Trans. Automatic Control*, vol. 34, pp. 831-847, 1989.

[38] Dufour, F. and P. Bertrand, "The filtering problem for continuous-time linear systems withMarkovian switching coefficients," *System & Control Letters*, vol. 23, no. 5, pp. 453–461, 1994.

[39] Dufour, F. and P. Bertrand, "An Image- Based Filter for Discrete-Time Markovian Jump Linear Systems," *Automatica*, vol. 32, no. 2, pp. 241–247, 1996.

[40] Dynkin, E. B., "*Markov Process*", Springer, New York, 1965.

[41] El-Ghaoui, L., and M. A. Rami, "Robust State-Feedback Stabilization of Jump Linear Systems," *Int. J. Robust and Nonlinear Control*, vol. 6 , pp. 1666-16671, 1997.

[42] Elliott, R. J., and D. D. Sworder, "Control of a Hybrid Conditionally Linear Gaussian Processes", *Journal of Optimization Theory and Applications*, vol. 74, pp. 75-85, 1994.

[43] Ezzine, J. and A. H. Haddad, "On the Controllability and Observability of Hybrid-Systems", *Int. J. Control*, vol. 49, pp. 2045-2055, 1989.

[44] Ezzine, J. and D. Karvanoglu, "On Almost-Sure Stabilization of Discrete-Time Parameter Systems: An LMI Approach", *Int. J. Control*, vol. 68, pp. 1129-1146, 1997.

[45] Feng, X. , K. A., Loparo, Y. Ji and H. J. Chizeck, "Stochastic Stability Properties of Jump Linear Systems," *IEEE Trans. Automat. Control*, vol. 37, no. 1, pp. 38– 53, 1992.

[46] Feng, Y., and H. Yan, "Optimal Production Control in a Discrte Manufacturing System with Unreliable Machines and Random Demand," *IEEE Trans. Automatic Control*, vol. 45, pp. 2280–2296, 2000.

[47] Fleming, W., S. Sethi and M. Soner, "An Optimal Stochastic Production Planning Problem with Randomly Fluctuating Demand", *SIAM J. Control and Optimization*, vol. 25, pp. 1494-1502, 1987.

[48] Fiagbedzi, Y. A. and A. E. Pearson, "A Multistage Reduction Technique for Feedback Stabilizing Distributed Time-Lag Systems", *Automatica*, vol. 23, pp. 311–326, 1987.

[49] Filar, J. A. and A. Haurie, "A Two-Factor Stochastic Production Model with Two-Time Scales," *Automatica*, vol. 37, pp. 1505–1513, 2001.

[50] Filar, J. A., V. G. Gaitsgory and A. Haurie, "Control of Singularly Perturbed Hybrid Stochastic Systems," *IEEE Trans. Automatic Control*, vol. 46, pp. 179–190, 2001.

[51] Fragoso, M. D, "On a Partially Observable LQG Problem for Systems with Markovian Jumping Parameters," *Systemsd & Control Letters*, vol. 10, pp. 349-356, 1988.

[52] Freeman, R.A., and P. V. Kokotovic, " A New Lyapunov Function for the Backstepping of 'Softer' Robust Nonlinear Control Law", in "*Nonlinear Control System Design Symposium*", Bordeaux, France, 1992.

[53] Fridman, E., and U. Shaked, "A Descriptor System Approach to $\mathcal{H}_\infty$ Control of Linear Time-Delay Systems", *IEEE Trans. Automatic Control*, vol. 47, no. 2, pp. 253-270, 2002.

[54] Fridman, E., and U. Shaked, "Delay-Dependent Stability and $\mathcal{H}_\infty$ Control: Constant and Time-Varying Delays", *Int. J. Control*, vol. 76, no. 1, pp. 48-60, 2003.

[55] Fu, M., C. E. de Souza and L. Xie, "H_∞-Estimation for Uncertain Systems", *Int. J. Robust and Nonlinear Control*, vol. 2, pp. 87-105, 1992.

[56] Gahinet, P., and P. Apkarian, "A Linear Matrix Inequality Approach to $\mathcal{H}_\infty$ Control," *Int. J. Robust and Nonlinear Control*, vol. 4, pp. 421–4480, 1994.

[57] Gahinet, P., A. Nemirovski, A. L. Laub and M. Chilali, "*LMI Control Toolbox*", The Math Works, Inc., Boston, MA, 1995.

[58] Geromel, J. C., J. Bernussou and P. L. D Peres, " Decentralized Control Through Parameter Space Optimization ", *Automatica*, vol. 30, pp. 1565-1578, 1994.

[59] Gershwin, S. B., "Hierarchical Flow Control: A Framework for Scheduling and Planning Discrete-Events in Manufacturing Systems", *Proceedings of the 28th Conference on Decision & Control*, Tampa-Florida, pp. 195– 209, 1989.

[60] Gershwin, S. B., "*Manufacturing Systems Engineering*", Prentice-Hall, New Jersey, 1994.

[61] Gorecki, H., S. Fuska, P. Garbowski and A. Korytowski,"*Analysis and Synthesis of Time-Delay Systems*", J. Wiley, New York, 1989.

[62] Hale, J., "*Functional Differential Equations*", Springer-Verlag, New York, 1977.

[63] Hale, J., and S. M. V. Lunel, "*Introduction to Functional Differential Equations*", Springer-Verlag, New York, 1991.

[64] Hu, Z., " Decentralized Stabilization of Large-Scale Interconnected Systems with Delays", *IEEE Trans. Automatic Control*, vol. AC-39, pp. 180- 182, 1994.

[65] Jeung, E. T., J. H. Kim and H. B. Park, "$\mathcal{H}_\infty$-Output Feedback Controller Design for Linear Systems with Time-Varying Delayed State", *IEEE Trans. Automatic Control*, vol. 43, pp. 713-718, 1998.

[66] Ji, Y. and H. J. Chizeck, "Controllability, Stabilizability and Continuous-Time Markovian Jump Linear-Quadratic Control", *IEEE Trans. Automatic Control*, vol. 35, pp. 777-788, 1990.

[67] Ji, Y. and H. J. Chizeck, "Jump Linear Quadratic Gaussian Control: Steady-State Solution and Testable Conditions," *Control-Theory and Advanced Technology*, vol. 6, no. 3, pp. 289–319, 1990.

[68] Kaliath, T., "A View of Three-Decades of Linear Filtering Theory", *IEEE Trans. Information Theory*, vol. IT-20, pp. 145-181, 1974.

[69] Khargonekar, P. P. and M. A. Rotea, "Mixed $\mathcal{H}_2/\mathcal{H}_\infty$ Control: A Convex Optimization Approach," *IEEE Trans. Automatic Control*, vol. 36, pp. 824-837, 1991.

[70] Kharitonov, V. L. and A. P. Zhabko, "Robust Stability of Time-Delay Systems" *IEEE Trans. Automatic Control* , vol. 39, pp. 2388–2397, 1994.

[71] Kharitonov, V. L. , "Robust Stability Analysis of Time-Deay Systems: A Survey" *Annual Reviews in Control* , vol. 23, pp. 185–196, 1999.

[72] Kim, J. H., "Robust Mixed $\mathcal{H}_2/\mathcal{H}_\infty$ Control of Time-Delay Systems," *Int. J. Systems Science*, vol. 32, pp. 1345-1351, 2001.

[73] Kolmannovskii, V. and A. Myshkis, "*Applied Theory of Functional Differential Equations*", Kluwer Academic Pub., New York, 1992.

[74] Komaroff, N., "Matrix Inequalities Applicable to Estimating Solution Sizes of Riccati and Lyapunov Equations," *IEEE Trans. Automatic Control*, vol. 34, pp. 97-98, 1989.

[75] Krasovskii, N. N. and E. A. Lidskii, "Analysis Design of Controllers in Systems with Random Attributes, Part I," *Automation and Remote Control*, vol. 22, pp. 1021–1025, 1961.

[76] Krasovskii, N. N. and E. A. Lidskii, "Analysis Design of Controllers in Systems with Random Attributes, Part II," *Automation and Remote Control*, vol. 22, pp. 1141–1146, 1961.

[77] Kreindler, E., and A. Jameson, "Conditions for Nonnegativeness of Partitioned Matrices," *IEEE Trans. Automatic Control*, vol. 17, pp. 147–148, 1972.

[78] Kushner, H., "Stochastic Stability and Control", Academic, New York, 1967.

[79] , H., " Robust Control and $\mathcal{H}_\infty$-Optimization: Tutorial Paper", *Automatica*, vol. 29, pp. 255-273, 1993.

[80] Lee, T. N. and U. L. Radovic, " Decentralized Stabilization of Linear Continuous and Discrete Time Systems with Delays in Interconnections", *IEEE Trans. Automatic Control*, vol. AC-33, pp. 757-760, 1988.

[81] Li, H., and M. Fu, "A Linear Matrix Inequality Approach to Robust H_∞ Filtering", *IEEE Trans. Signal Processing*, vol. 45, pp. 2338-2350, 1997.

[82] Logemann, H. and S. Townley, "The effect of Small Delays in the Feedback Loop in the Stability of Neutral Systems, *Systems and Control Letters*, vol. 27, 1996, pp. 267-274.

[83] M. S. Mahmoud and M. G. Singh, "*Large-Scale Systems Modelling*", Pergamon Press, London, 1981.

[84] Mahmoud, M. S., "Stabilizing Control of Systems with Uncertain Parameters and State-Delay", *Journal of the University of Kuwait (Science)*, vol. 21, pp. 185-202, 1994.

[85] Mahmoud, M. S. and N. F. Al-Muthairi, "Design of Robust Controllers for Time-Delay Systems", *IEEE Trans. Automatic Control*, vol. 39, no. 5, pp. 995-999, 1994.

[86] Mahmoud, M. S., "Output Feedback stabilization of Uncertain Systems with State Delay", *Advances in Theory and Applications*, C.T. Leondes (Ed.), vol. 63, pp. 197-257, 1994.

[87] Mahmoud, M. S., " Stabilizing Control for a Class of uncertain Interconnected Systems", *IEEE Trans. Automatic Control*, vol. AC-39, pp. 2484-2488, 1994.

[88] Mahmoud, M. S., " Guaranteed Stabilization of Interconnected Discrete-Time Uncertain Systems", *Int. J. Systems Science*, vol. 26, pp. 337-358, 1995.

[89] Mahmoud, M. S., "Dynamic Control of Systems with Variable State-Delays", *Int. J. Robust and Nonlinear Control*, vol. 6, pp. 123-146, 1996.

[90] M. S. Mahmoud, "Adaptive Stabilization of a Class of Interconnected Systems", *Int. J. Computers and Electrical Engineering*, vol.23, pp.225-238, 1997.

[91] M. S. Mahmoud and S. Kotob, "Adaptive Decentralized Model-Following Control", *J. Systems Analysis and Modelling Analysis*, vol. 27, pp. 169-210, 1997.

[92] Mahmoud, M. S. and S. Bingulac, "Robust Design of Stabilizing Controllers for Interconnected Time-Delay Systems", *Automatica*, vol. 34, pp. 795-800, 1998.

[93] Mahmoud, M. S. and M. Zribi, "Robust and $\mathcal{H}_\infty$ Stabilization of Interconnected Systems with Delays", *Proc. IEE Control Theory and Applications*, vol. 145, pp. 559-567, 1998.

[94] Mahmoud, M. S. , "Robust Stability and Stabilization of a Class of Uncertain Nonlinear Systems with Delays", *Int. J. Mathematical Problems in Engineering*, vol. 4, pp. 165-185, 1998.

[95] Mahmoud, M. S. and M. Zribi, "$\mathcal{H}_\infty$ Controllers for Time-Delay Systems using Linear Matrix Inequalities", *J. Optimization Theory and Applications*, vol. 100, pp. 89-123, 1999.

[96] Mahmoud, M. S., and L. Xie, "Stability and Positive Realness for Time-Delay Systems", *Int. J. of Mathematical Analysis and Applications*, vol. 239, pp. 7-19, 1999.

[97] Mahmoud, M. S., N. F. Al-Muthairi and S. Bingulac, "Robust Kalman Filtering for Continuous Time-Lag Systems", *Systems and Control Letters*, vol. 38, pp. 309-319, 1999.

[98] Mahmoud, M. S., "*Robust Control and Filtering of Time-Delay Systems*", Marcel-Dekker, New York, 2000.

[99] Mahmoud, M. S. ,"Robust $\mathcal{H}_\infty$ Control of Linear Neutral Systems", *Automatica*, vol. 36, pp. 757-764, 2000.

[100] Mahmoud M. S., "Robust $\mathcal{H}_\infty$ Control of Discrete Systems with Uncertain Parameters and Unknown Delays", *Automatica*, vol. 36, pp. 627-635, 2000.

[101] Mahmoud, M. S., L. Xie and Y. C. Soh, Robust Kalman Filtering for Discrete State-Delay Systems, *Proc. IEE Control Theory and Applications*, vol. 147, pp. 613-618, 2000.

[102] Mahmoud, M. S., "Robust $\mathcal{H}_\infty$ Control and Filtering for a Class of Nonlinear Dynamical Systems with Uncertainties and Unknown Delays", *J. Systems Analysis and Modelling Simulation*, vol. 39, pp. 393-421, 2000.

[103] Mahmoud, M. S. and M. Zribi "Stabilizing controllers using observers for uncertain systems with Delays", *Int. J. of Systems Science*, vol. 32, pp. 767-773, 2001.

[104] Mahmoud, M. S. and M. Zribi, "Guaranteed Cost Observer-Based Control of Uncertain Time-Lag Systems", *Int. J. Computers and Electrical Engineering*, vol. 27, pp. 530-544, 2001.

[105] Mahmoud, M. S., "Control of Uncertain State-Delay Systems: Guaranteed Cost Approach", *IMA J. Mathematical Control and Information*, vol. 18, pp. 109-128, 2001.

[106] Mahmoud, M. S., and L. Xie, "Passivity Analysis and Synthesis for Uncertain Time-Delay Systems", *Int. J. Mathematical Problems in Engineering*, vol. 7, pp. 455-484, 2001.

[107] Mahmoud M. S., and P. Shi, "Robust Control of Markovian Jumping Linear Discrete-Time with Unknown Nonlinerities", *IEEE Trans. Circuits and Systems-I*, vol. 49, pp. 538-542, 2002.

[108] Mahmoud M. S., "Robust $\mathcal{H}_\infty$ Control and Filtering for Time-Delay Systems", *Int. J. Systems Analysis and Modelling Simulation*, vol. 6, pp. 77-81, 2002.

[109] Mahmoud M. S., and L. Xie, "$\mathcal{H}_\infty$ Filtering for a Class of Nonlinear Time-Delay Systems", *Int. J. Systems Analysis and Modeling Simulation*, Vol. 42, No. 1, March 2002, pp. 85-102.

[110] Mahmoud, M. S., and M. Zribi, "Passive Control Synthesis for Uncertain Systems with Multiple-State Delays", *Int. J. Computers and Electrical Engineering (Special Issue on Time-Delay Systems)*, vol. 28, pp. 195-216, 2002.

[111] Mahmoud, M. S. and P. Shi, "Robust Control of Markovian Jumping Linear Discrete-Time Systems with Unknown Nonlinearities", *IEEE Trans. on Circuits and Systems-I*, vol. 49, pp. 538-542, 2002.

[112] Mahmoud, M. S., and M. Zribi, "Guaranteed Cost-Observer Control for Uncertain Time-Lag Systems", *Int. J. Computers and Electrical Engineering*, vol. 29, pp. 193-212, 2003.

[113] Mahmoud, M. S. and P. Shi, "Robust Kalman Filtering for Continuous Time-Lag Systems with Markovian Jump Parameters", *IEEE Trans. on Circuits and Systems-I*, vol. 50, pp. 98-105, 2003.

[114] Mahmoud, M. S. and P. Shi, "Robust Stability, Stabilization and $\mathcal{H}_\infty$ Control of Time-Delay Systems with Markovian Jump Parameters", *Int. J. Robust and Nonlinear Control*, Vol 13, 2003.

[115] Mahmoud, M. S., A. Ismail, P. Shi and R. K. Agrawal, "Transformation Methods for $\mathcal{H}_\infty$ Control of Uncertain Jumping Delayed Systems", *Proc. AIAA Guidance, Navigation and Control Conference*, Austin, Texas, August 11-14, 2003.

[116] Mahmoud, M. S. and A. Ismail, "Transformation Methods for Mixed $\mathcal{H}_2/\mathcal{H}_\infty$ Control of Jumping Time-Delay Systems", *Technical Report No. MsM-AI-ProjectI-5, UAE University*, submitted for publication

[117] Mahmoud, M. S. and A. Ismail, "A Descriptor Approach to Simultaneous $\mathcal{H}_2/\mathcal{H}_\infty$ Control of Jumping Time-Delay Systems", *Technical Report No. MsM-AI-ProjectI- , UAE University*, submitted for publication

[118] Mahmoud, M. S. and A. Ismail, "$\mathcal{H}_\infty$ Control of Uncertain Jumping Systems with Multi-State and Input Delays", *Technical Report No. MsM-AI-ProjectI-3, UAE University*, submitted for publication

[119] Mahmoud, M. S. and A. Ismail, "Robust $\mathcal{H}_\infty$ Analysis and Synthesis for Jumping Time-Delay Systems using Transformation Methods", *Technical Report No. MsM-AI-ProjectI- , UAE University*, submitted for publication

[120] Mahmoud, M. S. and A. Ismail, "Robust Performance Analysis and Synthesis of Multi-State-Delay Jumping Systems", *Technical Report No. MsM-AI-ProjectI-10, UAE University*, submitted for publication .

[121] Mahmoud, M. S. and A. Ismail, "Interconnected Jumping Tine-Delay Systems: Robust and $\mathcal{H}_\infty$ Control Schemes", *Technical Report No. MsM-AI-ProjectI-7, UAE University*, submitted for publication .

[122] Mahmoud M. S., "Robust Filtering for a Functional Time-Delay Jumping Systems", *Technical Report No. MsM-AI-02-01*, UAE University, 2002, submitted for publication.

[123] Mahmoud M. S., "LMI Approach to Robust Stability and $\mathcal{H}_\infty$ Control of Uncertain Neutral Jumping Systems", *Technical Report No. MsM-AI-02-02*, UAE University, 2002, submitted for publication.

[124] Mahmoud M. S., "Robust Estimation and Stabilization of Linear Uncertain Neutral Systems", *Technical Report No. MsM-AI-02-03*, UAE University, 2002, submitted for publication.

[125] Mahmoud M. S., and P. Shi, "Robust and $\mathcal{H}_\infty$ Control of Interconnected Systems with Markovian Jump Parameters", *Technical Report No. MsM-AI-02-04*, UAE University, 2002, submitted for publication.

[126] Malek-Zavarei, M. and M. Jamshidi, "*Time-Delay Systems: Analysis, Optimization and Applications*", North Holland, Amesterdam, 1987.

[127] Mao, X., "Stochastic Stabilization and Destabilization", *Systems and Control Letters*, vol. 23, pp. 279–290, 1994.

[128] Mao, X., "Robustness of Exponential Stability of Stochastic Differential Delay Equations", *IEEE Trans. Automatic Control*, vol. 41, pp. 442–447, 1996.

[129] Mao, X., "LaSalle-Type Theorems for Stochastic Differential Delay Equations", *J. Math. Anal Appl.*, vol. 236, 1999. pp. 350-369.

[130] Mariton, M. and P. Bertrand, "Output Feedback for a Class of Linear Systems with Stochastic Jump Parameters," *IEEE Trans. Automatic Control*, vol. 30, pp. 898-900, 1985.

[131] Mariton, M.“*Jump Linear Systems in Automatic Control* ,” Marcel Dekker, New York, 1990.

[132] Molchanov, A. P. and E. S. Pyatnitski, " Criteria of Asymptotic Stability of Differential and Difference Inclusions Encountered in Control Theory", *Systems and Control Letters*, vol. 13, pp. 59-64, 1989.

[133] Morozan, T. ,“Stability and Control for Linear systems with Jump Markov Perturbations,” *Stochastic Anal. and Appl.*, vol. 13, no. 1, pp. 91–110, 1995.

[134] Moheimani, S. O. R., and I. R. Petersen, “Optimal Quadratic Guaranteed Cost Control of a Class of Uncertain Time-Delay Systems”, *Proc. IEE, Part D*, vol. 144, (1997), pp. 183-188.

[135] Murray, D.“*Mathematical Biology,*” Springer, New York, 1989.

[136] Nagpal, K. M., and P. P. Khargonekar, “Filtering and Smoothing in H_∞ Setting”, *Systems and Control Letters*, vol. 22, pp. 123-129, 1994.

[137] O’Conner, D. and T. J. Tarn, "On the Stabilization by State Feedback for Neutral Differential Difference Equations", *IEEE Trans. Automatic Control*, vol. AC-28, pp. 615-618, 1983.

[138] Pan, Z. and T. Basar, “$\mathcal{H}_\infty$ Control of Large Scale of Jump Linear Systems via Averaging and Aggregation”, *Proceedings of the 34th Conference on Decision & Control*, New Orleans, pp. 2574– 2579, 1995.

[139] Pan, Z. and T. Basar, “$\mathcal{H}_\infty$ Control of Markovian Jump Systems and Solutions to Associated Piecewise-Deterministic differential Games”, *Analas of the International Society of Dynamic Games*, G. J. Olsder, (Ed.), Birkhauser, Boston, 1996.

[140] Park, P., "A Delay-Dependent Stability Criterion for Systems with Uncertain Time-Invariant Delays", *IEEE Trans. Automatic Control*, vol. 44, pp. 876–877, 1999.

[141] Petersen, I. R., B. D. O. Anderson and E. A. Jonckheere, "A First Principle Solution to the Non-Singular $\mathcal{H}$ Control Problem", *Int. J. Robust and Nonlinear Control*, vol. 1, pp. 171-185, 1991.

[142] Petersen, I. R., and D. C. McFarlane, "Optimal Guaranteed Cost Control and Filtering for Uncertain Linear Systems", *IEEE Trans. Automatic Control*, vol. 39, pp. 1971-1977, 1994.

[143] Rami, M. A., and L. El-Ghaoui, "LMI Optimization for Nonstandard Riccati Equations Arising in Stochastic Control," *IEEE Trans. Automatic Control*, vol. 41, pp. 1666-16671, 1996.

[144] Rotea, M. A. and P. P. Khargonekar, "Stabilization of Uncertain Systems with Norm-Bounded Uncertainty- A Control Lyapunov Function Approach ", *SIAM J. Control & Optimization*, vol. 27, pp. 1462-1476, 1989.

[145] Rotea, M. A. and P. P. Khargonekar, "$\mathcal{H}_2$-Optimal Control with $\mathcal{H}_\infty$ Constraint: The State Feedback Case," *Automatica*, vol. 27, pp. 307-316, 1991.

[146] Saberi, A., B. S. Chen, S. Peddapullaiah and U. L. Ly, "Simultaneous $\mathcal{H}_2/\mathcal{H}_\infty$ Optimal Control: The State Feedback Case," *Automatica*, vol. 29, pp. 1611-1614, 1993.

[147] Schmitendorf, W.E., " Designing Stabilizing Controllers for Uncertain Systems Using the Riccati Equation Approach", *IEEE Trans. Automatic Control*, vol.AC-33, pp. 376-379, 1988.

[148] Shaked, U. and C. E. de Souza, "Robust Minimum Variance Filtering", *IEEE Trans. Signal Processing*, vol. 43, pp. 2474-2483, 1995.

[149] Shi, P. and E. K. Boukas, "$\mathcal{H}_\infty$ Control for Markovian Jumping Linear Systems with Parametric Uncertainty," *J. Optimization Theory and Applications*, vol. 95, no. 1, pp. 75–99, 1997.

[150] Shi, P., "Filtering on Sampled-Data Systems with Parameteric Uncertainty," *IEEE Trans. Automat. Control*, vol. 43, pp. 1022–1027, 1998.

[151] Shi, P. , E. K. Boukas, and R. K. Agarwal, "Kalman filtering for continuous-time uncertain systems with Markovian jumping parameters", *IEEE Trans. Automat. Control*, vol. 44, no. 8, pp. 1592–1597, 1999.

[152] Shi, P. , E. K. Boukas, and R. K. Agarwal, "Control of Markovian Jump Discrete-Time Systems with Norm Bounded Uncertainty and Unknown delays," *IEEE Trans. Automat. Control*, vol. 44, no. 11, pp. 2139–2144, 1999.

[153] Soner, M. , "Singular Perturbations in Manufacturing", *SIAM J. Control and Optimization*, vol. 31, pp. 132-1476, 1993.

[154] Srichander, R. and B. K. Walker, "Stochastic Analysis for Continuous-Time Fault-Tolerant Control Systems", *Int. J. Control*, vol. 57, no. 2, pp. 433-452, 1989.

[155] Sworder, D. D. ,"Feedback Control of a Class of Linear Systems with Jump Parameters," *IEEE Trans. Automatic Control*, vol. 14, no. 1, pp. 9-14, 1969.

[156] Wang, Z., J. Lam and K. J. Burnham,"Stability Analysis and Observer Design for Neutral Delay Systems," *IEEE Trans. Automatic Control*, vol. 74, pp. 478-483, 2002.

[157] Willsky, A. S., "A survey of design methods for failure detection in dynamic systems," *Automatica*, vol. 12, no. 5, pp. 601–611, 1976.

[158] Willsky, A. S., H. J. Chizeck and D. Castanon, "Discrete-Time Markovian Jump Linear Quadratic Optimal Control", *Int. J. Control*, Vol. 43, pp. 213-231, 1986.

[159] Wonham, W. M., "Random Differential Equations in Control Theory", *Probabilistic Methods in Applied Mathematics*, A.T. Bharucha-Reid (Ed.), vol. 2, pp. 131-212, 1971.

[160] Xie, L. and Y. C. Soh, "Robust Kalman Filtering for Uncertain Systems", *Systems and Control Letters*, vol. 22, pp. 123-129, 1994.

[161] Xu, S., J. Lam and C. Yang, "$\mathcal{H}_\infty$ and Positive Real Control for Linear Neutral Delay Systems," *IEEE Trans. Automatic Control*, vol. 46, pp. 1321–13246, 2001.

[162] Zhang, Q., G. G. Yin and E. K. Boukas, "Optimal Control of a Marketing-Production Systems," *IEEE Trans. Automatic Control*, vol. 46, pp. 416–427, 2001.

[163] Zhou, K., K. Glover, B. Bodenheimer and J. Doyle, "Mixed $\mathcal{H}_2$ and $\mathcal{H}_\infty$ Performance Objectives I: Robust Performance Analysis," *IEEE Trans. Automatic Control*, vol. 39, pp. 1564-1574, 1994.

[164] Zhou, K., "*Essentials of Robust Control*", Prentice-Hall, New York, 1998.

[165] Zribi, M. and M. S. Mahmoud "$\mathcal{H}_\infty$-Control Design For Systems with Multiple delays", *Journal of Computers and Electrical Engineering*, Vol. 25, 1999, pp. 451-475.

[166] Zribi, M., M. S. Mahmoud, M. Karkoub and T. Li "$\mathcal{H}_\infty$-Controllers for Linearized Time-Delay Power Systems", *Proc. IEE Generation, Transmission and Distribution*, Vol. 147, pp. 401-408, 2000.

Index

About The Authors

MagdiSadek Mahmoud received the Ph.D. degree from Cairo University, Egypt in 1974. He has been a Professor of Systems Engineering since 1984. He served on the faculties of several universities world-wide including Cairo University; the American University in Cairo; the Egyptian Air Academy; MSA University; the Arab Academy of Sciences and Technology (**Egypt**), Kuwait University and KISR (**Kuwait**), UMIST (**UK**), Pittsburgh University and Case Western Reserve University (**USA**) and NTU (**Singapore**). He lectured in Europe (**UK, Germany, Switzerland**), (**Australia**) and (**Venezuela**). He has been actively engaged in teaching and research in the development of modern methodologies to computer control, systems engineering and information technology and has been a technical consultant on information, computer and systems engineering for numerous companies and agencies at all levels of government and the private sector.

Dr. Mahmoud is the principal author of nine (9) books, nine (9) book-chapters and the author/co-author of more than 300-refereed papers. He is the recipient of 1978, 1986 Science State Incentive Prizes for outstanding research in engineering (**Egypt**), of the Abdul-Hameed Showman Prize for Young Arab Scientists in engineering sciences, 1986 (**Jordan**) and of the Prestigious Award for Best Researcher at Kuwait University, 1992 (**Kuwait**), the State Medal of Science and Arts-first class, 1979 (**Egypt**) and the State Distinguished Award-first class, 1995 (**Egypt**).

He is listed in the 1979 edition of Who's Who in Technology Today (USA). He was the vice-chairman of the IFAC-SECOM working group on large-scale systems methodology and applications (1981-1986), and an associate editor of LSS Journal (1985-1988) and editor-at-large of the EEE series,

Marcel-Dekker, USA. He is an associate editor of the International Journal of Parallel and Distributed Systems of Networks, IASTED, since 1997. He is a member of the New York Academy of Sciences. He is a fellow of the IEE, a senior member of the IEEE, a member of Sigma Xi, the CEI (UK), the Egyptian Engineers society, the Kuwait Engineers society and a registered consultant engineer of information engineering and systems (Egypt).

Peng Shi received the B.S. degree in mathematics from Harbin Institute of Technology in 1982, the M.Sc. degree in modern control theory and applications from Harbin University of Engineering and Heilongjiang Institute of Applied Mathematics, China in 1985, the Ph.D. degree in electrical engineering from the University of Newcastle, Australia in 1994. He also has a doctor degree in mathematics from the University of South Australia in 1998.

Dr Shi lectured in Institute of Applied Mathematics at Heilongjiang University, China, from 1985-1989. He held a visiting fellow position in Department of Electrical and Computer Engineering, University of Newcastle, Australia from 1989-1990. He was a postdoctoral research associate at Centre for Industrial and Applied Mathematics, from 1995-1997, and a lecture at School of Mathematics, from 1997-1999, University of South Australia. he also held visiting fellow position in National Institute for Aviation Research, USA, and Department of Mechanical Engineering, Ecole Poly technique de Montreal, Canada. He joined in Defence Science and Technology Organisation, Australia in 1999 as a research scientist, and now as a task manager and senior research scientist. His research interests include operations research and analysis, robust control and filtering of sampled-data systems, hybrid systems, Markovian jump systems, fuzzy systems, time-delay systems, singularly perturbed systems, and mathematical modeling and optimization techniques and applications to defence industry.

Dr Shi has authored or co-authored over 90 journal publications. He is currently serving as Associate editor for the IEEE Control System Society Conference Editorial Board, and regional editor of Journal of Nonlinear Dynamics and Systems Theory. He is a senior member of IEEE and a member of SIAM. He also holds the adjunct positions of professor at Central Southern University, China, and research fellow at Centre for Industrial and Applied Mathematics, University of South Australia.